# THE PRINCIPLE OF TRUE REPRESENTATION

# THE PRINCIPLE OF TRUE REPRESENTATION

## MIND, MATTER AND GEOMETRY IN A SELF-CONSISTENT UNIVERSE

*by*

JOHN T. WINTHROP

iUniverse®

**THE PRINCIPLE OF TRUE REPRESENTATION**
**MIND, MATTER AND GEOMETRY IN A SELF-CONSISTENT UNIVERSE**



*iUniverse books may be ordered through booksellers or by contacting:*

*iUniverse*
*1663 Liberty Drive*
*Bloomington, IN 47403*
*www.iuniverse.com*
*1-800-Authors (1-800-288-4677)*



*ISBN: 978-1-5320-1201-3 (sc)*
*ISBN: 978-1-5320-1202-0 (hc)*
*ISBN: 978-1-5320-1236-5 (e)*

*Library of Congress Control Number: 2016919915*

*Print information available on the last page.*

*iUniverse rev. date: 12/07/2016*

To Riga

# Contents

## PART VI THE STRUCTURE OF SPACETIME



## PART VII THE HARD PROBLEM OF CONSCIOUSNESS

To get at the thing
Without gestures is to get at it as
Idea. She floats in the contention, the flux

Between the thing as idea and
The idea as thing. She is half who made her.
This is the final Projection, C.

WALLACE STEVENS

SO-AND-SO RECLINING ON HER COUCH

# Preface

**1.** Before 1700—and in particular during the eleven-hundred-year Christian era comprising the Middle Ages and Renaissance—the dynamics of the natural world were generally believed to be governed in detail, from moment to moment, by the hand of a divine Person: the Universe conceived as sensorium of God. But then, near the end of the seventeenth century, and coinciding more or less with the publication of Newton's *Principia* in 1687, the cumulative effect of the great scientific advances brought by Leonardo, Copernicus, Galileo and others during the Renaissance transformed almost overnight the old theological paradigm into a vision of the Universe as a great machine, set into motion perhaps by its Creator, but operating thereafter on its own on a principle of cause and effect. Thus was born the landmark period of European history known today as the Enlightenment or Age of Reason.

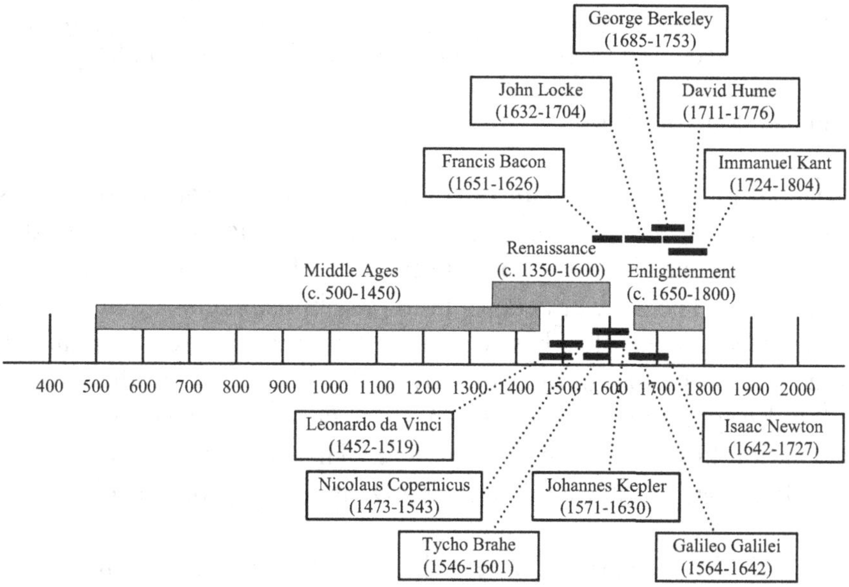


If before the Renaissance the inner workings of the natural world were hidden in divine mystery, with the Enlightenment there arose an attitude of confidence that the physical world, conceived as an ordered, law-abiding system, was in fact comprehensible, intelligible and ultimately knowable. The Enlightenment attitude might even be (and has been) said to mark the dawn of the Modern Mind. Oddly enough though, and perhaps ironically, the age of Enlightenment saw also the rise of modern Empiricism. According to Empiricism's main exponents Locke, Berkeley and Hume, what we deem to call "the world of nature" and claim to have knowledge of is actually no more than a construction of the mind derived from sense experience. On this skeptical view, there may or may not exist a real world external to the senses. But even if there is such a world, said these philosophers, it is quite inaccessible to rational inquiry. In other words, one need pay no attention to it; for the facts of the world as such, if there are any, are in principle beyond the grasp of science and inherently unknowable. The philosophers' doubt about the existence of an external world surely contradicts the otherwise optimistic spirit of the Enlightenment. Indeed, the philosopher giant of the age, Immanuel Kant, remarked that[1]

> …it still remains a scandal to philosophy and to human reason in general that the existence of things outside us (from which we derive the whole material of knowledge, even for our inner sense) must be accepted merely on *faith*, and that if anyone thinks good to doubt their existence, we are unable to counter his doubts by any satisfactory proof.

A scandal indeed. And Kant might have added: a scandal to physics, as well. For in the absence of knowledge of the facts of the world as such, how are we to know in our observation of nature how close we are to the truth? The answer according to the philosophers is that we have no way of knowing. The best we can do, they say, is create a plausible and consistent picture of the world by application of yet another product of the Enlightenment, namely, the steady interplay of theory and observation known as *scientific method.* That, in sum, is where things stood between science and philosophy at the end of the eighteenth century.

**2.** Now the question is, where do we stand today? On the side of physical science we have of course come a very long way since the Enlightenment. Indeed, the celebrated discoveries of electrodynamics, relativity in both its special and general formulations, and quantum mechanics rank not only among the greatest products of the scientific enterprise, but of human thought in general. On the philosophical side, though, it seems not much has changed in the last three hundred years.

Empiricism, while assuming a variety of guises and expressions in the more recent past—e.g., logical positivism, instrumentalism, verificationism, falliblism[2]—remains a background assumption in the day-to-day conduct of physical science. My guess is that most physicists, if pressed, would agree that to know something about the physical world is to be able to measure it; and if it is not measurable, then it is to be barred from the arena of science.

The present volume introduces a new direction in the pursuit of physical knowledge—a new way of doing physics and, beyond that, a new way of seeing and thinking about the natural world. Our ultimate aim is, of course, no different than that of science itself, namely, to uncover and explain the facts of the natural world. By that I mean we seek the *truth* about nature. However, as the Enlightenment philosophers have reminded us, physics as conventionally practiced is largely an empirical science. For our hypotheses, no matter how reasonable seeming or intellectually compelling, always need to be checked against observation. But if by general agreement observation must have the last word, then (to repeat the obvious) what we get by applying scientific method are not facts at all, but mere representations of the facts—"constructions of the mind derived from sense experience". Moreover, because one can never be sure how well the representations do represent the facts— i.e., how close they come to the truth—physics as conventionally conceived and practiced remains ever provisional and subject to refinement in the light of new observational data. The chapters below can be read as an attempt to close—or at least reduce—the allegedly unbridgeable gap between appearance and reality, i.e., between representation and fact. They argue for a move away from the seventeenth-century empiricism embraced even today to a rationalistic, less provisional, less tentative approach to the study of nature.

This would seem to be a tall order, at the very least an unlikely and perhaps impossible undertaking. We are, after all, talking about ascending to the God's eye view—the ability to see things as they are in themselves. Kant himself, while convinced of the existence of a transcendent world of noumenal beings, also believed such beings to be unknowable.[3] But, as we shall see in this book, Kant's pessimistic attitude towards noumenal access is almost certainly unwarranted. For not only can we demonstrate the existence of an external world, we can also describe at least some of its contents. There are three main steps. The first, taking a page from linear systems theory, is to treat the representation as a received message, that is to say, as a filtered and generally corrupted version of the noumenal fact. This step introduces a system operator whose function is to convert the fact into a representation. The second step is to demand exact correspondence between the fact and its representation. The resulting representation may then properly be said to be *true*. The third and final step is algebraically to combine the first two. There at once results a new universal law of physics—a Law of Laws—from which, after geometrization, particular dynamical laws may be obtained, all of which laws are true by definition.

Our second step—namely, the demand for exact correspondence between fact and representation—is actually nothing more than a formalized statement of Aristotle's correspondence theory of truth.[4] Because physical laws almost always derive from an underlying principle, for the purposes of this book I have decided to call the formalized correspondence theory the *Principle of True Representation*. The Law of Laws is a direct consequence of this principle. The striking thing about the Principle of True Representation is that it holds good even in the absence of observers. I interpret this to mean that the inanimate world all on its own is self-representing; it means that the universe at the most fundamental level comprises a self-consistent system. It also means that the P. of True Representation can corroborate the existence of an external world. For a material world whose structure depends upon its being self-representing may well differ in measurable ways from one that does not.

The main areas of application of the general theory just outlined are those suggested in the subtitle to the book: mind, matter and geometry, where in the category "matter" I include both the world of subatomic particles and the cosmos as a whole, and by "geometry" I mean the geometry of spacetime. This seems rather a lot for just one theory, but then that is what one wants in a comprehensive, connected account of reality: a theory vulnerable to falsification by a failure in any one of its areas of application. Even so, despite the breadth of subject matter covered, I make no claim to a theory of everything. Indeed, what I propose is a theory of physics *before* the Standard Model of particle physics, not beyond it. Quantum field theory, for instance, is hardly mentioned except as background material, not because it is unimportant, but because there is much to do before moving on to that arcane and difficult subject. The Principle of True Representation, in its role as a criterion of scientific truth, exposes not a few misconceptions in the (pre-Standard Model ) foundations of contemporary physics and science of mind, these having to do with, among other things, (1) the structure and dimensionality of spacetime, (2) the existence (or not) of the graviton, (3) the origin and values of the masses and electrical charges of the fundamental particles, (4) the origin of cosmic expansion and (5) the reality of free will. If I am right, if things really are as unsettled as this, then I suggest we are not at all ready to complete the Standard Model, let alone unify quantum mechanics and general relativity. We first need to get the foundations right. It seems to me especially obvious that any theory claiming to be a theory of everything yet failing to account for the possibility of mind and consciousness has been at best naïvely named. My hope is that the present theory will get us launched in the direction of something even larger, something having about it a legitimate air of finality.

The working out of these ideas has taken many years, decades in fact, and continues to this day. Early on (meaning late 'seventies, early 'eighties), I received helpful advice and criticism from Robert L. Mills, Jim Tough, Michael Polanyi, Robert L. Powell, Jim Young; and more recently from David Chalmers, John Searle, Martin Perl, John Learned and Robert Caldwell. David Caldwell's patient response to my questions concerning the behavior of the neutrino is particularly acknowledged. A good deal of this work was sent to the standard physics journals but was rejected for publication as being too far outside the mainstream. Hence the present all-inclusive volume. Undoubtedly the reviewers' comments led to a better product than would have resulted otherwise, and for that I am grateful.

## Notes and references

[1] Immanuel Kant, *Critique of Pure Reason*, Norman Kemp Smith, Trans. (Palgrave Macmillan, Houndmills, Basingstoke, Hampshire, UK, 1929), corrections added to the Preface to the Second Edition, pp. xxxix. Actually, on p. 245 of the *Critique*, Kant proposes a proof of the existence of the external world. The

argument seems to be this: (1) My own consciousness is experienced in time. (2) The reckoning of time presupposes a fixed temporal point of reference. (3) But this fixed point cannot be something in me, which is ever changing. (4) Therefore it derives from outside me, proving the existence of the external world. Whatever the merits of this existential argument it does not, of course, tell us in what the external world consists. One would not, after all, want to invoke comprehending minds when attempting to define the structure of mind-independent reality. See also Martin Heidegger, *Being and Time*, J. Macquarrie and E. Robinson, Trans. (HarperCollins, New York, 1962), p. 247.

[2] Karl Popper, *Realism and the Aim of Science* (Rowman and Littlefield, Totowa, New Jersey, 1983), p. xxxv.

[3] "Indeed when we rightly regard the objects of sense as mere phenomena we thereby admit that each such object is based upon a thing-in-itself of which we are not aware as it is constituted in itself, but only as known through its appearances, that is, by the manner in which our senses are affected by this unknown something. Therefore the intellect, by assuming phenomena, admits the existence of things-in-themselves. We may even say that the imagining of such beings underlying the phenomena, such mere intellectual beings, is not only permissible, but unavoidable." Immanuel Kant, *Prolegomena to Every Future Metaphysics That May Be Presented As a Science*, §32, reprinted in *The philosophy of Kant*, C. J. Friedrich, Ed. (Modern Library, New York, 1949), p. 87.

[4] Aristotle, *Metaphysics,* $1011^{b}$ 25, in *The Basic Works of Aristotle*, Richard McKeon, Ed., (Modern Library, New York, 2001). "To say of what is that it is not, or of what is not that it is, is false, while to say of what is that it is, and of what is not that it is not, is true;..."

# Conventions

Flat spacetime metric $g_{\mu\nu} = (+1, -1, -1, -1)$

Components of 3-vectors $\mathbf{x}$ denoted $x^j$, $(j = 1, 2, 3)$

Components of 4-vectors $x$ denoted $x^\mu$, $(\mu = 0, 1, 2, 3)$

Components of 5- or 6-vectors denoted $x^a$, $(a = 0, 1, 2, 3, 5, 6)$

Charge of the electron $q = Qe$, where $Q = -1$ and $e > 0$

Constants $c$ and $\hbar$ are *not* set to 1.

# Chapter 1

# Theme and motivation

> There is, I think, no theory-independent way to reconstruct phrases like 'really There'; the notion of a match between the ontology of a theory and its "real" counterpart in nature now seems to me illusive in principle.
>
> THOMAS S. KUHN[1]

## 1.1 Physics at an impasse

This book deals with the origin and structure of physical law. Our aim to find out why there are any laws at all, where the laws come from and why they have the forms they do. There are at least two reasons to do this. In the first place we should like to know what the world looks like in the absence of conscious observers. Observers, after all, are relative newcomers to the universe. Surely we want to know what goes on in the world in their absence or when they are not looking. We seek, in other words, an *objective* account of the world; a description, not of what it seems to be, but what in fact the world is like undisturbed by measurement. We are not, of course, supposed to be able to do this. From the logical positivist's point of view[2]— as expressed for instance in the above quotation of Thomas Kuhn—a theory is to be considered scientific (i.e., either verifiable or falsifiable) if and only if its claims are reducible to "terms of immediate experience"[3]. On this ground, for example, the theory that 'all swans are white' is legitimately scientific, whereas the theory that 'the Universe exists even if unperceived' is not. But such a viewpoint is surely lacking in imagination. For who is to say that one cannot have knowledge of the external world, even if it is unobservable? It may be that an unobservable external world leaves telltale traces of its existence in the part of the world that one *can* observe, just as our own privileged inner lives are revealed to others, in part, by facial expression and body language. That is just the sort of empirical evidence that we shall be looking for.

Secondly, it is obvious that natural scientists today face an imposing array of fundamental gaps in our knowledge of the world of nature. One gap concerns our inability to understand how subjective mental states could arise from objective brain states;[4] or more generally, how living, thinking beings could arise from dead matter.[5] Another is our inability to explain why the fundamental units of matter—leptons and quarks—are given to us in three families of successively greater mass;[6] why the masses have the oddly disparate values they do;[7] and why the neutrino—itself a species of lepton—appears to be almost, but not quite, massless.[8] A third gap concerns the structure of spacetime itself. Why, for instance, are there three spatial dimensions,[9] and could there exist additional unseen ones?[10] And finally, there are major gaps in our understanding of the universe as a whole. What, for example, is the source of dark energy—the energy of the vacuum thought to be responsible for the observed (and unexpected) accelerating cosmic expansion?[11] Why is its measured magnitude so much smaller (by a factor of perhaps $10^{-120}$!) than the value predicted by quantum field theory?[11] What are dark matter[12] and dark flow[13]? And the expected anti-baryonic content of the universe is nowhere to be found; where is it? We are thus confronted with a broad spectrum of deep mysteries of the natural world, each stubbornly resistant to solution—at least so far—and calling emphatically for resolution.

The various mysteries are connected, of course. For, as far as anyone knows, without material brain there can be no mind, and without space and time there can be neither mind nor matter. There is nonetheless a tendency among the professionals who think about such things to treat the mysteries separately. Thus, philosophers and neuroscientists actively pursue psychophysical theories of conscious experience without regard for what is happening in the world of particle physics; and physicists for their part construct Theories of Everything in extra spacetime dimensions oblivious to the phenomenon of consciousness. This tendency to see the world as a patchwork of unconnected elements is perhaps understandable, given in particular our current lack of understanding of the nature and origin of consciousness. But it has also led, in the writer's opinion, to egregious errors of fact and interpretation in both neuroscience and fundamental physics. On the evidence it seems that we have reached an impasse in our capacity to explain how the world works at its deepest level. Some have even seen in this predicament an end to foundational physical science[14] and science of mind[15] altogether. And yet, there may be a way forward. For if one could locate the ultimate source of physical law, as we shall attempt to do here, one might hope to find there not only the known laws, but also new or augmented ones with bearing on at least some of the unsolved mysteries of mind, matter and spacetime geometry. One might then have in hand at last the beginnings of a connected account of the workings of the natural world. That, in brief, is the motivation for the present work.

If discovering the source of physical law seems a task for philosophy as much as for physical science, that is because it is. Before modern times—that is, before the advent of the seventeenth-century Baconian idea that physical theories should be empirically tested[16]—the study of the natural world was considered to be a branch of philosophy, the study of first things. It was called, appropriately enough, *natural philosophy*,[17] a domain of knowledge one associates today with such names as Plato, Aristotle, Ptolemy, Galileo and Descartes. Like the first philosophers, we want to be able to say why the world is the way it is, and in this respect, our aim coincides with that of science itself: to explain the world around us. However, to meet this aim fully we suggest that one must first explain why the world is explainable in the first place, as only then can we begin to understand its inherent order. That is precisely where philosophy comes into play. Accordingly our work will take us to previously unexplored territory at the boundary between physics and metaphysics and from there to a consideration of some of the unsolved mysteries of particle physics, cosmology and the science of mind. As we shall see, it is philosophy not physics that furnishes the needed conceptual and analytical apparatus. In the end, perhaps surprisingly, and no doubt against the grain of conventional physical thought, we are going to get testable new physics from metaphysics.

The task of the remainder of this introductory chapter is twofold. First, we want to point out for a general audience some fairly obvious deficiencies in our current conception of the world of nature. The list presented here is far from exhaustive but is, one should think, of sufficient variety and depth to show just how far away we are from an adequate understanding of this world we inhabit. With these few examples we aim to convince the reader that the resolution of the mysteries is not to be found in mere subtleties of known formulations of physical theory. Rather it will require reformulation from the ground up, a radical rethinking of the world's underlying ontology. Second, so that we can see more or less where we are going, a technically-oriented, chapter-by-chapter summary of the book is provided.

## 1.2 Four lacunae in the scientific world-view

*1.2.1 The poverty of materialism.* Physical laws are summary expressions of the observed regularities of nature. Although there exist in nature many kinds of regularities, including social and psychological ones,[18] our interest here centers mainly on the regularities of inanimate nature, as those are universal in character, occurring everywhere and at all times, without exception. Thus Newton's law of universal gravitation expresses the observation that any two massive particles in the universe are attracted to each other with a force proportional to the square of the reciprocal distance between them. And Newton's second law of motion expresses the observation that the rate of change in the velocity of any material particle in the universe is proportional to the external force applied to it.

It is the business of physics, and of the physical sciences in general, to look for such laws and find ways to test them. The aim of this activity is to explain the world we live in and also to predict phenomena not yet observed. That said, we find ourselves faced with an obvious but crucial question: What, exactly, is meant by "the world," the object of our investigation? What is this vast thing whose behavior we are trying to explain? In what does it consist? It is easy to see that the prevailing *scientific* conception of the world—a view variously called *physicalism*, *philosophical materialism* or *scientific naturalism*[19]—cannot possibly be right—or at least not entirely so. As its name implies, materialism asserts that the world consists of matter and only matter, interacting with itself through various fields of force. On this view, everything we know or can know originates and abides in a purely material world. One can hardly quarrel with the successes of the naturalistic viewpoint. It has given us convincing explanations of everything from radioactive decay to population genetics, from stellar structure to the evolution of species. And yet, strictly speaking, it is false. For material particles and their interactions generally behave in a lawful manner, and the relevant laws, while not made of matter, are real enough. And so, contrary to the materialist doctrine, there do indeed exist things over and above the material world, namely the laws governing its behavior. But matter itself does not, and cannot, tell us where the laws come from or why they have the forms they do. On this ground alone, that is to say, from our ignorance of the ultimate source of physical law, we see that our understanding of the underlying structure of the world as a whole is incomplete.

Now the argument just given for a transcendent realm of physical law is unlikely to persuade the convinced advocate of scientific naturalism. He will counter that the laws of physics are not actually real; that they are just names given to regularities of the world discovered and documented by conscious beings (nominalism); or that they are mere concepts in the (material) mind—like the Pythagorean Theorem or the cardinal numbers—concepts that facilitate our talking about the world (conceptualism).[20] One understands the materialist's argument, of course, but it misses the point. Yes, of course we hold laws in the mind, and of course they do facilitate discussion amongst us. But unlike the Pythagorean Theorem or the cardinal numbers, the laws of nature very obviously have been here from the beginning, and will not go away when we do. They remain with the world, and the world continues under their control with or without us. That, in the author's view, guarantees the reality of the laws as well as anything could. Of course it is one thing to *say* that the laws are real and quite another to show that it is so. To do this we shall have to appeal to a higher principle, one going beyond materialism that reveals the transcendent (and mind-independent) source of the laws. This, as was said above, is the main aim of the present volume. If we are successful, the nominalist and conceptualist arguments against the reality of the laws can be considered refuted once and for all.

*1.2.2 The problem of consciousness.* Nothing is more mysterious than consciousness, not even quantum mechanics. Because, as far as we know, it arises only in association with living brains, one might be inclined to think of consciousness as a problem for neuroscience, cognitive science, psychology or even computer science, anything but physics. But that is to misunderstand the problem. The problem is not what consciousness is as such—something that, like the physical dimension *time*, may never be fully comprehended—but that it exists at all. One is reminded of Wittgenstein's oft-repeated dictum "It is not *how* things are in the world that is mystical, but *that* it exists."[21] The central problem of consciousness is this: How could *subjective* (private) conscious experience possibly arise in a material universe governed by *objective* (public) physical laws?[22] Or, put a little differently, how could privileged centers of experience arise in material world that is, in the eyes of science, centerless?[23] The answer in either case is clear: in an objective and centerless material world there can exist neither subjective experience nor the privileged centers required for its appearance. So either consciousness is an illusion and does not really exist[24]; or it does exist, but our current conception of the world is too narrow to accommodate it. The latter option, it seems reasonable to assume—given the obvious fact that we really are conscious—is the only plausible one. We can put the problem on sound physical footing as follows.

There exists in contemporary physical theory a powerful principle, due to Einstein, known as the (General) Principle of Relativity.[25] According to this principle, there are no privileged frames of reference in four-dimensional spacetime. This means that the laws of physics take the same form in all frames of reference in four-space, whatever their relative states of motion. The principle applies to all laws, not just those of mechanics. If this were not so, then the laws would undergo changes in form from frame to frame—a hopeless state of affairs for anyone trying to make sense of the regularities of nature. The power of this principle lies (among other things) in its critical function: if one thinks he has discovered a law of nature, but its form varies depending on the motion of the observer's frame of reference, then it must be judged wrong and without physical significance. Invariance with respect to frame of reference is a necessary condition for the truth of any proposed law of nature (though of course it is not sufficient).

Now suppose that, for simplicity, we restrict ourselves to so-called *inertial* frames of reference—frames moving at constant velocity relative to each other. Taking account of the three known spatial dimensions we can see that there exists a three-fold infinite set of inertial frames in four-space.[26] According to the Principle of Relativity, all such frames are equivalent and suitable for the formulation of the laws of physics in four-space. That is to say, none of them is privileged. But the frame of reference of individual conscious experience—the so-called *first-person* perspective[27]—is distinctly privileged, because the phenomena found within it—such as physical pain or hallucination—are unavailable for observation in any other frame of reference; the phenomena in question are observable from one, and only one, restricted point of view: that of the subject experiencing them. It follows that the seat of one's internal experiences, the first-person frame of reference, a place of *privileged access*, is not, and cannot be, a member of the three-fold set of inertial frames. The first-person perspective—the *sine qua non* of conscious experience—is thus incompatible with the Principle of Relativity; it fails to meet the criterion of equivalence with all other frames of reference with respect to the formulation of the laws of physics. From this we conclude that conscious experience lies outside the explanatory scope of standard physics in ordinary, four-dimensional spacetime.

The very existence of the phenomenon of consciousness shows that there is more to this world than meets the eye. When we ask, as we just did, How is consciousness even possible?, we see that we are actually asking a question about physical frames of reference, i.e., about the geometry of spacetime itself. Spacetime as presently understood simply does not support consciousness. But whatever the spacetime structure needed for consciousness may be, it exists whether we are here to take advantage of it or not. Thus the first question about consciousness is one of fundamental physics, not biology. To have any hope of understanding consciousness we claim that we must first identify those geometric and physical features of the world that have allowed consciousness to come into being in the first place. Very likely those features, like consciousness itself, are hidden from view. Moreover, those same features will almost certainly play a role in shaping the non-conscious part of the universe, thereby revealing their presence in an objective, public and observable manner

*1.2.3 The problem of sense data.* There is, of course, more to problem of consciousness than the fact of privileged access, definitive as it is. Equally problematic are such issues as the nature of sense data (qualia),[28] subjective vs. objective time,[29] free will[30] and mind-brain interaction[31]. We shall deal with each of these in some detail in **Chapters 17-19** of this book. For the moment, however, we want to draw particular attention to the problem of sense data, as this aspect of conscious experience, one that has puzzled scientists and philosophers alike at least since Locke and Hume, underscores most emphatically our incomplete understanding of the natural world.

We know that there exist—and have a good understanding of—such things as electromagnetic and acoustic radiation. These are physical things whose generation and behavior are described by known laws of physics. The ontology of these items is manifestly physical—they are publicly observable events situated in the four-dimensional spacetime continuum. We also know that there exist such things as light, color and sound. But, unlike the radiative phenomena giving rise to them, these are privately experienced qualities, existing only in the mind. Outside the mind (including the brain), all is dark and soundless. For light and sound are sense data, qualities experienced by the conscious being, the end result of a complex process of visual and auditory detection and cortical processing of incoming electromagnetic and acoustic radiation. Taste, smell, touch and pain, too, are sense data, the internally experienced results of applied external stimuli. What we do *not* know is the ontology of sense data. We do not know *what* they are, *where* they are in relation to the one experiencing them, or how they came to represent physical (radiative) stimuli. Nowhere in the scientific canon—the laws of spacetime physics—is there even a hint of explanation for the origin and existence of sense data.

Kant said it was a scandal that we could not prove the existence of the external world. (See the **Preface** to this book.) It seems equally scandalous that, not only do we continue to remain in the dark about sense data, one cannot find a single published list of open problems in physics that mentions the problem of sense data, let alone that of consciousness in general. The argument from the internal world of sense data underscores powerfully our ignorance of the basic makeup of the physical universe. For what sort of world can it be that manifests these?

*1.2.4 Where is the particle?* We said above that nothing is more mysterious than consciousness, not even quantum mechanics, and no doubt that is so. After all, we have in hand a well-developed and proven theory of quantum mechanics, whereas for consciousness we have not even a good definition.[32] On the other hand, just because we can make quantum calculations does not mean that we actually understand quantum mechanics. Nobel physicist Richard Feynman claimed that no one understood it, not even himself.[33] Now what exactly did Feynman mean by that? What is it about quantum mechanics that makes it at once computationally accessible yet difficult if not impossible to understand? The answer is, we believe, similar to the one just given for the intractability of consciousness: our current conception of reality is simply inadequate to explain why there is a quantum world at all, rather than the smoothly deterministic one underlying pre-twentieth-century physical thought. It is, in fact, not hard to see that our current formulation of basic quantum mechanics is radically incomplete, illustrating forcefully our ignorance of the overall makeup of the natural world and perhaps explaining why it is that, according to Feynman, no one understands quantum mechanics. To begin let us recall very briefly the origins of the quantum idea. By the end of nineteenth century it was obvious to everyone that matter consisted of particles, and light and radiant heat of waves. But this neat classical dichotomy of wave and particle would not survive the beginning decades of the twentieth century. In 1900 Max Planck was able to explain the spectrum of black body radiation (distribution of radiant energy vs. wavelength at a given temperature) by assuming that the material oscillator emits radiant energy, not continuously with wavelength as one assumes in classical theory, but in a discrete set of closely spaced energy levels.[34] In 1905 Albert Einstein, in an effort to explain the *photoelectric effect (photoemission)*, went even further and proposed to identify Planck's steps in radiant energy with actual corpuscles of light now called *photons*, discrete entities guided by the wave structure of the incident light.[35] Just as radically, in 1924 Louis de Broglie, inspired by the symmetry between electron and photon, suggested that the electron, under the right conditions, may exhibit undulatory properties.[36] de Broglie's insight was confirmed in 1927 by Davisson and Germer, who showed that the pattern of reflection of a beam of electrons from a crystal was similar to that obtained in the crystal diffraction of x-radiation.[37] Thus by end of the first quarter of the last century, *duality* of wave and particle was an established experimental fact: sub-atomic units of matter like electrons—objects one normally thinks of as comprising discrete particles—can sometimes act like waves; and electromagnetic radiation—something one normally thinks of as having a wave-like nature—can sometimes act like a particle. Why nature should behave in this strange manner we do not yet know. The fact that it does so comprises, in the writer's opinion, the central mystery of the physical microworld. In any case its confirmed existence drives the need for (a) a wave theory of matter, (b) a particulate theory of electromagnetic radiation and (c) a theory of the interaction of matter and radiation. Quantum mechanics was invented, at least in part, to address theoretically the various aspects and consequences of wave-particle duality.

The efforts of physicists over the last ninety years or so to build a quantum-mechanical theory of the microworld have been well rewarded. For unquestionably quantum mechanics is the most successful physical theory ever devised, a theory with unparalleled explanatory and predictive power. In particular, its most advanced form—the so-called Standard Model of particle physics[38-41]—describes with uncanny precision the interactions between all known particles of matter, e.g., the decay of one kind of particle into new particles, including radiation. Oddly enough, though, this remarkable theory says almost nothing about the makeup of the individual particle. It fails, for example, in what has come to be known as the *family* problem[42-45], to explain why nature provides three *families* or *generations* of leptons and quarks (the foundational building blocks of matter) when just one family would have sufficed to build a perfectly functional universe. Moreover it fails to explain the observed values of the masses of the particles comprising the three families; see **Fig. 1.1**.

To make the Standard Model work, these and a number of other unpredicted physical constants must first be put into it by hand.[46] The family problem alone is a clear sign that, despite its many successes, quantum mechanics as presently formulated is incomplete and in need of emendation. Something basic is missing and the question is: What is it?

To get at the answer, we shall need to go a little more deeply into the wave theory of matter.[47] For the moment let us forget about the creation and annihilation of particles and concentrate on the physics of the single particle. The simplest possible physical system is that of a single particle moving in a prescribed external field—an electron, for example, moving in the electrostatic field of a proton. When the external field is negligible or is turned off, the particle is said to be *free*. What the wave aspect of matter is telling us is that, prior to measurement, and contrary to common sense, the particle has no definite position or velocity.

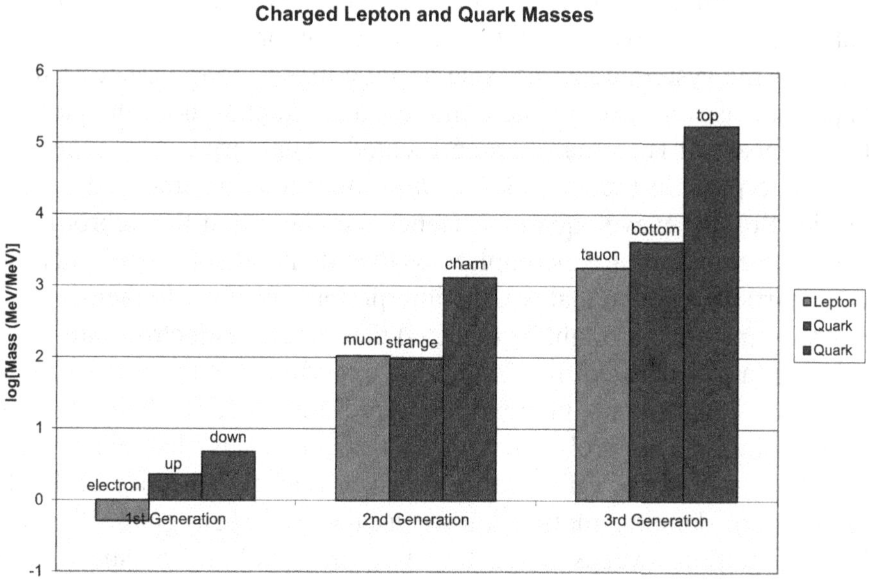


**Fig. 1.1** Logarithmic bar chart (Excel) of the masses of the charged leptons (electron, muon and tauon) and quarks (up, down, strange, charm, bottom, top), grouped according to generation. The masses, ranging over five orders of magnitude, form no obvious pattern. Mass values from J. Beringer *et al.* (Particle Data Group) Phys. Rev. **D86**, 010001 (2012). (URL: http://pdg.lbl.gov). Note: the up, down and strange quark masses are "current-quark masses" whereas the charm and bottom quark masses are "running" masses. The top quark mass is a "direct measurement".

Rather it has only the *probability* of being at a given location or of moving at a given velocity at a given time. This probabilistic behavior is specified by a certain function of space and time called the *wave function* or *probability amplitude* of the particle. It is almost always denoted by the symbol $\psi$ . The wave function $\psi$ determines the outcome of every conceivable experiment on the system—the expected values of the particle's position, velocity, energy and so on. For an electron moving slowly relative to the velocity of light *c*, the evolution of the wave function in time is described by the (non-relativistic) *Schrödinger wave equation*; and for electrons of any velocity not exceeding *c*, by the *Dirac wave equation.* Thus either wave equation, by having for its solution a wave function $\psi$ , determines for all times all that can be known about the dynamical behavior of the particle identified with $\psi$. We note, however—and this is the crucial point—that although the wave equation defines comprehensively the wave nature of the system under consideration, it makes no reference at all to its particle aspect. For nowhere in the wave equation does the particle itself appear. It is formally unrepresented. The corpuscular aspect of the single-particle theory is an *ad hoc* interpretation imposed by the physicist on the wave equation. Hence we claim that the Schrödinger and Dirac equations are incomplete, as they do not display explicitly the wave-particle dualism that we, the interpreters, attribute to them.

As against this, it might be objected that, because electrons are point particles (a fact derived from scattering experiments), their presence in the wave equation is inherently unrepresentable and their absence unsurprising. The problem with this statement is that even point particles exhibit an effective finite size due to an uncontrollable microscopic dancing motion known as the *Zitterbewegung*.[48] This dancing motion sweeps out a tiny but finite volume of space—the volume of the Zitterbewegung. As shown by Schrödinger, some features of the Zitterbewegung are derivable from the Dirac wave function $\psi$. But it is a description in terms of waves. Since it is not waves but the particle that is doing the dancing, there must exist a *second probability distribution* describing directly the point particle's Zitterbewegung. It is precisely this second distribution that is missing in the present formulation of quantum mechanics and whose absence renders it incomplete.

Our discussion so far has been limited to the quantum theory of single particles. In the real world, quantum systems may consist of many particles—particles that can interact, annihilate each other and thereby give rise to new ones. The extension of quantum mechanics to systems of many interacting particles is known as *quantum field theory*.[49] Perhaps quantum field theory resolves the problem of incompleteness just described. Let us see. To get from the single-particle theory to a theory of many particles, one makes use of a recipe called *second quantization*.[50] This is a procedure whereby the single-particle wave function $\psi$ is converted into a field *operator*, a formal entity that regulates the creation and annihilation of particles at every point of space. The relevance of second quantization for the present discussion is just this: Textbooks on quantum field theory commonly state, or imply, that second quantization of the Schrödinger or Dirac wave function "imparts to it some particle properties."[51] Now that is indeed true if all we are talking about are the *numbers* of particles created and annihilated. However, the description of the particle itself—the distribution of its Zitterbewegung—is no more present in quantum field theory than it is in the single-particle theory. In that sense, quantum field theory, like the single-particle theory, is incomplete.

As stated at the beginning of this subsection, modern quantum field theory culminates in the Standard Model of particle physics. The Standard Model is a *gauge theory*, meaning that the interactions it describes arise from the invariance of the theory under certain transformations known as *gauge transformations*.[49] There is much talk in the literature about physics beyond the Standard Model[52]—schemes involving the application of ever-more elaborate invariance principles—and with it the prospects of one day solving, among other things, the family problem. But that enterprise is, the author believes, doomed to failure. For in the end it fails to bring the particle itself explicitly into play, relying instead on operator representations of the particle. On the present view, what one needs to solve the family problem is not physics beyond the Standard Model, but physics *before* the Standard Model, namely, a completion of ordinary quantum mechanics, a revised formalism that expresses fully and symmetrically the dualism of wave and particle.

Going back to the single-particle theory, it should be understood that it is not only the wave equations of Schrödinger and Dirac that omit to show the particle. The equations of Maxwell (1873)[53] governing the propagation of electromagnetic radiation are similarly lacking. For these equations do not display explicitly dualism of wave and photon that we, the interpreters, attribute to them. If they did display the dualism, then it is perhaps Maxwell, not Planck and Einstein, who would today be credited with ushering in the era of quantum mechanics. As they stand, though, Maxwell's equations, like those of Schrödinger and Dirac, must be considered incomplete.

All of which brings us, to conclude this subsection, to what many consider the central outstanding problem of contemporary physics, the problem of *quantum gravity*—the problem of bringing quantum mechanics and Einstein's theory of gravitation—general relativity—together within a common formalism.[54] By somehow merging these two great pillars of physical science, physicists hope to forge a unified account of the four known fundamental forces of nature: electromagnetic, weak nuclear, strong nuclear and gravitational. In fact there exists already a wonderfully successful theory unifying the first two of these fundamental forces, namely the *electroweak sector* of the aforementioned Standard Model of particle physics.[55] The binding of nuclei by the strong force is in turn well described within the Standard Model by a separate yet similar formalism called *quantum chromodynamics*.[56] Theories that attempt to unify the electroweak theory and quantum chromodynamics are called grand unified theories (GUTs).[57] These have not been too successful, mainly because they predict an unobserved decay of the proton. And unifying these two individually-successful theories (electroweak and quantum chromodynamic) with general relativity has, for various technical reasons, proven even harder. Thus it remains to invent the all-encompassing theory, one that binds together the known four forces of nature within a common theoretical framework. Such a theory would—or so it is claimed—amount to a final Theory of Everything, leaving nothing of fundamental importance left to discover. In such a theory even the presently unpredicted constants of the Standard Model are expected to be accounted for. Many physicists think that some version of superstring theory in ten or eleven spacetime dimensions will turn out to be the final theory.[58]

What does all this have to do with the incompleteness problem under discussion here? Just this: It is a requirement of the quantum field theory of matter that it remain form-invariant under gauge transformations.[59] As a consequence of this requirement there arise in the theory an associated set of physical fields called *guage fields*, the presence of which is in fact necessary for the invariance of the theory. With each such field is associated (by way of second quantization) a new particle called a *gauge boson*, the quantum of the field. In quantum field theory, the gauge bosons arising in this manner become the carriers, or mediators, of the forces of interaction between leptons and quarks. In other words, in the quantum microworld the push of one thing on another takes place by means of an exchange of gauge bosons. Thus, within the Standard Model, the *photon* mediates the electromagnetic force; the *W and Z bosons*, the weak nuclear force; and *gluons*, the strong nuclear force; see **Table 1.1**.

| **Table 1.1** Force Mediating Gauge Bosons | | | | | |
|---|---|---|---|---|---|
| Force Mediated | Mediating Gauge Boson | Electric Charge | Spin | Mass (GeV/$c^2$)* | Observed |
| Electromagnetic | Photon ($\gamma$) | 0 | 1 | 0 | yes |
| Weak Nuclear | $W$ Boson ($W^{\mp}$) | $\mp e$ | 1 | 80.385 | yes |
| | $Z$ Boson ($Z^0$) | 0 | 1 | 91.188 | yes |
| Strong Nuclear | Gluon (g) | 0 | 1 | 0 | yes |
| Gravitational | Graviton (G) | 0 | 2 | $< 7\times10^{-41}$ | no |
| *Mass values from J. Beringer *et al.* (Particle Data Group), Phys. Rev. **D86**, 010001 (2012). | | | | | |

Significantly, each type of gauge boson occurring within the Standard Model not only was predicted theoretically but has been experimentally observed.[60-62] Now, what about the gravitational force? If it is to be united with the other three forces of nature in a Theory of Everything, then by symmetry it too should be mediated by a gauge boson. Physicists refer to this putative force-mediating particle as the *graviton.*[63] Does such a particle exist? We do not know. Even if it should exist, the interaction of the individual graviton with ordinary matter is far too feeble for it to be detected experimentally.[64] And, thanks to the incompleteness of quantum mechanics claimed here, neither do we have any good theoretical evidence for its existence. For the corpuscular aspect of the gravitational force—if there is such an aspect—is unrepresented in the gravitational field equations, just as it is unrepresented in the field equations of Dirac and Maxwell. The viability of string theory as an explanation of the world rests entirely on the existence or non-existence of the graviton. In a real sense, the completeness question raised here bears directly on the very future of theoretical particle physics.

## 1.3 Chapter abstracts

The book beyond this introductory chapter is divided into eight parts—various aspects of the physical world distinct yet connected by an underlying Principle of True Representation. To see in advance the scope of these connections, herewith are provided abstracts of the book's remaining nineteen chapters.

### PART I THE UNIFICATION OF APPEARANCE AND REALITY

**Ch. 2 The Principle of True Representation and the Law of Laws**. A world-model is postulated of sufficient generality to support the ontologically disparate phenomena of mind and matter. The model proposed is not one of any particular physical process. Rather it is a model of reality itself, in Wittgenstein's striking phrase, of all that is the case.[65] This we carry out by way of two time-honored concepts from the history of philosophy, namely, the Kantian dualism of noumenon and phenomenon; and the Cartesian dualism of mind and matter. Next we impose on our world-model a formal condition of scientific truth, a demand for self-consistency with respect to the totality of its parts. This core condition—from which the book takes its title—is designated the *Principle of True Representation.* The formal, symbolic expression of this principle comprises a new, comprehensive law of physics—a *Law of Laws*—from which we shall extract specific dynamical laws, each having the notable property of being true by definition.

**Ch. 3 Geometrization of the Law of Laws: $M^4 \times \bar{M}^2$ spacetime.** The Law of Laws is interpreted terms of a self-imaging wave field $\chi$ propagating in a six-dimensional product space $M^4 \times \bar{M}^2$, where $M^4$ is 4-D Minkowski space with the usual coordinates $(x^0, \mathbf{x})$; and $\bar{M}^2$ denotes a pseudo-Minkowski space consisting of two flat and infinite dimensions $x^5$ and $x^6$. Here the qualifier 'pseudo' signifies that, although $x^5$ is metrically time-like, it acts as if it were an additional spatial dimension; and although $x^6$ metrically space-like, it plays the role of an extra temporal dimension. Propagation of a field $\chi$ characterized by pseudofrequency $p^6 c/\hbar$ and pseudomomentum $p_5$ is directed parallel to $x^5$ under the action of a linear wave propagator $\hat{R}$. In a Fourier expansion of $\chi$ in coordinate $x^5$, the observable $n$th mode amplitude $\chi_n(x^0, \mathbf{x})$ is independent of both $x^5$ and $x^6$. Thus spacetime *looks* 4-dimensional. In a quantum-mechanical interpretation, propagator $\hat{R}$ represents the particle, and amplitude $\chi_n$ its guiding wave. Expansion index $-n$ represents the quantized values of electrical charge, $Q$, in the case of leptons, or quantized values of $Q+B$ in the case of quarks, where $B$ is the baryon number. In a multiparticle context, only those reactions in $M^4$ can occur that conserve pseudomomentum $p_5$.

## PART II SINGLE-PARTICLE LAWS OF PHYSICS

**Ch. 4 Self-imaging fermions: The spacetime origin of flavor.** The specific form of the geometrized Law of Laws is dictated by the behavior of field $\chi$ under spacetime transformations. In the case that $\chi$ is a Dirac spinor, the Law of Laws becomes a 6-D version of the Dirac equation. Thus it applies to single particles of spin ½: the fermion. The generic family $f$ of fermions is represented by $\chi \equiv \psi$, a self-imaging field propagating in $M^4 \times \overline{M}^2$. Expansion of $\psi$ in eigenstates of the fifth momentum operator $i\hbar\partial_5$ yields an infinite series of terms whose coefficients $\psi_n$ may be interpreted as fermion flavors belonging to family *f*. Their masses $m_n$ are given by $m_n c = (G_f^2 - p_{5n}^2)^{1/2}$, where $G_f = (p_a p^a)^{1/2}$ $(a = 0,1,2,3,5)$ is the radius of the five-momentum shell defining family *f*, and $p_{5n}$ is the pseudomomentum of the $n$th member of the family. Thus fermion mass is determined by spacetime geometry, not interaction with the vacuum Higgs field. In $M^4 \times \overline{M}^2$, (1) all flavor neutrinos are massless, yet may oscillate, and in fact *must* oscillate to exist; (2) neutrinos of different flavor are distinguished physically by their respective values of pseudomomentum $p_5$; and (3) baryon and lepton family numbers are strictly conserved.

**Ch. 5 Self-imaging massive bosons: Klein-Gordon and Maxwell-Proca equations in six spacetime dimensions.** Nature provides bosons of scalar, vector and antisymmetric tensor form. All three are captured in $M^4 \times \overline{M}^2$ by writing the general field $\chi$ as an expansion in the fifteen Dirac-Clifford matrices, plus the $4 \times 4$ unit matrix. The three boson fields arise as the coefficients of the expansion. Then, for massive scalar fields, the Law of Laws yields the scalar Klein-Gordon equation, describing free motion of the Higgs. For massive vector fields, it yields two Maxwell-Proca equations, one charge neutral ($p_5 = 0$) and the other charged ($p_5 \neq 0$). These two vectorial equations describe, respectively, free motion of the $Z^0$ and $W^{\pm}$ bosons. However, in $M^4 \times \bar{M}^2$, vectors have five components. Accordingly, the Law of Laws formalism generates two additional field equations, these describing the free motion of two spin-0, pseudoscalar bosons, $Z^5$ and $W^5$, with masses $m_Z$ and $m_W$, respectively. These predicted new particles interact with the rest of the world gravitationally only, and thus can be considered candidates for *dark matter*.

**Ch. 6 Self-imaging massless bosons: Maxwell's equations in six spacetime dimensions.** For massless tensor fields ($p_5 = p^6 = 0$), the Law of Laws yields in $M^4 \times \bar{M}^2$ two sets of Maxwell's equations. The first set entails the standard Maxwell tensor $F^{\mu\nu}$ and driving vector current $j^\mu$. Although the homogeneous equations of this first set (Bianchi identity) make their appearance unchanged, the inhomogeneous equations now contain the gradient of a constant scalar field $\phi$, a candidate for dark energy coming unexpectedly out of electromagnetic theory. The second set, analogous in structure to the first set, entails the four extra tensor components $F^{5\nu}$ and extra current component $j^5$. The $F^{5\nu}$ are derivable from a fifth vector potential component, $A^5$. However, $A^5$ is not a gauge field. Hence the longitudinally polarized photon coming from the second set, while energy bearing, is *dark*. As shown later in **Chapter 13**, the fifth component of the electromagnetic energy density, $\varpi^a$, when quantized, figures centrally in our calculation of the mass spectrum of the charged leptons.

## PART III GEOMETRODYNAMICS

**Ch. 7 General covariance.** For ready reference, a review is given of the formulas and conventions of tensor analysis, extended where necessary to six dimensions. The Law of Laws and the field equations proceeding from it described thus far in flat spacetime are shown to hold good in curved spacetime as well. In other words, the equations are generally covariant. The stage is set for new look at the interaction between spacetime and matter.

**Ch. 8 Geometrodynamics in 4 + 2 dimensions.** The Law of Laws has so far given us field equations of spinor, scalar, vector and rank-2 tensor form—laws describing the motion of free particles against the background stage of spacetime. The Law of Laws is a machine, now asked to describe the interaction between spacetime and matter. In response it generates a new set of field equations—a set analogous to Maxwell's equations—for the motion of a self-imaging rank-4 tensor, $F_{\alpha\beta\mu\nu}$. If this field tensor is assumed proportional to the Riemann-Cristoffel (R-C) curvature tensor, then the rank-3 matter current $j_{\nu\alpha\beta}$ appearing in the inhomogeneous equations of the set is conserved. Thus the R-C tensor enjoys the same degree of physical reality as the Maxwell $F_{\mu\nu}$. In this way 4-D spacetime $M^4$ is shown to be self-imaging. The Einstein gravitational field equations are found embedded in the homogeneous equations of the R-C set, i.e., the Bianchi relation.

With the R-C field $F_{\alpha\beta\mu\nu}$ is associated a Lorentz-like fifth force of Nature, a force distinct from the gravitational force of Einstein. Its strength is governed by a new fundamental constant, $\ell$, having the dimension of [Length]. Its magnitude is not given *a priori* but can be estimated theoretically. The first step is to solve the R-C field equations, a task most easily accomplished by the familiar potential method. To test its efficacy, the method is employed to find the metric interior and exterior to a spherical mass. The metric values obtained are comparable to those found by Schwartzschild, giving confidence not only in the potential method, but in the truth-value of the R-C field equations themselves. The next step, with the relevant potentials already in hand, is to calculate the R-C force exerted by the spherical mass on itself. It is found to be repulsive, counteracting the gravitational self-force of the sphere. The final step is to assume that the 'singularity' at the center of a black hole is not singular but finite owing to a balancing of the R-C and gravitational self-forces acting upon it. This results in an estimated value $\ell \sim \ell_P$, where $\ell_P$ is Planck's constant. It is suggested that the force of the Big Bang may have been the repulsive R-C self-force acting upon an initial singularity of radius $a << \ell$.

The energy of the R-C field does not gravitate. But there does exist a particle associated with the R-C field, one we call the *riemann*. This spin-1, photon-like object mediates the R-C force. However, because the Einstein field equations are embedded in the equations of Riemann-

Cristoffel, and propagator $\hat{R}$ is already identified with the riemann, no mediating particle is predicted for the gravitational force. The graviton, in other words, does not exist. This places string theory, in which the graviton plays a central role, in jeopardy. It throws doubt as well on the possibility of unifying gravity and quantum mechanics.

**Ch. 9 Cosmic expansion.** We know that a complete description of Nature requires the rank-2 Maxwell tensor $F_{\mu\nu}$, and possibly the rank-4 R-C tensor $F_{\alpha\beta\mu\nu}$ as well. The present chapter explores the potential relevance of a tensor field of rank intermediate between the other two, namely, the rank-3 object $F_{\alpha\mu\nu}$. If it is assumed antisymmetric between its latter two indices, then the Law of Laws yields for it a set of field equations analogous to those of Maxwell and Riemann-Cristoffel. The rank-2 matter current $j_{\nu\alpha}$ appearing in the inhomogeneous equations of the set is conserved in flat space, but not in curved space. Thus the new equations may apply to a description of the universe as a whole, which is already known to be flat within 0.04% margin of error.

With the rank-3 equations is associated a sixth force of nature. Its form, like that of the R-C force, is analogous to the Lorentz force of electromagnetism and depends linearly on the components of the rank-2 current density $j_{\nu\alpha}$. Current $j_{\nu\alpha}$ is proportional to the *gradient* of the convection current $j_\nu = \rho v_\nu$, where $\rho$ is the mass density and $v_\nu$ is the four-velocity. The sixth force is therefore an edge effect, and for that reason we designate it the *Cosmic Casimir (C-C) force.*

As gravitation is already accounted for by the R-C equations, the mass density $\rho$ cannot be gravitating. The only plausible candidate for it is the vacuum density $\rho_{\text{QFT}}$ given by quantum field theory, a density so great—~124 orders of magnitude larger than the mass density of the universe—that if it *did* gravitate, the universe would have collapsed soon after its inception. If $\rho_{\text{QFT}}$ is assumed to be uniformly distributed throughout the universe, it follows that if the C-C force exists anywhere, it is at the edge of a bounded universe, the only place where the gradient of $j_\nu$ would be non-vanishing. In that case, the C-C force becomes a candidate for the cause of cosmic expansion, provided it acts outwardly on the boundary of the universe.

Employing a potential formulation of the field equations, we find that the C-C force at the edge of the universe yields an equation equivalent to Friedman's for the time evolution of the scale factor $a$ in the

Robertson-Walker metric, provided that present-day non-gravitating mass density $\rho \sim (D_{\mathrm{U0}}/\ell)^2 \rho_0$, where $D_{\mathrm{U0}}$ is the present radius of the physical universe (as opposed to observable radius, $D_{\mathrm{OBS}}$), $\ell \sim \ell_P$, where $\ell_P$ is Planck's constant, and $\rho_0$ is the present-day observed mass density of the universe. If $\rho \sim \rho_{\mathrm{QFT}}$, then there results a prediction for the radius of the physical universe: $D_{\mathrm{U0}} \approx 3.75 D_{\mathrm{OBS}}$, a value very much smaller than that suggested by Guth inflation.

A comparison is given between three theories of cosmic expansion: the present one deriving from non-gravitating $\rho_{\mathrm{QFT}}$ in a bounded universe; the standard model of expansion induced by a cosmological constant; and one deriving from negative gravitating mass in a bounded universe.

## PART IV INTERNAL STRUCTURE OF THE FIELD QUANTA

**Ch. 10 Internal structure of the leptons and implications for baryon asymmetry.** Each flavor of lepton, whether charged or neutral, consists of a point particle *P* undergoing Zitterbewegung, giving to the lepton an effective finite size. The kinematics of point *P* can be described classically. In the case of the massless neutrino, point *P* is superluminal and follows a helical path about the mean direction of motion, generating the neutrino's intrinsic spin angular momentum. Apart from scale, this helical structure is the same for all three neutrino flavors. The charged leptons are slightly more complicated. In case of the electron, *P* is massless and follows a circular path at light speed *c*. In mu and tau, *P* has imaginary mass and follows at superluminal speed a helical path with circular orbital centerline. In all three cases, point *P*'s net motion generates spin. These classical depictions, crude as they are, indicate that mu and tau are not just more-massive versions of the electron.

To obtain a probabilistic picture of the internal geometry of any given lepton, we first integrate out the time variable in the corresponding wave propagator $\hat{R}$. This results in a probability distribution $D(\mathbf{x})$ for the position of point *P* within the volume of the Zitterbewegung. Then from the wave-mechanical, *external* charge five-current density $j^a$ one readily extracts the *internal* charge density $J^a$ dependent on distribution *D*. Similarly, from the external momentum five-current density $\varpi^a$ we extract the internal momentum current density, $\Pi^a$, also dependent on

*D*. These internal current densities, while representable as theoretical constructs, are not directly observable. But they do have observable effects, as is shown here in detail. To illustrate, calculating with these internal distributions, we obtain correct values of spin angular momentum and spin magnetic moment, as well as the form of the Dirac Hamiltonian of the hydrogen atom. Internal probability and momentum five-currents are obtained for the neutrino as well. From the latter we obtain correct spin vectors for the negative helicity neutrino and positive helicity antineutrino.

Probability distribution *D* contains negative densities along with the positive, a structural fact uninterpretable in Euclidian $E^3$. To accommodate this peculiar bipolar feature of the Zitterbewebung one is led to posit the presence of an internal, 2-D spacetime, $N^2$, a Minkowski-like geometry of one infinite temporal dimension, $\tau = u^0 / c$ and a bounded spatial dimension, *u*. With this addition, total spacetime enlarges to an 8-D product manifold $M^4 \times \bar{M}^2 \times N^2$. The Zitterbewegung now involves correlated motion of point *P* within both $E^3$ and $N^2$, where motion against positive *u* is encoded as negative values of *D*, and motion with positive *u* as positive values of *D*.

As $M^4 \times \bar{M}^2 \times N^2$ comprises a slab geometry, it is suggested that baryons reside on one of its bounding surfaces, and antibaryons on the other, thus potentially resolving the problem of missing antimatter.

**Ch. 11 Internal structure of the bosons.** Scalar, vector and tensor bosons are shown to have internal structure just as the leptons do. Like the leptons, their probability and energy densities are attributable to the Zitterbewegung of an internal point *P*. However, in contradistinction to the case of the leptons, the spin of the vector and tensor boson is not generated the Zitterbewegung. Rather it derives from an internal momentum density generated by the polarization of these objects. Thus the zero spin of a linearly polarized photon is attributable to a superposition of the oppositely-directed spins of its right- and left-hand circularly-polarized components.

## PART V THE FAMILY PROBLEM

**Ch. 12 Neutrino mixing and oscillation.** The theory of flavor presented in **Chapter 4** is applied to some unsolved problems in neutrino mixing and oscillation. In this theory, flavor neutrinos $|\nu_\ell\rangle$ arise in $M^4 \times \overline{M}^2$ as states proportional to eigenstates $|\ell\rangle$ of the fifth momentum operator $i\hbar\partial_5$. The $|\nu_\ell\rangle$ are in turn expandable in terms of eigenstates $|\ell j\rangle$ of the sixth momentum operator $i\hbar\partial_6$. The product state $|\ell\rangle|\ell j\rangle$ is the *j*th mass-squared eigenstate and interference between *N* of these gives rise to flavor oscillation. As the flavor neutrino carries a $U(1)$ charge, namely $p_5$, Majorana mass terms in the Lagrangian are forbidden. Thus flavor neutrinos are Dirac particles.

Against orthodoxy, flavor neutrinos are shown to be massless. Yet not only can they oscillate, but must do so to exist. And since they are massless, at least one of the mass-squared eigenstates must be tachyonic. And from *that* fact it follows that right-handed flavor neutrinos do not exist.

Flavor oscillation must occur while conserving lepton number, $L_\ell$. To ensure this, all mass-squared eigenstates carry lepton number. If there are 3 families and $3+n$ flavors, where $n$ is the number of sterile flavors, then the number of flavor states is $3\times(3+n)$. The extra flavor states beyond 3 are created by in-flight interference far from the production point and do not violate the known width of $Z^0$.

At least one sterile flavor is necessary, permitting the electron neutrino's mass to vanish. The LSND, MiniBooNE and reactor anomalies arise as side effects. Our predicted values of oscillation parameters $\Delta m_{41}^2$ and $\sin^2\theta_{24}$ conform to published values.

Absent a sterile flavor, mixing between $\nu_\mu$ and $\nu_\tau$ is predicted to be maximal, a consequence of the masslessness of the flavor neutrino. The presence of a sterile flavor pushes the mixing away from maximal, an effect confirmed by published values of $\sin^2\theta_{23}$.

**Ch. 13 Charged lepton mass formula.** It is commonly said, in light of the structure of the Standard Model of particle physics, that the quarks and charged leptons get their masses by interaction with the vacuum Higgs field. The SM fails, however, to the predict values of these masses. That is to be expected. For as was shown in **Chapter 4**, mass comes from extra dimensions, and the SM knows nothing of these. The present chapter undertakes to calculate the masses of the charged leptons by identifying those objects with the excited states of an electron trapped in a potential well in inner spacetime, $N^2$. We start by assuming that the extra rest energy of the generic lepton above that of the electron derives from the difference $\Delta U$ between the classical self energies of the particle as computed in two different reference frames, separated by a finite rotation angle in the $x^0$-$x^5$ plane of $M^4 \times \bar{M}^2$. Energy $\Delta U$ depends on both the particle's total mass $m$ and fifth component, $p_5^{\text{em}}$, of the electromagnetic momentum. Identifying the latter with the quantized energy spectrum $E_n$ of the trapped electron, we obtain an expression for $m$ given in terms of quantum number $n$. This yields calculated masses for $\mu$ and $\tau$ in good agreement with the accepted average experimental values. Also predicted is a hidden (dark) lepton $\sigma$ of mass intermediate between the masses of $\mu$ and $\tau$. All three charged leptons figure in defining the masses of the quarks, as will be shown in **Chapter 14**. The lepton series terminates at $\tau$ owing to the finite well depth. The mass of the electron is not predicted; its value is determined geometrically by the width of the well in which it is trapped.

**Ch. 14 Bare quark mass formulas.** Quark masses $m_q$ are calculated from the fermion mass formula $m_q c = (G_q{}^2 - p_5{}^2)^{1/2}$, where $G_q$ is the radius of the 5-momentum shell belonging to quark flavor $q$, and $p_5$ is its pseudomomentum. The $p_5$ value for quark $q$ is determined empirically by imposing conservation of $p_5$ on a suitable decay reaction $q \to \cdots$. Each radius $G_q$ depends on the mass of one of the charged leptons. Thus quarks $u$ and $d$ are coupled to $m_e$; $s$ to $m_\mu$; $c$ and $t$ to $m_\sigma$; and $b$ to $m_\tau$. Here $m_\sigma$ denotes the mass of the dark lepton $\sigma$ described in **Chapter 13**. The mass values given by these formulas agree well with those of the observed quark mass spectrum.

## PART VI THE STRUCTURE OF SPACETIME

**Ch. 15 Why are there three spatial dimensions?** To answer this, we calculate the Zitterbewegung of an electron in $N$-dimensional Euclidian $E^N$ and show that only when $N = 3$ is the Zitterbewegung—and hence the electron—physically viable. For even $N$ the radial probability distribution is in each case smeared out over the volume of Zitterbebegung, suggesting a degree of non-reentrant internal orbital motion, leading to eventual decay of the particle. In other words, the electron cannot exist in a universe of 2, 4, … spatial dimensions. For $N = 1$ the internal distribution consists of two isolated peaks, implying the absence of internal motion and hence failure to create rest mass. So a universe of one spatial dimension does not support the existence of the electron. For odd $N \geq 3$, the radial distribution consists of $(N-1)/2$ sets of sharply defined peaks, all falling on top of each other at the outer boundary of particle. Thus if $N > 3$, there will be two or more sets of peaks superimposed at the boundary. To each such set of peaks there correspond different values of spin and different values of charge, implying violations of both spin and charge conservation during the motion of internal point *P*. Thus, a universe of 5, 7,... spatial dimensions does not support the existence of the electron. For $N = 3$, however, the radial distribution entails a single sharply defined set of peaks at the electron's outer boundary, implying reentrant orbits and conservation of both spin and charge. A universe of three spatial dimensions thus supports the existence of the electron, and is the only universe able to do so.

**Ch. 16 The spacetime origins of electric charge and the fine structure constant.** In **Chapter 3** we saw that quantization of electric charge arises from an expansion of the family wave function in eigenstates of the fifth momentum operator, $i\hbar\partial_5$. But where does the physical attribute *charge* itself come from? Here we show that the $u^0$-$x^5$ plane of $\bar{M} \times N^2$ may be occupied by a doubly-periodic distribution of self-imaging charge, each period of which is charge neutral. To realize such a picture, an infinite periodic lattice of charge of amplitude $+e$ and period $a$ is assumed to occupy the $u^0$ axis of the $u^0$-$x^5$ plane. This is the source of a self-imaging charge field, $q(u^0, x^5)$, which propagates with wavelength $\lambda$ in direction $x^5$, driven by a two-dimensional Helmholz equation. By free-space propagation alone, the source field generates interference images of itself—so-called Fourier images—in planes spaced apart at intervals $\Delta x^5 \equiv D = 2a^2/\lambda$ perpendicular to the $x^5$ axis. In planes intermediate between any two Fourier-image planes there are formed additional periodic lattice structures, their amplitudes, together with the source amplitude, representing the integer and fractional charges of all fundamental particles: $\mp e$, $\pm 2e/3$, $\mp e/3$. The fractional charges are grouped just as they are found in the proton, neutron and their respective antiparticles. On these theoretical grounds, it appears at least plausible that charge is intrinsic to spacetime.

Also plausible is the possibility that the ratio $D/\lambda$ is an Eisenstein prime, 137, suggesting that the fine-structure constant $\alpha$ is ultimately a product of extra dimensions. If so, the value of $\alpha$ measured in $M^4$ would then derive from reciprocal prime $137^{-1}$ by an effective index of refraction $\nu = \alpha / 137^{-1} = 0.9997$.

**Ch. 17 Dual spacetime, dual time series and reduction of the state vector.** In physics, time is a dimension, on the same footing as space. But in conscious experience, it is a moving present moment. We show here how the two conceptions of time arise naturally out of a previously hidden world geometry, *dual spacetime*. To reveal the new geometry, we first transform the 6-D Dirac equation described in **Chapter 4** into one that *looks* four dimensional. We then examine the form of this equation under a real rotation $\vartheta$ in the $x^0$-$x^5$ plane. We find form invariance of the Dirac equation for two and only two rotation angles, $\vartheta = \pm\pi/2$. Denoting the original and rotated frames by $K$ and $K_D$, respectively, and assuming $\vartheta = -\pi/2$ so that $x^0$ and $x'^5$ point in the same direction, we conclude that $K$ and $K_D$ represent separate, mutually inaccessible realms of spacetime. Superimposed, they become dual spacetime, $K + K_D$.

In frame *K*, time $x^0$ represents one dimension of a static, 4-D block universe, a universe characterized by the so-called B-series of temporal succession: earlier-simultaneous-later. This same block universe exists as well in $K_D$ but with $x^0$ replaced by pseudospatial $x'^5$. In $K_D$ there exists neither time nor energy; the B-series in $K_D$ is one of pseudospatial succession: before-coincident-after.

Dimensions $x^0$ and $x'^5$ are connected by the rotation transformation $x'^5 = x^0 = ct$. This is also an equation of motion. It describes the motion in $K_D$ of a 3-D hyperplane $\mathbb{N}$ of infinite extent oriented perpendicularly to the $x'^5$ axis and moving in a direction parallel to that axis at speed *c*. Hyperplane $\mathbb{N}$ is a surface of simultaneity and a realization of the *moving present moment*, or *now*. In relation to it we have the so-called A-series of temporal succession: past-present-future.

Dual spacetime resolves the paradox of state vector collapse, i.e., the question of how entangled particles A and B know instantaneously how to behave when one of them is measured, even if widely separated at the time of measurement. In dual spacetime every event receives a dual description, one in *K*, the other in $K_D$. In *K*, the interval between A and B at the moment of collapse is space-like, and instantaneous communication between them impossible. In $K_D$, however, they are connected by moving present moment $\mathbb{N}$, a line of simultaneity, thus resolving the paradox.

## PART VII THE HARD PROBLEM OF CONSCIOUSNESS

**Ch. 18 Consciousness and the world.** The problem of consciousness is "hard" because it requires finding out how objective brain processes can be accompanied by subjective experience. We know that it is a problem in physics, because the first-person perspective is a frame of reference, and reference frames are the foundation stones of physics. The first-person frame is unique in two ways: it (1) is privileged and (2) expresses time in terms of a moving present moment. Dual frame $K_D$ is thus a logical candidate for the geometric home of conscious experience. The observer within this frame—the one who experiences—is commonly called the *Self*. Hence the following operational definition: *Consciousness* is the capacity of the individual or Self to read, understand and act upon internal representations of external events. It is, in other words, the capacity for perception and will, managed by understanding.

No man made machine, built out of ordinary matter, can be conscious. For conscious Selves occupy privileged frame $K_D$, an essential component inaccessible and unavailable to the builder of machines.

**Ch. 19 Mind-brain interaction in dual spacetime.** Two forms of mental operation are discussed in detail: perception and will. We find that neither operation can be understood in naturalistic, neurobiological terms alone. The naturalistic account of perception fails on three counts: (1) The binding operator required to construct from cortical data an image of the sensory input does not exist. (2) Even if it *did* exist, it should be unable to construct first-person mental images from third-person cortical data. (3) Nowhere in the brain are such qualia as color and pitch to be found; the brain is everywhere dark and silent. As for the will, according to modern neurobiology, free will is an illusion. But this cannot be so. For if volition is deterministic, then each voluntary act begins, absurdly, with an uncaused cause.

These defects disappear, however, when perception and will are interpreted in dual spacetime. While the conscious being's brain resides in frame $K$, her brain dual and mind occupy $K_D$. In perception, what she perceives are projections into the mind of the direct images of her environment formed by the perceptual apparatus, e.g., the retina or cochlea. These are presented to the mind as dual world lines in past block time, lines which, when scanned by the moving present moment, yield to the percipient the primary sensations of form and motion. The secondary sensations, or qualia, such as those of light, color and pitch, obtain as dualistic companions of the scanned worldlines. Parallel scanning of dual world line data from the cortex provides for understanding.

Will can be conceived as perception run in reverse. The intent of the will, i.e., the motor objective, is generated consciously as a dual world line in future block time. Because $K_D$ is parametrized by pseudo-spatial $x'^5$ rather than temporal $x^0$, this creation of the motor objective is carried out without expenditure of energy, undermining the conventional argument against dualism. Realization of the motor objective occurs sequentially as the moving present moment scans the motor objective's world line.

Our interactionist theory of mind and brain correctly predicts the unexpected results of the Libet experiment, corroborating mind-brain dualism and refuting the widely-held conclusion, drawn from that famous experiment, that free will is an illusion. A simple test of the connection between perception and will is presented. This test, performed on oneself, at once corroborates the present account of mind-brain dualism and falsifies biological naturalism.

## PART VIII CONCLUSION

**Ch. 20 From noumena to qualia.** This final chapter offers, in tabulated form, an annotated list of thirty-one problems dealt with in this book by the Principle of True Representation. Some of the results are explanatory in nature, while others make predictions, exposing the underlying Principle itself to falsification. Suggestions are made for future applications.

## Notes and references

[1] *The Structure of Scientific Revolutions, 2nd Ed.* (University of Chigago, Chicago, 1970), p. 206. Here Kuhn is saying that while theories are useful for 'puzzle-solving', this does not mean that improved theories get us closer to knowing what the world is "really" like. For Kuhn, the external world is unknowable in principle.

[2] For a compact account of logical positivism and related bibliography see *A Companion to Epistemology*, J. Dancy and E. Sosa, Eds. (Blackwell, Oxford, 1992), pp. 262-264. Although the influence of positivism in philosophy is not what it was in the 1920s and 1930s, the related debate between realists and antirealists on the status of the external world remains alive and well. See, for example, *Scientific Realism*, Jarrett Leplin, Ed. (University of California Press, Berkeley, 1984) and *Reality, Representation, and Projection*, J. Haldane and C. Wright, Eds. (Oxford University Press, Oxford, 1993). In the world of science, a positivist attitude still prevails; for most scientists (excepting, perhaps, string theorists), direct experience remains the sole source of knowledge of the physical world.

[3] Karl R. Popper, *Realism and the Aim of Science* (Rowman and Littlefield, Totowa, New Jersey, 1983), p. 215.

[4] For a sampling of the vast literature on the mind-body problem, see *Philosophy of Mind: Classical and Contemporary Readings*, David J. Chalmers, Ed. (Oxford University Press, Oxford, 2002). For a lucid critique of contemporary approaches to the problem see John R. Searle, *The Mystery of Consciousness* (New York Review, New York, 1997), supplemented by Searle's more recent article "Can Information Theory Explain Consciousness?" *New York Review of Books*, January 10, 2013.

[5] Robert Hazen, *Genesis: The Scientific Quest for Life's Origins* (Joseph Henry Press, Washington DC, 2005).

[6] Gordon Kane, *Modern Elementary Particle Physics* (Perseus, Cambridge, MA, 1993), Ch. 1. This volume remains relevant for its articulation of the foundations, which have changed little if at all in the years since the book's publication.

[7] Gordon Kane, "The Mysteries of Mass," Scientific American, June 27, 2005. Brings Ref. 6 (almost) up to date.

[8] Martin L. Perl, "The Leptons After 100 Years," Physics Today, pp. 34-40 (October 1997); W. C. Haxton and B. R. Holstein, Am. J. Phys. **68**, 15 (2000).

[9] For a thorough history of the problem from an anthropic perspective see J. D. Barrow and F. J. Tipler, *The Anthropic Cosmological Principle* (Oxford University Press, Oxford, 1986), pp. 258-276. See also M. Tegmark, "On the dimensionality of spacetime," Classical and Quantum Gravity **14**, L69-L75 (1997), [arXiv:gr-qc/9702052].

[10] B. Greene, *The Fabric of the Cosmos* (Alfred A. Knopf, New York, 2004).

[11] S. Perlmutter, "Supernovae, Dark Energy, and the Accelerating Universe," Physics Today 53-59 (April 2003).

[12] S. Weinberg, *Cosmology* (Oxford University Press, Oxford, 2008), pp. 185-200.

[13] A. Kashlinsky, F. Atrio-Barandela, D. Kocevski and H. Ebeling, "A measurement of large-scale peculiar velocities of clusters of galaxies: results and cosmological implications," arXiv:0809.3734.

[14] J. Horgan, *The End of Science* (Abacus, London, 1998).

[15] J. Horgan, *The Undiscovered Mind* (The Free Press, New York, 1999).

[16] Francis Bacon, *Novum Organum*, T. Fowler, Ed. (Oxford, 1878-1879).

[17] "Nature, Philosophical Ideas of," in *The Encyclopedia of Philosophy*, Volumes 5 and 6 (Macmillan, 1967); "Natural Philosophy," in *The Cambridge Dictionary of Philosophy* (Cambridge U. Press, Cambridge, U. K., 1995).

[18] Rudolph Carnap, *An Introduction to the Philosophy of Science* (Basic Books, New York, 1966), p. 8-9.

[19] "Physicalism," in *The MIT Encyclopedia of the Cognitive Sciences* (MIT Press, Cambridge, MA, 1999); E. B. Davis and R. Collins, "Scientific Naturalism," in *Science and Religion*, G. B. Ferngren, Ed. (Johns Hopkins U. Press, Baltimore, 2002), pp. 322-334.

[20] Owen Flanagan, *The Science of Mind* (MIT Press, Cambridge, 1991), p. 369.

[21] L. Wittgenstein, *Tractatus Logico-Philosophicus* (Routledge & Kegan Paul, London, 1961), Proposition 6.44.

[22] David J. Chalmers, *The Conscious Mind* (Oxford University Press, Oxford, 1996), p. xii. According to Chalmers, this is the "hard problem" of consciousness: "Why is all this [brain] processing accompanied by an experienced inner life?" See also Searle, *The Mystery of Consciousness*, *op. cit.*, p. 28; Thomas Nagel, *Mind and Cosmos: Why the Materialist Neo-Darwinism Conception of Nature Is Almost Certainly False* (Oxford U. Press, Oxford, UK, 2012), as well as a biologist's review of Nagel's book: H. Allen Orr, "Awaiting a New Darwin," *The New York Review of Books*, February 7, 2013, pp. 26-28.

[23] *A Dictionary of Philosophical Quotations*, A. J. Ayer and Jane O'Grady, Eds. (Blackwell, Oxford, 1992), p. 485.

[24] Chalmers, *op. cit.*, p. xii.

[25] A. Einstein, "The foundation of the general theory of relativity," Annalen der Physik **49**, 1916; reprinted in *The Principle of Relativity* (Dover), pp. 109-164.

[26] W. Pauli, *Theory of Relativity* (Pergamon, Oxford, 1958), p. 4.

[27] Jaegwon Kim, *Philosophy of Mind*, 2nd Ed. (Westview Press, Cambridge, MA, 2006), pp. 213-216.

[28] Roger Scruton, *Modern Philosophy* (Penguin, New York, 1994), pp. 330-331.

[29] Scruton, *op. cit.*, 365-377.

[30] Scruton, *op. cit.*, 248-250.

[31] Scruton, *op. cit.*, 209-226.

[32] "Consciousness cannot be defined: we may ourselves be fully aware what consciousness is, but we cannot without confusion convey to others a definition of what we ourselves clearly apprehend." Sir William Hamilton, *Lectures on Metaphysics*, **I**, 191. Quoted in D. Runes, *Dictionary of Philosophy* (Littlefield, Adams & Co. Ames, IA, 1959), p. 64.

[33] Richard Feynman, *The Character of Physical Law* (The M. I. T. Press, Cambridge, MA, 1965), p. 129.

[34] M. Planck, "On the Law of the Energy Distribution in the Normal Spectrum," Ann. Physik **4**, 553 (1901).

[35] A. Einstein, "On a heuristic viewpoint concerning the production and transformation of light," Ann. Physik **17**, 132 (1905).
[36] L. de Broglie, "A Tentative Theory of Light Quanta," Phil. Mag. **47**, 446 (1924); "Recherches sur la Théorie des Quanta," Ann. Phys. **3**, 22 (1925).
[37] Davison and Germer, "Diffraction of Electrons by a Crystal of Nickel," Phys, Rev. **30**, 705 (1927); "Reflection and Refraction of Electrons by a Crystal of Nickel," Proc. Nat. Acad. Sci. **14**, 619 (1928).
[38] S. L. Glashow, "Partial-symmetries of weak interactions," Nuc. Phys. **22**, 579 (1961).
[39] A. Salam and J. C. Ward, "Electromagnetic and weak interactions," Phys. Lett. **13**, 168 (1964).
[40] S. Weinberg, "A Model of Leptons," Phys. Rev. Lett. **19**, 1264 (1967).
[41] A. Salam, in *Elementary Particle Physics: Relativistic Groups and Analyticity (Nobel Symposium No. 8)*, Edited by N. Svartholm (Almqvist and Wiksell, Stockholm, 1968).
[42] A. Salam, "Gauge unification of fundamental forces," Rev. Mod. Phys. **52**, 525 (1980). K. Huang, *Quarks, Leptons and Gauge Fields* (World Scientific, Singapore), p.10.
[43] G. Kane, *Supersymmetry* (Perseus, Cambridge, MA, 2000), p. 26.
[44] G. L. Fitzpatrick, *The Family Problem* (Nova Scientific, Issaquah, WA, 1997).
[45] H. Georgi, *Lie Algebras in Particle Physics* (Benjamin, Menlo Park, CA, 1982), p. 247.
[46] W. N. Cottingham and D. A. Greenwood, *An Introduction to the Standard Model of Particle Physics* (Cambridge U. Press, Cambridge, UK, 1988), p. 184; Lee Smolin, *The Trouble with Physics* (Houghton Mifflin, Boston, 2006), pp. 12-13.
[47] L. I. Schiff, *Quantum Mechanics* (McGraw-Hill, New York, 1955).
[48] E. Schrödinger, Sitzber. Preuss. Akad. Wiss., Physik-math Kl. **24**, 418 (1930).
[49] F. Mandl and G. Shaw, *Quantum Field Theory* (John Wiley and Sons, Ltd., Chichester, U. K., 1984).

[50] Ref. 47, p. 348.

[51] Ref. 47, p. 341.

[52] Rabindra N. Mohapatra, *Unification and Supersymmetry: The Frontiers of Quark-Lepton Physics* (Springer-Verlag, New York, 2003), pp. 64-70.

[53] James Clerk Maxwell, *A Treatise on Electricity and Magnetism, Vol. 2* (Dover, New York, 1954), Ch. XX.

[54] Ref. 46, Smolin, *op. cit.*, pp. 3-5

[55] Ref. 10, pp. 263-266.

[56] Ref. 46, Smolin, *op. cit.*, pp. 62.

[57] Ref. 10, pp. 266-268.

[58] Ref. 10, pp. 327-412.

[59] Ref. 49, Ch. 12.

[60] The photon was hypothesized by Einstein to explain the photoelectric effect; see Ref. 35 and Henry Semat, *Introduction to Atomic and Nuclear Physics* (Rinehart & Co., Inc, New York, 1958), pp.110-113.

[61] The $W^{\mp}$ and $Z^0$ bosons predicted by the Standard Model were discovered at the CERN proton-antiproton collider in 1983. UA1 Collaboration, G. Arnison *et al.*, Phys. Lett. **122B**, 103 (1983); UA2 Collaboration, M. Banner *et al.*, Phys. Lett. **122B**, 476 (1983).

[62] The gluon predicted by quantum chromodynamics was first observed in three-jet events at the DESY facility (in Hamburg): R. Brandelik *et al.* (TASSO Collaboration), "Evidence for Planar Events in $e^+e^-$ Annihilation at High Energies," Phys. Lett. B **86,** 243-249 (1979).

[63] Ref. 52, p. 310.

[64] T. Rothman and S. Boughn, "Can Gravitons Be Detected?," Found. Phys. **36**, 1801-1825 (2006); arXiv:gr-qc/0601043.

[65] Ref. 21, Proposition 1.

# PART I

# THE UNIFICATION OF APPEARANCE AND REALITY

# Chapter 2

# The Principle of True Representation and the Law of Laws

...metaphysics is of no use in furthering output of experimental science. Discoveries and inventions in the land of phenomena? It can boast of none; its heuristic value, as they say, is absolutely nil in that area. From this point of view, there is nothing to be expected of it. There is no tilling of the soil in heaven.

JACQUES MARITAIN[1]

My thesis is that realism is neither demonstrable nor refutable.

KARL POPPER[2]

Experiment is the sole source of truth.

HENRI POINCARÉ[3]

## 2.1 Towards a new conception of the physical world: boundary conditions.

Against the claims of Maritain, Popper and Poincaré we shall in this chapter look to metaphysics to advance the cause of physical science. Our main goal as stated in the in the previous chapter is to locate the source of physical law. We began by asking about what constitutes the world and then entered into a critique of the materialist doctrine. We said that, despite its many explanatory successes, materialism cannot be the whole story because it does not, and cannot, account for the existence of physical law. To drive home the point we noted that the material world as presently conceived does not support the phenomenon of consciousness and conscious experience. In addition we noted that a number of our known laws consistent with materialism are incomplete as they fail to account for the dualism of field and particle. All of which now compels us to ask: "How does one get beyond materialism?" That the world consists at least in part of matter and energy is a clear given. Otherwise, how could one have knowledge of it (leaving aside the skeptical possibility that it is all a dream)? And so our question comes down to this: "What else could there be besides matter?" The answer, simply put, is *boundary conditions*. We are going to define the structure of the world by way of certain constraints placed upon it.

A simple analogy may help point the way. Suppose I am asked by a colleague from a desiccated planet, "What is a river?" I might answer simply, "A river is water—$H_2O$," and be technically correct. But such a response is plainly inadequate. To convey the nature of a river I need to add that it is water confined to a narrow and very long natural trough in the land, gradually changing in elevation to produce flow—or something to that effect. In other words, to define the system completely it is not enough to specify the material alone; one needs also to specify the conditions constraining it.

In a similar vein, the physical chemist and philosopher Michael Polanyi says this of the biological organism[4]:

> The structure and functioning of an organism is determined, like that of a machine, by constructional and operational principles which control boundary conditions left open by physics and chemistry. We may call this a *structural principle*, lying beyond the realm of physics and chemistry.

Our problem has thus narrowed somewhat. Our need now is to find boundary conditions sufficient to define the structure and functioning of the physical world—an application of Polanyi's structural principle writ large. The boundary conditions we seek, though, cannot be physical ones, like the banks and bottom of a river. Instead, they must transcend the physical world, just as physical laws do. In a word, they must be *meta*physical, having a mode of being different from that of the material world they constrain. This, obviously, is where philosophy enters the picture.

Materialism, from a philosophical point of view, is a species of metaphysical *monism*.[5] It admits to a world of one thing only—matter and its interactions. To get beyond materialism to a world of matter constrained by boundary conditions, we shall need to advance to a more complex philosophy, and this means *dualism*.[5] Dualism in the metaphysical sense is the theory that reality as a whole, or some domain of it, consists of two irreducible modes of being.[6] While the history of philosophy offers any number of dualisms for consideration,[6] the only ones of interest for us here are those having material substance as one of its modes. The companion mode then serves as the material world's boundary. In Western philosophy there are only two such: the Cartesian dualism of *mind* and body[7]; and the Kantian dualism of *reality* and *appearance* (*noumenon* and *phenomenon*)[8]. The author is well aware of the problematic and (to say the least) controversial status of both dualisms among philosophers and neuroscientists.[9] Physicists for their part show little interest in either dualism, except for a few theorists grappling with *measurement problem* of quantum mechanics.[10] For us, however, they are the way forward. In a moment we will describe the nature of the two theories and offer a defense of their relevance to physical theory. First, though, we want show how they fit together to form a graphic and plausible picture of reality as a whole—a material world with boundary conditions.

We may take Descarte's "body" and Kant's world of "appearance" to mean one and the same thing—the conventional material world, the everyday world of matter and energy that presents itself to the senses. This overlap of modes invites us to unite the two dualisms to form the composite model of reality shown in **Fig. 2.1**. The model consists of three contiguous areas which for convenience of discussion are labeled A, B and C. The interfaces between A and B and between B and C are labeled 1 and 2, respectively.

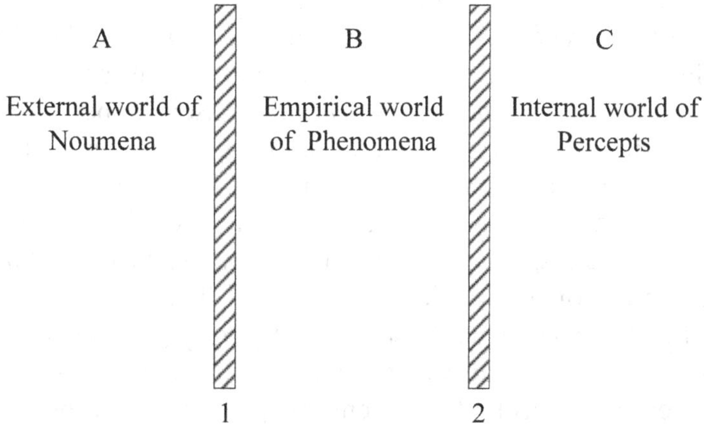


**Fig. 2.1** A three-part, ABC model of reality, with divisions 1 and 2.

Area A represents what philosophers call *the real external world*, a world brute facts existing independently of anyone's awareness or knowledge of it.[11] The things or facts of world A have properties and may interact, but those properties and interactions are by definition unobservable. Because they cannot be observed and are given to us only in thought, the facts in A were called by Kant *noumena*, or things-in-themselves (*Ding an sich*). The view that there exists such an unobservable yet real world is called *metaphysical realism*.[11] The belief that no such world exists, or is at least meaningless because unobservable, is called *anti-realism*.[11] The present work assumes the existence a real, external world. The problem of establishing its existence and its properties is known as the *problem of the external world*.[12]

Area B, conjoined with A at interface 1, is the *empirical world*, also known as the *world of appearance*. It is a fabricated world, made up of machine-derived images or *representations* of the facts of external world A. These representations in B are observable and knowable. They are manifest to the senses, but may or may not resemble the facts of A from which they derive. Kant called these observable representations *phenomena*, to distinguish them from the unobservable *noumena* of world A. We can think of the representations comprising world B as having been created by sensing apparatus located at interface 1. The view that physical knowledge derives from a polarity of external facts and empirically-derived representations is commonly referred to as *representative realism*,[13] or equivalently, *scientific realism*.[14]

Finally there is area C, connecting with B at interface 2. Area C represents the *private internal world* of conscious perception and awareness. In short, Area C is the *mind*. Descartes imagined it to consist of a mental substance different in kind from that of the material substance of world B. No one accepts that interpretation today, of course. Nevertheless, whatever its composition, on the present model it is where the act of observation takes place. The things of world C induced by the outside world are called *percepts* or *sense data*.[15] Percepts are internal *mental* representations of the images of the external world A created in world B by the sensing apparatus. Thus percepts are, in effect, mental representations of machine representations. We can think of percepts as being created (somehow) at interface 2.

In passing, we may note an appealing symmetry of the ABC model: boundary areas A and C are both hidden; area A because it is noumenal, and area C because it is privileged. Actually, Kant himself espoused such a three-part model, but seems not to have dwelt upon the ontological distinction between areas B and C. For us the distinction is crucial, as it offers the only hope of making scientific sense of the privileged nature of conscious experience.[16]

To illustrate the workings of this three-part model, consider the act of looking at a tree. (a) First, there is the noumenal tree, the primary visual object, rooted in world A. It is presumed to be there whether or not we happen to be looking or even exist. (b) When we look, optical photons scattered from the tree enter the eye and are focused by the lens of the eye to an image of the tree on the retina. The retinal image is then converted into electrical pulses, initiating a signal that is sent along the optic nerve and subsequent electrical and chemical pathways to the visual cortex. There, neurons are caused to fire. The firing neurons encode the image of the tree formed at the retina. One can think of the eye/brain combination as a machine creating a coded representation in B of the noumenal tree. (c) Finally, something completely mysterious happens. The coded representation in B, which, because it is dispersed widely throughout the volume of the brain and looks nothing like a tree, is somehow transformed into a structurally-coherent, mental image of the tree. This internal, mental image, in accordance with our model, resides in world C.

## 2.2 Derivation of the Law of Laws: the Projection Theory

We have set forth a three-part ABC model of the world by combining two metaphysical theories, the famous dualisms of Descartes and Kant. Our aim in doing so was to define the structure and lawful behavior of the material world by imposing upon it boundary conditions. Nevertheless, having combined the two theories, we find that what we have achieved thus far is not a theory of physics but a merely a theory of how one acquires *knowledge* of the physical world by successive representations, a machine-made representation at interface 1 and a mental one at interface 2. We set out to produce an *objective* description of the behavior of the facts of the physical world, but instead all we have got so far is a model of how those facts are transformed into *subjective* mental images. As such our model is as much epistemological as it is metaphysical. Clearly work remains if we are to achieve our objective of revealing the source of physical law. To meet our objective, we somehow have to eliminate from our model its subjective aspect. It means, among other things, removing from the picture the knowing subject, whose mental images are inherently subjective; it means moving from epistemology back to ontology.

This, as it turns out, is not hard to do. We have only to give formal expression to the connection between worlds A and B, and more particularly, to the manufacture of representations at interface 1. For this, the presence or absence of conscious observers and mental images in C are immaterial. All that matters is that there exists instrumentation at interface 1 capable of mapping the facts of A into representations in B. Of course an element of subjectivity remains in that scientific instruments are themselves the inventions and expressions of the intent of conscious minds. We are after all still immersed in an epistemic setting. But, as we shall see, this residue of subjectivity can be made to disappear, leaving in its place a picture of the world as it really is.

The connection we seek between a fact of world A and its representation in world B derives from a dual requirement of scientific realism, namely that there exist (a) *causal connection* and (b) *correspondence* between them.[14] Causal connection means that the properties of the representation originate in the facts of external reality; and correspondence means that the properties of the representation are those of the fact it represents. To see the significance of this, let us consider an arbitrary physical object $O$ of world A and let its behavior be fully specified by a descriptor, $\chi$. Here $\chi$ stands for an objective fact of physical reality—the Kantian noumenal object—knowledge of which provides one with all that can be known about $O$. It is thus the implicit aim of the scientific observer, using the best available instrumentation, to obtain as precise a representation of $\chi$ as possible. Let $\chi'$ denote such a representation, acquired by observation of $O$. A useful way of thinking about the causal linkage between fact and representation is to regard the representation $\chi'$ as a kind of projected or filtered version of the objective fact $\chi$. The problem of the external world can then be treated as a problem in *linear systems theory*. In the systems-theoretic context the fact $\chi$ and its representation $\chi'$ are to be regarded as, respectively, the input and output signals of a communications link—the measuring instrument—located at the boundary between worlds A and B and characterized by a system (instrumental) operator $\hat{R}'$. As shown schematically in **Fig. 2.2,** $\hat{R}'$ acts on the input signal $\chi$ to produce an output signal $\chi'$. In symbols we

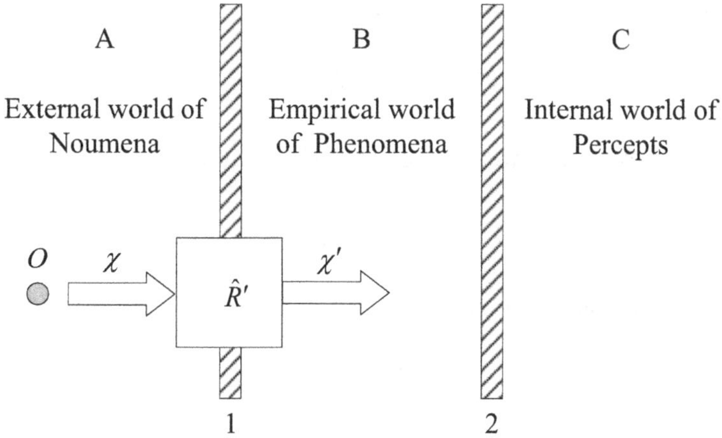


**Fig. 2.2** Operator $\hat{R}'$ operates on object $O$'s descriptor $\chi$, creating a representation $\chi'$.

have

$$\chi' = \hat{R}'\chi, \tag{2.1}$$

where the prime attached to $\chi'$ indicates the possible presence of error in the representation, and that attached to $\hat{R}'$ identifies the instrumental source of that error; it is a signal too of the *subjectivity* of the measurement process. Equation (2.1) provides one relation between $\chi$ and $\chi'$, that stemming from causal connection between worlds A and B. We shall refer to $\hat{R}'$ as the *projective transformation operator*, because it "projects" the descriptor $\chi$ into empirical world B. For now, the idea of projection is only a device for expressing the causal connection between worlds A and B. Later, however, the system operator and the projection process will turn out to have literal, geometric and physical meaning.

At this stage we know nothing about the structure of the operator $\hat{R}'$. Consequently, we do not know how well the representation $\chi'$ represents the fact $\chi$ to which it is causally related. If, for example, $\hat{R}'$ behaves as a low-pass filter, cutting out fine detail, then the agreement may be poor indeed. That is to say, in general, owing to the presence of error, $\chi' \neq \chi$. But now an interesting special case arises. Suppose that the measurement is error-free, that appearances do not lie, and that $\chi'$ corresponds precisely with the fact $\chi$. In that case we may remove the error-indicating prime from $\hat{R}'$, i.e.,

$$\hat{R}' \to \hat{R}, \tag{2.2}$$

giving (a) a revised expression of causal relation

$$\chi' = \hat{R}\chi \tag{2.3}$$

and (b) an expression of exact correspondence between $\chi$ and $\chi'$:

$$\chi' = \chi. \tag{2.4}$$

In addition, suppose we allow for the possibility that descriptor $\chi$ is algebraically complex, i.e., of the form $\chi = u + iv$, where $u$ and $v$ are the real and imaginary parts of $\chi$ and $i = \sqrt{-1}$. In that case, if we demand only the equality of the squared magnitudes of $\chi'$ and $\chi$, rather than that of $\chi'$ and $\chi$ themselves, then (2.4) can be replaced by the weaker relation

$$\chi' = r\chi, \; |r| = 1, \tag{2.5}$$

where $r$ is a unimodular constant [i.e., of the form $\exp(i\phi)$, $\phi$ real]. Equation (2.5), which includes (2.4) as a special case, gives us a second relation between $\chi'$ and $\chi$. We call it the *correspondence condition*, as it expresses error-free correspondence between the physical fact of world A and its representation in empirical world B. The representation $\chi'$ may now be eliminated between (2.3) and (2.5), yielding a single relation in the fact $\chi$:

$$\hat{R}\chi = r\chi. \tag{2.6}$$

This is our Law of Laws, the foundational result of this book. In mathematical terms it states that the descriptor $\chi$—which by definition provides a complete description of the object $O$—is an eigendescriptor of the projective transformation operator $\hat{R}$. The equation is unusual in that the eigenvalue $r$ is already known to have magnitude unity, a fact that severely constrains the structure of $\hat{R}$. Equation (2.6) applies to the description of any physical object $O$ for which there exist a causal relation of the form (2.3) and a condition of exact correspondence

(2.5). Our Law of Laws thus amounts to a general, if compressed, statement of the laws of physics, or at any rate, of those laws involving eligible descriptors $\chi$. Particular laws, as we shall soon see, are obtained by specifying $\hat{R}$ and the behavior of $\chi$ under spacetime transformations.

## 2.3 Truth, self-consistency and the Principle of True Representation

According to our derivation, the existence of the Law of Laws is entirely contingent on the correspondence condition (2.5). But to say that a representation $\chi'$ corresponds with reality is to say that it is *true* in the sense entailed by the *correspondence theory of truth*: truth as correspondence with the facts.[17] Normally one thinks of the correspondence theory as having to do with correspondence between a subjective utterance and the objective fact to which it refers. For example, a statement such as "the cat is on the mat" is said to be true if and only if the cat is actually on the mat; otherwise it is false. In the same way, a machine-derived representation $\chi'$ could be said to be true if and only if it represents exactly the fact $\chi$ to which it refers. The standard difficulty with correspondence as a criterion of truth is that the representation—whether human utterance or machine-derived image—differs ontologically from the fact to which it supposedly refers. The setting in which the utterance or measurement is made is, after all, epistemic, not ontological. Clearly, words are not the same as cats on mats; and photomicrographs are not the same as microorganisms. Moreover, real cats on mats and real microorganisms are empirically inaccessible; we know them in the epistemic setting by their representations alone. Hence the comparison between fact and representation necessary to ascertain truth or falsity normally cannot be made and for this reason the correspondence theory as conventionally employed appears logically flawed.

But the correspondence expressed by (2.5) sidesteps this difficulty, in effect rehabilitating the correspondence theory of truth. Consider what has happened here. We began with an assumption about the *subjective*, *epistemic* setting within which one acquires knowledge of an external fact $\chi$ by way of its instrumental representation $\chi'$. By demanding exact correspondence between fact and representation a formula was obtained, namely (2.6), that no longer refers to the knowing subject. This formula applies whether or not observers are present or have even existed. In a word, the description suddenly becomes *objective* and *ontological*. For the equality sign in the correspondence condition (2.5) means that the elements it connects are ontologically equivalent. It means that the fact $\chi$ and its representation $\chi'$ must both be considered objective elements of the physical world. In passing from epistemology to ontology we have arrived at a defensible implementation of the correspondence theory of truth. A theory conventionally judged to be logically flawed has suddenly risen to the level foundational principle—a formal criterion of truth. And that is not all. The passage from epistemology to ontology has uncovered a world of unexpected deep structure—a mind-independent, external world in which are present both facts and their corresponding representations. Now this is very odd, because in the absence of observing minds, how could there possibly be representations? The answer is that the world of inanimate nature manifests them all on its own. And of course the representations are necessarily true, for the world does not lie to itself. The physical world thus may be said to comprise a self-consistent system of correlated facts and representations. Our correspondence condition thus emerges as both a criterion of truth and an expression of self-consistency of the physical world. In recognition of its ontological status we propose to call our correspondence condition the *Principle of True Representation*. According to this principle, the existence of physical law, as expressed by the Law of Laws, derives from the fact that the world is truthfully self-representing. The Principle of True Representation ensures the objectivity (mind-independence) of the laws of physics, a fact expressed symbolically by the absence of primes in (2.6).

## 2.4 Unification

Now it may seem that in passing from epistemology to ontology we have got something for nothing. For it seems that, on paper, all we have really done is place an equality sign between $\chi$ and $\chi'$, and out comes the Law of Laws. We may well ask, what has happened physically? In particular, what has happened to the measuring instrument originally placed at the boundary between worlds A and B? The answer is this:

thanks to our criterion of truth, (2.5), the man-made instrument has been removed and replaced by a mind-independent state of unification: the unification of the worlds of appearance and reality. The correspondence condition (2.5) unifies the two worlds. This does not mean that the two worlds are now to be considered equivalent. Indeed they are *not* equivalent, and the fact that the phase factor $r$ need not be unity is an indication of this. Rather, we should think of them as complementary worlds, with "complementary" meant in the sense that Niels Bohr meant it: two mutually exclusive—but jointly necessary—ways of referring to the same phenomenon.[18] So we really have not got something for nothing. We have exchanged our epistemic instrumentation for an ontological state of unification. And as we have seen above, a unified reality is a self-consistent one, a world operating lawfully and independently of conscious observers.

There is another way of describing the exchange, one that sheds additional light on what it means to go from epistemology to ontology. We have said that by invoking our criterion of truth we have got rid of measuring instrument. The measurement process, however, invariably entails two things: (1) a spacetime gap between the object $O$ and its image $I$; and (2) a set of messenger particles carrying the object's features to a detector. In the case of the light microscope, to take a familiar example, the object to be studied is placed in the object plane of the instrument's objective lens and illuminated with a source of visible light, a stream of photons. A final image is formed by the eyepiece, through which the image may be viewed with the unaided eye or recorded photographically. In getting rid of the microscope, what we are really doing at the most fundamental level is getting rid of the spatial interval between object and image, as well as the messenger photons and the temporal interval it takes them to move through the instrument. The Law of Laws refers to none of these essentials of the measurement process. What it contains is a description of the object in itself, a description now free of epistemic associations, the result of a union between appearance and reality.

## 2.5 The aim of science

To conclude this chapter we want to show how our general Law of Laws fits into the arc of scientific progress. There are at least five points of entry.

(1) If the Principle of True Representation rehabilitates the correspondence theory of truth, it offers potentially an even greater service to physical science itself. We said earlier that the aim of science is to explain the world around us, and no doubt that is so. Scientific realism, however, in positing the existence of a real, external world, takes matters a step further: it implies that "the aim of science is the search for truth." But such a claim makes sense only if we have some way of knowing when we have found the truth. Unfortunately, on the standard view, the real state of affairs, which is external to us, is empirically inaccessible and hence unknowable. One could be in error (or not) and not know it. And so, on this standard view, scientific progress can at best be framed as *convergence* on the truth, each advance marked by a higher degree of verisimilitude; i.e., ability to explain previously explained facts as well as facts previously unexplained. Here are three standard examples of a convergence on scientific truth:

$$\left.\begin{array}{l}\text{Electrostatics}\\ \text{Magnetostatics}\end{array}\right\} \rightarrow \text{Maxwellian electromagnetics} \rightarrow ?$$

Newtonian gravity → Einsteinian gravity → ?

Schrödinger equation → Dirac equation → ?

Obviously the second term in each sequence improves on the first: The electromagnetics of Maxwell after all concern the *motion* of electricity and magnetism, whereas the term preceding it treats these phenomena separately and at rest. The gravitational equations of Einstein not only reduce to those of Newton for weak fields, but also predict new phenomena, e.g., the celebrated bending of light around the Sun. Similarly, the relativistic wave equation of Dirac not only reduces to Schrödinger's equation for slow particles, but among other things gives a natural explanation for intrinsic spin and predicts the existence of antiparticles. Now the question is, is the second term the end of each series, or could there be something more, some further elaboration of the Maxwell, Einstein or Dirac equations, as suggested by the question marks? On the conventional view, we don't know if we have reached the end or not. We can of course make guesses or *ad hoc* assumptions to explain some unexplained phenomenon, as for example Milgrom has done in modifying the gravitational inverse square law to explain the speeding up of galaxy clusters.[20] The Principle of True Prepresentation, however, yields a definite answer. By way of the Law of Laws, it conveniently circumvents the endless convergence process and gives the end result—the true state of affairs, as it were. As we shall see, none of the above sequences is presently complete. In each case the question mark is to be replaced by a new and more comprehensive theory, one that is true by definition.

(2) It is of course a staple of scientific method that theories are to be submitted to experimental test. If the Law of Laws is "true by definition," does that mean that one needn't bother testing it, that it is already beyond empirical falsification? No, it does not mean that, and for at least two reasons. First, to arrive at particular laws the general law (2.6) has to be expanded to exhibit at the very least relativistic covariance and one or two other elaborations having to do with extra dimensions. It is perfectly possible that these additional steps may spoil the truth-value of the Law of Laws, leading to a potential falsification of the theory. Second, even if the additional steps do no damage to the truth-value of the Law of Laws, it is possible that this law, even if true by definition, applies not to this world but to some other! We just do not know. Consequently the Law of Laws must still undergo the rigors of empirical test.

(3) It is often observed that, historically, physics has made its greatest strides via the unification of seemingly disparate and unconnected phenomena [cf. point (1) above]. Thus, for example, Newton's universal law of gravitation unified terrestrial gravity and celestial mechanics; and the electrodynamics of Maxwell unified the phenomena of electricity, magnetism and light. Today, when one speaks of a need for unification in physics, one usually has in mind the unification of quantum mechanics and general relativity—the problem of quantum gravity discussed briefly in the previous chapter of this book. The Principle of True Representation expresses a new and perhaps unexpected form of unification—that of the worlds of appearance and reality. It is a unification contingent on the existence of a truth connection between the two worlds. Of course whether the new unification bears fruit remains to be seen.

(4) Typically physical laws emerge from an underlying principle or specified condition. For example suppose that we wish to study the properties of circular planetary motion. We take as our guiding condition or principle that of equilibrium, i.e., zero radial acceleration. To meet this condition the gravitational and centrifugal forces acting on the orbiting body must balance, leading at once to Kepler's Third Law (for circular orbits).[22] In this book our purpose is to investigate not orbital motion but the structure of the universe as a whole. Our guiding principle or condition—a kind of equilibrium between fact and representation—is the Principle of True Representation, the demand that all representations image the external facts exactly. The result in this case is the Law of Laws. The two examples are compared in **Table 2.1**. We can see that the line of production of the laws in the

**Table 2.1** Physical laws and their underlying principles

| Context | Guiding Principle | Type of Principle | Resultant Law |
|---|---|---|---|
| Orbital physics | Equilibrium | Physical | Kepler's Third Law |
| Universe as a whole | True Representation | Metaphysical | Law of Laws |

two cases is similar. The main difference is that the equilibrium principle is "physical," whereas the Principle of True Representation is metaphysical: forces of equilibrium are observable (measurable) whereas the noumenal object is not. Oddly enough they both result in physical laws. The general message here is that one need not avoid a theory merely because it entails metaphysical assumptions. What matters is that it should have observable consequences.

(5) The Principle of True Representation helps us to answer our leading question, Why do the laws of physics have the forms they do? It helps, too, to answer another question, one that Karl Popper has referred to as *Newton's Problem*: Why are the laws the same at all points of the universe, no matter how far apart in space or time?[23] Unable or unwilling to explain this universality of law in terms of action at a distance, Newton was led to attribute it to the will of God. Today, thanks to Einstein, we know that action at a distance was never a viable

theory; that physical effects, the result of interactions between the physical constituents of the universe, spread spatially at finite speeds never exceeding that of light. But this only aggravates Newton's problem. The observable universe is about 92 billion light years across.[24] If it takes 92 billion years for the universe to communicate with itself between its most distant points (neglecting its accelerating rate of expansion!), how are we to understand that the laws of physics are same (if indeed the are!) at those and all other points of the universe? The theory of inflation[25] offers a possible answer: "during the inflationary era the part of the universe that we can observe would have occupied a tiny space, and there would have been plenty of time for everything in this space to be homogenized."[26] Although Weinberg is referring here to the homogenization (smoothing) of inhomogeneities in the cosmic microwave radiation background, one can at least imagine there occurring a homogenization of the laws as well. The Law of Laws suggests, perhaps, a more plausible answer. For the Law of Laws in the form (2.6) is a logical relation, expressing the self-consistency of the physical universe. Because it is independent of the variables of spacetime, it holds good everywhere and at all times. Thus it may be that the Principle of True Representation resolves Newton's problem.

## Notes and references


[1] Jacques Maritain, *The Degrees of Knowledge* (Notre Dame Press, Notre Dame, Indiana, 1998), p.p. 3-4.

[2] Karl Popper, *Objective Knowledge* (Oxford, Oxford, UK, 1972), p, 38.

[3] *Science and Hypothesis*, p. 140. Cited in *A Dictionary of Philosophical Quotations*, A. J. Ayer and Jane O'Grady, Eds. (Blackwell, Oxford, 1992), p. 355.

[4] Michael Polanyi, *Knowing and Being* (University of Chicago Press, Chicago, 1969), p. 219.

[5] Cf. the entry "Metaphysics," in *The Cambridge Dictionary of Philosophy*, Robert Audi, Ed. (Cambridge U. Press, Cambridge, UK, 1995).

[6] Cf. the entry "Dualism," in *Dictionary of Philosophy*, D. D. Runes (Littlefield, Adams & Co., Ames, Iowa, 1959).

[7] René Descartes, *Meditations* (Liberal Arts Press, New York, 1951), II and XI; reprinted in *Philosophy of Mind*, David J. Chalmers, Ed. (Oxford U. Press, Oxford, UK, 2002), pp. 10-21.
[8] Immanuel Kant, *Critique of Pure Reason*, Norman Kemp Smith, Trans. (Palgrave Macmillan, Houndmills, Basingstoke, Hampshire, UK, 1929), 266 ff., 291 ff.
[9] Owen Flanagan, *The Science of the Mind*, 2nd Ed. (MIT Press, Cambridge, Massachusetts, 1991), pp. 1-22.
[10] Max Jammer, *The Philosophy of Quantum Mechanics* (John Wiley & Sons, New York, 1974), Chapters 6 and 7.
[11] Cf. "Metaphyscial realism" in *The Cambridge Dictionary of Philosophy*, Robert Audi, Ed. (Cambridge U. Press, Cambridge, UK, 1995).
[12] Cf. entry "Problem of the External World," *A Companion to Epistemology*, J. Dancy and E. Sosa, Eds. (Blackwell, Malden, Massachusetts, 1993).
[13] Cf. entry "Representative realism," *A Companion to Epistemology, op. cit.*; also "Realism," in *The Encyclopedia of Philosophy, Volumes 7 and 8* (Macmillan, New york, 1967).
[14] E. McMullin, "A Case for Scientific Realism," in *Scientific Realism*, Jarrett Leplin, Ed. (U. of California Press, Berkeley, 1984), pp. 8-40.
[15] Cf. article "Perception," in *The Cambridge Dictionary of Philosophy, op. cit.*
[16] W. S. Sahakian and M. L. Sahakian, *Ideas of the Great Philosophers* (Barnes & Noble, 1966), p. 139.
[17] Cf. Ref. 4 in the Preface to this book, Ref. 14 above, and John R. Searle, *The Construction of Social Reality* (The Free Press, New York, 1995), pp. 199-226.
[18] Max Jammer, *op. cit.*, pp. 87, 95.
[19] Lee Smolin, *The Trouble with Physics* (Houghton Mifflin, Boston, 2006), pp. 31-33.
[20] M. Milgrom, "A Modification of the Newtonian Dynamics as a Possible Alternative to the Hidden Mass Hypothesis," Astrophys. J. **270**, 365-370 (1983).
[21] Cf. Smolin, *op. cit.* and S. Weinberg, "A Unified Physics by 2050?" Sci. Am., Dec. 1999, pp. 68-75.

[22] H. Goldstein, *Classical Mechanics* (Addison-Wesley, Reading, Massachusetts, 1959), p. 80.
[23] K. Popper, *Realism and the Aim of Science* (Rowman and Littlefield, Totowa, New Jersey, 1983), pp. 149-152.
[24] T. M. Davis and C. H. Lineweaver, "Expanding Confusion: Common Misconceptions of Cosmological Horizons and the Superluminal Expansion of the Universe," Astronomical Soc. of Australia **21,** 97-109 (2004).
[25] Alan H. Guth, *The Inflationary Universe* (Perseus, Reading, Massachusetts, 1997).
[26] S. Weinberg, *Cosmology*, (Oxford, Oxford, 2009), p. 205.

# Chapter 3

# Geometrization of the Law of Laws: $M^4 \times \bar{M}^2$ spacetime

Our main thesis may seem an unlikely one. For we suggest that the last frontier in physics is not physics at all, but metaphysics—that is to say, a program whose aim is to visualize and explain the world as it really is, not as it merely seems to be. The point of entry into this mind-independent reality is the projection theory set forth in the previous chapter and culminating in the Law of Laws, Eq. (2.6). In the present chapter we shall focus on the formal development of that Law. As it stands, Eq. (2.6) is simply a logical relation connecting facts and representations, a statement of self-consistency of the physical world. As such, it makes no direct reference to the dimensions of time and space. On the one hand there is merit in this for, as suggested in the previous chapter, the geometry-independence of the Law of Laws helps to resolve Newton's problem. But on the other hand, to get real physics from the Law of Laws, sooner or later we are going to have put it in terms of the geometry of spacetime. There is no obvious, *a priori* way of doing this. Nevertheless, a way does exist: one that makes use of a fortuitous similarity between the formal structure of the Law of Laws and that of the phenomenon of self-imaging[1] in ordinary wave optics. Accordingly, we begin with an account of optical self-imaging and then show how the optical *ansatz* not only points the way to a geometrization of the Law of Laws, but uncovers an unexpected extra-dimensionality of spacetime itself.

## 3.1 Self-imaging optical fields $u$ in $M^4$

We consider the propagation *in vacuuo* of a monochromatic, complex wave field $u$ in Minkowskian $M^4$ (3+1 spacetime). As shown in **Fig. 3.1**, we take a Cartesian frame of reference, $(x, y, z) = (\boldsymbol{\rho}, z)$, with $\boldsymbol{\rho}$ denoting the position vector in planes perpendicular to the z-axis, and orient its axes so that the source of the field lies in the half-space $z < 0$. Then, in the half-space $z \geq 0$, the field may be said to propagate principally in the direction of increasing $z$, that is to say, with no motion in the negative $z$ direction. In terms of these coordinates together with time $t$ the field $u$ may be written

$$u(\boldsymbol{\rho}, z, t) = v(\boldsymbol{\rho}, z)\exp(-i\omega t)\,, \tag{3.1}$$

where $v(\boldsymbol{\rho}, z)$ is the *complex amplitude* of the field and $\omega$ its angular frequency. We display the spacetime variables in this way to emphasize that, for given $z$, the amplitude $v(\boldsymbol{\rho}, z)$ varies as a function of the transverse position vector $\boldsymbol{\rho}$. Coordinate $z$ labels successive cross sections of the field.

Let us denote by $\Pi$ and $\Pi'$ two such cross sections, separated by distance $a$ in the half-space $z > 0$; see **Fig. 3.1**. We denote by $u$ and

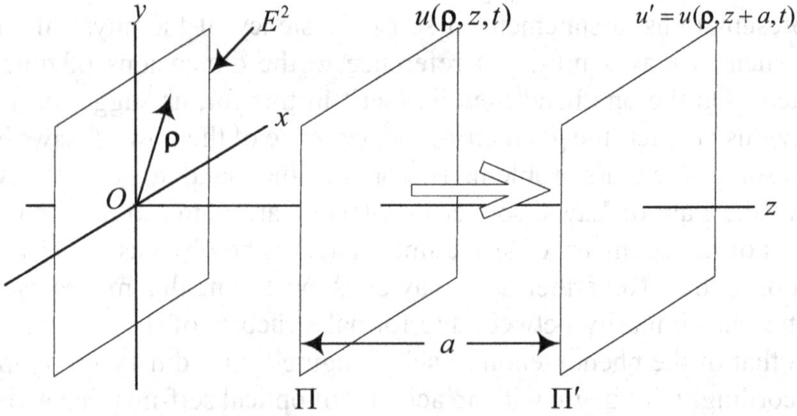


**Fig. 3.1.** Optical wave propagation in Euclidian $E^3$ with Cartesian coordinates $(x, y, z)$ and origin $O$. The field source lies in the halfspace $z < 0$. Vector $\boldsymbol{\rho} = (x, y)$ locates field points in $E^2$ transverse to the $z$-axis. The complex field amplitude at time $t$ and plane $\Pi$ is $u$; that at plane $\Pi'$, distance $a$ downstream from $\Pi$, is $u'$.

$u'$ the field distributions in planes $\Pi$ and $\Pi'$, respectively. Clearly the two distributions are causally connected, as $u'$ arises by propagation from $u$. The connection between them is expressed by an operator equation of the form

$$u' = \hat{\Gamma} u\,, \tag{3.2}$$

where $\hat{\Gamma}$ is the linear wave propagator. [By linear is meant that $\hat{\Gamma}(Au_1 + Bu_2) = A\hat{\Gamma}u_1 + B\hat{\Gamma}u_2$, where $A$ and $B$ are constants.] Propagator $\hat{\Gamma}$ operates on the complex distribution $u$, converting it into the distribution $u'$.

Now in general, owing to spatial dispersion (diffraction), the field amplitude at $\Pi'$ may bear little structural resemblance to the amplitude at $\Pi$. But now, just as in the projection theory of **Sec. 2.2**, an interesting special case arises: Suppose that the field structure at $\Pi'$ duplicates exactly that at $\Pi$. Fields that reproduce themselves in this manner by free-space propagation alone are said to be self-imaging.[1] If we demand only the equality of the field intensities at $\Pi$ and $\Pi'$, i.e., $|u'|^2 = |u|^2$, rather than that of $u'$ and $u$ themselves, then the condition for self-imaging can be written

$$u' = \gamma u\,, \qquad |\gamma| = 1\,, \tag{3.3}$$

where $\gamma$ is a unimodular constant. Eliminating $u'$ between Eqs. (3.2) and (3.3) we obtain

$$\hat{\Gamma}u = \gamma u\,. \tag{3.4}$$

The formal similarity between the process of propagation just outlined and the projection process of **Chapter 2** is evident. In particular, the general law of wave propagation (3.2) is analogous to the causal connection condition (2.3); the condition for self-imaging (3.3) is formally identical to our correspondence condition (2.5); and, finally, the law of propagation for self-imaging fields (3.4) is formally identical with our Law of Laws, Eq. (2.6). These formal similarities—in particular that between the unimodular eigenvalues $r$ and $\gamma$—strongly suggest an interpretation of the projection theory in terms of a self-imaging wave field. In such an interpretation, $\chi$ represents the self-imaging field, and $\hat{R}$ the wave propagator. To carry out this idea, we shall first work out the details of the optical case, which may then be appropriately recast giving analytical content to the projection theory.

We begin by noting that if a self-imaging field $u$ at plane $\Pi(z)$ images itself at plane $\Pi'(z+a)$, where $a = \text{Const.}$, then the field $u'$ at $\Pi'$ images itself at plane $\Pi''(z+2a)$. To show this, from (3.2) and (3.3) we have for the field amplitude $u''$ at $\Pi''$

$$u'' = \hat{\Gamma}u' = \gamma\hat{\Gamma}u = \gamma^2 u \,, \quad \left|\gamma^2\right| = 1 \,, \tag{3.5}$$

where we have made use of the linearity of $\hat{\Gamma}$. Hence $|u''|^2 = |u|^2$, as was to be shown. Generalizing this, we see that self-images are found at equally spaced intervals $a$ along the $z$-axis. The self-imaging field in question is, in other words, periodic in the coordinate $z$ with spatial period $a$. Thus the complex amplitude may be expressed as a Fourier series in the axial coordinate $z$:

$$v(\boldsymbol{\rho}, z) = \sum_{n=-\infty}^{+\infty} v_n(\boldsymbol{\rho}) \exp(ik_{zn} z) \tag{3.6}$$

and where

$$k_{zn} = K - \frac{2\pi n}{a} \,, \tag{3.7}$$

$a$ is the repeat interval, $K$ is an adjustable positive phase bias and the $v_n(\boldsymbol{\rho})$ are the $\boldsymbol{\rho}$-dependent coefficients of the expansion. If in (3.6) we replace $z$ by $z + a$, then we find for the constant $\gamma$ appearing in (3.3)

$$\gamma = \exp(iKa) \,. \tag{3.8}$$

The field $u$ defined by (3.4) is, as we have said, *periodically self-imaging*. In contrast, the individual modes of the expansion (3.6) labeled by index $n$ propagate without spatial dispersion and for that reason are called *non-diffracting beams*.[2-3] To demonstrate their non-dispersive behavior, let us write for the $n$th term of (3.1)

$$u_n(\boldsymbol{\rho}, z, t) = v_n(\boldsymbol{\rho}) \exp[-i(\omega t - k_{zn} z)] \,. \tag{3.9}$$

If in this expression we replace $z$ by $z + \zeta$, with $\zeta$ arbitrary, we find that

$$u_n(\boldsymbol{\rho}, z + \zeta, t) = \gamma_n u_n(\boldsymbol{\rho}, z, t) \,, \tag{3.10}$$

where

$$\gamma_n = \exp(ik_{zn}\zeta) \,. \tag{3.11}$$

Since $\zeta$ is arbitrary, the modal field $u_n$ indeed images itself in all planes $\Pi$ perpendicular to the $z$-axis

## 3.2 Representation of $\hat{\Gamma}$ in Fourier space

We now deduce the form of the propagation operator $\hat{\Gamma}$ in Fourier space. This is most simply done by seeing how $\hat{\Gamma}$ propagates the modal field $u_n$ from $z$ to $z+\zeta$: from (3.4) we have

$$\hat{\Gamma} u_n = \gamma_n u_n , \tag{3.12}$$

where $u_n$ is given by (3.9) and $\gamma_n$ by (3.11). We now represent the series coefficient $\upsilon_n(\boldsymbol{\rho})$ as a two-dimensional Fourier integral (infinite limits assumed)

$$\upsilon_n(\boldsymbol{\rho}) = \frac{1}{(2\pi)^2} \iint V_n(\boldsymbol{\kappa}) \exp(i\boldsymbol{\kappa}\cdot\boldsymbol{\rho}) d^2\boldsymbol{\kappa} , \tag{3.13}$$

where $V_n(\boldsymbol{\kappa})$ denotes the Fourier spectrum of $\upsilon_n(\boldsymbol{\rho})$ in $k$-space and where we have introduced the vector notation:

$$\boldsymbol{\kappa} = (k_x, k_y) ,\ d^2\boldsymbol{\kappa} = dk_x dk_y . \tag{3.14}$$

Inserting (3.13) into (3.9) we get from (3.12) the field at plane $\Pi(z+\zeta)$

$$\hat{\Gamma} u_n = \frac{1}{(2\pi)^2} \iint [V_n(\boldsymbol{\kappa}) \exp(ik_{zn} z)] \exp[-i(\omega t - \boldsymbol{\kappa}\cdot\boldsymbol{\rho} - k_{zn}\zeta)] d^2\boldsymbol{\kappa} . \tag{3.15}$$

The argument of the exponential containing the time can be interpreted as the scalar product $k\cdot x$ of two four-vectors, namely

$$k = (\frac{\omega}{c}, \boldsymbol{\kappa}, k_{zn}) \text{ and } x = (ct, \boldsymbol{\rho}, \zeta) , \tag{3.16}$$

where $c$ is the speed of light and where a metrical signature $g_{\mu\nu} = \text{diag}(+1,-1,-1,-1)$ is assumed. We demand that the scalar product $k\cdot k$ of four-vector $k$ with itself shall have constant value $k_C{}^2$:

$$\frac{\omega^2}{c^2} - \boldsymbol{\kappa}^2 - k_{zn}{}^2 = k_C{}^2 . \tag{3.17}$$

That is to say, $k$ is by definition an invariant four-vector in $k$-space. Eq. (3.17) is nothing but the relativistic energy-momentum equation for a quantum of the field, where $k_C = 2\pi / \lambda_C$ and $\lambda_C$ is the Compton w. l. of the field quantum. Multiplying through by $\hbar^2$ ($\hbar = h/2\pi$, $h$ being Planck's constant) we get equivalently

$$\frac{E^2}{c^2} - \varpi^2 - p_{zn}^{\ 2} = m^2 c^2 , \tag{3.18}$$

where $E$ is the energy of the quantum, $(\varpi, p_{zn}) = \mathbf{p} = (p_x, p_y, p_{zn})$ its three-momentum, and $m$ its mass. If $k_C > 0$, then (3.12) describes the quantum-mechanical self-imaging of a massive scalar field. This, however, is more generality than is needed for the demonstration of self-imaging in $M^4$. Since $\omega$ is constant one can always redefine the frequency to absorb $k_C$: $\omega' = (\omega^2 - k_C^{\ 2} c^2)^{1/2}$. Hence it suffices to take $k_C = 0$, corresponding to a particle of zero mass (unpolarized photon or Weyl neutrino). Condition (3.17) then reduces to

$$k_0^{\ 2} - \boldsymbol{\kappa}^2 - k_{zn}^{\ 2} = 0 , \tag{3.19}$$

where we have put wave number $\omega / c \equiv k_0 = 2\pi / \lambda$ and $\lambda$ is the radiation wavelength *in vacuuo*. Eq. (3.19) defines a circle in the $k_x$-$k_y$ plane of radius

$$\kappa_n = \sqrt{k_0^{\ 2} - k_{zn}^{\ 2}} . \tag{3.20}$$

Since $|\boldsymbol{\kappa}| = \kappa_n =$ Const., the radiation vectors $\mathbf{k} = (\boldsymbol{\kappa}, k_{zn})$ form a cone in $k$-space, as indicated in **Fig. 3.2**. We call (3.19) the *on-shell condition*: it states that the only non-vanishing values of the Fourier spectrum $V_n(\boldsymbol{\kappa})$ are those lying on a circle of radius $\kappa_n$ as shown in **Fig. 3.2**. The spectrum $V_n$ is thus of the form

$$V_n(\boldsymbol{\kappa}) = \Phi(\phi)\delta(\boldsymbol{\kappa}^2 - \kappa_n^{\ 2}) , \tag{3.21}$$

where $\delta$ denotes the Dirac delta function and $\Phi$ is a function of the angular coordinate $\phi$ in the transverse plane of $k$-space. Thanks to the $\delta$-function, we may replace the *constant* $k_{zn}$ appearing in the

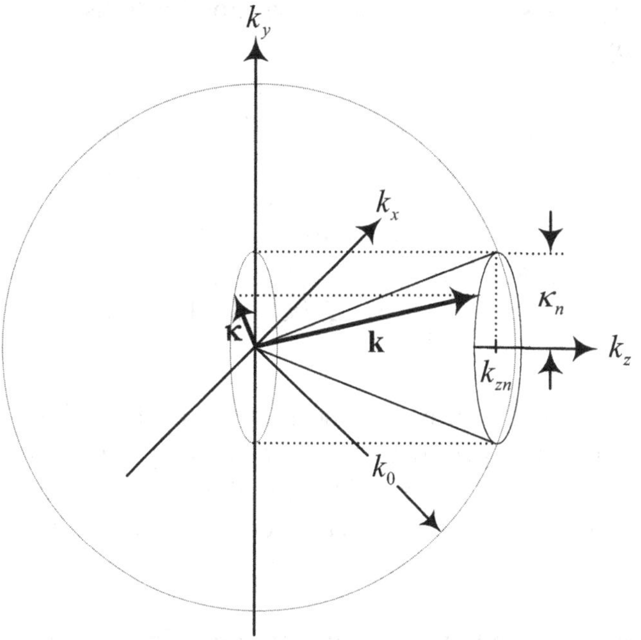


**Fig. 3.2.** The transverse components $\boldsymbol{\kappa}$ of the propagation vectors $\mathbf{k}$ for non-diffracting fields $u_n$ have constant magnitude $\kappa_n$. Thus vectors $\mathbf{k} = (\boldsymbol{\kappa}, k_{zn})$ lie on a conical surface of height $k_{zn}$ and base radius $\kappa_n$ in *k*-space. The tips of the transverse components end on a circle of diameter $\kappa_n$ in the $k_x, k_y$ plane. The region outside a disk of radius $|\boldsymbol{\kappa}| = k_0$—where $k_0$ is the wave number—is forbidden for self-imaging; fields for which $\kappa_n > k_0$ are evanescent and do not propagate.

argument of the exponential of (3.15) containing the time with the *variable* function of $\boldsymbol{\kappa}$ obtained by solving (3.19) for $k_{zn}$:

$$k_{zn} = \begin{cases} \sqrt{k_0^{\,2} - \boldsymbol{\kappa}^2}, & \boldsymbol{\kappa}^2 \le k_0^{\,2} \\ i\sqrt{\boldsymbol{\kappa}^2 - k_0^{\,2}}, & \boldsymbol{\kappa}^2 > k_0^{\,2} \end{cases} \quad . \tag{3.22}$$

Inserting this into (3.15) we obtain

$$\hat{\Gamma} u_n = \frac{1}{(2\pi)^2} \iint F(k_0, \boldsymbol{\kappa}, \zeta) U_n(\boldsymbol{\kappa}, z, t) \exp(i\boldsymbol{\kappa} \cdot \boldsymbol{\rho}) d^2\boldsymbol{\kappa}, \tag{3.23}$$

where $U_n(\boldsymbol{\kappa}, z, t)$ is the two-dimensional Fourier transform of the field at $z$ given by (3.9):

$$U_n(\boldsymbol{\kappa}, z, t) = V_n(\boldsymbol{\kappa}) \exp[-i(\omega t - k_{zn} z)], \tag{3.24}$$

and

$$F(k_0, \boldsymbol{\kappa}, \zeta) = \begin{cases} \exp\left(i\zeta\sqrt{k_0^{\,2} - \boldsymbol{\kappa}^2}\right), & \boldsymbol{\kappa}^2 \le k_0^{\,2} \\ \exp\left(-\zeta\sqrt{\boldsymbol{\kappa}^2 - k_0^{\,2}}\right), & \boldsymbol{\kappa}^2 > k_0^{\,2} \end{cases}. \tag{3.25}$$

Thus we see that the effect of the propagator $\hat{\Gamma}$ acting on $u_n$ is to multiply its Fourier spectrum $U_n(\boldsymbol{\kappa}, z, t)$ by a spatial dispersion factor $F(k_0, \boldsymbol{\kappa}, \zeta)$. Note that for self-imaging the on-shell radius $\kappa_n$ cannot exceed $k_0$; if it does, the signal becomes evanescent and fails to propagate.

## 3.3 $\hat{\Gamma}$ as convolution operator in configuration space

Utilizing the convolution theorem, the $k$-space representation (3.23) can be expressed as a convolution integral in configuration space:

$$\begin{aligned} \hat{\Gamma} u_n &= f(k_0, \boldsymbol{\rho}, \zeta) * u_n(\boldsymbol{\rho}, z, t) \\ &= \iint f(k_0, \boldsymbol{\rho} - \boldsymbol{\rho}', \zeta) u_n(\boldsymbol{\rho}', z, t) d^2\boldsymbol{\rho}' \end{aligned} \tag{3.26}$$

where

$$\begin{aligned} f(k_0, \boldsymbol{\rho}, \zeta) &= \frac{1}{(2\pi)^2} \int F\left(k_0, \boldsymbol{\kappa}, \zeta\right) \exp\left(i\boldsymbol{\kappa} \cdot \boldsymbol{\rho}\right) d^2\boldsymbol{\kappa} \\ &= -\frac{1}{2\pi} \frac{\partial}{\partial \zeta} \left[ \frac{\exp(ik_0 r)}{r} \right] \end{aligned}, \tag{3.27}$$

$r^2 = \boldsymbol{\rho}^2 + \zeta^2$ and $d^2\boldsymbol{\rho}' = dx'dy'$. Here the function $f$ is the familiar spherical wave function of wave optics. Equation (3.26) expresses Huygens' Principle: it describes the field at plane $\Pi(z+\zeta)$ as a superposition of spherical wavelets emanating from all points of the field at plane $\Pi(z)$. Huygens' Principle in the convolution form (3.26) is known as the Rayleigh-Sommerfeld diffraction integral. It allows us to express our propagation operator in the succinct form

$$\hat{\Gamma} = f\, *, \tag{3.28}$$

where $*$ denotes convolution.

We have derived the form of the operator (3.28) through its action on the non-diffracting beam $u_n$. But $\hat{\Gamma}$ is linear. Hence in (3.26) we may sum over index *n*, giving

$$\hat{\Gamma} u = f * u\,, \tag{3.29}$$

where *u* is the periodically self-imaging field (3.1), with $v(\boldsymbol{\rho}, z)$ given by (3.6): the propagator $\hat{\Gamma}$ is the same for both general self-imaging and non-diffracting fields.

## 3.4 $\hat{\Gamma}$ as differential operator

We define the time derivative and the two-dimensional gradient operator as follows:

$$\partial_0 = \partial / \partial(ct) \quad \text{and} \quad \boldsymbol{\partial} = (\partial / \partial x, \partial / \partial y)\,. \tag{3.30}$$

Then utilizing the Fourier correspondences

$$k_0 \rightleftharpoons i\partial_0 \quad \text{and} \quad \boldsymbol{\kappa} \rightleftharpoons -i\boldsymbol{\partial} \tag{3.31}$$

we have from (3.23), after summing over *n* to obtain the periodically self-imaging field *u* and its Fourier transform *U*

$$\begin{aligned} \hat{\Gamma} u &= F(i\partial_0, -i\boldsymbol{\partial}, \zeta) \frac{1}{(2\pi)^2} \iint U(\boldsymbol{\kappa}, z, t) \exp(i\boldsymbol{\kappa} \cdot \boldsymbol{\rho}) d^2\boldsymbol{\kappa} \\ &= F(i\partial_0, -i\boldsymbol{\partial}, \zeta) u(\boldsymbol{\rho}, z, t) \end{aligned}\,. \tag{3.32}$$

Thus from (3.25), (3.31) and (3.32) we obtain $\hat{\Gamma}$ in the form of a differential operator:

$$\hat{\Gamma} = \exp\left(-\zeta\sqrt{\partial_0{}^2 - \partial^2}\right). \tag{3.33}$$

Now $\hat{\Gamma}$ is also a displacement operator, as it displaces the field $u$ from $z$ to $z+\zeta$. Expanding the left hand side of (3.10) in a Taylor series about $z$ we find

$$\hat{\Gamma}u_n = [\exp(\zeta\partial_z)]u_n, \tag{3.34}$$

where $\partial_z = \partial / \partial z$, so in addition to (3.33) we have

$$\hat{\Gamma} = \exp(\zeta\partial_z). \tag{3.35}$$

We thus obtain the differential equation

$$\left[\exp\left(-\zeta\sqrt{\partial_0{}^2 - \partial^2}\right) - \exp(\zeta\partial_z)\right]u = 0. \tag{3.36}$$

Expanding each exponential in a power series in $\zeta$, we obtain to first and second order in $\zeta$, respectively,

$$\left(\sqrt{\partial_0{}^2 - \partial^2} - \partial_z\right)u = 0 \tag{3.37}$$

and

$$\left(\partial_0{}^2 - \partial^2 - \partial_z{}^2\right)u = 0. \tag{3.38}$$

The second of these equations is the familiar wave equation of scalar wave optics. Taking the time derivative, we see that the complex amplitude $v$ satisfies a three-dimensional Helmholz equation:

$$\left(\partial^2 + \partial_z{}^2 + k_0{}^2\right)v(\boldsymbol{\rho}, z) = 0 \tag{3.39}$$

If we now insert into (3.39) the Fourier series representation (3.6) we obtain a set of *two*-dimensional Helmholz equations in the $v_n$:

$$\left(\partial^2 + \kappa_n{}^2\right)v_n(\boldsymbol{\rho}) = 0, \tag{3.40}$$

where $\kappa_n$ is given by (3.20). One can think of the index $n$ as generating a *family* of non-diffracting beam *flavors* $\upsilon_n$ belonging to the *generation parameter* $k_0$; i.e., to the same three-momentum shell of radius $k_0$. In this way the optical theory outlined here anticipates the theory of fermion flavor to be developed in **Chapter 4**.

Wave equation (3.38) corresponding to second order in $\zeta$ is itself of second order. Apparently wave equation (3.37) corresponding to first order in $\zeta$ *wants* to be of first order, but can't get there, as it still depends quadratically on the differential operators $\partial_0$ and $\boldsymbol{\partial}$. This defect is easily corrected, following a now-familiar argument of Dirac. The aim is to produce a symmetrical relation that puts all four differential operators $\partial_\mu = (\partial_0, \boldsymbol{\partial}, \partial_z)$ of (3.37) on the same footing in $3+1$ spacetime. The relation should be linear in the $\partial_\mu$ because (3.37) depends linearly on $\partial_z$. In addition it must be consistent with (3.19), from which condition the differential equations ultimately derive. Let us therefore return to (3.19) and replace it with the linear relation

$$Ik_0 - \left(\mathbf{s}\cdot\boldsymbol{\kappa} \pm \sigma_z k_{zn}\right) = 0 \tag{3.41}$$

where $\mathbf{s} = (\sigma_x, \sigma_y)$, the $\sigma_i$ are the Pauli matrices and $I$ is the unit matrix:

$$\sigma_x = \begin{bmatrix} 0 & 1 \\ 1 & 0 \end{bmatrix}, \sigma_y = \begin{bmatrix} 0 & -i \\ i & 0 \end{bmatrix}, \sigma_z = \begin{bmatrix} 1 & 0 \\ 0 & -1 \end{bmatrix}, I = \begin{bmatrix} 1 & 0 \\ 0 & 1 \end{bmatrix}. \tag{3.42}$$

Because the $\sigma_i$ anticommute and $\sigma_i^2 = 1$ ($i = x, y, z$), Eq. (3.41) is indeed consistent with (3.19). Then in place of (3.22) we have from (3.41)

$$Ik_{zn} = \pm\sigma_z (Ik_0 - \mathbf{s}\cdot\boldsymbol{\kappa}) \tag{3.43}$$

and in place of (3.33) and (3.35) we obtain differential forms appropriate to matrix wave functions

$$\hat{\Gamma} = \exp\left[\pm i\zeta\sigma_z\left(Ii\partial_0 + \mathbf{s}\cdot i\boldsymbol{\partial}\right)\right] = \exp\left[\mp\zeta\sigma_z\left(I\partial_0 + \mathbf{s}\cdot\boldsymbol{\partial}\right)\right] \tag{3.44}$$

and

$$\hat{\Gamma} = \exp\left(I\zeta\partial_z\right), \tag{3.45}$$

Equating the first-order terms in the expansions of these forms, we get two linear wave equations, corresponding to the upper and lower signs in (3.44):

$$\left(\partial_0 + \mathbf{s}\cdot\boldsymbol{\partial} + \sigma_z\partial_z\right)u_L = 0 \tag{3.46}$$

and

$$\left(\partial_0 + \mathbf{s}\cdot\boldsymbol{\partial} - \sigma_z\partial_z\right)u_R = 0 \ . \tag{3.47}$$

These linearized versions of (3.37) are in fact the Weyl equations for left- and right-handed massless neutrinos (a mirror inverts $z$, changing left into right). Taking the time derivatives, we have for the complex amplitudes $\upsilon_L$ and $\upsilon_R$

$$\left(k_0 + s\cdot i\partial + \sigma_z i\partial_z\right)\upsilon_L = 0 \tag{3.48}$$

and

$$\left(k_0 + s\cdot i\partial - \sigma_z i\partial_z\right)\upsilon_R = 0 . \tag{3.49}$$

Utilizing (3.6) we obtain for the equations of the non-diffracting neutrino fields

$$\left(k_0 + \mathbf{s}\cdot i\partial - \sigma_z k_{zn}\right)\upsilon_{Ln} = 0 \tag{3.50}$$

and

$$\left(k_0 + \mathbf{s}\cdot i\partial + \sigma_z k_{zn}\right)\upsilon_{Rn} = 0 , \tag{3.51}$$

in which $k_{zn} = \sqrt{k_0^{\,2} - \kappa_n^{\,2}}$ .

## 3.5 Self-imaging fields $\chi$ in $M^4 \times \bar{M}^2$

We are now ready to put into geometric context our Law of Laws, Eq. (2.6). For this we have two powerful tools at our disposal: first, the theory of optical self-imaging detailed above and whose general structure so closely resembles that of the projection theory of **Sec. 2.2**; and second, Lorentz covariance, which demands form-invariance of the laws of nature under Lorentz transformations, i.e., transformations that connect coordinate systems moving rectilinearly and at constant rates relative to one another and leave invariant the quadratic form

$x^\mu x_\mu = c^2t^2 - x^2 - y^2 - z^2$.

To begin, let us look at our optical self-imaging equations with an eye to their invariance properties. By inspection we see that wave equations (3.40), (3.50) and (3.51) are invariant under circular rotations in the $x$-$y$ plane, but *not* under Lorentz transformations. Indeed, the $z$ and $t$ dimensions do not even appear in these equations, as they have been differentiated away. The wave equations for non-diffracting optical fields $v_n(\boldsymbol{\rho})$ thus do not qualify as laws of nature in the sense demanded by Lorentz covariance: they are frame-dependent. But on the other hand, it is perfectly clear what it would take to render (3.40), for example, Lorentz invariant: there would have to appear in place of the two coordinates of the $x$-$y$ plane the complete set of coordinates of 4-space; and in place of the two-dimensional gradient the complete set of gradient operators of 4-space, i.e.,

$$\left.\begin{aligned} \boldsymbol{\rho} &= (x, y) \rightarrow x^\mu = (x^0, x^1, x^2, x^3) \\ \boldsymbol{\partial} &= (\partial_x, \partial_y) \rightarrow \partial_\mu = (\partial_0, \partial_1, \partial_2, \partial_3) \end{aligned}\right\} . \tag{3.52}$$

Thus, for the projection theory, we see that the plane perpendicular to the principal direction of propagation of the descriptor $\chi$, with $\chi$ now interpreted as a self-imaging field, cannot be the simple $x$-$y$ plane sufficient to optical self-imaging. Instead, the plane perpendicular to the principal direction of propagation must be the full 4-dimensional, $M^4$ hyperplane. This of course stretches the imagination but is unavoidable if the Law of Laws is to be a legitimate law of nature.

But if $M^4$ defines cross sections of the propagating field $\chi$, then what possibly can be the principal direction of propagation? Clearly it cannot be along one of the dimensions of ordinary $M^4$—the $z$-axis, for example—for it would then define a privileged direction in $M^4$, violating the Principle of Relativity. And in any case the dimensions $x^\mu$ are already used up forming the cross sectional $M^4$ hyperplane. Thus, to geometrize the projection theory, one is compelled to postulate a fifth spacetime dimension, $x^5$, playing in the projection theory the role that the $z$-dimension played in the optical theory, i.e.,

$$\left.\begin{array}{l} z \rightarrow x^5 \\ \partial_z \rightarrow \partial_5 \end{array}\right\}. \tag{3.53}$$

Dimension $x^5$ is now to be considered the principal direction of propagation of the field $\chi$ ; see **Fig. 3.3.** The new dimension is flat,

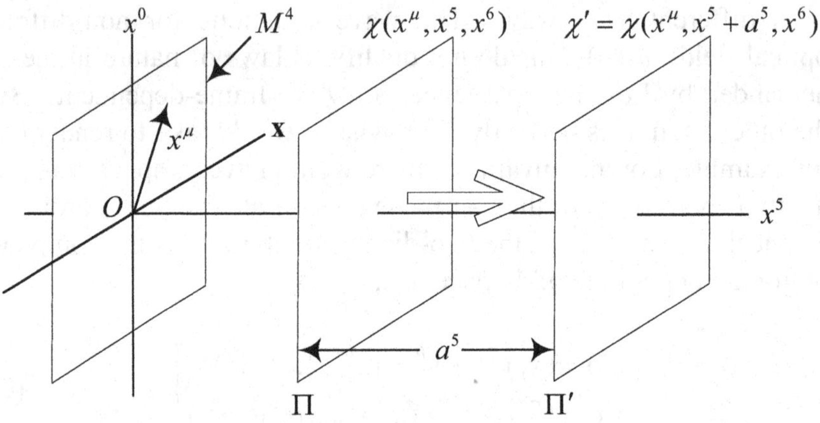


**Fig. 3.3.** Propagation of descriptor $\chi$ in six-dimensional product space $M^{4+2} = M^4 \times \bar{M}^2$ with coordinates $(x^\mu, x^5, x^6)$ and origin $O$. The field source lies in the half-space $x^5 < 0$. Vector $x^\mu = (x^0, x^1, x^2, x^3)$ locates field points in Minkowskian $M^4$, transverse to the $x^5$ axis. The complex field amplitude at (pseudo)time $\tau = x^6 / c$ and plane Π is $\chi$ ; that at plane Π′, distance $a^5$ downstream, is $\chi'$.

infinite in extent and perpendicular to $M^4$, just as in the optical case the flat and infinite $z$ axis is perpendicular to the $x$-$y$ plane. Although $x^5$ behaves as a spatial dimension, it is, as we shall see in a moment, metrically time-like ($g_{55} = +1$). It is not seen because the observable non-diffracting complex amplitudes comprising $\chi$ are independent of it, just as the optical non-diffracting fields $v_n(\mathbf{\rho})$ are independent of $z$. Infinitesimal rotations in the $x^\mu$-$x^5$ planes are forbidden, as such rotations would cause the $x^\mu$ to participate in the principal direction of propagation, thereby violating the Principle of Relativity and Lorentz covariance.

We are not quite done uncovering the structure and dimensionality of spacetime implied by the projection theory. We recall that in our theory of optical self-imaging the field $u$ was assumed to be monochromatic, i.e., driven by a single temporal angular frequency $\omega$. If we are to maintain the parallel with the optical theory, we should likewise assume the field $\chi$ to be monochromatic. However, as time $t = x^0 / c$ now occupies the $M^4$ cross section, and is to be considered on the same geometric footing as the three spatial dimensions of $M^4$, $\chi$'s temporal frequency spectrum need not be limited to the single frequency $\omega$. The field $\chi$, in other words, need not be monochromatic in the usual temporal sense. How, then, are we to understand the requirement of monochromaticity? The answer, one is forced to conclude, is that there must exist a second time dimension, here to be designated $\tau = x^6 / c$, a dimension playing the same role in the projection theory as ordinary time $t$ played in the optical theory, and carrying with it the single angular frequency $\Omega = k^6 c$. That is, in making the transition from the optical theory to the projection theory we make the role replacements:

$$\left.\begin{aligned} x^0 &\to x^6 \\ k^0 &\to k^6 \\ \partial_0 &\to \partial_6 \end{aligned}\right\}. \tag{3.54}$$

Like $x^5$, dimension $x^6$ is flat, infinite in extent and perpendicular to $M^4$. It is also perpendicular to $x^5$. Oddly enough, $x^6$, while serving as a legitimate second time dimension, is, as we shall soon see, metrically space-like ($g_{66} = -1$).

Taken together, the new extra dimensions, $x^5$ and $x^6$, comprise a second Minkowskian space, which we designate $\bar{M}^2$, and where the overbar reminds us that the metrical signatures of the two dimensions comprising it are just opposite to what one would expect. The total spacetime implied by the projection theory is thus the six-dimensional product space

$$M^{4+2} = M^4 \times \bar{M}^2 \tag{3.55}$$

with metrical signature

$$g_{ab} = \mathrm{diag}(g_{00}, g_{11}, g_{22}, g_{33}; g_{55}, g_{66}) = \mathrm{diag}(+1, -1, -1, -1; +1, -1).$$

In close analogy with (3.1), with the role substitutions indicated in (3.52-54), and defining for $M^4$ the coordinates $x^\mu \equiv x$, we have for the general field $\chi$ of the projection theory

$$\chi(x,x^5,x^6) = w(x,x^5)\exp(-ik_6 x^6), \tag{3.56}$$

where $w(x,x^5)$ is the complex amplitude, and $-k_6 x^6 = k^6 x^6 = \Omega\tau$ .

We now suppose the field $\chi$ to be self-imaging in $M^{4+2}$, imaging itself periodically in the principal direction of propagation $x^5$ with period $a^5$. The complex amplitude can then be written as a Fourier series in the coordinate $x^5$:

$$w(x,x^5) = \sum_{-\infty}^{+\infty} w_n(x)\exp(-ik_{5n}x^5), \tag{3.57}$$

where

$$k_{5n} = K_5 - \frac{2\pi n}{a^5}, \tag{3.58}$$

$a^5$ is the repeat interval, $K_5$ is a tunable positive phase bias and the $w_n(x)$ are the $x$-dependent coefficients of the expansion.

Now, as depicted in **Fig. 3.3**, let us consider the propagation of field $\chi$ from cross sectional plane $\Pi = \Pi(x^5)$ to plane $\Pi' = \Pi(x^5 + a^5)$. The correspondence condition (2.5) reads

$$\chi' = \chi(x,x^5 + a^5,x^6) = r\chi(x,x^5,x^6). \tag{3.59}$$

with $\chi$ given by (3.56). Substituting (3.57) we find that

$$r = \exp(-iK_5 a^5). \tag{3.60}$$

This is the eigenvalue of $\hat{R}$ proper to *periodically self-imaging* fields $\chi$. As in the optical case, the individual modes of the Fourier expansion propagate without spatial dispersion and are thus to be considered *non-diffracting beams* in $M^{4+2}$ spacetime. Their non-dispersive behavior is shown by first writing for the *n*th term of (3.56)

$$\chi_n(x,x^5,x^6) = w_n(x)\exp\left[-i(k_6 x^6 + k_{5n}x^5)\right]. \tag{3.61}$$

If in this expression we replace $x^5$ by $x^5 + \xi^5$, with $\xi^5$ arbitrary, we get

$$\chi_n(x, x^5 + \xi^5, x^6) = r_n \chi_n(x, x^5, x^6), \tag{3.62}$$

where

$$r_n = \exp(-ik_{5n}\xi^5). \tag{3.63}$$

Since $\xi^5$ is arbitrary, the modal field $\chi_n$ does indeed image itself in all planes $\Pi$ normal to the $x^5$ axis.

## 3.6 Representation of $\hat{R}$ in Fourier space

We now define the operator $\hat{R}$ in Fourier space. For the modal field $\chi_n$ the Law of Laws (2.6) reads

$$\hat{R}\chi_n = r_n \chi_n, \tag{3.64}$$

where $\chi_n$ is given by (3.61) and $r_n$ by (3.63). To put this in Fourier form we first express the series coefficient $w_n(x)$ as a 4-dimensional Fourier integral (infinite limits assumed)

$$w_n(x) = \frac{1}{(2\pi)^4} \iiiint W_n(k) \exp(-ik \cdot x)\, d^4k, \tag{3.65}$$

where $W_n(k)$ is the Fourier spectrum of $w_n(x)$,

$$k \equiv k^\mu = (k^0, k^1, k^2, k^3),\ d^4k = dk^0 dk^1 dk^2 dk^3, \tag{3.66}$$

and $k \cdot x = k_\mu x^\mu$. Putting (3.65) into (3.61) we get from (3.64) the field at plane $\Pi(x^5 + \xi^5)$

$$\begin{aligned} R\chi_n = \frac{1}{(2\pi)^4} \int \left[ W_n(k) \exp(-k_{5n} x^5) \right] \\ \times \exp\left[ -i(k \cdot x + k_{5n}\xi^5 + k_6 x^6) \right] d^4k \end{aligned} \tag{3.67}$$

The argument of the second exponential can be interpreted as the scalar product $k_a x^a$ of two six-vectors, namely,

$$k^a = \left(k, k_n^5, k^6\right) \text{ and } x^a = \left(x, \xi^5, x^6\right). \tag{3.68}$$

In complete analogy with the optical theory we shall require the scalar product $k_a k^a$ of the six-vector $k_a$ with itself to vanish:

$$k^2 + k_{5n}{}^2 - k_6{}^2 = 0, \tag{3.69}$$

where we have used the fact that $x^5$ is time-like and $x^6$ is space-like: $k_5 = k^5$ and $k_6 = -k^6$. In terms of momentum this can be written

$$p_6{}^2 - p^2 - p_{5n}{}^2 = 0, \tag{3.70}$$

where we have defined $p^a = \hbar k^a$ . This is the energy-momentum relation for the quantum of the field in six dimensions, with $p^6 c$ representing the energy of the quantum. As such it says that in six dimensions the mass of the quantum is zero. We might, of course, have assumed a non-vanishing mass. However, since $p^6 = \hbar\Omega / c$ is constant, the non-vanishing mass can be absorbed in a redefined frequency $p'^6$, just as in the optical analogue. Thus with no loss of generality it suffices to assume vanishing $k_a k^a$, as we have done in (3.69). Now we know already that in $M^4$ [see (3.18)] $p$ is an invariant four-vector:

$$p^2 = m^2 c^2. \tag{3.71}$$

Hence (3.70) becomes

$$p_6{}^2 - p_{5n}{}^2 = m^2 c^2. \tag{3.72}$$

Since $p_6$ is constant, we see at once that *the fifth momentum* $p_5$ *is invariant*. In $M^{4+2}$ spacetime we have *three* relativistic invariants to work with—$m$, $p^5$ and $p^6$—any two of which can be considered independent.We also now see why the time $x^6$ must be metrically space-like and $x^5$ time-like. For if $x^6$ and $x^5$ had metrical signatures opposite to those assumed in (3.67), then instead of (3.72) one would have

$$p_6^{\,2} - p_{5n}^{\,2} = -m^2c^2\,, \tag{3.73}$$

differing from (3.72) by a minus sign on the r. h. s. Since $p_6^{\,2} \geq p_{5n}^{\,2}$, one would then have $m^2 \leq 0$, implying that all quanta arising in the projection theory are either massless or tachyonic. As tachyonic particles are neither causal nor observed, we must reject as unphysical the metrical assignments leading to (3.73).

Just as (3.19) defines in the 2-dimensional $\boldsymbol{\kappa}$-plane a circle of radius $\kappa_n$, Eq. (3.69) defines in the 4-dimensional $k$-plane a hyperbola of transverse semi-axis $k_{Cn}$, where

$$k_{Cn} = \sqrt{k_6^{\,2} - k_{5n}^{\,2}} = 2\pi / \lambda_{Cn}\,, \tag{3.74}$$

and $\lambda_{Cn}$ is the Compton wavelength of the quantum belonging to the $n$th non-diffracting field, $\chi_n$; see **Fig. 3.4.** Eq. (3.69) is the on-shell condition for non-diffracting fields in $M^4$. It states that the only non-vanishing values of the Fourier spectrum $W_n(k)$ are those lying on the hyperbola defined by (3.69). Thus, in direct analogy with (3.21),

$$W_n(k) \propto \delta\left(k^2 - k_{Cn}^{\,2}\right). \tag{3.75}$$

Owing to the $\delta$ function, we are free to replace the *constant* $k_{5n}$ appearing in the second exponential of (3.67) with the *variable* function of $k$ obtained by solving (3.69) for $k_{5n}$:

$$k_{5n} = \begin{cases} \sqrt{k_6^{\,2} - k^2}\,, & k^2 \leq k_6^{\,2} \\ -i\sqrt{k^2 - k_6^{\,2}}\,, & k^2 > k_6^{\,2} \end{cases}\,, \tag{3.76}$$

and where we have assumed propagation in the direction of increasing $x^5$ ($\xi^5 > 0$). Inserting this into (3.67) we obtain

$$\hat{R}\chi_n = \frac{1}{(2\pi)^4}\iiiint F(k,\xi^5,k^6)\mathrm{X}_n(k,x^5,x^6)\exp(-ik\cdot x)d^4k\,, \tag{3.77}$$

where $\mathrm{X}_n(k, x^5, x^6)$ is the four-dimensional Fourier transform of the field $\chi_n(x, x^5, x^6)$ given by (3.61):

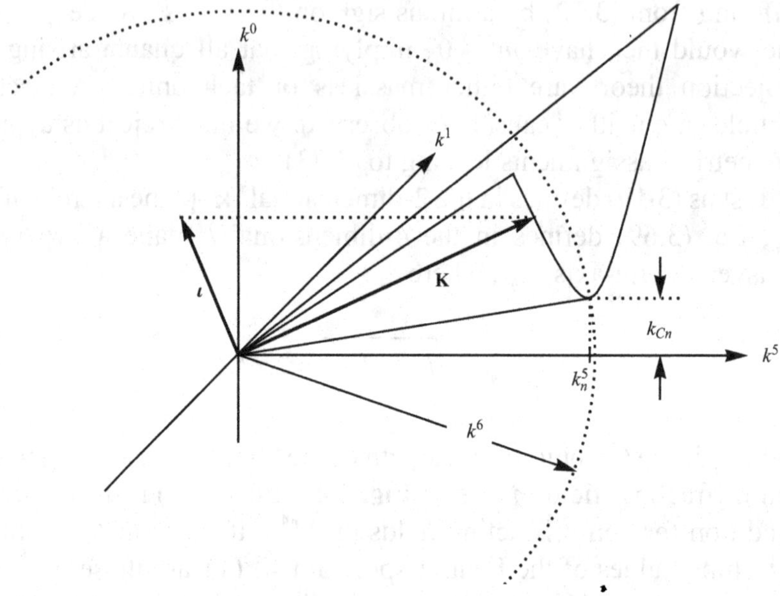


**Fig. 3.4.** The transverse components $\boldsymbol{\iota} = \mathbf{e}_\mu k^\mu$ of the propagation vectors $\mathbf{K} = \boldsymbol{\iota} + \mathbf{e}_5 k_n^5$ for non-diffracting fields $\chi_n$ have constant magnitude $k_{Cn} = m_n c / \hbar$, where $m_n$ is the mass of the field quantum [see Eq. (3.69)]. (Here unit vectors $\mathbf{e}_a$ obey the relation $\mathbf{e}_a \cdot \mathbf{e}_b = g_{ab}$ .) Thus vectors **K** lie on a conical surface of height $k_n^5$ and hyperboloidal base of semi-transverse axis $k_{Cn}$ in 5-dimensional $k$-space. The tips of the transverse components end on a hyperpoloid of semi-transverse axis $k_{Cn}$ in the $M^4$ hyperplane. Vector **K** as depicted represents a particle in motion in the negative $x^1$ direction. Fields for which $k_{Cn} > k^6$ are evanescent and do not propagate.

$$\mathrm{X}_n(k, x^5, x^6) = W_n(k) \exp[-i(k_6 x^6 + k_{5n} x^5)], \qquad (3.78)$$

and

$$F(k, \xi^5, k^6) = \begin{cases} \exp\left(-i\xi^5 \sqrt{{k_6}^2 - k^2}\right), & k^2 \le {k_6}^2 \\ \exp\left(-\xi^5 \sqrt{k^2 - {k_6}^2}\right), & k^2 > {k_6}^2 \end{cases}, \qquad (3.79)$$

Eq. (3.79) defines the projection operator $\hat{R}$ in Fourier space. The effect of $\hat{R}$ operating on $\chi_n(x, x^5, x^6)$ is to multiply the latter's Fourier spectrum (3.78) by the dispersion factor (3.79). For propagation distances $\xi^5 > 0$ this factor effectively suppresses all frequencies lying outside the region of the Fourier plane bounded by the hyperpola $k^2 = k_6^{\ 2}$ (See **Fig. 3.4**). Therefore, for sustained propagation, one must have $k_{Cn} < k^6$.

## 3.7 $\hat{R}$ as convolution operator in configuration space

By the convolution theorem Fourier integral (3.77) can be expressed as a four-dimensional convolution in configuration space:

$$\begin{aligned}\hat{R}\chi_n &= f\left(x;\xi^5,k^6\right) * \chi_n\left(x;x^5,x^6\right) \\ &= \iiiint f\left(x-x';\xi^5,k^6\right)\chi_n\left(x';x^5,x^6\right)d^4x'\end{aligned} \tag{3.80}$$

where $f$ is the inverse Fourier transform of the spectral filter (3.79),

$$\begin{aligned}f\left(x;\xi^5,k^6\right) &= \frac{1}{(2\pi)^4}\iiiint F\left(k,\xi^5,k^6\right)\exp\left(-ik\cdot x\right)d^4k \\ &= \frac{1}{2\pi^2}\frac{\partial}{\partial\xi^5}\frac{1}{s}\frac{\partial}{\partial s}\begin{cases}\dfrac{\exp(-ik^6 s)}{s}, & s^2 \ge 0 \\ \dfrac{\exp(-k^6 s)}{s}, & s^2 < 0\end{cases}\end{aligned}, \tag{3.81}$$

where $s^2 = x^2 + (\xi^5)^2$. Eq. (3.80) is nothing but Huygens' principle in $M^{4+2}$ spacetime: it expresses the field at plane $\Pi(x^5 + \xi^5)$ in terms of a superposition of elementary wavelets, $f\left(x;\xi^5,k^6\right)$, diverging from the source plane $\Pi(x^5)$. These wavelets have the $\exp(ik_0 r)/r$ form familiar from wave optics—see (3.27)—appropriately generalized to $M^{4+2}$ spacetime. This result reinforces our depiction of the projection theory as a process of propagation. It also allows us to write our propagation operator $\hat{R}$ in the compact form

$$R = f\,* \tag{3.82}$$

where $*$ denotes four-dimensional convolution in the $M^4$ hyperplane.

Because $\hat{R}$ is a linear operator, we may in (3.80) sum over *n*, giving

$$\hat{R}\chi = f * \chi, \tag{3.83}$$

where $\chi$ is the periodically self-imaging field (3.56), with *w* given by (3.57). Exactly as in the optical theory, the propagation operator is the same for both general self-imaging and non-diffracting fields.

## 3.8 $\hat{R}$ as differential operator

We have in $M^{4+2}$ these Fourier correspondences:

$$k \rightleftharpoons i\partial,\ k_5 \rightleftharpoons i\partial_5,\ \text{and}\ k_6 \rightleftharpoons i\partial_6. \tag{3.84}$$

Inserting them into (3.79) we have from (3.77), after summing over *n* to obtain the periodically self-imaging field $\chi$ and its Fourier transform X:

$$\hat{R}\chi = F(i\partial, \xi^5, i\partial_6)\chi(x, x^5, x^6), \tag{3.85}$$

where

$$\chi(x, x^5, x^6) = \frac{1}{(2\pi)^4}\iiiint \mathrm{X}(k, x^5, x^6)\exp(-ik\cdot x)d^4k. \tag{3.86}$$

Thus from (3.85) and (3.79) we find $\hat{R}$ in differential-operator form

$$\hat{R} = \exp\left[\xi^5\sqrt{\partial_6{}^2 - \partial^2}\right], \tag{3.87}$$

where $\partial^2 = \partial_\mu\partial^\mu$. But $\hat{R}$ is also a displacement operator, taking the field $\chi$ from plane $\Pi(x^5)$ to plane $\Pi(x^5 + \xi^5)$. Expanding the left hand side of (3.62) in a power series in $\xi^5$ we obtain

$$\hat{R}\chi = \left[\exp(\xi^5\partial_5)\right]\chi, \tag{3.88}$$

so that, in addition to (3.87),

$$\hat{R} = \exp(\xi^5\partial_5), \tag{3.89}$$

giving the differential equation

$$\left[\exp\left(\xi^5\sqrt{\partial_6{}^2-\partial^2}\right)-\exp\left(\xi^5\partial_5\right)\right]\chi\left(x,x^5,x^6\right)=0. \quad (3.90)$$

Expanding this in powers of $\xi^5$, we obtain from the first and second order terms, respectively,

$$\left(\sqrt{\partial_6{}^2-\partial^2}-\partial_5\right)\chi=0 \quad (3.91)$$

and

$$\left(\partial^2+\partial_5{}^2-\partial_6{}^2\right)\chi=0. \quad (3.92)$$

Eq. (3.92) is the analog in $M^{4+2}$ of wave equation (3.38) of scalar wave optics. Taking the derivative with respect to time $x^6$ we obtain

$$\left(\partial^2+\partial_5{}^2+k_6{}^2\right)w(x,x^5)=0, \quad (3.93)$$

a five-dimensional Helmholz equation in the complex amplitude, $w$. If in this equation we insert the Fourier series representation (3.57) we obtain a set of *four*-dimensional Helmholz (Klein-Gordon) equations in the non-diffracting beam amplitudes $w_n$:

$$\left(\partial^2+k_{Cn}{}^2\right)w_n(x)=0, \quad (3.94)$$

where $k_{Cn}$ is defined by (3.74). In parallel with the optical theory, index $n$ generates a family of non-diffracting beam flavors belonging to the generation parameter, $k^6$.

Eqs. (3.92)-(3.94) are of second order. As a first-order equation, (3.91) obviously is unsatisfactory. To obtain a first-order equation, we return to (3.69) and replace it with the linear relation

$$\gamma\cdot k\pm\gamma^5k_{5n}+Ik_6=0, \quad (3.95)$$

where the five $\gamma$s, namely $\gamma\equiv\gamma^\mu=(\gamma^0,\gamma^1,\gamma^2,\gamma^3)$ and $\gamma^5=i\gamma^0\gamma^1\gamma^2\gamma^3$ are the Dirac matrices and $I$ is the $4\times4$ identity matrix. With

$g_{\mu\nu} = \text{diag}(+1,-1,-1,-1,)$, we have

$$\gamma^\mu\gamma^\nu + \gamma^\nu\gamma^\mu = 2g^{\mu\nu}, \; \gamma^\mu\gamma^5 + \gamma^5\gamma^\mu = 0, \; (\gamma^5)^2 = 1, \tag{3.96}$$

ensuring consistency of (3.95) with (3.69). Then instead of (3.76) we have from (3.95)

$$Ik_{5n} = \mp\gamma^5\left(\gamma\cdot k + Ik_6\right) \tag{3.97}$$

and in place of (3.87) and (3.89) we obtain

$$\hat{R} = \exp\left[\mp i\gamma^5\xi_5\left(\gamma\cdot i\partial + Ii\partial_6\right)\right] \tag{3.98}$$

and

$$\hat{R} = \exp\left(I\xi^5\partial_5\right). \tag{3.99}$$

The first order terms in the series expansion of these differential forms yield two linear wave equations corresponding to the upper and lower signs in (3.98):

$$\left(\gamma\cdot i\partial + \gamma^5 i\partial_5 + i\partial_6\right)\chi_P(x,x^5,x^6) = 0 \tag{3.100}$$

and

$$\left(\gamma\cdot i\partial - \gamma^5 i\partial_5 + i\partial_6\right)\chi_Q(x,x^5,x^6) = 0, \tag{3.101}$$

whereas the second order terms in the expansion again yield (3.92). The two linearized versions of (3.91) are the analogs of the Weyl equations (3.46) and (3.47) obtained in the optical theory. In the projection theory, however, the wave functions $\chi_P$ and $\chi_Q$ do not refer to the handedness of massless particles. Rather, in the case that $\chi_P$ and $\chi_Q$ are Dirac spinors, they represent, respectively, the probability and charge amplitudes of the Dirac particle.

For completeness, taking the derivatives with respect to time $x^6$, we obtain for the complex amplitudes $w_P$ and $w_Q$

$$\left(\gamma\cdot i\partial + \gamma^5 i\partial_5 - k^6\right)w_P(x,x^5) = 0 \tag{3.102}$$

and

$$\left(\gamma\cdot i\partial - \gamma^5 i\partial_5 - k^6\right)w_Q(x,x^5) = 0, \tag{3.103}$$

and for the non-diffracting fields

$$\left(\gamma \cdot i\partial + \gamma^5 k_{5n} - k^6\right) w_{Pn}(x) = 0 \qquad (3.104)$$

and

$$\left(\gamma \cdot i\partial - \gamma^5 k_{5n} - k^6\right) w_{Qn}(x) = 0\,, \qquad (3.105)$$

and where $k_{5n}$ and $k^6 = -k_6 = \Omega / c$ are related by (3.74). Again, index $n$ generates a family of non-diffracting beam flavors belonging to the generation parameter, $k^6$.

## 3.9 Comments and observations

*3.9.1 Reality of the new coordinates and of* $\chi$. Guided by the optical self-imaging *ansatz* we have succeeded in placing our Law of Laws (2.6) on the stage of physical spacetime. A theory that began in **Chapter 2** with projection as a metaphor for the conversion of fact into representation has in this chapter evolved into a literal, physical theory of wave propagation. Surprisingly enough, we have in the evolutionary process uncovered two new spacetime dimensions, the time-like spatial dimension $x^5$ and the space-like temporal one $x^6$. We emphasize that, despite their ultimate origin in speculative (metaphysical) thought, the extra coordinates are to be considered physically real—quite as real as the familiar coordinates of Minkowskian $M^4$—and no mere mathematical device. Without them it is our belief that none of the open problems of physics mentioned in **Chapter 1** has any hope of resolution. With them, as we shall see in the subsequent chapters of this book, their resolution becomes reasonably accessible.

Also, like the dimensions $x^5$ and $x^6$, the wave function $\chi(x, x^5, x^6)$ is to be considered objectively real. Here one recalls the perennial debate over the interpretation of the quantum-mechanical wave function, $\psi$, whose time development is described by the Schrödinger wave equation.[4] On Born's probabilistic interpretation, $\psi$ is not physically real but represents only our *knowledge* of the physical system with which it is associated. Heisenberg, on the other hand, noting that $\psi$ satisfies a deterministic wave equation, concluded that $\psi$ must be physically real and not merely the expression of a state of knowledge.

Heisenberg's argument applies equally well to our descriptor $\chi$: it satisfies a deterministic wave equation and so must be physically real. But we already knew that. For we saw in **Sec. 2.3** that, as a consequence of the correspondence condition (2.5), the fact $\chi$ is to be considered an objective element of the physical world. The present physical theory is, in other words, ontological, not epistemic. It deals with real things, not merely our knowledge of them.

*3.9.2 The Projection Theory is a single-particle theory.* By assumption, the wave function $\chi$ resulting from geometrization of the projection theory is a function of a single 6-vector of position, $x^a = (x, x^5, x^6)$. Thus $\chi$ is a single-particle descriptor and the theory under discussion here is to be considered a single-particle theory. That is, it describes the properties and behavior in external fields of single, non-composite particles such as muons and photons, but is applicable only indirectly to composite objects like protons and mesons. But even the single-particle theory offers much of interest. In particular, one notes that despite the unqualified success of quantum field theory—the Standard Model— in describing the interactions of all known subatomic particles, it is unable to predict the masses of those particles. The projection theory, on the other hand, for all its limitations, can among other things readily account for the masses of all leptons and free quarks. The point is that there is much to learn before embarking on a theory of many particles, provided we get the geometry right.

*3.9.3 Origin of wave-particle dualism.* At the heart of the physical microworld there lies duality of wave and particle. The idea that a physical entity can be one thing or another depending on point of view arguably comprises the main distinction between the classical and quantum world views. And yet, as emphasized in *Sec. 1.2.4*, it is an idea whose truth has been supported mainly by experimental observation: for the particle as such appears nowhere in the standard formalism of quantum mechanics. Fortunately, and perhaps unexpectedly, the developments of the present chapter offer a new and completed quantum formalism, one in which the particle itself makes an explicit appearance. To see this, we may combine (3.64) and (3.82) to yield

$$\chi_n = r_n^{-1} f * \chi_n . \tag{3.106}$$

This expresses the non-diffracting wave function $\chi_n$ as a self-weighted superposition of invariant propagator functions $f$. In view of the meaning of convolution, the propagator $f$ may be said to be simultaneously present at all spacetime points of the wave field, with weighting at those points determined by the wave field itself. This suggests that we identify $f$ with the spacetime distribution of the *particle*, and $\chi_n$ with its guiding wave field, a wave-particle picture consistent with the one proposed by de Broglie.[5] In fact, invariant distributions similar to $f$ were proposed long ago by McManus[6] and by Feynman.[7]

Occasionally one reads that the particle—conceived as a kind of microscopic billiard ball—is a fiction; that it is actually a wave of some sort.[8] That claim is in a sense confirmed by (3.106), for the entity representing the particle is the invariant wave propagator $f$. However, in the present instance something remarkable happens when we integrate out the time variation in $f$: there then results a purely spatial distribution—one we can interpret to represent the time-averaged motion of a point particle. The distribution so formed is that of the particle's Zitterbewegung. Later, we are going to derive the internal distributions of a variety of particles: electron, neutrino, photon, and a new particle we shall call the riemann. For now, we wish only to emphasize that dualism of wave and particle is a direct and observable consequence of wave propagation along coordinate $x^5$. Where there is propagation of a wave function $\chi_n$ there necessarily appears a particle, represented by the wave propagator, $f$.

*3.9.4 Invisibility of the extra coordinates.* In **Sec. 3.5** mention was made that the new spatial dimension $x^5$ is not seen because the observable non-diffracting fields comprising $\chi$ are independent of it. Inspection of Eqs. (3.104) and (3.105) shows that indeed the non-diffracting fields $w_{Pn}(x)$ and $w_{Qn}(x)$ depend on the coordinates of $M^4$ alone, dimensions $x^5$ and $x^6$ having been differentiated away in what amounts to a process of dimensional reduction. Thus spacetime *looks* four-dimensional. We note, however, that although the non-diffracting fields are independent of $x^5$ and $x^6$, these fields are still embedded in a 6-dimensional spacetime. This means that, from a 6-dimensional perspective, the physical objects represented by $w_{Pn}(x)$ and $w_{Qn}(x)$ are actually infinite cylinders or, more precisely, infinite line currents extending from $x^5 = -\infty$ to $x^5 = +\infty$, as depicted in **Fig. 3.5**. The 4-D cross sections of these infinite cylinders in the $M^4$ hyperplane are the microscopic objects (such as electrons and photons) observed in particle accelerators and in everyday life.

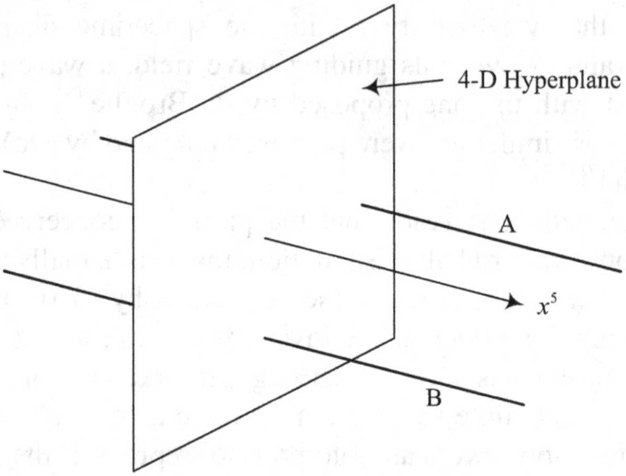


**Fig. 3.5** Physical objects in $M^{4+2}$ take the form of infinite line currents like A and B running parallel to the $x^5$ axis. Their cross sections in the 4-D hyperplane $M^4$ are the objects (such as electrons, photons, table and brains) observed in particle physics and in the everyday world.

*3.9.5 All single particles are massless in six dimensions.* This picture of infinite line currents intercepting $M^4$ may at first glance seem implausible. In particular, because the mass $m$ of the current's cross section in $M^4$ is in general non-vanishing, one might conclude that the total mass of the infinite line current is infinite. Such a conclusion would be incorrect. For in fact, as we observed in connection with Eq. (3.70), the mass of the quantum in six dimensions is exactly zero. And it is zero for precisely the same reason that the quanta observed in $M^4$ are non-tachyonic: the metrical signatures of $x^5$ and $x^6$ are opposite to those attached to the standard spatial and temporal dimensions of $M^4$. Thus while the picture of a universe comprised of infinite line currents is decidedly odd, it is at least not physically absurd. It is a natural result of our demand for consistency between the worlds of appearance and reality.

*3.9.6 Pseudomomentum and pseudoenergy.* In **Sec. 3.6** we argued that $x^5$ must be metrically time-like and $x^6$ metrically space-like. However, since the field descriptor $\chi$ propagates in the direction of $x^5$, it is clear that $x^5$ does *not* represent a second time dimension, but rather is to be interpreted as an extra spatial dimension, a coordinate like $x$, $y$ and $z$ denoting spatial location, but one whose metrical signature happens to be time-like. Similarly, $x^6$ does not represent an extra spatial dimension, but is to be interpreted as an extra temporal dimension in the restricted sense that it supplies the angular frequency $k^6 = \Omega / c$ associated with propagation along $x^5$. The free particle thus may be said to possess a time-like fifth component of momentum, $p_5 = \hbar k_5$; a time-like fifth component of velocity $\upsilon_5 = p_5 c / p_0$; and a space-like component of energy, $p^6 = \hbar k^6$. In view of their reversed metrical signatures, $p_5$ and $p^6$ will be called the *pseudomomentum* and *pseudoenergy*, respectively.

*3.9.7 Conservation of* $p_5$. As is well known, conservation of the Lagrangian under spacetime translations implies conservation of energy-momentum.[9] In $M^{4+2}$ spacetime invariance under translations in $x^5$ implies conservation of pseudomomentum $p_5$.[10] Indeed, we have already proved the invariance of $p_5$ from (3.72). Because infinitesimal rotations in the $\mu$-5 planes are forbidden, conservation of $p_5$ holds separately from conservation of the four $p_\mu$. In the case of a single particle, for which $p_5 = \text{inv.}$, the law of conservation of $p_5$ tells us nothing new. However, in a many-particle context involving interactions, it is distinctly non-trivial. For it implies that *only those reactions can occur that conserve the overall* $p_5$. As we shall see in the next chapter [**Sec. 4.13**], it is this law, combined with conservation of electric charge, that underlies conservation of lepton and baryon numbers.

*3.9.8 The neutrino in six dimensions.* In **Sec. 3.8** we noted that Dirac spinor pairs $\chi_P$ and $\chi_Q$ distinguish not between right and left handedness but between probability and electric charge. In the case of a charged particle, such as the electron, the distinction is one without a difference: the currents deriving from the two spinors are directly proportional. The neutrino, however, is another matter. For although it possesses a non-vanishing probability density, the neutrino is also electrically neutral. Thus for the neutrino the distinction between $\chi_P$ and $\chi_Q$ could not be sharper. And it is a distinction unavailable to the Dirac theory of the neutrino in four dimensions. Why does this matter? It matters because it provides a new foundation for the study of the neutrino, one yielding surprising and perhaps heretical results. For instance, later in **Secs. 4.7** and **12.4**, we will show that the charge density of the neutrino deriving from $\chi_Q$ is directly proportional to square of its mass. Since the charge of the neutrino is zero, it follows that the neutrino is massless, just as the Standard Model assumes it to be. Oscillation between different flavors of neutrino nevertheless occurs owing to very slight departures from monochromaticity in the $k_6$ frequency domain; see **Chapter 12**. With these results we have still another example of new physics coming out of the projection theory.

*3.9.9 Origin of quantized electric charge.* We have seen in some detail how flavor arises out of self-imaging in $M^{4+2}$. We start with a general wave function $\chi(x, x^5, x^6)$ propagating in direction $x^5$ with pseudotemporal frequency $k^6$. Next, because $\chi$ is self-imaging, we take it to be periodic in $x^5$ and expand it in a Fourier series. Expansion index *n* then generates a family (or generation) of non-diffracting beam amplitudes (flavors) $w_n(x)$ belonging to generation parameter $k^6$. Thus flavor can be considered a physical attribute defined by two parameters: $k^6$ and *n*. Now we have a good idea of the meaning of parameter $k^6$; it represents (when multiplied by $\hbar c$) the maximum rest energy of any member of the family of particles belonging to it; see **Fig. 3.4**. But what about parameter *n*? Could it have a physical significance beyond that of generator of terms in a Fourier series? To explore this, let us take the $w_n(x)$ to describe a family of fundamental fermions. The constants $k_{Cn}$ defined by (3.74) would then represent (when multiplied by $\hbar / c$) their masses. Then, if in (3.58) we were to take $K_5 = k^6$, the fermion corresponding to index $n = 0$ would be massless. The fermion so obtained is of course the neutrino. But the neutrino is electrically neutral. Thus we conclude that the integer parameter $-n$ represents quantized values of electrical charge, $Q$, in the case of leptons; or quantized values of $Q + B$ in the case of quarks, where *B* denotes baryon number. Of course we have not *proved* that $-n$ represents what we just have said it represents. We will later provide such a proof (see **Sec. 4.9**). Meanwhile we shall simply state that self-imaging in $M^{4+2}$ not only provides a physical definition of flavor, but potentially explains quantization of electrical charge.

## Notes and references


[1] W. D. Montgomery, "Self-Imaging Objects of Infinite Aperture," J. Opt. Soc. Am. **57**, 772 (1967); "Self-Imaging Objects of Finite Aperture," J. Opt. Soc. Am., **58**, 1112 (1968).

[2] J. Durnin, J. J. Miceli and J. H. Eberly, "Diffraction-free beams," Phys. Rev. Lett. **58**, 1499 (1987).

[3] J. Durnin, "Exact solutions for nondiffracting beams. I. The scalar theory," J. Opt. Soc. Am. A **4**, 651 (1987).

[4] M. Jammer, *The Philosophy of Quantum Mechanics* (John Wiley, New York, 1974), pp. 38-44.

[5] Jammer, *op cit.,* pp. 44-49.

[6] H. McManus, Proc. Roy. Soc. (London) **A195**, 323 (1948).

[7] R. P. Feynman, Phys. Rev. **74**, 1430 (1948).

[8] John Duffield, *Relativity+* (Corella Ltd., UK, 2009), p. 113.

[9] S. Schweber, *An Introduction to Relativistic Quantum Field Theory* (Row, Peterson, Evanston, Illinois, 1961) p. 186ff, 208.

[10] If the $5\mu$ component of the energy-momentum tensor $T^{ab}$ does not depend on the spacetime coordinates explicitly then

$$\int\left(\frac{dT^{50}}{dx^0}+\sum_{k=1}^{3}\frac{dT^{5k}}{dx^k}\right)dx^3=\frac{d}{dx^0}\int T^{50}dx^3=\dot{p}_5=0,$$

provided the $T^{5k}$ vanish at large distances.

# PART II

# SINGLE-PARTICLE LAWS OF PHYSICS

# Chapter 4

# Self-imaging fermions: the spacetime origin of flavor

In the previous chapter, by analogy with optical self-imaging, we obtained both linear and second-order equations for the single-particle descriptor $\chi$ in six-dimensional spacetime $M^4 \times \bar{M}^2$; see (3.92), (3.100) and (3.101). In this and the following two chapters (altogether making up **PART II** of the book) we shall derive particular single-particle propagation laws, the structure of each dictated by the form of the propagator $\hat{R}$ and the behavior of $\chi$ under spacetime transformations. The field descriptors $\chi$ to be considered are those for:

Fermions (this chapter): $\chi \rightarrow$ Dirac spinor $\psi$
Bosons (**Chapters 5** and **6**): $\chi \rightarrow$ Klein-Gordon scalar $\phi$
Maxwell-Proca 5-vector $V^a$
Maxwell $5\times 5$ tensor $F^{ab}$

Note that these fermionic and bosonic fields are the very ones which, in a 4-dimensional context, enter the description of the electro-weak interaction. However, we are not immediately interested in the details of interactions, as these are already more than adequately described by the Standard Model. Instead, we seek to describe the behavior of *free* fields and particles in six spacetime dimensions. It perhaps bears repeating, we embark here on physics *before* the Standard Model, not beyond it.

The propagation laws we are about to discuss are of course *wave* equations, the descriptor $\chi$ in each case being the wave function. But the propagator $\hat{R}$ does more than propagate $\chi$ in the direction of $x^5$. For as noted in *Sec. 3.9.3*, $\hat{R}$ also represents the *particle* itself, the quantum of the field denoted by $\chi$. We will describe in detail the structure of the quanta later in **PART IV** of this book. For now it is enough to know that, for the fields under discussion here, where there is propagation along $x^5$, there exists a particle; and if no propagation, then there is no particle.

It should be noted as well that, in the formulation of these single-particle laws, the role played by spacetime is a limited one: it furnishes the parameters required to express those laws. Spacetime provides the stage on which the laws operate, but introduces no dynamical effects of its own. Of course the laws must still hold in curved spacetime, and later in **PART III** we will reformulate them allowing them to act on the stage of non-Euclidian geometry. Also in **PART III**, we shall take up the problem of geometrodynamics, where spacetime reveals its own equations of motion and dynamical properties.

We begin our study of the propagation laws with those relating to fermions. Not surprisingly we shall reproduce the Dirac equation. However, in six spacetime dimensions, the Dirac equation properly distinguishes between the different flavors of fermion. And so we begin here our attack on an elusive physical problem already alluded to in *Sec. 1.2.4*, namely, the family problem of the fundamental fermions. The driving idea is that these particles and their masses arise from higher dimensions, not gauge symmetry. The present chapter addresses those components of the problem relating to flavor, family replication and mass generation. The analysis, as we shall see, sheds surprising new light on the mass of the neutrino. In addition we find that in all processes baryon number $B$ and lepton family numbers $L_\ell$ ($\ell = e, \mu, \tau$) are separately conserved. In **PART V** of the book we shall apply the theory of flavor developed here to problems of neutrino mixing and oscillation, and to obtain formulas for the masses of the quarks and charged leptons.

## 4.1 The family problem

There are at present known to exist six lepton flavors and six quark flavors, all of which have been observed experimentally, either directly or indirectly.[1, 2] These twelve fundamental fermions are conventionally organized into three families, or generations, whose members are differentiated by electric charge $Q$ as shown in **Table 4.1**.[1] It is often said that the first generation of leptons and quarks, accompanied by an appropriate set of force-mediating bosons, would have sufficed to build a working universe practically indistinguishable from the one we live in.[3-5] From that point of view, the second and third generations appear to be superfluous.[6] The family problem[7-9],

**Table 4.1** Conventional arrangement of leptons and quarks of electric charge $Q$ into three generations $G$.

| Particle Type | Charge $Q$ | Generation $G$ | | |
|---|---|---|---|---|
| | | I | II | III |
| Lepton | 0 | $\nu_e$ | $\nu_\mu$ | $\nu_\tau$ |
| | $-1$ | $e$ | $\mu$ | $\tau$ |
| Quark | 2/3 | $u$ | $c$ | $t$ |
| | $-1/3$ | $d$ | $s$ | $b$ |

or flavor puzzle[10], is the dual problem of explaining the origin of these superfluous generations and why there are two of them rather than some other number.

A closely related problem is that of calculating the mass spectra of the charged leptons and bare quarks.[11, 12] Because the masses of the charged leptons increase monotonically with generation, as do the bare masses of quarks of like charge (see **Fig. 1.1**), one suspects that the solution to the replication problem may lead to a solution of the mass problem. That is in fact the case, as we shall later see (**Chapters 13** and **14**.)

The family problem has thus far easily resisted the many attempts to solve it.[9] There are, we think, two reasons for this. First, and most importantly, nowhere in the Standard Model (SM) of electroweak interactions[13-16] is the concept of flavor physically defined or represented. At present one has, not a physical theory of flavor, but an empirically-based scheme of classification.[17] Thus, for example, the flavor neutrinos $\nu_e$ and $\nu_\mu$ are labeled as they are to reflect the fact that each always gives rise to its own charged lepton in charged-current interactions like $\nu_\ell n \to \ell^- p$. But the act of naming obviously does not explain how these neutrinos differ physically. It is clear that a true theory of flavor, if it is to result in a solution to the family problem, will have to go deeper to the source of that fundamental attribute.

The second impediment to progress against the family problem has been, in the writer's estimation, the fixed idea that fermions acquire their masses by interaction with the vacuum Higgs field, $v$.[15, 18-20] True, fermions have masses, so the SM Lagrangian must contain fermion mass-like terms, and the only known way to get them and still respect gauge symmetry is by spontaneous symmetry breaking. But the appearance of mass-like terms in the Lagrangian subsequent to spontaneous breakdown of the underlying $SU(2)_L \times U(1)_Y$ gauge symmetry by no means proves that fermion masses are generated by coupling to the Higgs field. On the contrary, if it is true that mass is tied to flavor, then, because flavor is unaccounted for in the SM, it seems unlikely that spontaneous symmetry breaking can be the true source of fermion mass. A more plausible scenario is one in which the fermions come to the SM Lagrangian *already endowed* with their intrinsic masses, $m_f$. These preëxisting masses then determine the strengths of the Higgs-fermion interactions through the Yukawa coupling constants $g_f = \sqrt{2}m_f / v$. (A similar argument applies to the masses of the gauge bosons, which masses, thanks to the Higgs mechanism, may be said to generate the value of the weak mixing angle, $\theta_W$.) In short, we suggest that the Higgs mechanism does *not* generate the fermion masses as is generally supposed, but simply provides a gauge invariant way of getting preëxisting masses into the Lagrangian.

But if spontaneous symmetry breaking is not the source of fermion mass, then what is? Even more to the point, if it is true that mass is linked to flavor, then what exactly is "flavor"? Our approach to the family problem—one leading to explanations of both flavor and mass generation—is very simple. The familiar grouping of leptons and quarks into generations as depicted in **Table 4.1** suggests the possibility of treating flavor not as a mere label, but as a *physical* attribute with two degrees of freedom, generation $G$ and electric charge $Q$. Our proposed theory of flavor should then do two things: it must (1) furnish physical manifestations of these parameters and (2) exhibit them in quantized form. The idea is that, in a well-formed theory, quantized values of $G$ will lead to distinct families of fermions, and quantized values of $Q$ to distinct flavors within a family, in accordance with the structure of **Table 4.1**.

But of course we already know of such a theory. We first encountered it in our study of self-imaging optical fields in Minkowskian $M^4$; with reference to Helmholtz equation (3.40), we pointed out that expansion index $n$ generates a family of non-diffracting beam flavors $\upsilon_n$ belonging to generation parameter $k^0$. We encountered it again, even more relevantly, in our study of self-imaging fields in $M^4 \times \bar{M}^2$, motivated by the Law of Laws (2.6); with respect to linear wave equations (3.104) and (3.105), we noted that index $n$ generates a family of non-diffracting beam flavors $w_{Pn}$ and $w_{Qn}$ belonging to generation parameter $k^6$. Thus to develop a good theory of flavor for fermions, we need only formulate our theory of self-imaging in $M^4 \times \bar{M}^2$ in terms of Dirac spinors. This we now proceed to do.

## 4.2 Towards a Dirac equation in six dimensions

Equations for fermions—particles of spin one-half—are for us formally more telling than those for bosons. This is because they make contact with the Lie group generated by the fifteen Dirac-Clifford matrices, $\Gamma_i$. The significance of this is that *fifteen* is also the number of independent rotations—whether circular or relativistic—in a space of six dimensions. The Clifford algebra thus provides assurance in the reality of 6-dimensional spacetime; that is to say, to believe in the reality of our two extra dimensions one need not rely entirely on the analogy with optical self-imaging. Moreover, rather fortuitously, the Clifford algebra formally confirms our choice of metrical signature for the extra dimension $x^6$ (but not of $x^5$).

*4.2.1 Invariants of motion.* In accordance with the projection theory, we assume a six-dimensional product space $M^4 \times X^2$, where $M^4$ denotes four-dimensional Minkowski space and $X^2$ is a two-dimensional space consisting of flat, infinite dimensions $x^5$ and $x^6$. We assume the metric $g_{ab} = \mathrm{diag}(+1,-1,-1,-1; g_{55}, g_{66})$, where the signs of $g_{55}$ and $g_{66}$ are to be determined. We will find them to be + and −, respectively, so that $M^4 \times X^2$ becomes $M^4 \times \bar{M}^2$ as assumed in the previous chapter.

We start by finding the invariants of motion in $M^4 \times X^2$. For this we look to the inhomogeneous Lorentz group in six dimensions. There are six generators of translations, $\hat{k}_a = i\partial_a$ ($a = 0,1,2,3,5,6$), which, after multiplication by $\hbar$, can be identified with the components of the energy-momentum six-vector, $(p_\mu, p_5, p_6)$ ($\mu = 0,1,2,3$). In addition, there are 15 Hermitian generators of rotations, $\hat{M}_{ab} = x_a\hat{k}_b - x_b\hat{k}_a$. The commutation rules between these generators are completely analogous to those in four dimensions:[21]

$$\left[\hat{k}_a,\, \hat{k}_b\right] = 0\,, \tag{4.1}$$

$$\left[\hat{M}_{ab},\, \hat{k}_c\right] = i\left(g_{bc}\hat{k}_a - g_{ac}\hat{k}_b\right), \tag{4.2}$$

$$\left[\hat{M}_{ab},\, \hat{M}_{cd}\right] = -i\left(g_{ac}\hat{M}_{bd} + g_{bd}\hat{M}_{ac} - g_{ad}\hat{M}_{bc} - g_{bc}\hat{M}_{ad}\right), \tag{4.3}$$

In four dimensions the scalar invariant of the homogeneous group is known to be $\hat{k}_\mu \hat{k}^\mu$. But in six dimensions $\hat{k}_\mu \hat{k}^\mu$ need not be invariant. In six dimensions, the scalar invariant operator of the homogeneous group is[22]

$$\hat{k}_a \hat{k}^a = \hat{k}_\mu \hat{k}^\mu + g_{55}\hat{k}_5^{\;2} + g_{66}\hat{k}_6^{\;2}\,. \tag{4.4}$$

(It is invariant because it commutes separately with each of the $\hat{k}_a$ and $\hat{M}_{ab}$.) In momentum space (4.4) becomes the invariant scalar product

$$k_a k^a = M^2 c^2 / \hbar^2\,, \tag{4.4a}$$

which is the energy-momentum relation for a mass *M* in six dimensions. Now of course we know very well that in the real world $\hat{k}_\mu \hat{k}^\mu$ is invariant: spacetime *appears* to be four-dimensional. Thus, if our theory is to have to do with the real world, we must be able to prove separately that

$$g_{55}\hat{k}_5^{\;2} = \text{invariant} \tag{4.5}$$

and

$$g_{66}\hat{k}_6^{\;2} = \text{invariant.} \tag{4.6}$$

This is accomplished as follows.

*4.2.2 A linearized wave equation.* Here we seek a six-dimensional form of the single-particle Dirac equation applicable to fermions. This equation should of course be linear in the six momentum operators $\hat{p}_a = \hbar \hat{k}_a$. Thanks to the $\gamma^5$ matrix [see Eq. (3.95)] the invariant operator (4.4) indeed factorizes, but to render the equation entirely linear we must have in (4.4a) $M = 0$: as noted already in *Sec. 3.9.5*, mass vanishes in six dimensions. We then obtain[23]

$$\overline{\gamma}^a \hat{p}_a \chi = \left(\gamma^\mu \hat{p}_\mu + \overline{\gamma}^5 \hat{p}_5 + \gamma^6 \hat{p}_6\right)\chi = 0, \quad a = 0,1,2,3,5,6 \quad (4.7)$$

where $\chi = \chi\left(x, x^5, x^6\right)$ is a four-component spinor and where we have defined

$$\overline{\gamma}^a = (\gamma^\mu, \overline{\gamma}^5, \gamma^6), \text{ with}$$

$$\overline{\gamma}^5 = \begin{cases} -\overline{\gamma}_5 = i\gamma^5 \text{ if } x^5 \text{ is space-like} \\ +\overline{\gamma}_5 = \gamma^5 \text{ if } x^5 \text{ is time-like} \end{cases}$$

and

$$\gamma^6 = -i\left(g_{66}\right)^{1/2} I = \begin{cases} -\gamma_6 = +I \text{ if } x^6 \text{ is space-like} \\ \\ +\gamma_6 = -iI \text{ if } x^6 \text{ is time-like} \end{cases}.$$

We see at once that $x^6$ cannot be time-like. For if it were, then the Hamiltonion $-\gamma^0\left(\gamma^k \hat{p}_k + \overline{\gamma}^5 \hat{p}_5 + \gamma^6 \hat{p}_6\right)c$ associated with (4.7) would contain the non-hermitian term $i\gamma^0 \hat{p}_6 c$. *Thus, in a six-dimensional universe containing fermions, the sixth axis must be space-like:* $g_{66} = -1$. Interestingly, this choice of signature is confirmed by considering the fifteen Dirac matrices $\Gamma_i$ as generators of a Lie group. These may be arranged in the form[24] $(a,b = 0,1,2,3,5,6)$

$$M_{ab} = -M_{ba} = \frac{1}{2}\begin{pmatrix} 0 & \gamma_0\gamma_1 & \gamma_0\gamma_2 & \gamma_0\gamma_3 & \gamma_0\overline{\gamma}_5 & \gamma_0 \\ & 0 & \gamma_1\gamma_2 & \gamma_1\gamma_3 & \gamma_1\overline{\gamma}_5 & \gamma_1 \\ & & 0 & \gamma_2\gamma_3 & \gamma_2\overline{\gamma}_5 & \gamma_2 \\ & & & 0 & \gamma_3\overline{\gamma}_5 & \gamma_3 \\ & & & & 0 & \overline{\gamma}_5 \\ & & & & & 0 \end{pmatrix}, \quad (4.8)$$

with the commutation relations

$$\left[M_{ab}, M_{cd}\right] = -g_{ac}M_{bd} - g_{bd}M_{ac} + g_{ad}M_{bc} + g_{bc}M_{ad} \quad (4.9)$$

and metrical signature $g_{ab} = \text{diag}(+1,-1,-1,-1; g_{55}, g_{66})$. The $M_{ab}$ form a group in that the commutator of any pair yields, by (4.9), another member of the group. Now suppose we form the commutator of any pair taken from the sixth column, say $M_{06}$ and $M_{56}$. We have

$$[M_{06}, M_{56}] = [\gamma_0, \overline{\gamma}_5]/4 = \gamma_0\overline{\gamma}_5/2 = M_{05}. \qquad (4.10)$$

But from (4.9),

$$[M_{06}, M_{56}] = -g_{66}M_{05}. \qquad (4.11)$$

Hence, $g_{66} = -1$, QED. Unfortunately, the group does not reveal the sign of $g_{55}$, as is readily shown by forming the commutator of any pair from the fifth column of (4.8).

*4.2.3 Lorentz invariance in six dimensions.* Now if (4.7) is to be a physically valid equation of motion in $M^4 \times X^2$, it must be form invariant under six-dimensional Lorentz transformations. We define Lorentz transformations in $M^4 \times X^2$ by the rotation transformation

$$x'^a = \Lambda^a{}_b x^b, \quad a,b = 0,1,2,3,5,6 \qquad (4.12)$$

where invariance of the quadratic form $x'^a x'_a$ under rotation requires that

$$\Lambda^a{}_b \Lambda_a{}^c = g_b{}^c. \qquad (4.13)$$

Form invariance of (4.7) under (4.12) means that there exists a $4\times4$ transformation matrix $S(\Lambda)$ with inverse $S^{-1}(\Lambda)$ ($S^{-1}S = I$) such that, operating on $\chi$, in the primed coordinate system (4.7) takes the form

$$\overline{\gamma}^b \hat{p}'_b \chi' = 0, \qquad (4.14)$$

where

$$\chi'(x') = S\chi(x), \qquad (4.15)$$

$x = \Lambda^{-1}x'$, and the gamma matrices remain unchanged, retaining the forms defined in (4.7). To find *S* we substitute for $\hat{p}_a$ and $\chi$ in (4.7) their representations in the primed frame, giving

$$S\overline{\gamma}^a \Lambda^b{}_a S^{-1} \hat{p}'_b \chi' = 0 \; . \qquad (4.16)$$

Comparing this with (4.14) we see that $S$ must be such that

$$S\overline{\gamma}^a \Lambda^b{}_a S^{-1} = \overline{\gamma}^b, \qquad (4.17)$$

or, alternatively,

$$\overline{\gamma}^a \Lambda^b{}_a = S^{-1} \overline{\gamma}^b S. \qquad (4.18)$$

Let us now look at rotations in the 6-$\mu$ and 6-5 planes. From (4.18), since $\gamma^6$ is the identity matrix,

$$\overline{\gamma}^a \Lambda^6{}_a = S^{-1} \gamma^6 S = \gamma^6. \qquad (4.19)$$

Hence

$$\Lambda^6{}_a = g^6{}_a. \qquad (4.20)$$

In other words, rotations in the 6-$\mu$ and 6-5 planes are forbidden, thus establishing (4.6): *the sixth component of momentum* $p^6$ *is invariant.*

Thus the number of possible rotations in $M^4 \times X^2$ has come down from fifteen to ten. To find the form of $S$ corresponding to these remaining ten,[25] we consider rotations through an infinitesimal angle $\varepsilon$ defined by coefficients

$$\Lambda^a{}_b = g^a{}_b + \varepsilon \lambda^a{}_b, \quad a, b = 0,1,2,3,5 \qquad (4.21)$$

where $\lambda^{ab} = -\lambda^{ba}$. Note that (4.21) excludes index 6 since, in accordance with (4.20), rotations in all planes containing axis $x^6$ are forbidden. Matrix $S(\Lambda)$ and its inverse now read

$$\left. \begin{aligned} S(\Lambda) &= I + \varepsilon T \\ S^{-1}(\Lambda) &= I - \varepsilon T \end{aligned} \right\}. \qquad (4.22)$$

Inserting (4.21) and (4.22) into (4.18) we obtain to first order in $\varepsilon$

$$\overline{\gamma}^a T - T \overline{\gamma}^a = \lambda^a{}_b \overline{\gamma}^b, \quad a, b = 0,1,2,3,5 \qquad (4.23)$$

A set of matrices $T$ satisfying (4.23) indeed exists and is given by

$$T = \lambda^{ab} M_{ab}, \tag{4.24}$$

where

$$M_{ab} = \frac{1}{2}(\bar{\gamma}_a \bar{\gamma}_b - g_{ab}) \quad , a,b = 0,1,2,3,5. \tag{4.25}$$

The ten $M_{ab}$ may be arranged in the form,

$$M_{ab} = \frac{1}{2}\begin{pmatrix} 0 & \gamma_0\gamma_1 & \gamma_0\gamma_2 & \gamma_0\gamma_3 & \gamma_0\bar{\gamma}_5 \\ & 0 & \gamma_1\gamma_2 & \gamma_1\gamma_3 & \gamma_1\bar{\gamma}_5 \\ & & 0 & \gamma_2\gamma_3 & \gamma_2\bar{\gamma}_5 \\ & & & 0 & \gamma_3\bar{\gamma}_5 \\ & & & & 0 \end{pmatrix}, \tag{4.26}$$

and are in fact the generators of a *five*-dimensional subgroup of (4.8) with metric $(+1,-1,-1,-1,g_{55})$. They obey commutation rules

$$[M_{ab}, M_{cd}] = -g_{ac}M_{bd} - g_{bd}M_{ac} + g_{ad}M_{bc} + g_{bc}M_{ad}. \tag{4.27}$$

The algebra of the Dirac matrices thus readily accommodates and supports our proposed six-dimensional geometry with infinite fifth and sixth dimensions.

## 4.3 Dirac equation for probability amplitude

Let us now consider the effect of the sign of $g_{55}$ on the equation of motion (4.7). If $x^5$ is space-like, then $\bar{\gamma}^5 = i\gamma^5$ and the Hamiltonian associated with (4.7) contains the extra term $-i\gamma^0\gamma^5\hat{p}_5$. Because this extra term is Hermitian, wave equation (4.7) exhibits at least the formal characteristics required of a true description of nature. However, because all components of the five-momentum $(\hat{p}_\mu, \hat{p}_5)$ are on an equal footing, permitting infinitesimal rotations in all four planes $x^5$-$x^\mu$, such a description does not include or imply the crucial condition $\hat{p}_5{}^2 = \text{inv.}$ We conclude that, if $x^5$ is space-like, then the corresponding equation of motion (4.7), although viable in a formal sense, bears no relation to the real world.

On the other hand, if $x^5$ is time-like, then $\overline{\gamma}^5 = \gamma^5$ and the extra term $-\gamma^0\gamma^5\hat{p}_5 c$ in the Hamiltonian is anti-Hermitian. In that case it seems that (4.7) cannot be considered a valid equation of motion, except in the special case that $\hat{p}_5\chi = 0$. But this assessment is wrong. For if we multiply (4.7) from the left by the metric operator $\gamma^5\hat{p}_5 + I\hat{p}^6$, we obtain (recalling that $\gamma^6 = I$),

$$\left[\gamma^\mu\left(\hat{p}^6 - \gamma^5\hat{p}_5\right)\hat{p}_\mu + \hat{p}_5{}^2 - \hat{p}_6{}^2\right]\chi = 0\,, \tag{4.28}$$

and for this equation the associated Hamiltonian $H$, given by

$$H \propto \left(\hat{p}^6 - \gamma^5\hat{p}_5\right)\boldsymbol{\alpha}\cdot\hat{\mathbf{p}} + \gamma^0\left(\hat{p}_6{}^2 - \hat{p}_5{}^2\right), \tag{4.29}$$

is Hermitian. (Here we have put $\gamma^0\boldsymbol{\gamma} = \boldsymbol{\alpha}$.) From the point of view of hermiticity of the Hamiltonian, (4.28) can be considered a good equation of motion. But let us check for invariance under five-dimensional Lorentz transformations. To obtain (4.28) we multiplied (4.7) by the factor $I\hat{p}^6 + \gamma^5\hat{p}_5$. Thus form invariance of (4.28) requires that

$$S\left(I\hat{p}^6 + \gamma^5\hat{p}_5\right)S^{-1} = I\hat{p}_6' + \gamma^5\hat{p}_5'\,. \tag{4.30}$$

From (4.18) we see that, when $\hat{p}_5\chi \neq 0$, (4.30) is satisfied if and only if $\Lambda^\mu{}_5 = g^\mu{}_5$ $(\mu = 0,\ 1,\ 2,\ 3)$. This means that infinitesimal rotations in the $\mu$-5 planes are forbidden. It also means that, in momentum space, $\hat{p}_5{}^2 = \text{inv}$. Hence if $x^5$ is time-like, i.e., $g_{55} = +1$, then, with (4.30) in force, (4.28) is not only formally valid but meets criterion (4.5) for relevance in the real world. Equation (4.28) [as well as (4.7), with $\overline{\gamma}^5 = \gamma^5$ and $\gamma^6 = I$] therefore represents a valid equation of motion.

*4.3.1 The Dirac probablility amplitude is self-imaging.* Let us review: By combining (a) invariance of the homogeneous group of translations in six dimensions, (b) the group algebra of the Dirac matrices $(\gamma^\mu, \gamma^5)$ and (c) invariance under Lorentz transformations in six dimensions, we are led to a linear equation for fermions in six-dimensional spacetime $M^4 \times \overline{M}^2$, namely Eq. (4.7). With the metric assignments $(+,-,-,-;+,-)$ that equation may now be written

$$\left(\gamma^{\mu}\hat{p}_{\mu}+\gamma^{5}\hat{p}_{5}+\hat{p}_{6}\right)\chi=0. \tag{4.31}$$

But this equation is structurally identical with (3.100), the equation of motion of the self-imaging descriptor, $\chi_P$. Thus, with the identification $\chi_P \equiv \chi$, we conclude that the Dirac probability amplitude $\chi$ is self-imaging. As in (3.56)-(3.58), $\chi$ takes the form of a Fourier series[26] in the axial coordinate $x^5$:

$$\chi(x,x^5,x^6)=\psi(x,x^5)\exp(-ip_6x^6/\hbar), \tag{4.32}$$

with

$$\psi(x,x^5)=\sum_{n=-\infty}^{\infty}\psi_n(x)\exp(-ip_{5n}x^5/\hbar), \tag{4.33}$$

$$p_{5n}=P_5-2\pi n\frac{\hbar}{a^5}, \quad (p_{5n}\geq 0) \tag{4.34}$$

$a^5$ is the repeat interval, $P_5$ is a constant phase bias, and the four-component spinors $\psi_n(x)$ are the coefficients of the expansion. Note that, as in (3.58), propagation in the positive $x^5$ direction is assumed. The motion of the amplitude $\chi$ in six dimensions is given by (4.31) and, after multiplication by $I\hat{p}^6+\gamma^5\hat{p}_5$, Eq. (4.28). Our theory of self-imaging fields in six dimensions is thus not as radical as it might have seemed one chapter ago. For it appears now to be consistent with, and indeed embedded in, the familiar group algebra of the Dirac matrices.

*4.3.2 Interpretation of coordinates $x^5$ and $x^6$* . We have shown on purely formal grounds that $x^5$ must be time-like and $x^6$ must be space-like.[27] However, as we said in *Sec. 3.9.5*, despite these metrical assignments, $x^5$ does not represent a second time dimension, nor does $x^6$ represent a fourth spatial dimension. It is easy to see how one might think otherwise: In (4.31) the operators $\hat{p}_0$ and $\hat{p}_5$ appear symmetrically in the sum $\gamma^0\hat{p}_0+\gamma^5\hat{p}_5$, suggesting that $x^5$ should be considered a second temporal dimension on the same footing as $x^0$;[28] and the operators $\hat{p}_k$ and $\hat{p}_6$ appear together in the sum $\gamma^k\hat{p}_k+I\hat{p}_6$, suggesting that $x^6$ should be considered a fourth spatial dimension on the same footing as the $x^k$. However, multiplication of (4.31) by the operator $I\hat{p}^6+\gamma^5\hat{p}_5$, giving rise to (4.28), breaks the symmetry between $x^0$ and $x^5$ and between $x^k$ and $x^6$ as well. In the Hamiltonian (4.29) the quadratic operator $-\gamma^0 \hat{p}_5^{\,2}$ plays a role similar to that of the operator $\hat{\mathbf{p}}^2$ in non-relativistic quantum theory, except that its sign is negative, reflecting its time-like character; and the quadratic operator $\gamma^0 \hat{p}_6^{\,2}$ plays the role of rest energy, even though $\hat{p}_6$ is space-like. These formal considerations serve to justify our previous claims as to the meaning of the new dimensions, namely, that $x^5$ is to be interpreted as an extra spatial dimension whose metrical signature happens to be time-like; and $x^6$ is to be interpreted as an extra temporal dimension whose metrical signature happens to be space-like. All of this serves to justify our referring to $p^5$ as the *pseudomomentum*; and to $p^6$ as the *pseudoenergy*.

## 4.4 Fermion flavor and mass generation by self-imaging

*4.4.1 The mass-squared operator.* We look now into the meaning of the quadratic operator $\hat{p}_6^{\,2} - \hat{p}_5^{\,2}$ appearing in our physically-relevant equation of motion (4.28). Having defined completely the metric $g_{ab}$ in six dimensions, we can now write from (4.7) (with $a, b = 0,1,2,3,5$)

$$(\gamma^a \hat{p}_a + \hat{p}_6)(\gamma^b \hat{p}_b - \hat{p}_6)\chi = (\hat{p}_\mu \hat{p}^\mu - \widehat{m^2} c^2)\chi = 0\,, \qquad (4.35)$$

where

$$\widehat{m^2} = \left(\hat{p}_6^{\,2} - \hat{p}_5^{\,2}\right)/c^2\,. \qquad (4.36)$$

Clearly $\widehat{m^2}$ is to be interpreted as a *mass-squared operator.* This operator replaces the constant $m^2$ of conventional Dirac theory in four dimensions.

Let us now insert expansion (4.32) into (4.28). This yields a set of Dirac equations in four dimensions

$$\left[\gamma^\mu \left(p^6 - \gamma^5 p_{5n}\right)\hat{p}_\mu - m_n^{\,2} c^2\right]\psi_n(x) = 0\,, \qquad (4.37)$$

corresponding to a set of fermions with quantized mass-squared values

$$m_n^{\,2} = (p_6^{\,2} - p_{5n}^{\,2})/c^2\,. \qquad (4.38)$$

We have thus arrived at the beginnings of a theory of fermion flavor: for each definite value of the *generation parameter* $p^6 = -p_6$, the expansion index $n$ generates a *family* of fermion *flavors* belonging to $p^6$ with corresponding fifth momenta $p_{5n}$, masses $m_n$ and probability amplitudes

$$\chi_n(x, x^5, x^6) = \psi_n(x)\exp\left[-i\left(p_{5n}x^5 + p_6 x^6\right)/\hbar\right]. \qquad (4.39)$$

We define *fermion family f* as the set of all particles belonging to the same 5-momentum shell of radius

$$p^6(f) \equiv G_f \qquad (4.40)$$

The *flavor* of the $n$th member of such a set is jointly defined by the parameters $(G_f, n)$. The probability amplitudes (4.39) can be considered non-diffracting beams in $M^4 \times \bar{M}^2$ [see Eq. (3.61)]. Ultimately we shall study two types of fermion family: the leptons $\ell, \nu_\ell$ (where $\nu_\ell$ is the neutrino belonging to charged lepton $\ell$) and quarks $q$. Remarkably, expansion (4.33) not only generates families of leptons but also explains the quantization of electric charge: for (as we shall show in **Sections 4.9** and **4.11**) we may take the integers $-n$ to represent the quantized values of electric charge $Q$ in the case of leptons; or the quantized values of $Q + B$ in the case of quarks (where $B$ denotes baryon number).

For antiparticles, in a hole-theoretic interpretation, $p_{5n} \to -p_{5n}$. Thus, corresponding particles and antiparticles belong to the same family, but travel in opposite directions along the $x^5$ axis (except in the case of the electron, for which $p_5 = 0$; see *Sec. 4.8.1* below).

Note also that, unless $p_5 = 0$, wave equation (4.37) is not invariant under spatial inversion, nor is it invariant under time reversal. This signifies, in all fundamental fermions other than the electron, the presence of an internal helical motion (*Zitterbewegung*), the handedness of which cannot be reversed by Lorentz boost. We shall return to this point later in **Chapter 10**.

*4.4.2 Accounting for neutrino oscillation and the charged-fermion mass spectrum.* We must point out that our theory of flavor as developed thus far is deficient in two important respects. In the first place, the non-diffracting beams $\chi_n(x,x^5,x^6)$ are eigenstates of the mass-squared operator with eigenvalues given by (4.38). In other words, the $\chi_n$ are all states of definite mass. Therefore (4.37), as it stands, does not properly describe flavor neutrinos $\nu_\ell$, which are known experimentally *not* to be states of definite mass (because they appear to oscillate between one flavor and another). In the second place, our theory so far offers no explanation for the replication of families of fermions (see *Sec. 1.2.3* and especially **Fig. 1.1**). To do so it must provide a means for quantizing parameter $G_f$, thereby generating as many families as there are allowed values of $G_f$.

The first defect is easy to correct. As we shall show in **Chapter 12**, one need only replace the monochromatic wave function $(p^6 = G)$ with a polychromatic one $(p^6 = G + g_j,\ j = 1,2,\ldots)$. Interference between component wave functions of differing frequency $p^6$ produces neutrino flavor oscillation.

Correcting the second defect is more complicated. It requires replacing $\hat{p}_6$ with an operator that acts not only on $x^6$, but on the coordinates of a two-dimensional inner spacetime, $N^2$, as well. As we shall see in **Chapter 13**, the mass spectrum of the charged leptons is generated by the energy eigenvalues of the Hamiltonian of an electron trapped in a well in $N^2$. In **Chapter 14** we show that each quark mass $m_q$ derives from coupling to one of the charged lepton masses $m_\ell$.

4.4.3 *A note on the origin of mass.* According to the Standard Model of particle physics, the quarks and charged leptons acquire their inertial masses by interaction with a scalar Higgs field present throughout all space. In this picture, mass arises as a kind of drag force on the particle as it accelerates through the Higgs field. The recent discovery of the Higgs particle,[29, 30] validating the existence of its parent field, serves to reinforce this story of the origin of inertial mass—a story now universally accepted among physicists and—thanks to the popular media—lay people alike. However, a glance at Eq. (4.38) shows that this story is wrong. Mass is *not* created by interaction with the Higgs field. Rather, it is a product of (pseudo) momentum and energy coming from the extra dimensions $x^5$ and $x^6$, respectively. And the best evidence of this claim is that in extra dimensions we can do something that one cannot do with the Higgs theory, namely, calculate the charged lepton and quark mass spectra.

But if the Higgs field is not directly responsible for mass, then what good is it? Why does it even exist? A plausible explanation is this. The Higgs field arises as a byproduct of symmetry breaking in the Standard Model. Without symmetry breaking all particles in the model would be massless. With symmetry breaking they are allowed non-vanishing masses, though the Standard Model cannot tell us their values. Thus one may say that the presence of the Higgs field correlates with, or signals, the presence of finite particle masses. Put more strongly, the Higgs field allows or authorizes the presence of massive fundamental particles. And that, we claim, is the sole purpose of the Higgs field. The particles themselves come to the Standard Model already endowed with mass. What the interaction *does* create are the values of the Yukawa constants coupling the Higgs field to the preëxisting masses; it does not create the masses themselves.

## 4.5 Dirac equation for electric charge amplitude

In the previous chapter we obtained a companion equation to Eq. (3.100), namely (3.101), representing motion of the charge amplitude. The companion equation for fermions is derivable from (4.28) by rewriting it in the form

$$\left[\gamma^{\mu}\left(\hat{p}^{6}-\gamma^{5}\hat{p}_{5}\right)\hat{p}_{\mu}-\left(\hat{p}^{6}+\gamma^{5}\hat{p}_{5}\right)\left(\hat{p}^{6}-\gamma^{5}\hat{p}_{5}\right)\right]\chi=0, \quad (4.41)$$

from which we obtain an equation for charge amplitude $\chi_Q$

$$\left[\gamma^{\mu}\hat{p}_{\mu}-\gamma^{5}\hat{p}_{5}-\hat{p}^{6}\right]\chi_Q=0, \quad (4.42)$$

where

$$\chi_Q=\frac{\hat{p}^{6}-\gamma^{5}\hat{p}_{5}}{mc}\chi, \quad (4.43)$$

and $m$ is the mass of the charged particle of definite mass associated with the family of particles defined by (4.37). From (4.43) we see that probability amplitude $\chi$ acts as a potential for charge amplitude $\chi_Q$. If we now multiply (4.42) from the left by the factor $-\gamma^{5}\hat{p}_{5}+I\hat{p}^{6}$ we obtain

$$\left[\gamma^{\mu}\left(\hat{p}^{6}+\gamma^{5}\hat{p}_{5}\right)\hat{p}_{\mu}+\hat{p}_{5}{}^{2}-\hat{p}_{6}{}^{2}\right]\chi_{Q}=0\,, \tag{4.44}$$

an equation which, with its hermitian Hamiltonian, represents the charge amplitude companion to (4.28).

## 4.6 Conservation of probability and charge

Here we seek conserved probability and electric-charge currents, $s^a$ and $j^a$, for the Dirac particle in six dimensions ($a=0,1,2,3,5,6$). As already noted in *Sec. 3.9.7*, for a charged particle such as the electron, the two currents are directly proportional. For the neutrino, however, they differ from each other entirely. The two currents are thus central to a proper study of the neutrino.

*4.6.1 Probability current density.* From (4.28) we have for the equations of motion of the probability amplitude $\chi$ and its adjoint $\tilde{\chi}=\chi^{\dagger}\gamma^{0}$ (where the dagger $^{\dagger}$ denotes Hermitian conjugate)

$$\gamma^{\mu}\left(\partial_{6}+\gamma^{5}\partial_{5}\right)\partial_{\mu}\chi+\left(\partial_{6}{}^{2}-\partial_{5}{}^{2}\right)\chi=0 \tag{4.45}$$

$$\partial_{\mu}\tilde{\chi}\left(\overleftarrow{\partial}_{6}-\gamma^{5}\overleftarrow{\partial}_{5}\right)\gamma^{\mu}+\left(\partial_{6}{}^{2}-\partial_{5}{}^{2}\right)\tilde{\chi}=0\,, \tag{4.46}$$

where $\overleftarrow{\partial}_{a}$ denotes an operation on the symbol to its left. Multiplying (4.45) from the left by $\tilde{\chi}$, (4.46) from the right by $\chi$, and then subtracting we obtain

$$\begin{aligned}&\partial_{\mu}\left[\tilde{\chi}\left(\overleftarrow{\partial}_{6}-\gamma^{5}\overleftarrow{\partial}_{5}\right)\gamma^{\mu}\chi-\tilde{\chi}\gamma^{\mu}\left(\partial_{6}+\gamma^{5}\partial_{5}\right)\chi\right]\\&\quad+\partial_{5}\left(\tilde{\chi}\partial_{5}\chi-\partial_{5}\tilde{\chi}\cdot\chi\right)\\&\quad+\partial_{6}\left(\partial_{6}\tilde{\chi}\cdot\chi-\tilde{\chi}\partial_{6}\chi\right)\\&\quad=\tilde{\chi}\left(\overleftarrow{\partial}_{6}-\gamma^{5}\overleftarrow{\partial}_{5}\right)\gamma^{\mu}\partial_{\mu}\chi-\partial_{\mu}\tilde{\chi}\gamma^{\mu}\left(\partial_{6}+\gamma^{5}\partial_{5}\right)\chi\end{aligned} \tag{4.47}$$

But from (4.31) we have

$$\gamma^{\mu}\partial_{\mu}\chi = -\left(\gamma^{5}\partial_{5} + \partial_{6}\right)\chi \tag{4.48}$$

and the adjoint equation

$$\partial_{\mu}\tilde{\chi}\gamma^{\mu} = \tilde{\chi}\left(\gamma^{5}\overleftarrow{\partial}_{5} - \overleftarrow{\partial}_{6}\right). \tag{4.49}$$

The r. h. s. of (4.47) thus vanishes identically, giving an equation of continuity form. Multiplying (4.47) by $i\hbar c/2$, we can write it as

$$\partial_{\mu}s^{\mu} + \partial_{5}s^{5} + \partial_{6}s^{6} = 0, \tag{4.50}$$

where the $s^{a}$ are the components of the conserved probability current density:

$$s^{\mu} = \frac{c}{2}\left[\tilde{\chi}\gamma^{\mu}\left(\hat{p}^{6} - \gamma^{5}\hat{p}_{5}\right)\chi - \tilde{\chi}\gamma^{\mu}\left(\overleftarrow{\hat{p}}^{6} - \gamma^{5}\overleftarrow{\hat{p}}_{5}\right)\chi\right], \tag{4.51}$$

$$s^{5} = \frac{c}{2}\left(\tilde{\chi}\hat{p}^{5}\chi - \hat{p}^{5}\tilde{\chi}\cdot\chi\right), \tag{4.52}$$

$$s^{6} = \frac{c}{2}\left(\tilde{\chi}\hat{p}^{6}\chi - \hat{p}^{6}\tilde{\chi}\cdot\chi\right). \tag{4.53}$$

Note that since operators $\hat{p}_5$ and $\hat{p}_6$ are invariant, the current components $s^5$ and $s^6$ are separately conserved (and are independent of $x^6$). Note also that no Dirac fermion, including the neutrino, can be an eigenstate of $\gamma^5$. For if it were, then one would have $\chi = (1 \pm \gamma^5)\chi/2$, which, when substituted into (4.52) and (4.53), gives $s^5 = 0 = s^6$, signifying "no particle." In this theory all probability amplitudes are four-component spinors.

Let us now substitute for $\chi$ in these expressions using the expansion (4.32) and (4.33). This yields separate continuity laws for the amplitudes $\psi_n(x)$ defined by (4.37):

$$\partial_{\mu}s_n{}^{\mu} + \partial_{5}s_n{}^{5} + \partial_{6}s_n{}^{6} = 0, \tag{4.54}$$

where

$$s_n{}^{\mu} = c\tilde{\psi}_n\gamma^{\mu}\left(p^{6} - \gamma^{5}p_{5}\right)\psi_n, \tag{4.55}$$

$$s_n{}^{5} = cp^{5}\tilde{\psi}_n\psi_n, \tag{4.56}$$

$$s_n{}^{6} = cp^{6}\tilde{\psi}_n\psi_n. \tag{4.57}$$

*4.6.2 Normalization and expectation value.* In accordance with (4.55), the states $\psi_n$ are normalized such that

$$1 = \langle \psi_n | (p^6 - \gamma^5 p_5) | \psi_n \rangle . \tag{4.58}$$

The expectation value of an operator $\hat{O}$ is defined by

$$\langle \hat{O} \rangle = \langle \psi_n | \hat{O} (p^6 - \gamma^5 p_5) | \psi_n \rangle , \tag{4.59}$$

where it is assumed that $\hat{O}$ commutes with the metric operator $p^6 - \gamma^5 p_5$.

*4.6.3 Charge current density.* Just as (4.28) and its adjoint lead to a law of conservation of probability current density $s^a$, Eq. (4.44) and its adjoint provide a law of conservation of electric charge current density $j^a$. To obtain the latter, comparison of (4.44) and (4.28) shows that in (4.54)-(4.57) one need only make the replacements $s_n{}^a \to j_n{}^a$, $\psi_n \to \psi_{Qn}$ and $\gamma^5 \to -\gamma^5$. This yields

$$\partial_\mu j_n{}^\mu + \partial_5 j_n{}^5 + \partial_6 j_n{}^6 = 0 , \tag{4.60}$$

where

$$j_n{}^\mu = -e \tilde{\psi}_{Qn} \gamma^\mu (p^6 + \gamma^5 p_5) \psi_{Qn} , \tag{4.61}$$

$$j_n{}^5 = -e p^5 \tilde{\psi}_{Qn} \psi_{Qn} , \tag{4.62}$$

$$j_n{}^6 = -e_6 p^6 \tilde{\psi}_{Qn} \psi_{Qn} . \tag{4.63}$$

and where, from (4.43) and (4.32),

$$\psi_{Qn} = \frac{p^6 - \gamma^5 p_{5n}}{mc} \psi_n . \tag{4.64}$$

Note that, as in the case of probability, the currents $j_n{}^5$ and $j_n{}^6$ are separately conserved. Note also that we have included in the currents $j_n{}^\mu$ and $j_n{}^5$ the unit of electric charge $-e = -|e|$. However, in $j_n{}^6$ we have written $e_6$ rather than $e$. Why do we do this? At this point the role of the sixth dimension seems simply to be to provide momentum (pseudoenergy) $p^6$. Geometrically one pictures $p^6$, not as a line current like $p^5$, but as the radius (half-transverse axis) of the momentum shell in five-space as depicted in **Fig. 3.4**. In other words, one suspects that, although the sixth dimension does yield a component of probability current, it may not generate a corresponding charge current; writing (4.63) as we have done allows for the possibility that $e_6 = 0$. This suspicion will be borne out in our study of Maxwell's equations in six dimensions (**Chapter 6**).

Now whatever the status of $j_n^{\ 6}$, the five-vector $j_n^{\ a}$ clearly represents charge-current density. For if we substitute for $\psi_{Qn}$ using (4.62), we obtain

$$j_n^{\ a} = -e\frac{m_n^{\ 2}}{m^2}s_n^{\ a}, \qquad (4.65)$$

where $m_n^{\ 2}$ is defined by (4.38) and $m$ is the mass of the charged lepton of family $\ell$. In the case of the charged lepton, where for some $n$, $m_n = m$, we have from (4.65)

$$j_n^{\ a} = -es_n^{\ a}, \quad (a = \mu, 5). \qquad (4.66)$$

Thus we are indeed justified in identifying $j_n^{\ a}$ with charge-current density and $\psi_{Qn}$ with charge amplitude: in the case of the charged lepton, $j_n^{\ a}$ and $s_n^{\ a}$ are directly proportional, as they must be.

## 4.7 Neutrino mass and the oscillation imperative

However, in the case of the neutrino, where for some $n$, $m_n = m_\nu$, Eq. (4.65) reads

$$j_n^{\ a} = -e\frac{m_\nu^{\ 2}}{m^2}s_n^{\ a}, \qquad (4.67)$$

where $m_\nu$ is the neutrino mass. Integrating over all space we obtain for the charge $q_\nu$ of the neutrino:

$$\langle j_n^{\ 0}\rangle / c = q_\nu = -e\frac{m_\nu^{\ 2}}{m^2}. \qquad (4.68)$$

But the neutrino is electrically neutral: $q_\nu = 0$. Hence, from (4.68)

$$m_\nu^2 = 0\,. \tag{4.69}$$

That is, on the present theory, *all neutrinos are massless*. In addition, from (4.38) we see that for the neutrino belonging to generation $p^6 = G$

$$p_5 = G\,. \tag{4.70}$$

As compelling a result (4.69) may be, there is a difficulty. The neutrinos under discussion are at this point objects of definite mass. As such their wave functions $\psi_n \equiv \nu$ are defined by Dirac equation (4.37):

$$\left(1-\gamma^5\right)\left[\gamma^\mu \hat{p}_\mu - G\left(1-\gamma^5\right)\right]\nu(x) = 0\,. \tag{4.71}$$

As this is a zero mass equation, wave function $\nu$ must be of the form $\nu = (1+\gamma^5)\phi$, where $\phi$ is a four-component Dirac spinor. Consequently neutrino state $\nu$ is an eigenstate of $\gamma^5$, and the probability currents $s^5$ and $s^6$ of (4.56) and (4.57) both vanish identically. But this contradicts the finite values of $p_5$ and $p^6$ defining the massless neutrino of definite mass. The conclusion is clear: neutrinos of definite mass cannot exist in six-dimensional $M^4 \times M^2$. The structures of $s^5$ and $s^6$ do not support them.

But neutrinos do exist, and to give them life the state of definite mass discussed so far must be replaced by a more-general state consisting of a superposition of mass-squared eigenstates. Such a state will no longer be an eigenstate of $\gamma^5$ and therefore can describe real neutrinos. This superposition of states, as we know, leads to oscillation between the different flavors of neutrino. Thus the *oscillation imperative: in $M^4 \times M^2$ neutrinos must oscillate to exist.* Now the interesting point is that real flavor neutrinos, like (the non-existent) neutrinos of definite mass, must still have zero mass. This can be seen as follows. By definition flavor neutrino $\nu$ belongs to family $G$ if and only if its 5-momentum vector terminates on the momentum shell of radius $G$, i.e., $p_\mu p^\mu + p_{5\nu}^{\ 2} = G^2$. However, the state $\nu$ now consists of a series of terms, where, in accordance with the remarks made in *Sec. 4.4.2*, the squared length of the five-momentum vector of the *j*th term is $(G+g_j)^2$, with $g_j$ a small departure from $G$. Hence, in order that $\nu$ belong to family $G$, we demand that

$$\left\langle \hat{p}_6{}^2 \right\rangle = G^2,$$

(4.72)

where the average value entails a summation over the internal index $j$. But, according to (4.70), the propagating wave vector $\nu$ is already an eigenstate of $\hat{p}_5$ with eigenvalue $G$. Hence (4.72) can be rewritten as

$$\left\langle \hat{p}_6{}^2 - \hat{p}_5{}^2 \right\rangle = \left\langle \widehat{m^2 c^2} \right\rangle = m_\nu{}^2 c^2 = 0 . \qquad (4.73)$$

Thus we see that if neutrino $\nu$ is to belong to family $G$, it must have vanishing mass. This applies all flavors of neutrino.

The subject of neutrino mixing and oscillation represents today an enormous industry, theoretical and experimental. We shall deal with it further in some detail in **Chapter 12**, giving there a direct proof of the zero-mass condition (4.73) and showing, among other things, that neutrinos need not have mass to oscillate, and moreover may do so while conserving lepton number.

## 4.8 Evaluation of constants $P_5$, $a^5$ and $p^6$ for the three known families of leptons

In **Sec. 4.4** we showed how fermion flavors arise as terms in a Fourier expansion of the beam amplitude in the axial coordinate, $x^5$. In particular we saw how, for a given value of the generation parameter $p^6 \equiv G$, the expansion index $n$ generates a family of fermion flavors belonging to $G$, with corresponding fifth momenta $p_{5n} = P_5 - 2\pi n\hbar / a^5$, masses $m_n = (G^2 - p_{5n}{}^2)^{1/2} c$ and probability amplitudes given by (4.39). We are now in a position to determine, for the three known families of leptons, the values of the constants $P_5$ and $a^5$ and the effective range of the expansion index $n$ appearing in (4.34). This will lead to numerical values for parameter $G$ for each of the three families.

*4.8.1 The electron family of leptons.* We may begin by assuming for the electron, with $n = n_e$, vanishing pseudomomentum, $p_5$:

$$p_{5en_e} \equiv p_5(e^-) = 0 . \qquad (4.74)$$

Note that the subscript *e* in $p_{5en_e}$ denotes the electron family. Then we have from (4.38), with $p^6 \equiv G_e$ and $m_{n_e} \equiv m_e$ (mass of the electron)

$$G_e = m_e c\,. \tag{4.75}$$

Thus (4.37) reduces to the standard Dirac equation for the electron. This is perfectly reasonable. For had we assumed non-vanishing pseudomomentum, the equation of motion (4.37) would then fail to conserve parity, implying handedness the electron does not have. Using (4.74) in (4.34) we have for the constant bias $P_5$:

$$P_5 = n_e \frac{h}{a^5}, \tag{4.76}$$

where for brevity we have written $2\pi\hbar = h$ (Planck's constant).

Now without loss of generality we may assign to the neutrino the index value $n = n_\nu = 0$. This gives from (4.34)

$$p_{5en_\nu} \equiv p_5(\nu_e) = P_5\,. \tag{4.77}$$

Moreover, we know from (4.73) that the neutrino has zero mass. Hence, from (4.38)

$$p_5(\nu_e) = G_e\,. \tag{4.78}$$

From this and (4.77),

$$P_5 = G_e\,. \tag{4.79}$$

Combining this with (4.76) we obtain

$$\frac{h}{a^5} = \frac{G_e}{n_e}. \tag{4.80}$$

We now substitute (4.79) and (4.80) into (4.34) to obtain

$$p_{5en} = G_e\left(1 - \frac{n}{n_e}\right). \tag{4.81}$$

What is $n_e$? Clearly $n_e = 1$. For if $n_e > 1$, then (4.38) generates $n_e - 1$ unobserved charged leptons with masses intermediate between 0 and $m_e$; and $n_e < 0$ yields unobserved charged particles of imaginary mass. Thus we have from (4.80) and (4.75)

$$\frac{h}{a^5} = m_e c, \tag{4.82}$$

and for the electron family

$$p_{5en} = m_e c(1-n), \quad (n = 0,1) \tag{4.83}$$

which may also be written in the form

$$p_{5en} = G_e - n m_e c, \quad (n = 0,1) \tag{4.84}$$

where $p_{5e0} = p_5(\nu_e)$ and $p_{5e1} = p_5(e^-)$. Inserting (4.83) into (4.38) and using the fact that $n = n^2$ we obtain the mass formula for the electron family

$$m_{en} = n m_e, \tag{4.85}$$

where $m_{e0} = m_{\nu_e} = 0$ and $m_{e1} = m_e$.

The five-momentum vectors $p^a = (p^\mu, p^5)$ differentiating the electron and its neutrino are now completely determined. The invariant resultants of these vectors are depicted in **Fig. 4.1**, together with those of the positron and electron antineutrino. The sixth momentum component $p^6$ is represented by the momentum-shell cross-sectional radius $G_e = m_e c$.

*4.8.2 The mu and tau families of leptons.* Having found the five-momentum vectors defining the electron family of leptons, we shall find it easy to derive those defining the families mu and tau. To do this we first apply the law of conservation of $p_5$ (see *Sec. 3.9.7*) to the decay process $\ell^- \to \nu_\ell + e^- + \bar{\nu}_e$, with $\ell = \mu, \tau$. The $p_5$ balance for this reaction reads

$$p_5(\ell^-) = p_5(\nu_\ell) + p_5(e^-) + p_5(\bar{\nu}_e). \tag{4.86}$$

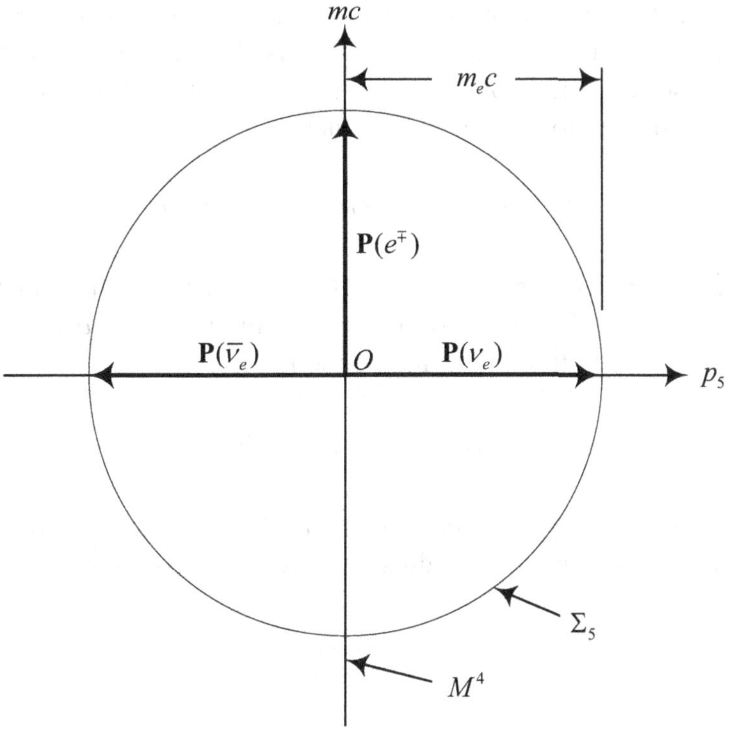


**Fig. 4.1.** Five-momentum vectors $\mathbf{P} = \mathbf{e}_a p^a = \mathbf{e}_\mu p^\mu + \mathbf{e}_5 p^5$ differentiating the electron and its neutrino in $M^4 \times \overline{M}^2$. Unit vectors $\mathbf{e}_a$ satisfy $\mathbf{e}_a \cdot \mathbf{e}_b = g_{ab}$ $(a = 0,1,2,3,5)$. As shown, all resultant vectors $\mathbf{P}$ and their transverse components $\mathbf{e}_\mu p^\mu$ lie in the plane of the paper. Vector $\mathbf{e}_\mu p^\mu$ is of length *mc*, and the vertical axis is so labeled. All vectors $\mathbf{P}$ are of length $p^6 = G_e$, their tips lying in the 5-D momentum shell $\Sigma_5$. $M^4$ = 4-dimensional hyperplane. In the case of the electron $p^a(e^-) = (p_0, \mathbf{p}, 0)$ and for its (massless) neutrino $p^a(\nu_e) = (|\mathbf{p}|, \mathbf{p}, m_e c)$. Also shown for comparison are vectors for the positron $e^+$ and antineutrino $\overline{\nu}_e$: $p^a(e^+) = (p_0, \mathbf{p}, 0)$ and $p^a(\overline{\nu}_e) = (|\mathbf{p}|, \mathbf{p}, -m_e c)$.

Now from (4.38) and definition (4.40) we have for the massless neutrino $\nu_\ell$

$$p_5(\nu_\ell) = p_\ell{}^6 \equiv G_\ell. \tag{4.87}$$

Then substituting (4.87), (4.74) and (4.78) into (4.86) we have

$$p_5(\ell^-) = G_\ell - m_e c, \tag{4.88}$$

where for $p_5(\overline{\nu}_e)$ we have used the fact that, in a hole-theoretic context, the fifth momentum of an antiparticle is the negative of that of the particle [see comment following Eq. (4.40)]. Note that (4.87) and (4.88) can be combined to form a single equation

$$p_{5\ell n} = G_\ell - n m_e c, \quad (n = 0,1) \tag{4.89}$$

where $p_{5\ell 0} = p_5(\nu_\ell)$ and $p_{5\ell 1} = p_5(\ell^-)$. This is identical in form to (4.84). Thus (4.89) holds for all three families: $\ell = e, \mu, \tau$. Now for $n = 1$, mass definition (4.38) may be written

$$G_\ell^2 - p_5(\ell^-)^2 = m_\ell^2 c^2 . \tag{4.90}$$

Eliminating $p_5(\ell^-)$ between this and (4.88) we obtain the connection between momentum shell radius $G_\ell$ and the mass $m_\ell$ of the charged member of lepton family $\ell$:

$$G_\ell = \frac{m_\ell^2 + m_e^2}{2m_e} c . \tag{4.91}$$

Inserting (4.89) into (4.38) and making use of (4.91) and the fact that $n = n^2$ we obtain the mass formula for lepton family $\ell$:

$$m_{\ell n} = n m_\ell , \quad \ell = e, \mu, \tau \tag{4.92}$$

where $m_{\ell 0} = m_{\nu_\ell} = 0$ and $m_{\ell 1} = m_\ell$. Numerically, using the experimentally-known values of the $m_\ell$, we obtain from (4.91) the values of $G_\ell$ listed in **Table 4.1**. These values will later prove decisive in establishing conservation of lepton family and baryon numbers (see **Sec. 4.13**).

**Table 4.1.** Values of momentum shell radius $G_\ell$ calculated from Eq. (4.87)

| $\ell$ | Experimental $m_\ell$ [MeV/$c^2$] | $G_\ell / c$ [MeV/$c^2$] |
|---|---|---|
| $e$ | 0.511 | 0.511 |
| $\mu$ | 105.658 | $1.1\times10^4$ |
| $\tau$ | 1777.0 | $3.1\times10^6$ |

The five-momentum vectors $p^a = (p^\mu, p^5)$ defining all lepton flavors are now completely determined. Those for the electron family were depicted already in **Fig. 4.1**. Vectors $p^a$ differentiating members of lepton families $\ell = \mu, \tau$ are shown in **Fig. 4.2.**

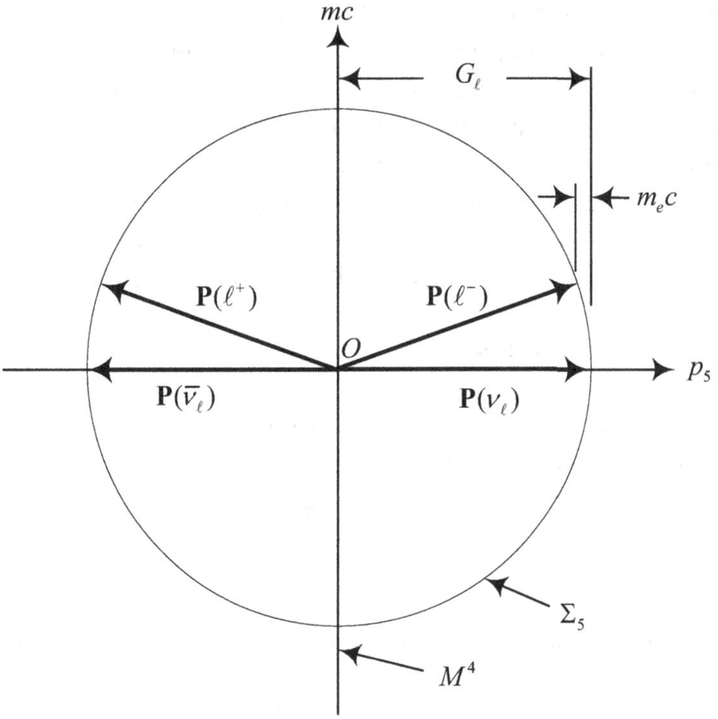


**Fig. 4.2.** Five-momentum vectors $\mathbf{P} = \mathbf{e}_a p^a = \mathbf{e}_\mu p^\mu + \mathbf{e}_5 p^5$ differentiating the charged leptons $\ell^-$ and their corresponding neutrinos $\nu_\ell$ ($\ell = \mu, \tau$) in $M^4 \times \bar{M}^2$. Unit vectors $\mathbf{e}_a$ satisfy $\mathbf{e}_a \cdot \mathbf{e}_b = g_{ab}$ $(a = 0,1,2,3,5)$. As shown, all resultant vectors $\mathbf{P}$ and their transverse components $\mathbf{e}_\mu p^\mu$ lie in the plane of the paper. Vector $\mathbf{e}_\mu p^\mu$ is of length *mc*, and the vertical axis is so labeled. All vectors $\mathbf{P}$ are of length $p^6 = G_\ell$, their tips lying in the 5-D momentum shell $\Sigma_5$. $M^4$= 4-dimensional hyperplane. For the charged lepton $\ell^-$, $p^a(\ell^-) =$ $(p_0, \mathbf{p}, G_\ell - m_e c)$ and for the (massless) neutrino $p^a(\nu_\ell) = (|\mathbf{p}|, \mathbf{p}, G_\ell)$. Also shown for comparison are resultant vectors for the charged lepton $\ell^+$ and antineutrino $\bar{\nu}_\ell$: $p^a(\ell^+) = (p_0, \mathbf{p}, -G_\ell + m_e c)$ and $p^a(\bar{\nu}_e) = (|\mathbf{p}|, \mathbf{p}, -G_\ell)$.

*4.8.3 Distinguishing charged and neutral leptons of different flavor.* We now see, both graphically and numerically, that the mu and tau charged leptons are not simply more massive versions of the electron. They in fact differ markedly from the electron, and from each other, in virtue of their different values of pseudomomentum, (4.88). Similarly, flavor neutrinos of different families, though massless and electrically neutral, are nevertheless physically distinguished from each other by their respective values of pseudomomentum (4.87).

*4.8.4 On the geometric origin of the electron mass.* In **Sec. 4.4** we said that to get replication of families we need to quantize parameter $G$. It is worth noting that the ground state $G$-value, namely $G_e$, is nothing but electron mass $m_e$ (multiplied by $c$). This mass is to considered a universal constant, one determined, remarkably enough, by an invariant interval of the fifth dimension, namely the repeat interval, $a^5$. According to (4.84) this interval is numerically equivalent to, and determines, the Compton wavelength of the electron, $h/m_e c$.

## 4.9 Charge quantization

Integrating over all space we obtain from (4.65) the charge $q_{\ell n}$ of the $n$th member of lepton family $\ell$

$$q_{\ell n} = -e\frac{m_{\ell n}^{2}}{m_{\ell}^{2}}. \tag{4.93}$$

Then from (4.92), and again using the fact that $n = n^2$, we obtain

$$q_{\ell n} = -ne, (n = 0,1) \tag{4.94}$$

where $q_{\ell 0} = q_{\nu_\ell} = 0$ and $q_{\ell 1} = q_\ell = -e$. Thus we have proved that index $n$ represents electric charge. The mystery of charge quantization can be considered solved: quantization of electric charge is a direct consequence of the phenomenon of self-imaging in six spacetime dimensions.

## 4.10 Connection between $p_5$, electric charge $Q$ and lepton number $L_\ell$

We can now eliminate the index $n$ in (4.89) using (4.94):

$$p_{5\ell n} = G_\ell + \frac{q_{\ell n}}{e} m_e c. \tag{4.95}$$

On hole-theoretic grounds, the signs of both $p_5$ and $q$ are reversed for antiparticles. An expression for $p_5$ valid for both particles and antiparticles can therefore be written in the form

$$p_5 = L_\ell G_\ell + Q m_e c\,, \tag{4.96}$$

where $L_\ell$ is the lepton family number and $Q \equiv q/e$; for simplicity the indices $\ell n$ attached to $p_5$ and $q$ have been suppressed. By definition,

$$\left.\begin{aligned} L_\ell &= +1 \text{ for } \nu_\ell \text{ and } \ell^- \\ &= -1 \;\text{ for } \bar{\nu}_\ell \text{ and } \ell^+ \\ &= 0 \quad \text{for leptons and antileptons} \\ &\qquad \text{not in family } \ell \end{aligned}\right\} . \tag{4.97}$$

## 4.11 Quark flavor generation and family replication

As indicated in **Sec. 4.4**, quark flavors are generated by an expansion of the free quark wave function in eigenstates of the mass squared operator $m^2c^2 = G_q{}^2 - \hat{p}_5{}^2$, where $G_q$ is the radius of the five-momentum shell defining quark family $q$. To compute quark masses from (4.38) and to study processes involving both hadrons and leptons, we shall need a formula, analogous to (4.96), connecting the $p_5$ and $Q$ values of the hadrons. To derive this formula we first determine empirically the $p_5$ values of the quarks by imposing conservation of $p_5$ on the following three reactions:

(i) $\beta$-decay:

$$n(udd) \rightarrow p(uud) + e^- + \bar{\nu}_e$$

(ii) Decay of the charmed meson[31]:

$$D^0(c\bar{u}) \rightarrow K^-(s\bar{u}) + \pi^+(u\bar{d})$$

(iii) Decay of the top quark[32]:

$$t \to \begin{cases} W^+ b \\ W^+ s \quad \left(W^+ \to \ell^+ \nu_\ell\right) \\ W^+ d \end{cases}$$

Since $p_5(e^\mp) = 0$ and $p_5(\nu_e / \overline{\nu}_e) = +m_e c / - m_e c$ [see (4.83)], reactions (i), (ii) and the first channel of (iii) yield, respectively,

$$\left.\begin{aligned} p_{5u} &= p_{5d} + m_e c \\ p_{5c} &= p_{5s} + m_e c \\ p_{5t} &= p_{5b} + m_e c \end{aligned}\right\}. \tag{4.98}$$

In addition, the three channels of reaction (iii) imply

$$p_{5d} = p_{5s} = p_{5b} \equiv R/3, \tag{4.99}$$

where $R$ is a phenomenological constant whose value will be estimated momentarily. From (4.98) and (4.99) we obtain

$$p_{5u} = p_{5c} = p_{5t} \equiv R/3 + m_e c. \tag{4.100}$$

Equations (4.99) and (4.100), and corresponding expressions for antiquarks, are special cases of the general formula

$$p_5 = BR + (B + Q) m_e c, \tag{4.101}$$

where $B$ is the baryon number ($= 1/3$ for quarks), $Q_{u,\,c,\,t} = +2/3$ and $Q_{d,\,\mathrm{s},\,\mathrm{b}} = -1/3$. Since hadrons are built up from quarks, (4.101) applies to hadrons as well as quarks and thus represents our required formula, the analog of formula (4.96) applicable to leptons. Comparing (4.101) and (4.34), we see that, for quarks, the constant $P_5 = BR$ and the integer flavor index $-n = B + Q$. Furthermore, we still have $h/a^5 = m_e c$, as determined in **Sec. 4.8** for leptons [Eq. (4.82)]. Remarkably, the electron mass appears in both lepton formula (4.96) and quark formula (4.101), demonstrating unequivocally the pre-Standard Model etiology of $m_e$.

## 4.12 Constant *R*

To estimate the value of *R*, let us first recall that (4.101) was derived empirically by demanding conservation of $p_5$ in selected decay processes, e.g., $\beta$-decay. Equation (4.101) thus represents the $p_5$ values of quarks *in confinement*—despite that expression's relation to the free-particle modal expansion (4.33). The value of *R*, then, should reflect the fact that real quarks are not free, but are confined inside hadrons.

To see how to get such a value, let us write down from (4.96) the $p_5$ value for the lowest order *leptonic* mode, namely, the electron:

$$p_5 = L_e G_e + Q m_e c\,, \tag{4.102}$$

where $G_e c = m_e c^2$, the rest energy of the electron. Now the rest energy $G_e c$ can be considered the energy of confinement of a massless point circulating about the mean position of the electron; this circulating motion is the electron's Zitterbewegung.[33] By direct analogy with (4.102), we propose to identify the energy *Rc* appearing in the quark formula (4.101) with the energy of confinement of quarks trapped in the lowest hadronic mode, namely, the $\pi$ meson. There are at least two ways of specifying this energy of confinement.

(1) If it is assumed that the mass of the $\pi$ meson is due principally to the energy of confinement of its constituent quarks, then

$$R \sim 135\ \mathrm{MeV}/c\,. \tag{4.103a}$$

(2) At the end of **Sec. 4.4** it was mentioned that the masses $m_\ell$ of the charged leptons can be identified with the energetic modes of an electron trapped in inner spacetime $N^2$. The first excited state of the trapped electron is the muon. Its rest energy $m_\mu c^2$ is an energy of confinement and as such becomes a candidate for the energy of confinement of the trapped quark:

$$R \sim m_\mu c = 105.7\ \mathrm{MeV}/c\,. \tag{4.103b}$$

This second possibility for $R$ seems implausible until we realize—as we shall see in **Chapter 14**—that each one of the quark masses $m_q$ is coupled to one of the lepton masses $m_\ell$. On that ground it is not unreasonable to identify $R$ with a known and calculable energy of confinement, namely, $m_\mu c^2$. In any case, estimates (4.103a) and (4.103b) are reasonably close in value, despite their resting on different arguments, and either one will serve the purposes of the following section.

## 4.13 Conservation of lepton family and baryon numbers

In the minimal Standard Model of electroweak interactions, neutrinos are assumed to be massless and lepton family numbers $L_\ell$ are separately conserved. Conservation of lepton number follows from $U(1)$ symmetry in the SM Lagrangian. The theory under discussion here is consistent with the SM in that it predicts massless neutrinos. Thus one might expect family lepton number to be conserved as well. That is in fact the case. We shall now show that conservation of both lepton family number and baryon number follows directly from the conservation of electric charge $Q$ and pseudomomentum $p_5$.

Consider an arbitrary process $i \to f$, and let the total changes in $Q$, $p_5$, $B$ and $L_\ell$ be denoted $\Delta Q$, $\Delta p_5$, $\Delta B$ and $\Delta L_\ell$ respectively. For example, by $\Delta L_\ell$ is meant

$$\Delta L_\ell = \sum_f L_\ell^f - \sum_i L_\ell^i \qquad (\ell = e,\, \mu,\, \tau), \tag{4.104}$$

where the summations run over all initial and final state leptons and antileptons belonging to family $\ell$. Then from (4.96) and (4.101), we have, after summing over all lepton families $\ell$,

$$\Delta p_5 = \sum_\ell G_\ell \Delta L_\ell + (R + m_e c)\Delta B + m_e c \Delta Q\,. \tag{4.105}$$

But $p_5$ and $Q$ are conserved in any process. Hence $\Delta p_5$ and $\Delta Q$ vanish identically and (4.105) reduces to

$$\sum_\ell G_\ell \Delta L_\ell + (R + m_e c)\Delta B = 0\,. \tag{4.106}$$

Now from **Table 4.1** and Eq. (4.103) we have $G_e \ll R \ll G_\mu \ll G_\tau$. Hence each term of (4.106) must separately vanish, giving

$$\Delta B = 0 \tag{4.107}$$

and

$$\Delta L_\ell = 0 \quad (\ell = e,\ \mu, \tau). \tag{4.108}$$

*In all processes, baryon number B and lepton family numbers* $L_\ell$ $(\ell = e,\ \mu, \tau)$ *are separately conserved.* According to this theory, proton decay is forbidden, as is neutrinoless double $\beta$-decay—to mention just two processes under active experimental investigation. It does not, however, forbid neutrino oscillation, as we shall see in **Chapter 12.**

## 4.14 Summary and experimental prospects

We now summarize our main results and point out ways in which our general theory of flavor may be empirically disproved.

1. Each family $f$ of fundamental fermions is represented by a self-imaging wave function $\chi(x^\mu, x^5, x^6) = \psi(x^\mu, x^5)\exp(-ip_6 x^6 / \hbar)$ propagating in six-dimensional spacetime $M^4 \times \bar{M}^2$, where the fifth and sixth dimensions $x^5, x^6 \in \bar{M}^2$ are infinite, flat and metrically time-like and space-like, respectively. The momentum representation of the amplitude $\psi(x^\mu, x^5)$ vanishes everywhere except on the five-momentum shell of radius $p^6 = G_f$. The coefficients $\psi_n(x^\mu)$ of the expansion of $\psi$ in eigenstates $\exp(-ip_{5n}x^5 / \hbar)$ of the fifth momentum operator $\hat{p}_5$ are identified with the fermion flavors comprising family *f*. Thus the attribute *flavor* requires two parameters for its specification: the shell radius $G_f$ and the integer expansion index *n*. In the case of leptons, $-n$ represents electric charge *Q* (in units of $e$) and, in the case of quarks it represents electric charge *Q* plus baryon number *B*.

2. Due to the exponential structure of the eigenstates of $\hat{p}_5$, the Lagrangian density is independent of $x^5$, ensuring invisibility of the extra coordinate.

3. Invariance of the Lagrangian under translations in $x^5$ implies conservation of the total pseudomomentum, $p_5$. Because infinitesimal rotations in the $\mu$-5 planes of $M^4 \times \bar{M}^2$ are forbidden, conservation of $p_5$ holds separately from conservation of the four $p_\mu$. Hence only those reactions can occur that conserve the total $p_5$.

4. Neutrino mass-squared eigenstates $\nu_{\ell j}$ arise as simultaneous eigenstates of the fifth momentum operator $\hat{p}_5$ and $\hat{p}_6$. Interference between states of different frequency $p^6 = G + g_j, j = 1,2,...$ produces flavor oscillation. (See **Chapter 12**.)

5. The mass $m$ of any charged fermion $f$ in this theory is given by $m_f^2 c^2 = G_f^2 - p_5^2$. The observed masses $m_\ell$ of the charged leptons are identified with the energy eigenvalues of the Hamiltonian of an electron trapped in a one-dimensional well in a two-dimensional, internal spacetime, $N^2$. From the $m_\ell$ one obtains discrete values of the momentum shell radii $G_\ell$, thus explaining the replication of lepton families. (See **Chapter 13**.) The quark masses $m_q$ are determined by way of coupling to the charged lepton masses. (See **Chapter 14**.)

6. All flavor neutrinos are massless and must oscillate to exist. *Our general theory of flavor would be falsified if neutrinos were shown by direct (kinematic) measurement to have non-zero mass. At present such measurements provide only upper bounds on $m_\nu$ and are consistent with $m_\nu = 0$.*[34-36] Note that the conclusion drawn from oscillation experiments that neutrinos have mass[37] does not constitute evidence against the present theory. For that conclusion rests on an untested (and unwarranted) assumption, namely, that neutrinos must have mass to oscillate.

7. Electrically neutral and massless, neutrinos of different flavor are physically distinguished by their respective values of pseudomomentum $p_5$.

8. From the conservation of electric charge $Q$ and pseudomomentum $p_5$, it follows that, in all processes, lepton family number $L_\ell$ and baryon number $B$ are separately conserved. *Our general theory of flavor would thus be falsified if neutrinoless double beta* decay[34] *were to be*

*observed; if protons were observed to* decay[38]*; or if experimental evidence were found for the any of the decays* $\mu \to e\gamma$ [30] , $\tau \to \mu\gamma$ [40] *or* $\tau^- \to \ell^-\ell^+\ell^-$ [41].

**Notes and references**


[1] F. Halzen and A. D. Martin, *Quarks and Leptons: An Introductory Course in Modern Particle Physics* (Wiley, New York, 1984), p. 285.

[2] CDF Collaboration, F. Abe *et al.*, Phys. Rev. D **50**, 2966 (1994); Phys. Rev. Lett. **73**, 224 (1994).

[3] H. Primakoff, in *Nuclear and Particle Physics at Intermediate Energies*, edited by J. B. Warren (Plenum, New York, 1976), p. 1.

[4] F. E. Close, *An Introduction to Quarks and Partons* (Academic, London, 1979), p. 443.

[5] G. Kane, *Modern Elementary Particle Physics* (Perseus, Cambridge, Mass., 1993), p. 76.

[6] R. N. Mohapatra, *Unification and Supersymmetry* (Springer-Verlag, New York, 1986), p. 165.

[7] A. Salam, Rev. Mod. Phys. **52**, 525 (1980).

[8] K. Huang, *Quarks, Leptons and Gauge Fields* (World Scientific, Singapore), p. 10; G. Kane, *Supersymmetry* (Perseus, Cambridge, MA, 2000), p. 26.

[9] G. L. Fitzpatrick, *The Family Problem* (Nova Scientific, Issaquah, WA, 1997).

[10] H. Georgi, *Lie Algebras in Particle Physics* (Benjamin, Menlo Park, CA, 1982), p. 247.

[11] S. L. Glashow, Comments Nucl. Part. Phys. **8**, 105 (1978).

[12] A. O. Barut, Phys. Rev. Lett. **42**, 1251 (1979).

[13] S. L. Glashow, Nucl. Phys. **22**, 579 (1961).

[14] A. Salam and J. C. Ward, Phys. Lett. **13**, 168 (1964).

[15] S. Weinberg, Phys. Rev. Lett. **19**, 1264 (1967).

[16] A. Salam, in *Elementary Particle Physics: Relativistic Groups and Analyticity (Nobel Symposium No. 8)*, edited by N. Svartholm (Almqvist and Wiksell, Stockholm, 1968).

[17] M. L. Perl, Physics Today **50**, 34 (1997)

[18] Huang, *loc. cit.*, pp. 113-117

[19] Ref. [1], p. 338.

[20] Z. Berezhiani, hep-ph/9602325.
[21] S. Schweber, *An Introduction to Relativistic Quantum Field Theory* (Row, Peterson, Evanston, Illinois, 1961) p. 45.
[22] A. O. Barut, "Dynamical Symmetry Group Based on Dirac Equation and its Generalization to Elementary Particles," Phys. Rev. **135**, B839 (1964).
[23] An equation of this form without the extradimensional interpretation was introduced by A.O. Barut, "Reformulation of the Dirac Theory of the Electron," Phys. Rev. Lett. **20**, 893 (1968); A. O. Barut and W. Thacker, "Covariant generalization of the *Zitterbewegung* of the electron and its SO(4,2) and SO(3,2) internal algebras," Phys. Rev. D **31**, 1386 (1985).
[24] A similar group, arranged somewhat differently, was introduced by Barut, Ref. 22.
[25] Here we follow Schweber, *op cit.*, p. 77, merely extending the analysis presented there to five dimensions.
[26] C. Wetterich in *Physics in Higher Dimensions*, edited by T. Piran and S. Weinberg (World Scientific, Singapore, 1986), p. 208. Wetterich describes a similar approach to getting fermion masses from higher dimensions. He starts, however, with a massless single-particle equation rather than one incorporating mass explicitly from the outset.
[27] The possibility that spacetime might entail a flat, time-like fifth dimension was considered long ago by A. S. Eddington, *Fundamental Theory* (Cambridge, 1953), Ch. 6, p. 126. Eddington, however, rejected the idea as "scarcely imaginable."
[28] Two- and three-dimensional time domains have been proposed by many authors. P. Demers, Can. J. Phys. **53**, 1687 (1975); R. Mignani and E. Recami, Lett. Nuovo Cimento **16**, 449 (1976); P. T. Pappas, Lett. Nuovo Cimento **22**, 429 (1978); E. A. Cole, Nuovo Cimento **A60**, 1 (1980); W. E. Hagston and I. D. Cox, Found. Phys. **15**, 773 (1985).
[29] ATLAS Collaboration, *Combined search for the Standard Model boson in pp collisions at* $\sqrt{s} = 7\,TeV$ *with the ATLAS detector*, PH-EP-2012-167 (2012). CMS Collaboration, *Observation of a new boson with a mass near 125 GeV*, CMS-PAS-HIG-12-020 (2012).
[30] But doubts about the identity of the discovered particle have

been raised. See A. Belyaev, M. S. Brown, R. Foadi and M. T. Frandsen, "Technicolor Higgs boson in the light of LHC data," Phys. Rev. D **90**, 035012 (13 August 2014).

[31] F. E. Close, *An Introduction to Quarks and Partons* (Academic Press, London, 1979), p. 383.

[32] D. E. Groom, *et al.* (Particle Data Group), Eur. Phys. Journ. **C15**, 1 (2000); http//pdg.lbl.gov.

[33] K. Huang, "On the Zitterbewegung of the Dirac Electron," Am. J. Phys. **20**, 479 (1952).

[34] C. Weinheimer and K. Zuber, "Neutrino Masses," arXiv:1307.3518 [hep-ex] (4 Sep 2013).

[35] G. Drexlin *et al.*, "Current Direct Neutrino Mass Experiments," arXiv:1307.0101 [physics.ins-det].

[36] E. W. Otten and C. Weinheimer, "Neutrino mass limit from tritium beta decay," arXiv:0909.2104 [hep-ex].

[37] P. Fisher, B. Kayser and K. S. McFarland, "Neutrino Mass and Oscillation," arXiv:hep-ph/9906244 (4 Jun 1999).

[38] Ref. [6], pp. 103-114.

[39] M. Ahmed *et al.*, Phys. Rev. D **65**, 112002 (2002).

[40] K. Abe *et al.*, Phys. Rev. Lett. **92**, 171802 (2004).

[41] B. Aubert *et al.*, Phys. Rev. Lett. **92**, 121801 (2004).

# Chapter 5

## Self-imaging *massive* bosons: Klein-Gordon and Maxwell-Proca equations in six spacetime dimensions

The wave functions of massive spin 0 and spin 1 elementary bosons are, respectively, scalar and vector objects, whereas that of the massless spin 1 boson (photon) is an antisymmetric tensor. Bosons as a whole are thus formally more diverse than the leptons, which are representable in terms of a single species of mathematical object, the Dirac spinor. On the other hand, their equations of motion are quadratic and consequently much simpler than the linear ones of the leptons. Accordingly it is perfectly possible to treat the different forms boson separately utilizing quadratic equation (3.92), with $\chi$ taking on separately the scalar, vector or tensor forms of descriptor. This however, would be a mistake. For there exists a formalism, *viz*., the Dirac-Clifford algebra, that allows us to treat the bosons as a whole, as members of a family not unlike that of the fermions. To carry out this unified treatment of the bosons one simply writes down the Dirac equation in six dimensions, with the Dirac spinor replaced by a quaternion matrix array

$$\mathrm{X} = \begin{bmatrix} \underline{A} & \underline{B} \\ \underline{C} & \underline{D} \end{bmatrix} = \lambda I + \sum_{c,d=0,1,2,3,5,6} \lambda^{cd} M_{cd},$$

where quaternions $\underline{A},\ldots,\underline{D}$ derive from an expansion of matrix X in the 15 Dirac matrices $M_{cd}$ [see (4.8)], plus the unit matrix, $I$. The expansion, with field coefficients $[\lambda(x),\ \lambda_{cd}(x)]$, is of sufficient generality to represent all forms of boson: scalar, vector and tensor.

The chief benefit of this unified approach is that it brings to light two previously unnoticed, stable, spin 1 pseudoscalar bosons, one having the mass of the known $Z^0$ boson, the other that of the $W^{\pm}$ boson, each capable of interacting with one, and only one, external field, namely the field of gravity. The new bosons, which we designate $Z^5$ and $W^5$, thus can be considered candidates for *dark matter*.

The present chapter concentrates on the properties and behavior of the massive bosons—particles associated with the wave equations of Klein-Gordon and Maxwell-Proca. The massless bosons—and Maxwell's equations in six dimensions—are dealt with separately in **Chapter 6.**

## 5.1 Constants of the motion

Bosons, like the leptons, propagate in 6-dimensional spacetime. As such they are endowed with a six-momentum vector $p^a = (p^0, \mathbf{p}, p^5, p^6)$, where $p^5$ and $p^6$ denote invariant pseudomomentum and pseudoenergy, respectively. As was shown in **Sec. 3.6** [see Eq. (3.72)], the latter two extra-dimensional momenta determine particle mass $m$ in accordance to the formula

$$m^2c^2 = p_6^{\ 2} - p_5^{\ 2}. \tag{5.1}$$

We shall need the values of the constants in this formula for a proper description of the equations of motion of the bosons. However, none of these constants is known *a priori* and must be found empirically.

*5.1.1 Scalar boson.* We take the Higgs boson $H$ to represent the scalar case. The $p_5$ value of the Higgs is readily deduced from the electron-positron collision process

$$e^+ + e^- \rightarrow H + \ell^+ + \ell^- \ . \tag{5.2}$$

On hole-theoretic grounds the signs of both $p_5$ and charge are reversed for antiparticles (see *Secs. 4.4.1 and 4.8.2*). Thus the $p_5$ contributions of the sums $e^+ + e^-$ and $\ell^+ + \ell^-$ are both zero. But the $p_5$ values on the two sides must balance. Thus we have for the Higgs

$$p_5(H) = 0\,, \tag{5.3}$$

and then from (5.1):

$$m_H c = p^6(H)\,. \tag{5.4}$$

According to (5.4) the mass of the Higgs particle is defined by a single parameter from the sixth spacetime dimension: the pseudoenergy $p^6(H)c$ (=125 GeV).[1] On the present theory, this extra-dimensional explanation of the origin of the Higgs mass replaces the presumption of Higgs mass being due to self-interaction.

*5.1.2 The photon.* We next consider the photon $\gamma$, which mediates the electromagnetic force. As the photon is massless, we have at once from (5.1)

$$p^6(\gamma) = p^5(\gamma). \tag{5.5}$$

From the electron-positron annihilation reaction $e^-e^+ \to 2\gamma$ we find for the photon's pseudomomentum

$$p^5(\gamma) = 0 \tag{5.6}$$

and then from (5.5),

$$p^6(\gamma) = 0. \tag{5.7}$$

That is, the photon's pseudomomentum $p^5(\gamma)$ and pseudoenergy $p^6(\gamma)$ are both zero. As shown in **Fig. 5.1**, the photon's momentum vector lies on the surface of the momentum light cone in reciprocal $M^4$ spacetime. This is to be constrasted with the situation depicted in **Fig. 3.4** for a particle of non-vanishing mass $m_n$; the propagating field of such a particle, irrespective of its $p_5$ value, is always driven by a non-vanishing pseudoenergy $p^6$. We might point out that, although the photon's momentum vector is confined to ordinary reciprocal 4-space, the photon nevertheless forms an infinite line current in $M^{4+2}$, as depicted in **Fig. 3.5.** The current velocity along $x^5$, however, is zero.

*5.1.3 The $Z^0$ and $W^\pm$ bosons.* We turn now to the weak force, mediated by the three gauge bosons $W^-$, $Z^0$ and $W^+$. These are massive particles having electric charge $-e$, 0, and $+e$, respectively. From the decay reaction $Z^0 \to e^+e^-$ we see at once that that

$$p_5(Z^0) = 0. \tag{5.8}$$

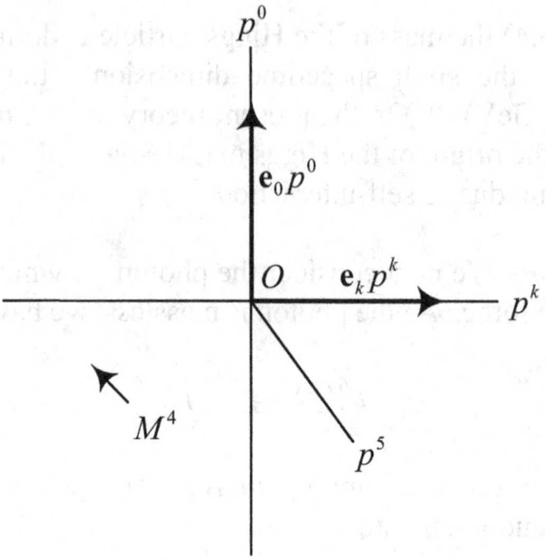


**Fig. 5.1.** Shown are components of the massless photon's momentum 5-vector $\mathbf{P} = \mathbf{e}_a p^a = \mathbf{e}_0 p^0 + \mathbf{e}_k p^k$, where unit vectors $\mathbf{e}_a$ satisfy $\mathbf{e}_a \cdot \mathbf{e}_b = g_{ab}$. In the case of the photon, $p^5 = 0 = p^6$. Hence resultant $\mathbf{P}$ is a null vector, a point at the origin. Note that, although the neutrino is massless, its 5-momentum vector $\mathbf{P}$ is not a null vector, because for the neutrino $p_5 = p^6 \neq 0$. See **Figs. 4.1** and **4.2**.

Consider next the decay reaction $Z^0 \to W^+ e^- \overline{\nu}_e$. In the light of (4.74), (4.78) and (5.8), the $p_5$-balance of this reaction reads $0 = p_5(W^+) + 0 - m_e c$. Hence

$$p_5(W^+) = m_e c\,. \tag{5.9}$$

Similarly, from the decay $Z^0 \to W^- e^+ \nu_e$ we obtain

$$p_5(W^-) = -m_e c\,. \tag{5.10}$$

Thus, with integer $-n$ $(n = 1, 0, -1)$ denoting electric charge, we may write

$$p_{5(n)} = -n m_e c\,, \tag{5.11}$$

which, with $p_{5(0)} \equiv p_5(Z^0)$ and $p_{5(\mp 1)} \equiv p_5(W^\pm)$, encompasses all three $p_5$ values (5.8)-(5.10).

It is of interest to compare (5.11) with the corresponding formula (4.92) for the electron family of leptons. Writing $Q = -n$ we have for the electron family

$$p_{5(n)} = (L_e - n)m_e c\,, \tag{5.12}$$

where $L_e$ is the lepton number. On comparing the two formulas we see that

$$p_5(Z^0) = p_5(e^{\mp}) = 0\,, \tag{5.13}$$

$$p_5(W^+) = p_5(\nu_e) = m_e c\,, \tag{5.14}$$

and

$$p_5(W^-) = p_5(\overline{\nu}_e) = -m_e c\,. \tag{5.15}$$

The law of conservation of pseudomomentum applied to the above decay processes has thus revealed an unexpected connection between the *e*-family of leptons and the three-member set of (heavy) gauge bosons. Namely, the members of each of the particle pairs $(e^{\mp}, Z^0)$, $(\nu_e, W^+)$ and $(\overline{\nu}_e, W^-)$ have the same $p_5$-value. It appears that the three massive bosons do indeed comprise a particle family, in the sense that their $p_5$-values, like those of the electron family of leptons, depend linearly on the electric charge. **Fig. 5.2** shows for comparison the 5-momentum vectors $p^a$ belonging to the gauge boson family and those of the electron family of leptons. The similarity between the two family structures is evident. The pair-wise symmetry seen here between lepton and boson families might well be considered a realization of the supersymmetry idea prevalent in some extensions of the Standard Model (including superstring theory).[2] Of course only in extra dimensions does the symmetry reveal itself.

*5.1.4 More on the origin of mass.* Further to our comments of *Sec. 4.7.2*, inspection of (5.1) shows that the masses of the gauge bosons derive not from interaction with the Higgs field, but from invariant parameters from the extra dimensions $x^5$ and $x^6$; specifically, the pseudoenergies $p^6(Z^0), p^6(W^{\pm})$ and the pseudomomenta $p_5(W^{\pm})$. Even the weak mixing angle[3] $\theta_W$ is revealed here to have an extra-dimensional origin. For we may write, noting that $m_e c = p_e^6$,

$$\cos\theta_W = \frac{m_W}{m_Z} = \frac{\sqrt{p_{6W}^{\ 2} - p_{6e}^{\ 2}}}{p_{6Z}}, \tag{5.16}$$

giving $\theta_W$ in terms of pseudoenergies coming from the sixth spacetime dimension.

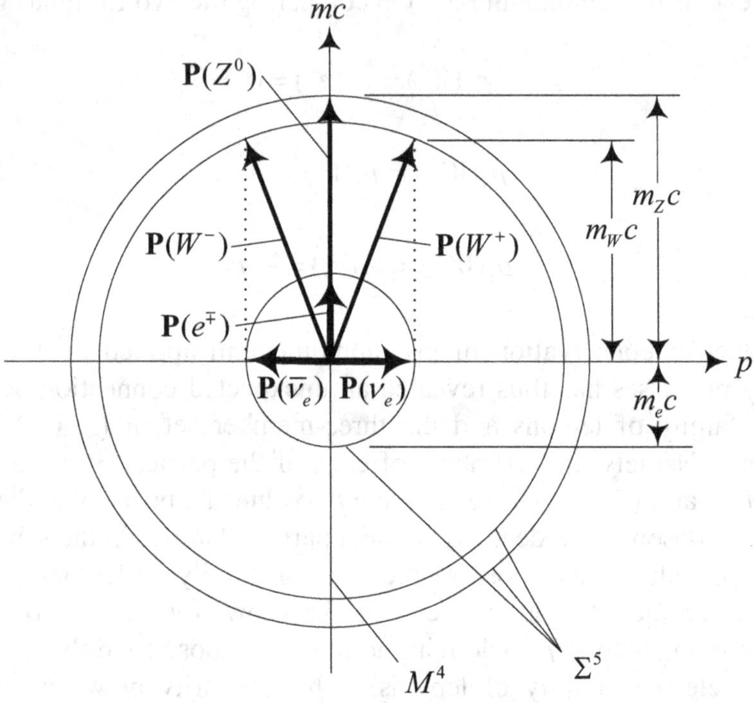


**Fig. 5.2** Five-momentum vectors $\mathbf{P} = \mathbf{e}_a p^a = \mathbf{e}_\mu p^\mu + \mathbf{e}_5 p^5$ differentiating the gauge bosons $Z^0$ and $W^\pm$ and, for comparison, the electron family $e^\mp$, $\nu_e$ and $\overline{\nu}_e$, in $M^4 \times \overline{M}^2$. Unit vectors $\mathbf{e}_a$ satisfy $\mathbf{e}_a \cdot \mathbf{e}_b = g_{ab}$, ($a = 0,1,2,3,5$). As shown, all resultant vectors $\mathbf{P}$ and their transverse components $\mathbf{e}_\mu p^\mu$ lie in the plane of the paper. Vector $\mathbf{e}_\mu p^\mu$ is of length *mc*, and the vertical axis is so labeled. Vectors $\mathbf{P}(e^\mp, \nu_e, \overline{\nu}_e)$, $\mathbf{P}(Z^0)$ and $\mathbf{P}(W^\pm)$ terminate on 5-D momentum shells $\Sigma_5$ of radii $p^6 = m_e c$, $m_Z c$ and $(m_W^{\ 2} + m_e^{\ 2})^{1/2} c$, respectively. $M^4$ = 4-dimensional hyperplane. For $Z^0$, $p^a(Z^0) = (p_0, \mathbf{p}, 0)$ and for $W^\pm$, $p^a(W^\pm) = (p_0, \mathbf{p}, \pm m_e c)$. Similarly for the electron, $p^a(e^\mp) = (p_0, \mathbf{p}, 0)$ and for the electron's neutrinos $p^a(\nu_e / \overline{\nu}_e) = (|\mathbf{p}|, \mathbf{p}, \pm m_e c)$. Noteworthy is the similarity of the group of three bosons to that of the three leptons, suggesting a real world instance of supersymmetry.

The minimal Standard Model is often criticized for its dependence on a substantial number of undetermined constants (18 or 19 of them, it is said), the idea being that a better and deeper theory—one going "beyond the Standard Model"—would provide the means for calculating the values of these undetermined constants. Given the results of this and the previous chapter, that hope would seem to be in vain. For many of these constants—the masses of the fundamental fermions and bosons and the weak mixing angle, for example—are simply features of spacetime geometry and—like $c$ and $h$—fundamentally incalculable. The Standard Model is perhaps closer to a final theory of interactions than one might think.

## 5.2 General equation of motion for bosons

As recalled at the beginning of this chapter, bosons come in three forms: scalar, vector and tensor. Consequently, to treat the bosons as a whole we shall need a new form of descriptor $\chi$, one having within it scalar, vector and tensor components. The form we seek is a $4\times 4$ *matrix*, $\mathrm{X}$, more specifically a $2\times 2$ array of *quaternions*. We shall first write down the equation of motion of our new descriptor $\chi$ and then derive the projection operator $\hat{R}$ and eigenvalue $r$ required to generate that equation. The latter step verifies that the general boson equation of motion is self-imaging, and thus an expression of the Principle of True Representation and the Law of Laws.

We adopt as our equation of motion Eq. (3.100) with a source term, $\sigma$:

$$\left(\gamma^\mu \hat{k}_\mu + \gamma^5 \hat{k}_5 + \hat{k}_6\right)\chi\left(x^a\right) = \sigma\left(x^a\right), \quad a = 0,1,2,3;5,6. \quad (5.17)$$

Since $\chi$ is to be self-imaging, both $\chi$ and $\sigma$ must be eigenstates of $\hat{k}_5$ and $\hat{k}_6$. They are therefore of the form

$$\chi\left(x^a\right) = \mathrm{X}(x^\mu)\exp[-i(k_5 x^5 + k_6 x^6)] \quad (5.18)$$

and

$$\sigma\left(x^a\right) = i\Sigma(x^\mu)\exp[-i(k_5 x^5 + k_6 x^6)]. \quad (5.19)$$

In terms of the field function Χ and source function Σ Eq. (5.17) reduces to an equation in ordinary $M^4$ :

$$\left(\gamma^{\mu}\hat{k}_{\mu}+\gamma^{5}k_{5}-k^{6}\right)\mathrm{X}\left(x^{\mu}\right)=i\Sigma\left(x^{\mu}\right). \tag{5.20}$$

Then, in explicit differential form, (5.20) reduces to

$$\left(\gamma^{\mu}\partial_{\mu}-i\gamma^{5}k_{5}+ik^{6}\right)\mathrm{X}\left(x^{\mu}\right)=\Sigma\left(x^{\mu}\right). \tag{5.21}$$

This is our equation of motion of the general boson field. In it Χ and Σ are taken to be square matrices

$$\mathrm{X}=\begin{bmatrix}\underline{A} & \underline{B}\\ \underline{C} & \underline{D}\end{bmatrix} \tag{5.22}$$

and

$$\Sigma=\begin{bmatrix}\underline{M} & \underline{N}\\ \underline{P} & \underline{Q}\end{bmatrix}, \tag{5. 23}$$

where the underlined elements are quaternions of the form $t+\boldsymbol{\sigma}\cdot\mathbf{x}$, $t$ is a complex scalar, $\mathbf{x}$ is a complex three-vector and $\boldsymbol{\sigma}$ denotes the Pauli spin matrices (3.42).

We now show how (5.21) proceeds from our Law of Laws (3.64), which, with Χ as descriptor, reads:

$$\hat{R}\mathrm{X}=r\mathrm{X}. \tag{5.24}$$

Let us first rewrite (5.20) in the form

$$\left(\gamma^{\mu}\hat{k}_{\mu}-k^{6}-i\Sigma\mathrm{X}^{-1}\right)\mathrm{X}=-\gamma^{5}k_{5}\mathrm{X} \tag{5.25}$$

where $\mathrm{X}^{-1}$ is the matrix inverse of Χ:

$$\mathrm{X}^{-1}\mathrm{X}=\mathrm{X}\mathrm{X}^{-1}=I. \tag{5.26}$$

If we now multiply (5.25) from the left by $i\gamma^5\xi_5$ then exponentiate the prefactor of X on either side, we obtain precisely (5.24), with

$$\hat{R} = \exp\left[i\gamma^5\xi_5\left(\gamma^\mu \hat{k}_\mu - k^6 - i\Sigma\mathrm{X}^{-1}\right)\right] \tag{5.27}$$

and

$$r = \exp\left(-ik_5\xi^5\right) \; . \tag{5.28}$$

The projection operator $\hat{R}$ is just (3.98) with the lower (+) sign and now including a source-dependent term $i\Sigma\mathrm{X}^{-1}$. The form of the eigenvalue $r$ is that required of all non-diffracting fields in 6-space; see (3.63). Expansion of the exponentials of (5.24) yields to all orders in $\xi^5$ our field equation (5.20). Having found $\hat{R}$ and its eigenvalue we have thereby demonstrated the self-imaging property of matrix X. We should emphasize that matrix X represents a infinite line current running parallel to the $x^5$ axis, and having the same cross section in all hyperplanes $M^4$ perpendicular to it. See **Fig. 3.5**.

Physical fields are introduced as follows. As is well known, the fifteen Dirac-Clifford matrices $\gamma^\mu$, $\sigma^{\mu\nu} = i\left(\gamma^\mu\gamma^\nu - \gamma^\nu\gamma^\mu\right)/2$, $\gamma^5\gamma^\mu$ and $\gamma^5$, together with the unit matrix $I$, are linearly independent and form a basis for the vector space of all $4\times 4$ matrices.[4] [For the algebraic properties of these matrices, see Eq. (3.96).] An arbitrary $4\times 4$ matrix may be expanded in these sixteen basis matrices. The physical fields we are looking for are the coefficients of the expansion.[5] We write for the general field X and source $\Sigma$

$$\mathrm{X} = aI + a_\mu\gamma^\mu + a_{\mu\nu}\sigma^{\mu\nu} + a_{5\nu}\gamma^5\gamma^\nu + a_5\gamma^5 \tag{5.29}$$

and

$$\Sigma = sI + s_\mu\gamma^\mu + s_5\gamma^5, \tag{5.30}$$

where $a(x)$ and $s(x)$ are scalars, $a_\mu(x)$ and $s_\mu(x)$ are vectors, $a_{\mu\nu}(x)$ is an antisymmetric tensor, $a_{5\nu}(x)$ is a pseudovector, and $a_5(x)$ and $s_5(x)$ are pseudoscalars. Because the equations of motion coming out of (5.21) are going to be either of rank 0 or 1 (scalar, pseudoscalar or vector), we exclude from $\Sigma$ rank-2 sources of the form $s_{ab}$.

We now substitute for X and Σ in (5.21) using (5.29) and (5.30). Then by equating coefficients of the basis matrices formed on the two sides of the equation, we obtain the desired equation of motion, as well as various subsidiary conditions, e.g, the Bianchi relation.

## 5.3 Scalar boson: the Klein-Gordon equation

Let us try this method for the simplest possible case, the scalar boson. For the field and source matrices it is convenient to write

$$\mathrm{X} = -ik^6\phi I + a_\mu \gamma^\mu \tag{5.31}$$

$$\Sigma = sI \tag{5.32}$$

where $\phi$ is the scalar field function. In the chiral representation,[6] the $\gamma$-matrices are

$$\gamma^0 = \begin{bmatrix} 0 & -I \\ -I & 0 \end{bmatrix},\ \gamma^i = \begin{bmatrix} 0 & \sigma^i \\ -\sigma^i & 0 \end{bmatrix},\ \gamma^5 = \begin{bmatrix} I & 0 \\ 0 & -I \end{bmatrix}, \tag{5.33}$$

where the $\sigma^i$ are the Pauli matrices defined in (3.42) and $I$ is the $2\times 2$ identity matrix. In this representation, X and Σ assume the $4\times 4$ matrix forms

$$\mathrm{X} = \begin{bmatrix} -ik^6\phi & a_0 - \boldsymbol{\sigma}\cdot\mathbf{a} \\ a_0 + \boldsymbol{\sigma}\cdot\mathbf{a} & -ik^6\phi \end{bmatrix} \tag{5.34}$$

$$\Sigma = \begin{bmatrix} sI & 0 \\ 0 & sI \end{bmatrix} \tag{5.35}$$

Our field and source matrices do indeed exhibit the quaternion arrays indicated in (5.22) and (5.23).

Putting (5.31) and (5.32) into (5.21) [with $k_5 = 0$; see (5.3)] and equating coefficients of like Dirac-Clifford matrices on the two sides we obtain:

$$\text{Coefficients of } I\text{: } \partial_\mu a^\mu + k_6{}^2\phi = s \tag{5.36}$$

$$\text{Coefficients of } \gamma^\mu\text{: } a_\mu = \partial_\mu \phi \tag{5.37}$$

$$\text{Coefficients of } \sigma^{\mu\nu}\text{: } \partial_\mu a_\nu - \partial_\nu a_\mu = 0 \tag{5.38}$$

The first of these is the equation of motion, expressed in terms of the vector field $a_\mu(x)$ and scalar field $\phi(x)$; this is a scalar analogue of Maxwell-Proca. The second equation defines vector $a_\mu$ as the gradient of $\phi$; thus $\phi$ is acting as a potential for $a_\mu$. The third equation is analogous to the Bianchi identity and is satisfied identically by (5.37). Putting (5.37) into (5.36) we obtain the scalar Klein-Gordon equation

$$\left(\partial_\mu \partial^\mu + k_H{}^2\right)\phi = s, \tag{5.39}$$

complete with scalar source, *s*. Note that we have defined from (5.1) and (5.4)

$$k^6 \equiv k_H = m_H c/\hbar,$$

where $k_H$ represents the reciprocal reduced Compton wavelength of the Higgs particle.

Deriving (5.39) was our main goal, but from equations (5.36)-(5.38) we can derive also the symmetrical stress energy tensor $T^{\mu\nu}$ associated with field $\phi$. To do this we first define the vector object $K^\mu = a^\mu s$, an obvious analogue of the Lorentz force density. Then, substituting for source *s* using (5.36) we obtain

$$\begin{aligned} K^\mu &= a^\mu \partial_\nu a^\nu + k_H{}^2 a^\mu \phi \\ &= \partial_\nu (a^\mu a^\nu) - a_\nu (\partial^\nu a^\mu) + k_H{}^2 a^\mu \phi \end{aligned} \tag{5.40}$$

Then making use of (5.37) and (5.40), we readily obtain

$$K^\mu = \partial_\nu T^{\mu\nu} \quad , \tag{5.41}$$

where

$$T^{\mu\nu} = \partial^\mu\phi\,\partial^\nu\phi + \frac{1}{2}g^{\mu\nu}\left(k_H{}^2\phi^2 - \partial_\lambda\phi\,\partial^\lambda\phi\right). \tag{5.42}$$

From this we get the well-known energy and momentum densities for the scalar field. Define $T^{0\nu} \equiv (\varpi^0, \boldsymbol{\varpi})c$, where $\varpi^\nu \equiv (\varpi^0, \boldsymbol{\varpi})$ has the units and appearance of a momentum density 4-vector (though not actually a vector). Then

$$\varpi^0 c = \frac{1}{2}\left[\left(\partial^0\phi\right)^2 + \left(\nabla\phi\right)^2 + k_H{}^2\phi^2\right] \tag{5.43}$$

$$\boldsymbol{\varpi} c = -\partial_0\phi\,\nabla\phi \tag{5.44}$$

Typically one derives equation (5.39) by way of the Lagrangian formalism.[7] But this is not a real derivation because the Lagrangian density employed is picked beforehand to give the desired result. In contrast, the present derivation makes no prior assumptions about the sought-for equation of motion, except that it should be scalar. One merely feeds into (5.5) the basic components $\phi$, $a_\mu$ and $s$, and the Dirac-Clifford algebra then assembles them automatically (one might say magically) into an equation of motion and a set of auxiliary relationships. This has worked well for the scalar case. One hopes that the same process applied to field components of higher order will generate formulas not seen before. That hope is to be realized in the following two sections.

## 5.4 Charge-neutral vector bosons $Z^0$ and $Z^5$

*5.4.1 Expansion in Dirac-Clifford matrices.* We now seek a vectorial equation of motion. From our extra-dimensional point of view, the concept of vector now includes a fifth component, $V^5$; that is, 5-vector $V^a = (V^\mu, V^5)$. Accordingly, our source expansion $\Sigma$ will have 5 independent terms; and our field expansion X will contain the maximum 16 terms, including the scalar term $\phi$. For the charge-neutral vector equations these expansions are conveniently written

$$\mathrm{X} = \phi + \frac{k^6}{i}\gamma^\mu Z_\mu + \frac{1}{2i}\sigma^{\mu\nu}Z_{\mu\nu} + i\gamma^5\gamma^\nu Z_{5\nu} + k^6\gamma^5 Z_5 \qquad (5.45)$$

and

$$\Sigma = \gamma^\mu s_\mu + i\gamma^5 s_5\,, \qquad (5.46)$$

where $Z_\mu(x)$ and $s_\mu(x)$ are vectors, $Z_{\mu\nu}(x)$ is an antisymmetric tensor, $Z_{5\nu}(x)$ is a pseudovector, and $Z_5(x)$ and $s_5(x)$ are pseudo-scalars. These expansions can be written in matrix form. Let us define, in analogy with electromagnetism,

$$\left.\begin{aligned} Z_{0j} &= \frac{E_Z{}^j}{c}, \\ Z_{ij} &= -\varepsilon_{ijk}B_Z{}^k, \\ Z_{50} &= -\frac{E_5}{c}, \\ Z_{5j} &= B_5{}^j \end{aligned}\right\} \qquad (5.47)$$

Equations (5.45) and (5.46) become

$$\mathrm{X} = \begin{bmatrix} \phi + k^6 Z_5 + \boldsymbol{\sigma}\cdot\left(-\dfrac{\mathbf{E}_Z}{c} + i\mathbf{B}_Z\right) & -i\left(\dfrac{E_5}{c} + k^6 Z_0\right) + i\boldsymbol{\sigma}\cdot\left(-\mathbf{B}_{Z5} + k^6\mathbf{Z}\right) \\ -i\left(-\dfrac{E_5}{c} + k^6 Z_0\right) - i\boldsymbol{\sigma}\cdot\left(\mathbf{B}_{Z5} + k^6\mathbf{Z}\right) & \phi - k^6 Z_5 + \boldsymbol{\sigma}\cdot\left(\dfrac{\mathbf{E}_Z}{c} + i\mathbf{B}_Z\right) \end{bmatrix} \qquad (5.48)$$

and

$$\Sigma = \begin{bmatrix} iL_5 & L_0 - \boldsymbol{\sigma}\cdot\mathbf{L} \\ L_0 + \boldsymbol{\sigma}\cdot\mathbf{L} & -iL_5 \end{bmatrix} \qquad (5.49)$$

Our field and source matrices once again exhibit the quaternion arrays indicated in (5.22) and (5.23).

Putting (5.45) and (5.46) into (5.21) [with $\hbar k_5 = 0$; see (5.8)] and equating coefficients of like Dirac-Clifford matrices we obtain: [8]

Coefficients of $I$: $$k^6\left(\partial_\mu Z^\mu - \phi\right) = 0 \tag{5.50}$$

Coefficients of $\gamma^\mu$: $$\partial_\mu Z^{\mu\nu} + \partial^\nu \phi + k_6^{\,2} Z^\nu = s^\nu \tag{5.51}$$

Coefficients of $\sigma_{\mu\nu}$: $$k^6\left[Z^{\mu\nu} - \left(\partial^\mu Z^\nu - \partial^\nu Z^\mu\right)\right] - i\varepsilon^{\mu\nu\rho\sigma}\partial_\rho Z_{5\sigma} = 0 \tag{5.52}$$

Coefficients of $\gamma^5\gamma_\sigma$: $$\frac{1}{2}\varepsilon^{\sigma\lambda\mu\nu}\partial_\lambda Z_{\mu\nu} + k^6\left(Z^{5\sigma} + \partial^\sigma Z^5\right) = 0 \tag{5.53}$$

Coefficients of $\gamma^5$: $$-\partial_\mu Z^{5\mu} + k_6^{\,2} Z^5 = s^5 \quad . \tag{5.54}$$

If $\partial_\nu Z^\nu = 0$ ( Lorentz condition), Eq. (5.50) yields $\phi = 0$. This equation, and with it the evaluation of $\phi$, disappears altogether when $k^6 = 0$ (Maxwell's equations).

Equation (5.51) is our 4-vectorial equation of motion, expressed in terms of the scalar $\phi$, vector $Z^\nu$ and tensor $Z^{\mu\nu}$. It is ultimately to become the vectorial equation of Maxwell-Proca.

Equation (5.52) is actually two equations owing to the presence of the imaginary unit, *i*:

$$k^6\left[Z^{\mu\nu} - \left(\partial^\mu Z^\nu - \partial^\nu Z^\mu\right)\right] = 0 \tag{5.55}$$

and

$$\partial_\rho Z_{5\sigma} - \partial_\sigma Z_{5\rho} = 0 \tag{5.56}$$

In (5.55) $Z^\nu$ plays the role of vector potential, defining tensor $Z^{\mu\nu}$; this equation disappears when $k^6 = 0$ (Maxwell's equations). Equation (5.56) is the Bianchi identity for the pseudovector $Z_{5\nu}$ [cf. equation (5.38)]. Note that in (5.52) the imaginary unit *i* is essential; otherwise one would have an inadmissible mixture of vector and pseudovector. This explains the presence and placement of the factors *i* in (5.45) and (5.46).

Like (5.52), equation (5.53) is actually two equations. The first term vanishes identically owing to (5.55):

$$\partial_\lambda Z_{\mu\nu} + \partial_\mu Z_{\nu\lambda} + \partial_\nu Z_{\lambda\mu} = 0\,. \tag{5.57}$$

This is the true Bianci identity. Provided $k^6 \neq 0$ the second equation reads

$$Z^{5\sigma} = -\partial^\sigma Z^5\,. \tag{5.58}$$

In this equation, which is analogous to (5.55), $Z^5$ plays the role of pseudoscalar potential, defining pseudovector $Z^{5\sigma}$. This definition satisfies (5.56) identically. Although (5.58) disappears when $k^6 = 0$, the definition of $Z^{5\sigma}$ still holds good because of (5.56).

Equation (5.54) is a fifth equation of motion, expressed in terms of the pseudoscalar $Z^5$ and pseudovector $Z^{5\mu}$.

*5.4.2. Maxwell-Proca equation for* $Z^0$. *Polarization.* We now show how (5.51) reduces to Maxwell-Proca. First eliminate $Z^{\mu\nu}$ from (5.51) using (5.55). This gives

$$\partial_\mu \partial^\mu Z^\nu - \partial^\nu (\partial_\mu Z^\mu - \phi) + k_6^{\;2} Z^\nu = s^\nu\,. \tag{5.59}$$

But the term in parentheses vanishes owing to (5.50). Hence we get Maxwell-Proca:

$$\left(\partial_\mu \partial^\mu + k_Z^{\;2}\right) Z^\nu = s^\nu\,, \tag{5.60}$$

complete with source $s^\nu$. Note that we have defined, from (5.1) and (5.8),

$$k^6 \equiv k_Z = \frac{m_Z c}{\hbar}\,, \tag{5.61}$$

where $k_Z$ is the reciprocal reduced Compton wavelength of $Z^0$ and $m_Z$ is its mass.[9] If the source is conserved then we have from (5.60) the Lorentz condition

$$\partial_\nu Z^\nu = 0\,, \tag{5.62}$$

and then from (5.50), $\phi = 0$.

We should point out that while $Z^\nu$ has four components, they are not all independent. The Lorentz condition (5.62) reduces the degree of freedom from four to three. This translates to three orthonormal states of polarization: two transverse and one longitudinal. Suppose the motion is in the $x^3$ direction. Three possible polarization states are

$$\left.\begin{array}{l}\left(0,Z^1,0,0\right)\\ \left(0,0,Z^2,0\right)\\ \left(Z^0,0,0,Z^3\right)\end{array}\right\} \qquad (5.63a)$$

The first two are linearly-polarized transverse modes and may be combined to form two mutually-orthogonal, elliptically-polarized states. The third state is the longitudinal one and its components must satisfy (5.62). Let $\partial^\nu\varphi$ be a solution of the free wave equation (5.60). Then for longitudinal polarization we may pick

$$Z^0 = -\frac{\partial_3\varphi}{k_Z},\ Z^3 = \frac{\partial_0\varphi}{k_Z}. \qquad (5.63b)$$

In the rest frame of this particle, $\partial_3\varphi = 0$, $\partial_0\varphi = k_Z\varphi'$ and the field vector becomes

$$(0,0,0,\varphi'). \qquad (5.63c)$$

*5.4.3 Maxwell-Proca equation for $Z^5$: Dark matter.* Now what about the fifth equation of motion, (5.54)? Eliminating $Z^{5\sigma}$ by means of (5.58) we obtain

$$\left(\partial_\mu\partial^\mu + k_Z{}^2\right)Z^5 = s^5. \qquad (5.64)$$

And so we have a new equation of motion, and with it a new particle, one we shall call $Z^5$. The $Z^5$ is a charge-neutral, pseudoscalar particle and, like its scalar counterpart $\phi$, has 0 spin (because there is no polarization vector). It has the same mass as $Z^0$, namely $m_Z$. But unlike $Z^0$, the $Z^5$ is not a gauge particle. It does *not* arise out a demand for invariance of a theory of interaction against local phase changes of the form $\chi \to \chi\exp[if(x^5)]$. Such a demand is impermissible, as it would destroy the periodic dependence of the wave function on $x^5$ as indicated in (4.39), and with it the possibility of self-imaging. Instead $Z^5$ arises simply as a fifth component of the five vector $(Z^{\mu}, Z^5)$, a straightforward product of the Dirac-Clifford formalism, and one not subject to gauge ambiguity. Now it is the function of a *gauge* boson to mediate forces; in other words, a gauge boson (by definition) interacts. Thus the $Z^5$, a non-gauge boson, mediates no forces. And in return, nothing interacts with it. It is therefore stable, as nothing can mediate its decay. Nor, for the same reason, can it be the product of a decay; it must have begun its life as an elemental product of the Big Bang. The $Z^5$, however, presumably responds to gravity, and if abundant could with its considerable mass affect the large-scale structure of the universe. It is thus an obvious candidate for *dark matter*.[10]

## 5.5 Electrically charged vector bosons $W^{\pm}$ and $W^5$

*5.5.1 Expansion in Dirac-Clifford matrices.* We seek now the equations of motion of the charged vector bosons. This case is a little more complicated than the neutral case because now $\hbar k_5 \neq 0$ [see (5.14) and (5.15)]. The question is, what will the Dirac-Clifford algebra do with it? To find out, we first define the field and source expansions

$$\mathrm{X} = \phi + \frac{k^6}{i}\gamma^{\mu}W_{\mu} + \frac{1}{2i}\sigma^{\mu\nu}W_{\mu\nu} + i\gamma^5\gamma^{\nu}W_{5\nu} + k^6\gamma^5 W_5 \tag{5.65}$$

and

$$\Sigma = \gamma^{\mu}l_{\mu} + i\gamma^5 l_5 , \tag{5.66}$$

where $W_{\mu}(x)$ and $l_{\mu}(x)$ are vectors, $W_{\mu\nu}(x)$ is an antisymmetric tensor, $W_{5\nu}(x)$ is a pseudovector, and $W_5(x)$ and $l_5(x)$ are pseudoscalars. Putting these into (5.21) and equating like coefficients of the Dirac-Clifford matrices we obtain:

Coefficients of $I$: $$k^6\left(\partial_{\mu}W^{\mu} + k_5 W^5 - \phi\right) = 0 \tag{5.67}$$

Coefficients of $\gamma^{\mu}$: $$\partial_{\mu}W^{\mu\nu} + k_5 W^{5\nu} + \partial^{\nu}\phi + k_6{}^2 W^{\nu} = l^{\nu} \tag{5.68}$$

$$\text{Coefficients of } \sigma_{\mu\nu}\text{:}\quad k^6\left[W^{\mu\nu} - \left(\partial^\mu W^\nu - \partial^\nu W^\mu\right)\right] - i\varepsilon^{\mu\nu\rho\sigma}\left(\partial_\rho W_{5\sigma} + \frac{1}{2}k_5 W_{\rho\sigma}\right) = 0 \tag{5.69}$$

$$\text{Coefficients of } \gamma^5\gamma_\sigma\text{:}\quad \frac{1}{2}\varepsilon^{\sigma\lambda\mu\nu}\partial_\lambda W_{\mu\nu} + k^6\left(W^{5\sigma} + \partial^\sigma W^5 + k^5 W^\sigma\right) = 0 \tag{5.70}$$

$$\text{Coefficients of } \gamma^5\text{:}\quad -\partial_\mu W^{5\mu} - k^5\phi + k_6{}^2 W^5 = l^5 \tag{5.71}$$

If $\partial_\mu W^\mu = 0$ (Lorentz condition), Eq. (5.67) yields $k_5 W^5 - \phi = 0$ . This equation, and with it the evaluation of $k_5 W^5 - \phi$, disappears when $k^6 = 0$ (Maxwell's equations).

Equation (5.68) is our 4-vectorial equation of motion, expressed in terms of the scalar $\phi$, vector $W^\nu$ , pseudovector $W^{5\nu}$ and tensor $W^{\mu\nu}$ . It is ultimately to become the Maxwell-Proca equation for the $W^\pm$ bosons.

Equation (5.69) is actually two equations owing to the presence of the imaginary unit, *i*:

$$k^6\left[W^{\mu\nu} - \left(\partial^\mu W^\nu - \partial^\nu W^\mu\right)\right] = 0 \tag{5.72}$$

and

$$\partial_\rho W_{5\sigma} - \partial_\sigma W_{5\rho} = -k_5 W_{\rho\sigma} \tag{5.73}$$

In (5.72) $W^\nu$ plays the role of vector potential, defining tensor $W^{\nu\mu}$; this equation disappears when $k^6 = 0$ . Equation (5.73) is the Bianchi identity for the pseudovector $W_{5\nu}$ . Interestingly, this identity, unlike (5.56), is inhomogeneous: the r.h.s. is not zero but proportional to the tensor field.

Equation (5.70), like (5.69), is actually two equations. The first term vanishes identically owing to (5.72):

$$\partial_\lambda W_{\mu\nu} + \partial_\mu W_{\nu\lambda} + \partial_\nu W_{\lambda\mu} = 0. \tag{5.74}$$

This is the true Bianci identity for the $W^{\pm}$ bosons. Provided $k^6 \neq 0$ the second equation reads

$$W^{5\sigma} = -\partial^{\sigma} W^5 - k^5 W^{\sigma} \qquad (5.75)$$

In this equation $W^5$ plays the role of pseudoscalar potential, defining pseudovector $W^{5\sigma}$. But vector $W^{\sigma}$ also figures on the r.h.s. This definition satisfies inhomogeneous Bianchi (5.73) identically.Equation (5.71) is a fifth equation of motion, expressed in terms of the scalar $\phi$, pseudoscalar $W^5$ and pseudovector $W^{5\mu}$.

*5.5.2. Maxwell-Proca equation for $W^{\pm}$.* We now show how (5.68) reduces to Maxwell-Proca. First eliminate $W^{\mu\nu}$ from (5.68) using (5.72). This gives

$$\partial_{\mu}\partial^{\mu} W^{\nu} + k_5 W^{5\nu} - \partial^{\nu}(\partial_{\mu} W^{\mu} - \phi) + k_6{}^2 W^{\nu} = l^{\nu}. \qquad (5.76)$$

With (5.67) this becomes

$$\partial_{\mu}\partial^{\mu} W^{\nu} + k_5 (W^{5\nu} + \partial^{\nu} W^5) + k_6{}^2 W^{\nu} = l^{\nu}. \qquad (5.77)$$

Then finally, making use of (5.75), we get Maxwell-Proca

$$\left(\partial_{\mu}\partial^{\mu} + k_W{}^2\right) W^{\nu} = l^{\nu}, \qquad (5.78)$$

where we have defined

$$\sqrt{k_6{}^2 - k_5{}^2} \equiv k_W = \frac{m_W c}{\hbar}. \qquad (5.79)$$

Here $k_W$ is the reciprocal reduced Compton wavelength of $W^{\pm}$ and $m_W$ is its mass.[9]

If the source is conserved then we have from (5.78) the Lorentz condition

$$\partial_{\nu} W^{\nu} = 0, \qquad (5.80)$$

and then from (5.67), the scalar field

$$\phi = k_5 W^5 . \tag{5.81}$$

The $W^{\pm}$ fields have of course the same polarization structure as that of $Z^0$.

*5.5.3 Maxwell-Proca equation for $W^5$: More dark matter.* Let us now look at our fifth equation of motion, (5.71). Eliminating $W^{5\sigma}$ and $\phi$ by means of (5.75) and (5.81) we obtain

$$\partial_\mu \partial^\mu W^5 + k^5 \partial_\mu W^\mu + (k_6{}^2 - k_5{}^2) W^5 = l^5 . \tag{5.82}$$

Then finally from (5.79) and (5.80)

$$\left(\partial_\mu \partial^\mu + k_W{}^2\right) W^5 = l^5 . \tag{5.83}$$

And so we have yet another new equation of motion, and with it a new spin-0, pseudoscalar particle, one we shall call $W^5$. Its mass is evidently $m_W$. The $W^5$ is nominally charged; both $+e$ and $-e$ forms should obtain. However, the $W^5$, like $Z^5$ is not a gauge particle. It does not interact electromagnetically or in any other way except gravitationally. Thus the $W^5$ should behave in either form as if it were a less massive $Z^5$. Like $Z^5$, the $W^5$ seems a likely candidate for dark matter[10]—if sufficiently abundant. As the two forms of $W^5$ have opposite $k_5$ values, they should have arisen primordially in equal numbers to conserve the fifth momentum.

## 5.6 Stress-energy tensor for the vector boson

Let us consider the interaction of current $s^\nu$ with an external field $Z^{\mu\nu}$, as expressed by the force density $K^\mu = Z^\mu{}_\nu s^\nu$. Then eliminating source $s^\nu$ using (5.35) (with $\phi = 0$) we have in analogy to (5.40)

$$\begin{aligned} K^\mu &= Z^{\mu\lambda}\left(\partial_\nu Z^\nu{}_\lambda + k_Z{}^2 Z_\lambda\right) \\ &= \partial_\nu\left(Z^{\mu\lambda} Z^\nu{}_\lambda\right) - Z^\nu{}_\lambda \partial_\nu Z^{\mu\lambda} + k_Z{}^2\left[\frac{1}{2}\partial^\mu\left(Z^\lambda Z_\lambda\right) - \partial^\lambda\left(Z_\lambda Z^\mu\right)\right] \end{aligned} \tag{5.84}$$

where we have used Lorentz condition (5.62). Then, after some manipulation of indices and use of Bianchi identity (5.57) we obtain

$$K^{\mu} = -\partial_{\nu} T^{\mu\nu} \tag{5.85}$$

where

$$T^{\mu\nu} = -Z^{\mu\lambda} Z^{\nu}{}_{\lambda} + k_Z{}^2 Z^{\mu} Z^{\nu} + \frac{1}{2} g^{\mu\nu} \left( \frac{1}{2} Z^{\rho\sigma} Z_{\rho\sigma} - k_Z{}^2 Z^{\rho} Z_{\rho} \right). \tag{5.86}$$

From this we get the momentum density components $T^{0\nu}/c = \varpi^{\nu} = (\varpi^0, \boldsymbol{\varpi})$ [see discussion following (5.42)], where

$$\begin{aligned} \varpi^0 c &= \frac{1}{2}\left[ \left(Z^{0j}\right)^2 + \left(Z^{ij}\right)^2 + k_Z{}^2 \left(Z_0{}^2 + Z_i{}^2\right) \right] \\ &= \frac{1}{2}\left[ \frac{\mathbf{E}_Z{}^2}{c^2} + \mathbf{B}_Z{}^2 + k_Z{}^2 \left(Z_0{}^2 + \mathbf{Z}^2\right) \right] \end{aligned} \tag{5.87}$$

$$\boldsymbol{\varpi} c = \frac{\mathbf{E}_Z}{c} \times \mathbf{B}_Z + k_Z{}^2 Z^0 \mathbf{Z}. \tag{5.88}$$

These components reduce according to state of polarization. Referring to (5.63a-c), we have for:

*Transverse polarization:*

$$\varpi^0 c = \frac{1}{2} \left( \frac{\mathbf{E}_Z{}^2}{c^2} + \mathbf{B}_Z{}^2 + k_Z{}^2 \mathbf{Z}^2 \right) \tag{5.89}$$

$$\boldsymbol{\varpi} c = \frac{\mathbf{E}_Z}{c} \times \mathbf{B}_Z . \tag{5.90}$$

*Longitudinal polarization* [cf. (5.43), (5.44)]:

$$\varpi^0 c = \frac{1}{2}\left[ (\partial_0 \varphi)^2 + (\nabla \varphi)^2 + k_Z{}^2 \varphi^2 \right] \tag{5.91}$$

$$\boldsymbol{\varpi} c = -\partial_0 \varphi \, \nabla \varphi . \tag{5.92}$$

*Particle at rest:*

$$\varpi^0 c = \frac{k_Z^{\ 2}}{2}\left[(\varphi')^2 + \varphi^2\right] \tag{5.93}$$

$$\boldsymbol{\varpi} c = 0\,. \tag{5.94}$$

Exactly similar equations hold for the $W^{\pm}$ bosons.

The equations of motion for $Z^5$ and $W^5$ are identical in structure to scalar equation (5.39). Their energy and momentum densities are thus immediately obtainable from (5.43) and (5.44).

## Notes and references

[1] Lucas Taylor (4 July, 2012), "Observation of a New Particle with a Mass of 125 GeV" (http://press.web.cern.ch/news/observation-new-particle-mass-125-gev).

[2] Gordon Kane, *Supersymmetry* (Perseus Publishing, Cambridge, Massachusetts, 2000); *Supersymmetry and Beyond: From the Higgs Boson to the New Physics* (Basic Books, New York, 2013).

[3] F. Mandl and G. Shaw, *Quantum Field Theory* (John Wiley, New York, 1984), pp. 271, 301, 314.

[4] S. Schweber, *An Introduction to Relativistic Quantum Field Theory* (Row, Peterson, Evanston, IL, 1961), pp. 70-74.

[5] A Dirac-algebraic formulation similar to the present one, with application to supersymmetry, has been given by F. Gürsey, "A Dirac Algebraic Approach to Supersymmetry," in *Quantum, Space and Time—the Quest Continues*, edited by A. O. Barut and A. van der Merwe (Cambridge U. Press, Cambridge, 1984), pp. 539-546.

[6] C. Itzykson and J. B. Zuber, *Quantum Field Theory* (McGraw-Hill, New York, 1980), p. 694.

[7] F. Mandl and G. Shaw, *op. cit.*, p. 32.

[8] To carry this out, in addition to definitions (3.96), the following relationships may prove helpful:

$$\gamma^\mu \gamma^\nu = g^{\mu\nu} - i\sigma^{\mu\nu}$$

$$\gamma^\nu \gamma^\lambda \gamma^\mu - \gamma^\mu \gamma^\lambda \gamma^\nu = 2\varepsilon^{\lambda\mu\nu\sigma}\gamma^5 \gamma_\sigma$$

$$\gamma^5 \sigma^{\mu\nu} = \frac{i}{2} \varepsilon^{\mu\nu\rho\sigma} \sigma_{\rho\sigma}$$

where

$$\sigma^{\mu\nu} = \frac{i}{2}\left(\gamma^\mu \gamma^\nu - \gamma^\nu \gamma^\mu\right)$$

and $\varepsilon^{\mu\nu\rho\sigma}$ is the totally asymmetric Levi-Civita tensor

$$\varepsilon^{\mu\nu\rho\sigma} = \begin{cases} +1 & \text{if } \{\mu,\nu,\rho,\sigma\} \text{ is an even permutation of } \{0,1,2,3\} \\ -1 & \text{if it is an odd permutation} \\ 0 & \text{otherwise} \end{cases}$$

quoting directly from C. Itzykson and J-B. Zuber, *Quantum Field Theory* (McGraw-Hill, New York, 1980), p. 692.

[9] The masses of the three gauge bosons are known experimentally to be:

$$m_Z = 91.1876 \pm 0.0021 \text{ GeV}/c^2$$
$$m_W = 80.385 \pm 0.015 \text{ GeV}/c^2.$$

J. Beringer *et. al.* (Particle Data Group), Phys. Rev. **D86**, 010001 (2012) (URL: http://pdg.lbl.gov).

[10] M. Drees and G. Gerbier, "Mini-Review of Dark Matter: 2012," (arXiv:1204.2373 [hep-ph] 11 Apr 2012).

# Chapter 6

# Self-imaging *massless* bosons: Maxwell's equations in six spacetime dimensions

In the previous chapter we showed how the Law of Laws, by way of an expansion of the wave function in Dirac-Clifford matrices, generates in a natural way the equations of motion of the massive bosons $H$, $Z^0$ and $W^{\pm}$. As a bonus it gave also the equations of motion of two new fundamental particles, $Z^5$ and $W^5$: massive, stable, pseudoscalar particles responsive to the gravitational field alone and thus candidates for dark matter.

In the present chapter we propose to develop in a similar way the equations of motion of massless bosons. The equations in question are of course Maxwell's equations, but enhanced, because we are now working in six spacetime dimensions. The enhancements mainly are these: (1) Maxwell's homogeneous equation (Bianchi identity) appears in its usual form, but in the inhomogeneous equation there now appears the gradient of a constant scalar field, a field one may tentatively identify with dark energy (cosmological term of Einstein). (2) There appears a second set of equations of the Maxwell form induced by the fifth charge current density, $j^5$. This second set will prove decisive in our later calculation of the mass spectra of the charged leptons and quarks. (3) This second set of equations yields a new form of radiation, a form we call *dark radiation*, because, like dark matter, it interacts with nothing but gravity. And associated with it is a new massless boson, the *dark photon*, $A^5$.

## 6.1 Expansion in Dirac-Clifford matrices.

For the massless field, we have $k_5 = k^6 = 0$ [see (5.6) and (5.7)]. The general Law of Laws for bosons (5.21) then reduces to

$$\gamma^{\mu}\partial_{\mu}\mathrm{X}\left(x^{\mu}\right)=\Sigma\left(x^{\mu}\right). \tag{6.1}$$

The field and source matrices appropriate to this expression are readily obtained from (5.45) and (5.46) by putting $k^6 = 0$ and replacing $Z_{ab}$ and $s_a$ with the familiar $F_{ab}$ and $j_a$ notation of Maxwell theory:

$$\mathrm{X} = I\phi + \frac{i}{2}\sigma^{\mu\nu}F_{\mu\nu} + i\gamma^5\gamma^\nu F_{5\nu} \tag{6.2}$$

and

$$\Sigma = \mu_0\gamma^\mu j_\mu + i\mu_5\gamma^5 j_5\,, \tag{6.3}$$

where $\phi(x^\mu)$ is a scalar, $j_\mu(x^\mu)$ is a vector, $F_{\mu\nu}(x^\mu)$ is an antisymmetric tensor, $F_{5\nu}(x^\mu)$ is a pseudovector, and $j_5(x^\mu)$ is a pseudoscalar. The real constants $\mu_0$, $\mu_5$ represent vacuum magnetic permeabilities; because $j^5$ is separately conserved—rotations in the $x^5$-$x^\mu$ plane being forbidden (see *Sec.4.4.3*) — manifest covariance need not be respected where the fifth axis is concerned.

With the definitions

$$\left.\begin{aligned} F_{0j} &= \frac{E^j}{c}, \\ F_{ij} &= -\varepsilon_{ijk}B^k, \\ F_{50} &= -\frac{E_5}{c}, \\ F_{5j} &= B_5^{\,j} \end{aligned}\right\} \tag{6.4}$$

(6.2) and (6.3) become in matrix form

$$\mathrm{X} = \begin{bmatrix} \phi + \boldsymbol{\sigma}\cdot\left(-\dfrac{\mathbf{E}}{c} + i\mathbf{B}\right) & i\left(-\dfrac{E_5}{c} + \boldsymbol{\sigma}\cdot\mathbf{B}_5\right) \\ i\left(\dfrac{E_5}{c} + \boldsymbol{\sigma}\cdot\mathbf{B}_5\right) & \phi + \boldsymbol{\sigma}\cdot\left(\dfrac{\mathbf{E}}{c} + i\mathbf{B}\right) \end{bmatrix} \tag{6.5}$$

and

$$\Sigma = \begin{bmatrix} i\mu_5 j_5 & \mu_0\left(j_0 - \boldsymbol{\sigma}\cdot\mathbf{j}\right) \\ \mu_0\left(j_0 + \boldsymbol{\sigma}\cdot\mathbf{j}\right) & -i\mu_5 j_5 \end{bmatrix}. \tag{6.6}$$

Here we see explicitly the field and source quaternion arrays defining electromagnetism in six dimensions. We note that a quaternion formulation of Maxwell's equations in four spacetime dimensions has been described by Edmonds.[1] If in (6.2) and (6.3) we omit all terms except those involving $F_{\mu\nu}$ and $j_\mu$, the resulting matrices, when substituted into (6.1), yield Maxwell's equations, with **E** and **B** representing as usual the electric and magnetic fields, respectively.

## 6.2 The field equations in covariant form

If in Eqs. (5.50)-(5.54) we put $k^6 = 0$, we obtain—after converting to Maxwell notation—following two sets of equations:

Set I (8 component equations)

$$\partial_\mu F^{\mu\nu} + \partial_\mu (g^{\mu\nu}\phi) = \mu_0 j^\nu \tag{6.7}$$

$$\varepsilon^{\sigma\lambda\mu\nu}\partial_\lambda F_{\mu\nu} = 0 \tag{6.8}$$

Set II (7 component equations)

$$\partial_\mu F^{5\mu} = -\mu_5 j^5 \tag{6.9}$$

$$\varepsilon^{\rho\sigma\mu\nu}\partial_\mu F_{5\nu} = 0 \tag{6.10}$$

As a formal matter the first thing we notice about these component equations is that there are 15 of them, 1 less than the maximum number possible. The shortfall is due to the fact that, given (6.2) and (6.3), Eq. (6.1) contains no term proportional to the identity matrix, *I*.

Secondly we see that Eqs. (6.7) are just Maxwell's inhomogeneous equations, modified by the divergence of a symmetric tensor field $g^{\mu\nu}\phi$. Now what about this additional element? Do Maxwell's inhomogeneous equations really require alteration in six dimensions? To explore this, take the divergence of (6.7). Since $F^{\mu\nu}$ is antisymmetric, the first term automatically vanishes. Then, to conserve charge ($\partial_\nu j^\nu = 0$), we must have

$$\partial_\nu \partial^\nu \phi = 0, \tag{6.11}$$

which is to hold everywhere, throughout all space and time. So either $\phi$ is an eternally existing, massless plane wave, which is hardly reasonable, or it is a constant. Assuming

$$\phi = \text{const.} \tag{6.12}$$

we have $\partial_\mu (g^{\mu\nu}\phi) = 0$ and (6.7) goes over to the conventional Maxwell form.

Equation (6.8) is the Bianchi relation, giving the four homogeneous equations of Maxwell. It is satisfied identically by expressing $F_{\mu\nu}$ in terms of a vector potential $A_\mu$:

$$F_{\mu\nu} = \partial_\mu A_\nu - \partial_\nu A_\mu . \tag{6.13}$$

Note that the case of the massive bosons, the sequence is reversed: definitions (5.55) and (5.72) are generated automatically by the Dirac-Clifford formalism and used subsequently to yield the Bianchi relations (5.57) and (5.74).

The equations of Set I thus represent Maxwell's original equations in their covariant formulation. Interestingly enough, with these equations—thanks to the extra dimensions $x^5$ and $x^6$ — there enters a new universal constant, $\phi$, the significance of which will be discussed in **Sec. 6.7**.

Let us turn now to the equations of Set II. Their formal similarity to the equations of Set I is evident. In (6.9) we have a fifth inhomogeneous equation, generated by the pseudoscalar fifth current density $j^5$. And in (6.10) we have a relation of the Bianchi form, yielding four additional homogeneous equations in the pseudovector $F_{5\mu}$. This latter relation is satisfied identically by writing $F_{5\mu}$ in terms of the gradient of a pseudoscalar potential $A_5(x^\mu)$:

$$F_{5\mu} = -F_{\mu 5} = -\partial_\mu A_5 \,. \tag{6.14}$$

Note that this definition contains no term $\partial_5 A_\mu$: to render $A_\mu$ self-imaging and $x^5$ invisible, $A_\mu$ (and all other physical fields) must be independent $x^5$, i.e., $\partial_5 A_\mu = 0$; see *Sec. 3.9.3*. The physical meaning of Set II will be discussed in **Sec. 6.4**.

We note that a set of equations similar to (6.9) and (6.10) were obtained by Hagston and Cox by extending the indices on the Maxwell field equations.[2] Their equations differ from ours, in both form and effect, because their fields can depend on the extra coordinate $x^5$, whereas ours cannot.

## 6.3 The field equations in vector form

By way of summary, let us write out our two sets of Maxwell equations in vector form using definitions (6.4).

Set I (8 component equations )

$$\nabla \cdot \left( \frac{\mathbf{E}}{c} \right) = \mu_0 j_0, \qquad \text{(Gauss)} \tag{6.15}$$

$$-\partial_0 \left( \frac{\mathbf{E}}{c} \right) + \nabla \times \mathbf{B} = \mu_0 \mathbf{j}, \quad \text{(Ampère)} \tag{6.16}$$

$$\partial_0 \mathbf{B} + \nabla \times \left( \frac{\mathbf{E}}{c} \right) = 0, \qquad \text{(Faraday)} \tag{6.17}$$

$$\nabla \cdot \mathbf{B} = 0 \qquad \text{(Supplementary)} \tag{6.18}$$

Set II (7 component equations)

$$\partial_0\left(\frac{E_5}{c}\right)+\nabla\cdot\mathbf{B}_5=\mu_5 j_5\,,\quad \text{(Analogue of Ampère)} \tag{6.19}$$

$$\partial_0\mathbf{B}_5+\nabla\left(\frac{E_5}{c}\right)=0\,,\quad \text{(Analogue of Faraday)} \tag{6.20}$$

$$\nabla\times\mathbf{B}_5=0 \quad \text{(Supplementary)} \tag{6.22}$$

Here $\boldsymbol{E}$ and $\boldsymbol{B}$ are the usual electric and magnetic fields, and $E_5$ and $\mathbf{B}_5$ are their analogues generated by the fifth current, $j_5$. In terms of potentials, we have from (6.4)

$$\left.\begin{aligned}&\frac{\mathbf{E}}{c}=-\partial_0\mathbf{A}-\nabla A_0\\&\mathbf{B}=\nabla\times\mathbf{A}\\&\frac{E_5}{c}=\partial_0 A_5\\&\mathbf{B}_5=-\nabla A_5\end{aligned}\right\} \tag{6.23}$$

To these sets of equations we add the material equations

$$\varepsilon_0\mu_0=\varepsilon_5\mu_5=\frac{1}{c^2}\,, \tag{6.24}$$

where $\varepsilon_0$ and $\varepsilon_5$ represent vacuum dielectric constants.

## 6.4 Energy-momentum tensor in five dimensions

Now what are we to make of the second set of Maxwell's equations? Do the fields $E_5$ and $\mathbf{B}_5$ really exist? And if they do, then how do they manifest themselves physically? To answer this one might begin by writing down the Lorentz force in five dimensions. This is easy to do and presumably would reveal the forces exerted by $E_5$ and $\mathbf{B}_5$ on external currents, $j^\mu, j^5$. The effort, however, would be in vain. For just as the massive fields $Z^5$ and $W^5$ are not gauge fields [see (5.64) and (5.83)], neither is $A_5$ a gauge field. It arises solely as a product of the Dirac-Clifford formalism, *not* out of a demand that the theory remain invariant against local phase changes of the form $\chi \to \chi \exp[if(x^5)]$. Such a demand is impermissible, as it would destroy the periodic dependence of the wave function on $x^5$ as indicated in (4.39), and along with it the possibility of self-imaging. Now if the field $A_5$ exists, then there exists a massless boson—one we shall designate $A^5$—to go with it. But it is not a gauge particle; $A^5$ interacts with nothing except gravity, and no external Lorentz force derives from it. One might reasonably call $A^5$ a *dark photon* and the field it belongs to *dark radiation*. Whatever we choose to call it, the upshot is this: because $A^5$ mediates no forces, $E_5$ and $\mathbf{B}_5$ have no effect at all on external currents.[3]

But that does not mean that $E_5$ and $\mathbf{B}_5$ are without physical significance. For although these fields cannot interact with external currents, they nevertheless carry energy (energy that does not do work!) and momentum. At least one of their effects is, in fact, to provide the charged leptons $\mu^\mp$ and $\tau^\mp$ with an electromagnetic mass unavailable to the electron. We shall later find that the mass contribution from potential $A_5$ is quantized, yielding the observed spectrum of charged lepton masses. It is therefore of interest, anticipating our later studies of lepton (and quark) mass, to develop now an expression for electromagnetic self-energy in five dimensions. We shall define the energy-momentum tensor $T^{\mu\nu}$ of the electromagnetic field as that tensor whose divergence yields the electromagnetic *self-force* density[4], $K^\mu$:

$$K^\mu = -\partial_\nu T^{\mu\nu}. \tag{6.25}$$

The fields comprising $T^{\mu\nu}$ are those generated by *internal* charge-current densities $J^\mu$ and $J^5$. These are the currents associated, *not* with the particle's global probability cloud $\psi$, but with the particle itself—microscopic currents created by the *Zitterbewegung*. The four-vector $K^\mu$ thus represents the density of force that the fields generated by $J^\mu$ and $J^5$ exert on the currents themselves. It may be written in the form

$$K^{\mu} = F^{\mu\nu} J_{\nu} + F^{5\mu} J_{5}, \tag{6.26}$$

where the second term on the right represents the contribution to the self-force from the fifth potential, $A_5$. We have emphasized that $A_5$ is not a gauge field. The self-force (6.26) thus must be viewed as simply a formal device for obtaining the energy momentum tensor, $T^{\mu\nu}$.

In terms of internal currents, Maxwell's equations (6.7) and (6.9) now read

$$\partial_{\mu} F^{\mu\nu} = \mu_0 J^{\nu} \tag{6.27}$$

and

$$\partial_{\mu} F^{5\mu} = -\mu_5 J^{5}. \tag{6.28}$$

Inserting these into (6.26) and making use of (6.8) and (6.10) we obtain from (6.25)

$$T^{\mu\nu} = \frac{1}{\mu_0}\left(-F^{\mu\sigma} F^{\nu}{}_{\sigma} + \frac{1}{4} g^{\mu\nu} F^{\rho\sigma} F_{\rho\sigma}\right) - \frac{1}{\mu_5}\left(-F^{5\mu} F_{5}{}^{\nu} + \frac{1}{2} g^{\mu\nu} F^{5\rho} F_{5\rho}\right). \tag{6.29}$$

The first two terms comprise the familiar energy-momentum tensor associated with the classical electromagnetic field, $F_{\nu\mu}$. The last two terms are due to the fifth dimension $x^5$ and the new Maxwell field $F_{5\mu}$. The electromagnetic self-energy density (in units of momentum density)≡ $\varpi^0$ $T^{00}/c$ is found from (6.29) to be

$$\varpi^0 \equiv \frac{T^{00}}{c} = \frac{1}{2\mu_0 c}\left(\frac{\mathbf{E}^2}{c^2} + \mathbf{B}^2\right) + \frac{1}{2\mu_5 c}\left(\frac{E_5{}^2}{c^2} + \mathbf{B}_5{}^2\right). \tag{6.30}$$

Note that the contribution to the self-energy from $A_5$ is positive when $\mu_5 > 0$.

The electromagnetic momentum density is given by the components $\varpi^n \equiv T^{0n}/c$:

$$\boldsymbol{\varpi} = \varepsilon_0 \mathbf{E}\times\mathbf{B} + \varepsilon_5 E_5 \mathbf{B}_5, \tag{6.31}$$

where we have made use of the material equations (6.24).[5] Here we see that, by way of their normal product, $E_5$ and $\mathbf{B}_5$ do contribute to the momentum density vector. The photon $A^5$, though dark, carries both energy and momentum.

We may also define a fifth component, $\varpi^5$, of the electromagnetic momentum density. In analogy with (6.25), we define the energy-momentum *vector* $T^{\mu 5}$ as that vector whose divergence yields the self-force density *scalar*, $K^5$:

$$K^5 = -\partial_\mu T^{\mu 5}. \tag{6.32}$$

Also, in analogy with (6.26) we write

$$K^5 = F^{5\nu} J_\nu. \tag{6.33}$$

Inserting (6.27) into (6.33) and making use of (6.10) we obtain from (6.32)

$$T^{\mu 5} = \frac{1}{\mu_0} F^{\nu 5} F^{\mu}{}_{\nu}. \tag{6.34}$$

Thus we have $\varpi^5 = T^{05}/c$, or

$$\varpi^5 = \varepsilon_0 \boldsymbol{E} \cdot \boldsymbol{B}_5. \tag{6.35}$$

As we shall see in **Chapter 13**, it is precisely $\varpi^5$ which, when integrated over all space, becomes quantized, yielding the charged lepton mass spectrum.

## 6.5 The retarded potentials

To evaluate the momentum density 5-vector $\varpi^a = (\varpi^0,\ \boldsymbol{\varpi},\ \varpi^5)$ one needs expressions for the potentials $A^\mu$ and $A^5$. To find these we must solve (6.27) and (6.28) expressed in terms of potentials. We shall demand that the potentials $A^\mu$ satisfy the constraint (Lorentz condition)

$$\partial_\mu A^\mu = 0, \tag{6.36}$$

in which case (6.27) becomes

$$\partial_\mu \partial^\mu A^\nu = \mu_0 J^\nu \tag{6.37}$$

In terms of $A^5$, (6.28) automatically reads [using (6.14)]

$$\partial_\mu \partial^\mu A^5 = \mu_5 J^5 , \tag{6.38}$$

where $J^\nu$ and $J^5$ are *internal* current densities. The solutions to these equations are the retarded potentials

$$A^\nu(\mathbf{x},t) = \frac{\mu_0}{4\pi}\int \frac{J^\nu(\boldsymbol{\xi}, t - r/c)}{r} d^3\boldsymbol{\xi} \tag{6.39}$$

and

$$A^5(x,t) = \frac{\mu_5}{4\pi}\int \frac{J^5(\boldsymbol{\xi}, t - r/c)}{r} d^3\boldsymbol{\xi} , \tag{6.40}$$

where $r = |\mathbf{x} - \boldsymbol{\xi}|$. For the currents $J^\nu$ and $J^5$ we specifically avoid employing currents associated with the Dirac theory of the electron,[6] first because they result in divergent self-energy, and second because in the Dirac theory $J^5$ does not exist. Instead, we shall employ currents derivable from the particle's *Zitterbewegung*. Finding and displaying those currents form the subject matter of **Chapter 10**.

## 6.6 Radiative states of polarization

Here we identify the polarization states of electromagnetic plane waves *in vacuuo*. We do this for the radiation associated with both Maxwell sets I and II.

*6.6.1 Maxwell Set I.* The massive Maxwell-Proca fields $Z^\mu$ and $W^\mu$ studied in the previous chapter are 4-vectors, and as such potentially may exhibit four states of polarization. However, because the 4-divergences $\partial_\mu Z^\mu$ and $\partial_\mu W^\mu$ necessarily vanish (Lorentz condition), only three of the four polarization states of $Z^\mu$ and $W^\mu$ are independent: two transverse and one longitudinal.

In contrast, the massless Maxwell field $F^{\mu\nu}$ studied in the present chapter is an antisymmetric tensor and so potentially may have *six* states of polarization, corresponding to the six components of field vectors $\mathbf{E}$ and $\mathbf{B}$. However, of the six, only two are independent and physically significant. To show this let us first write down Maxwell's equations (6.15)-(6.18), with the source terms set to zero:

$$\nabla\cdot\left(\frac{\mathbf{E}}{c}\right)=0, \qquad \text{(Gauss)} \qquad (6.41)$$

$$-\partial_0\left(\frac{\mathbf{E}}{c}\right)+\nabla\times\mathbf{B}=0\,, \qquad \text{(Ampère)} \qquad (6.42)$$

$$\partial_0\mathbf{B}+\nabla\times\left(\frac{\mathbf{E}}{c}\right)=0\,, \qquad \text{(Faraday)} \qquad (6.43)$$

$$\nabla\cdot\mathbf{B}=0 \qquad \text{(Supplementary)} \qquad (6.44)$$

The 3-divergence Gauss and Supplementary laws can be considered analogues of the Lorentz condition. Together they reduce the number of independent components of $\mathbf{E}$ and $\mathbf{B}$ to two each. The Ampère law then supplies two relations between the remaining four independent components, these two relations corresponding to the two independent components of vector $\mathbf{E}$. Alternatively, the Faraday law yields the same two relations, these corresponding to the two independent components of vector $\mathbf{B}$. Either way, we see that the original number of independent components, six, has been reduced to two, as anticipated. Their equations of motion, readily obtained from (6.42) and (6.43) are

$$(\partial_0{}^2-\nabla^2)\mathbf{E}=0 \qquad (6.45)$$

and

$$(\partial_0{}^2-\nabla^2)\mathbf{B}=0\,. \qquad (6.46)$$

To identify the two independent components, let us take a plane wave traveling in the positive $x^3$ direction. The $\mathbf{E}$ and $\mathbf{B}$ field components take the form

$$E^i = e^i \sin(\phi + \alpha^i) \tag{6.47}$$

and

$$B^i = b^i \sin(\phi + \beta^i), \tag{6.48}$$

where $\phi = k_0 x^0 + k_3 x^3$ , $e^i$ and $b^i$ are constant coefficients, and $\alpha^i$ and $\beta^i$ are constant phase angles. With these definitions, from either (6.45) or (6.46) we have at once that

$$k_0 = -k_3 \equiv k, \tag{6.49}$$

and from the Gauss and Supplementary laws (6.41) and (6.44)

$$e^3 = b^3 = 0. \tag{6.50}$$

Now in the Ampère law (6.42) write for $\mathbf{B}$ its equivalent from the r.h.s. of (6.48). This gives

$$\partial_0\left(\frac{\mathbf{E}}{c}\right) = k\left[b^2\cos(\phi+\beta^2),\ -b^1\cos(\phi+\beta^1),\ 0\right]. \tag{6.51}$$

Taking the time derivative of $\mathbf{E}$ directly from (6.47) gives

$$\partial_0\left(\frac{\mathbf{E}}{c}\right) = k\left[\frac{e^1}{c}\cos(\phi+\alpha^1),\ \frac{e^2}{c}\cos(\phi+\alpha^2),\ 0\right]. \tag{6.52}$$

On comparing the latter two expressions we find that

$$\left.\begin{aligned} b^1 &= -\frac{e^2}{c} \\ b^2 &= \frac{e^1}{c} \\ \beta^1 &= \alpha^2 \\ \beta^2 &= \alpha^1 \end{aligned}\right\}. \tag{6.53}$$

And so our propagating $\mathbf{E}$ and $\mathbf{B}$ fields (6.47) and (6.48) become, in terms of the *two* coefficients $e^1$ and $e^2$,

$$\mathbf{E} = \left[ e^1 \sin(\phi + \alpha^1),\ e^2 \sin(\phi + \alpha^2),\ 0 \right] \tag{6.54}$$

and

$$\mathbf{B} = \frac{1}{c}\left[ -e^2 \sin(\phi + \alpha^2),\ e^1 \sin(\phi + \alpha^1),\ 0 \right]. \tag{6.55}$$

We conclude as follows:

(1) All components of $\mathbf{E}$ and $\mathbf{B}$ are transverse to the direction of motion. There are no longitudinal components, thanks to the Gauss and Supplementary conditions (6.41) and (6.44).

(2) Thus there are two modes of polarization, both transverse. In terms of $\mathbf{E}$ they are

$$\left.\begin{array}{c} \left(0, E^1, 0, 0\right) \\ \left(0, 0, E^2, 0\right) \end{array}\right\} \tag{6.56}$$

By appropriate choice of phase angles $\alpha^1$ and $\alpha^2$ these two linearly polarized states may be combined to form two mutually-orthogonal elliptically-polarized states.

(3) From (6.54) and (6.55) we see that $\mathbf{E}$ and $\mathbf{B}$ are mutually perpendicular:

$$\mathbf{E} \cdot \mathbf{B} = 0. \tag{6.57}$$

It may not be superfluous to mention that our deductions about polarization of the electromagnetic field were accomplished here without reference the vector potential $A^\mu$ or gauge invariance. Nor should that be necessary, as the $\mathbf{E}$ and $\mathbf{B}$ fields are by nature invariant against choice of gauge. Nevertheless, perhaps as a consequence of the centrality of gauge invariance in the theory of interactions, polarization is now often explained in terms of gauge theory rather than Maxwell's equations directly. In that approach, one first imposes the Lorentz condition, which involves replacing the old $A^\mu$ with a new one

$A'^{\mu} = A^{\mu} + \partial^{\mu}\chi$, with scalar field $\chi$ so defined that $\partial_{\mu}A'^{\mu} = 0$. One then introduces a *second* gauge transformation $A''^{\mu} = A'^{\mu} + \partial^{\mu}\Lambda$, involving a second scalar field $\Lambda$ so defined that $\partial_{\mu}A''^{\mu} = 0$, with the aim of forcing $A''^{0}$ and $A''^{3}$ to vanish, leaving behind the non-vanishing transverse components $A''^{1}$ and $A''^{2}$. Now while this procedure does succeed in deriving the transversality of the electromagnetic field, waiting upon successive gauge transformations supplied by physicists cannot be the way that Nature works. The old textbook approach, based solely on Maxwell's equations is, in the writer's opinion, much to be preferred.

*6.6.2 Maxwell Set II.* With the source $j_5$ set to zero, Maxwell's Set II becomes

$$\partial_0\left(\frac{E_5}{c}\right) + \nabla\cdot\mathbf{B}_5 = 0, \quad \text{(Ampère analog)} \tag{6.58}$$

$$\partial_0\mathbf{B}_5 + \nabla\left(\frac{E_5}{c}\right) = 0, \quad \text{(Faraday analog)} \tag{6.59}$$

$$\nabla\times\mathbf{B}_5 = 0 \quad \text{(Supplementary)} \tag{6.60}$$

The equations of motion of fields $E_5$ and $\mathbf{B}_5$ follow from (6.58) and (6.59):

$$(\partial_0^{\,2} - \nabla^2)E_5 = 0 \tag{6.61}$$

and

$$(\partial_0^{\,2} - \nabla^2)\mathbf{B}_5 = 0. \tag{6.62}$$

Let us again assume propagation in the $x^3$ direction and write

$$E_5 = e_5\sin(\phi + \alpha_5) \tag{6.63}$$

and

$$B_5^{i} = b_5^{i}\sin(\phi + \beta_5^{i}), \tag{6.64}$$

Given (6.61) and (6.62), Eq. (6.49) still obtains, and from Supplementary (6.60) we have at once that

$$B_5^1 = B_5^2 = 0 . \tag{6.65}$$

Now in the Analog of Ampère law (6.58), write for $\mathbf{B}_5$ its value on the r.h.s. of (6.64). We get

$$\partial_0\left(\frac{E_5}{c}\right) = kb_5^3 \cos(\phi + \beta_5^3) . \tag{6.66}$$

But by direct time differentiation of (6.63),

$$\partial_0\left(\frac{E_5}{c}\right) = k\frac{e_5}{c}\cos(\phi + \alpha_5) . \tag{6.67}$$

Hence,

$$\left.\begin{aligned} b_5^3 &= \frac{e_5}{c} \\ \beta_5^3 &= \alpha_5 \end{aligned}\right\} . \tag{6.68}$$

And so our two dark radiation fields (6.63) and (6.64) become, in terms of the single coefficient $e^3$ and single phase angle $\alpha_3 \equiv 0$

$$E_5 = e_5 \sin\phi \tag{6.69}$$

and

$$\mathbf{B}_5 = \frac{1}{c}\left(0,\ 0,\ e_5 \sin\phi\right) . \tag{6.70}$$

In contrast to the normal "light" fields $\mathbf{E}$ and $\mathbf{B}$, our dark field $\mathbf{B}_5$ is longitudinally polarized. **Fig. 6.1** summarizes the geometric relationship of all three vectors, with $\mathbf{k}$ denoting the direction of propagation.

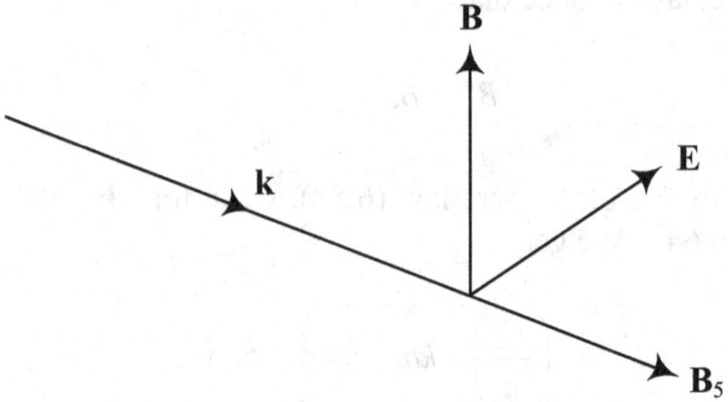


**Fig. 6.1** Showing the relationship between mutually perpendicular field vectors $\mathbf{E}$, $\mathbf{B}$ and $\mathbf{B}_5$. Vector $\mathbf{k}$ denotes direction of propagation.

## 6.7 Energy of the vacuum

To conclude this chapter—somewhat speculatively it should be said—let us return to the constant $\phi$ appearing in Maxwell's inhomogeneous equation, (6.7). If $\phi$ is non-vanishing, as in theory it could be, then it exists whether or not charged matter is present. In other words, our quaternion formulation of electromagnetism in extra dimensions predicts the possible existence of an ubiquitous vacuum energy density, $u_\Lambda$, a prediction, it might be noted, arising here without the help of quantum fields, superstrings or Einstein gravity. From (6.5) we see that $\phi$ has the same units as $\mathbf{E}/c$ and $\mathbf{B}$. This suggests that we can derive an algebraic expression for the universal vacuum energy $u_\Lambda$ in a manner analogous to the one used to find the electromagnetic self-energy density (6.30) of a single particle. In the absence of an electromagnetic field (6.7) becomes

$$\partial_\mu (g^{\mu\nu}\phi) = \mu_0 j^\nu , \tag{6.71}$$

both sides of which actually equal 0. Let us now define, taking (6.26) as a model, the *self-force of the vacuum* (which also equals zero!)

$$K^\mu = -(g^{\mu\nu}\phi) j_\nu . \tag{6.72}$$

Then, using (6.71) to eliminate $j_\nu$, we obtain

$$K^{\mu} = -\partial_{\lambda}\left(\frac{1}{2\mu_0} g^{\mu\lambda}\phi^2\right). \tag{6.73}$$

We now suppose $K^{\mu}$ to be derivable from a symmetric energy-momentum tensor $T_{\Lambda}^{\mu\nu}$, as in (6.25). This gives from (6.73)

$$T_{\Lambda}^{\mu\lambda} = \frac{1}{2\mu_0} g^{\mu\lambda}\phi^2, \tag{6.74}$$

the 00 component of which nicely parallels the self-energy expression (6.30). Indeed, with $u_{\Lambda} = T_{\Lambda}^{00}$, we have the background energy density

$$u_{\Lambda} = \frac{\phi^2}{2\mu_0}. \tag{6.75}$$

If we identify $u_{\Lambda}$ with the density $\Omega_{\Lambda}\rho_{\text{crit}}c^2$ of cosmic dark energy,[7,8] then

$$\phi = \left(2\Omega_{\Lambda}\rho_{\text{crit}} / \varepsilon_0\right)^{1/2} \tag{6.76}$$

where the critical density (in terms of mass) $\rho_{\text{crit}} \sim 10^{-29}$ g/cm$^3$ $= 10^{-26}$ kg/m$^3$, the permittivity of free space $\varepsilon_0 = 8.85 \times 10^{-12}$ coulomb$^2$/newton m$^2$, and $\Omega_{\Lambda} \sim 0.683$, yielding an observationally-defined numerical value of the constant $\phi$ appearing in (6.75).

Now the cosmic dark energy density is an attribute of empty space itself.[9] We know that dark energy is gravitationally repulsive, but other than that its ultimate origin and detailed properties remain a mystery. In particular, it may or may not actually be constant. Thus the possibility that dark energy ultimately derives, as we have shown here, from classical electrodynamics in extra dimensions is of no little interest:

(1) For if our $u_{\Lambda}$ really is the dark energy density, then, as we can see from (6.75), dark energy is constant.
(2) From (6.76) we see that the units of $\phi$ include the coulomb, the unit of electric charge. This suggests that, unless there is something even deeper going on, the vacuum may be the source for yet another

physical property of presently unknown origin: that of electric charge. Later in **Chapter 15** we are going to show that in a lattice representation of time, six-dimensional spacetime itself can supply this fundamental property in both its integral and one-third integral manifestations.

## Notes and references


1. J. D. Edmonds, Jr., "Maxwell's eight equations as one quaternion equation," Am. J. Phys. **46**, 430 (1978).
2. W. E. Hagston and I. D. Cox, "An Extended Theory of Relativity in a Six-Dimensional Manifold," Found. Phys. **15**, 773 (1985).
3. In contrast, the Hagston and Cox formulation of Ref. 2 predicts new Lorentz forces due to the extra Maxwell fields.
4. W. Heitler, *The Quantum Theory of Radiation* (Oxford University Press, London, 1954), pp. 27-34.
5. The electromagnetic energy flux density $[\text{energy} \times \text{vol}^{-1} \times \text{velocity}]$, also known as the Poynting vector $\mathbf{S} = \mathbf{E} \times \mathbf{H}$, is related to our momentum density $[\text{momentum} \times \text{vol}^{-1}]\ \boldsymbol{\varpi}$ in this way: $\mathbf{S} = c^2 \boldsymbol{\varpi}$.
6. V. Weiskopf, "On the self-Energy and the Electromagnetic Field of the Electron," Phys. Rev. **56**, 72 (1939). Reprinted in *Selected Papers on Quantum Electrodynamics*, edited by J. Schwinger (Dover, New York, 1958).
7. Hans C. Ohanian, *Gravitation and Spacetime* (W. W. Norton & Co., New York, 1976), pp. 267-271; H. A. Atwater, *Introduction to General Relativity* (Pergamon Press, Oxford,1974), pp. 169-275.
8. S. Weinberg, *Cosmology* (Oxford University Press, Oxford, UK, 2008), pp. 38, 50.
9. S. Perlmutter, "Supernovae, Dark Energy, and the Accelerating Universe," Physics Today, 53 (April 2003).

# PART III

# GEOMETRODYNAMICS

# Chapter 7

# General covariance

Whatever may be the structure and dimensionality of spacetime, the coordinates employed to represent it are not, of course, observable entities like electrons and photons. Thus one demands that, on paper, the dynamical laws of physics should be expressible in terms independent of any particular system of coordinates. Laws that can be written in this coordinate-free way are said to be *generally covariant*.[1] General covariance ensures that transformation from one coordinate system to another leaves unaltered the physical content of the laws of physics. We have previously referred to frame-invariance of the laws as the (General) Principle of Relativity; see *Sec. 1.2.2*.

Our formulation of the Law of Laws has up to this point assumed an inertial frame of reference, i.e., a coordinate frame selected from an infinite set of such frames, each moving uniformly relative to one another; see **Sec. 3.5**. Consequently the various single-particle laws generated thus far by the Law of Laws are invariant against Lorentz transformation but are not generally covariant. In this chapter we shall extend our formulation to render those laws generally covariant. Technically this entails replacing the ordinary derivative $\partial_\mu$ wherever it occurs with a covariant derivative $\nabla_\mu$ whose form depends on the structure of the object on which it operates.

There are at least two good reasons to proceed from special (Lorentz) covariance to general covariance. First, and most obviously, we have from the beginning claimed the Law of Laws to be the source of all dynamical laws governing the behavior of single particles. Because all physical laws are required to be generally covariant, we are natually obliged to show that the Law of Laws too exhibits general covariance.

Second, general covariance is the key to describing the interaction between gravity and matter-generated fields of probability and electromagnetism. Einstein proposed that one must demand invariance of the laws of physics not just between frames in uniform relative motion, but between accelerated frames as well—general covariance. But by the Principle of Equivalence, Einstein reasoned, one cannot tell whether the apparent acceleration of a massive body is due to having observed it from an accelerated frame of reference or to an applied force of gravitation.[1] Hence generally covariant single-particle laws entail automatically the effects of gravitation. The law of gravitation itself requires separate attention. In **Chapter 8** we shall show how Einstein's field equations arise out of the Law of Laws, appearing embedded in a larger law of geometrodynamics.

## 7.1 Elements of tensor analysis

To facilitate our work in this and subsequent chapters, we review now the basic formulas and notational conventions of tensor analysis, extended, where necessary, to six dimensions.[2]

*7.1.1 Metric tensor.* In six dimensional spacetime $M^{4+2} = M^4 \times \bar{M}^2$ the metric tensor is denoted $g_{ab}$ $(a,b = 0,1,2,3;5,6)$. The portion of it spanning subspace $M^4$ is denoted, as usual, $g_{\mu\nu}$ $(\mu,\nu = 0,1,2,3)$. In flat $M^4$ the diagonal components of the metric are $(+1,-1,-1,-1)$; those of $\bar{M}^2$—which is always flat—are $(+1,-1)$. Now if the physical phenomena occurring within $M^4$ are to be self-imaging, as depicted in **Fig 3.3**, then $M^4$ must be everywhere perpendicular to $x^5$ and $x^6$. That is, whatever the $g_{\mu\nu}$ within $M^4$ may be doing, subspace $M^4$, viewed as a hypersurface in $M^{4+2}$, appears to be flat. Since $\bar{M}^2$ is also flat and perpendicular to $M^4$, we have

$$g_{\mu 5} = g_{\mu 6} = 0. \tag{7.1}$$

Also, for self-imaging, all physical phenomena occurring within $M^4$ must be independent of $x^5$ and $x^6$. Hence

$$g_{\mu\nu,5} = g_{\mu\nu,6} = 0, \tag{7.2}$$

where the comma denotes ordinary derivative: i.e., for any function $f(x^a)$, $\partial f / \partial x^a = \partial_a f = f_{,a}$. In matrix form, the complete (symmetrical) metric tensor looks like this

$$g_{ab} = \begin{pmatrix} g_{00} & g_{01} & g_{02} & g_{03} & 0 & 0 \\ & g_{11} & g_{12} & g_{13} & 0 & 0 \\ & & g_{22} & g_{23} & 0 & 0 \\ & & & g_{33} & 0 & 0 \\ & & & & +1 & 0 \\ & & & & & -1 \end{pmatrix}. \tag{7.3}$$

If $A_a$ and $A^a$ are covariant and contravariant vectors, respectively, then

$$A_a = g_{ab} A^b \tag{7.4}$$

and

$$A^a = g^{ab} A_b . \tag{7.5}$$

Here $g^{ab}$ is the cofactor of $g_{ab}$ in the determinant of $g_{ab}$, divided by the value of the determinant. Thus we find

$$g^{55} = g_{55} = +1 \tag{7.6}$$

and

$$g^{66} = g_{66} = -1 . \tag{7.7}$$

One has also

$$g_{ab} g^{bc} = g_a^{\ c} = 1 \text{ for } a = c \text{ and } = 0 \text{ for } a \neq c , \tag{7.8}$$

so that

$$g_a^{\ b} A^a = A^b . \tag{7.9}$$

*7.1.2 Christoffel symbols.* These are defined in terms of the metric as follows.

1. Christoffel symbol of the *first kind*:

$$\Gamma_{\mu\nu\sigma} = \frac{1}{2}(g_{\mu\nu,\sigma} + g_{\mu\sigma,\nu} - g_{\nu\sigma,\mu}) . \tag{7.10}$$

It is symmetrical in the last two indices:

$$\Gamma_{\mu\nu\sigma} = \Gamma_{\mu\sigma\nu} . \tag{7.11}$$

From the definition it follows that:

$$\Gamma_{\mu\nu\sigma} + \Gamma_{\nu\mu\sigma} = g_{\mu\nu,\sigma} \tag{7.12}$$

2. Christoffel symbol of the *second kind*:

$$\Gamma^{\mu}_{\nu\sigma} = g^{\mu\lambda}\Gamma_{\lambda\nu\sigma} . \tag{7.13}$$

It is symmetrical in the two lower indices:

$$\Gamma^{\mu}_{\nu\sigma} = \Gamma^{\mu}_{\sigma\nu} \tag{7.14}$$

Note that these definitions extend to six dimensions, e.g.,

$$\Gamma_{abc} = \frac{1}{2}(g_{ab,c} + g_{ac,b} - g_{bc,a}) . \tag{7.15}$$

However, in the light of (7.1) and (7.2), these symbols vanish whenever $a$, $b$ or $c = 5$ or $6$. For example $\Gamma_{\mu\nu 5} = (g_{\mu\nu,5} + g_{\mu 5,\nu} - g_{\nu 5,\mu})/2 = 0$. As we shall see in the next chapter, the vanishing of these symbols limits the amount of new gravitation physics one can expect to get from the extra dimensions.

*7.1.3 Covariant derivative of a tensor.* The *covariant derivative* of a *tensor* of order $\alpha...\omega$ is expressed by a semi-colon: $T_{\alpha...\omega;\mu}$. For tensors in general:

scalar $S$: $$S_{;\mu} = S_{,\mu} \tag{7.16}$$

covariant vector $A_\mu$: $$A_{\mu;\sigma} = A_{\mu,\sigma} - \Gamma^{\alpha}_{\mu\sigma}A_{\alpha} \tag{7.17}$$

tensor $T_{\mu\nu}$: $$T_{\mu\nu;\sigma} = T_{\mu\nu,\sigma} - \Gamma^{\alpha}_{\mu\sigma} T_{\alpha\nu} - \Gamma^{\alpha}_{\nu\sigma} T_{\mu\alpha} \tag{7.18}$$

contravariant vector $A^{\mu}$: $$A^{\mu}_{;\sigma} = A^{\mu}_{,\sigma} + \Gamma^{\mu}_{\alpha\sigma} A^{\alpha} \tag{7.19}$$

tensor $T^{\mu\nu}$: $$T^{\mu\nu}_{;\sigma} = T^{\mu\nu}_{,\sigma} + \Gamma^{\mu}_{\alpha\sigma} T^{\alpha\nu} + \Gamma^{\nu}_{\alpha\sigma} T^{\mu\alpha}. \tag{7.20}$$

For some special tensors:

metric $g_{\mu\nu}$: $$g_{\mu\nu;\sigma} = 0 \tag{7.21}$$

$$g^{\mu\nu}_{;\sigma} = 0 \tag{7.22}$$

Dirac $\gamma_a$: $$\gamma_{\mu;\sigma} = \gamma_{\mu,\sigma} - \Gamma^{\alpha}_{\mu\sigma} \gamma_{\alpha} \tag{7.23}$$

$$\gamma^{\mu}_{;\sigma} = \gamma^{\mu}_{,\sigma} + \Gamma^{\mu}_{\alpha\sigma} \gamma^{\alpha} \tag{7.24}$$

$$\gamma_{5;\sigma} = \gamma^{5}_{;\sigma} = 0 \tag{7.25}$$

*7.1.4 Covariant differentiation of spin-dependent quantities.* The connection between spin and spacetime curvature is expressed by the anticommutation relation

$$\gamma_{\mu}\gamma_{\nu} + \gamma_{\nu}\gamma_{\mu} = I g_{\mu\nu} \tag{7.26}$$

In flat space, where the $\gamma_{\mu}$ are constant matrices, one has $\partial_{\sigma}\gamma_{\mu} = 0$. The corresponding expression in curved space is written[3]

$$\nabla_{\sigma}\gamma_{\mu} = \gamma_{\mu;\sigma} + \gamma_{\mu}\Gamma_{\sigma} - \Gamma_{\sigma}\gamma_{\mu} = 0. \tag{7.27}$$

Here $\nabla_{\sigma}$ is a covariant derivative which when applied to the $\gamma_{\mu}$ in curved space yields zero. The *spin coefficients* $\Gamma_{\sigma}$ are given by[4]

$$\Gamma_{\sigma} = \frac{1}{4}\gamma_{\nu}\gamma^{\nu}_{;\sigma}. \tag{7.28}$$

Covariant differentiation by $\nabla_\sigma$ has the property

$$\nabla_\sigma(AB)=(\nabla_\sigma A)B+A(\nabla_\sigma B). \tag{7.29}$$

Thus the covariant derivative $\nabla_\sigma$ of a product $\gamma_\alpha\gamma_\beta ... A_{\mu\nu...}$ of Dirac matrices and an ordinary tensor yields:

$$\nabla_\sigma(\gamma_\alpha\gamma_\beta ... A_{\mu\nu...})=\gamma_\alpha\gamma_\beta ... A_{\mu\nu...;\sigma} \tag{7.30}$$

For example,

$$\nabla_\sigma(\sigma^{\mu\nu}F_{\mu\nu})=\sigma^{\mu\nu}F_{\mu\nu;\sigma}, \tag{7.31}$$

where $\sigma^{\mu\nu}$ is the spacetime-dependent version of the spin tensor appearing in (6.2).

The covariant derivative of a four-component (Dirac) spinor $\chi$ is denoted by $D_\mu$. This operation gives:[3]

$$D_\mu\chi=\chi_{;\mu}+\Gamma_\mu\chi, \tag{7.32}$$

which may be compared with (7.27). The operators $\nabla_\sigma$ and $D_\mu$ differ in accordance to different types of spin-dependent quantity they are applied to. In either case $\Gamma_\sigma$ is defined by (7.28).

*7.1.5 The curvature tensor.* The Riemann-Christoffel curvature tensor $R^\mu_{\nu\rho\sigma}$ is defined by

$$R^\mu{}_{\nu\rho\sigma}=\Gamma^\mu{}_{\nu\sigma,\rho}-\Gamma^\mu{}_{\nu\rho,\sigma}+\Gamma^\kappa{}_{\nu\sigma}\Gamma^\mu{}_{\kappa\rho}-\Gamma^\kappa{}_{\nu\rho}\Gamma^\mu{}_{\kappa\sigma}. \tag{7.33}$$

It is asymmetric in the last two lower indices:

$$R^\mu{}_{\nu\rho\sigma}=-R^\mu{}_{\nu\sigma\rho}. \tag{7.34}$$

It obeys also the cyclic relationship

$$R^\mu{}_{\nu\rho\sigma}+R^\mu{}_{\rho\sigma\nu}+R^\mu{}_{\sigma\nu\rho}=0. \tag{7.35}$$

Lowering the upper index to the first position below we obtain

$$R_{\mu\nu\rho\sigma} = \frac{1}{2}(g_{\mu\sigma,\nu\rho} - g_{\nu\sigma,\mu\rho} - g_{\mu\rho,\nu\sigma} + g_{\nu\rho,\mu\sigma}) \\ + \Gamma_{\beta\mu\sigma}\Gamma^{\beta}{}_{\nu\rho} - \Gamma_{\beta\mu\rho}\Gamma^{\beta}{}_{\nu\sigma} \quad . \tag{7.36}$$

This form of the curvature tensor is antisymmetric in both the first and last pair of indices:

$$R_{\mu\nu\rho\sigma} = -R_{\nu\mu\rho\sigma} = R_{\nu\mu\sigma\rho} . \tag{7.37}$$

It is symmetric with respect to interchange of the first and last pair of indices:

$$R_{\mu\nu\rho\sigma} = R_{\rho\sigma\mu\nu} = R_{\sigma\rho\nu\mu} . \tag{7.38}$$

To quote Dirac,[5] "The result of all these symmetries is that, of the 256 components of $R_{\mu\nu\rho\sigma}$, only 20 are independent."

In general, covariant derivatives do not commute. The degree to which they do not commute is governed by the curvature tensor. The following relations, of which we will make much use, are readily established:[6]

$$A_{\nu;\rho;\sigma} - A_{\nu;\sigma;\rho} = A_{\beta} R^{\beta}{}_{\nu\rho\sigma} \tag{7.39}$$

$$A_{\mu\nu;\rho;\sigma} - A_{\mu\nu;\sigma;\rho} = A_{\alpha\nu} R^{\alpha}{}_{\mu\rho\sigma} + A_{\mu\alpha} R^{\alpha}{}_{\nu\rho\sigma} \tag{7.40}$$

$$A_{\lambda\mu\nu;\rho;\sigma} - A_{\lambda\mu\nu;\sigma;\rho} = A_{\alpha\mu\nu} R^{\alpha}{}_{\lambda\rho\sigma} + A_{\lambda\alpha\nu} R^{\alpha}{}_{\mu\rho\sigma} + A_{\lambda\mu\alpha} R^{\alpha}{}_{\nu\rho\sigma} \tag{7.41}$$

$$A_{\kappa\lambda\mu\nu;\rho;\sigma} - A_{\kappa\lambda\mu\nu;\sigma;\rho} = A_{\alpha\lambda\mu\nu} R^{\alpha}{}_{\kappa\rho\sigma} + A_{\kappa\alpha\mu\nu} R^{\alpha}{}_{\lambda\rho\sigma} + A_{\kappa\lambda\alpha\nu} R^{\alpha}{}_{\mu\rho\sigma} + A_{\kappa\lambda\mu\alpha} R^{\alpha}{}_{\nu\rho\sigma} \tag{7.42}$$

The curvature tensor satisfies the *Bianci relations*:

$$R^{\mu}{}_{\nu\rho\sigma;\tau} + R^{\mu}{}_{\nu\sigma\tau;\rho} + R^{\mu}{}_{\nu\tau\rho;\sigma} = 0 \tag{7.43}$$

$$R_{\mu\nu\rho\sigma;\tau} + R_{\mu\nu\sigma\tau;\rho} + R_{\mu\nu\tau\rho;\sigma} = 0. \tag{7.44}$$

*7.1.6 The Ricci tensor.* Contraction of the first and last indices of $R_{\mu\nu\rho\sigma}$ yields the *Ricci tensor*:

$$R^{\mu}{}_{\nu\rho\mu} = R_{\nu\rho} = R_{\rho\nu}, \tag{7.45}$$

the symmetry of which follows from (7.38). Contracting again, we obtain the *scalar curvature*, *R*:

$$R^{\nu}{}_{\nu} = R. \tag{7.46}$$

## 7.2 Extension of the spinor, scalar, vector and tensor field equations to curved spacetime

In curved spacetime our Law of Laws (2.6) has the same general form it has in flat spacetime. The picture of self-imaging in hyperplanes $M^4$ perpendicular to $x^5$ remains unchanged and exactly as illustrated in **Fig. 3.3**. Thus eigenvalue *r*, too, is unchanged, because $k_5$ is invariant. What does change is the content of projection operator $\hat{R}$. To make the move to curved spacetime we must replace in $\hat{R}$ the ordinary derivative with an appropriate covariant derivative. Applied to descriptor $\chi$, the revised operator $\hat{R}$ generates a new field equation in $\chi$, one that holds good in curved spacetime.

*7.2.1 The Dirac equation.* In flat space the operator $\hat{R}$ for fermions is given by (3.98). When applied to spinors $\chi$, the upper sign yields an equation for probability amplitude (4.31), the lower sign an equation for charge amplitude (4.42). In curved space (3.98) goes over to

$$\hat{R} = \exp\left[\pm i\gamma^5 \xi_5 \left(\gamma^\sigma i D_\sigma + I i \partial_6\right)\right], \tag{7.47}$$

where the structure of the covariant derivative $D_\sigma$ is defined by (7.32). The ordinary derivative $\partial_6$ remains ordinary because it is invariant. Similarly, the displacement operator form of $\hat{R}$, (3.99), is unchanged because $\partial_5$ is invariant. In curved space the probability and charge amplitude equations become

$$\left(\gamma^{\mu} i\hbar D_{\mu} + \gamma^{5}\hat{p}_{5} - I\hat{p}^{6}\right)\chi = 0\,, \tag{7.48}$$

and

$$\left(\gamma^{\mu} i\hbar D_{\mu} - \gamma^{5}\hat{p}_{5} - I\hat{p}^{6}\right)\chi_{Q} = 0\,. \tag{7.49}$$

*7.2.2 The equations of Klein-Gordon and Maxwell-Proca.* In flat space the operator $\hat{R}$ for both scalar and vector boson descriptor X is given by (5.27). This form of the operator contains X itself, which in turn consists of an expansion in the Dirac-Clifford matrices, with the field tensors as coefficients of the expansion. Thus to obtain $\hat{R}$ in curved space we must replace the ordinary derivative in (5.27) by the covariant derivative $\nabla_{\sigma}$ defined by Eqs. (7.27)-(7.31).

$$\hat{R} = \exp\left[\gamma^{5}\xi_{5}\left(-\gamma^{\mu}\nabla_{\mu} - ik^{6} + \Sigma \mathrm{X}^{-1}\right)\right]. \tag{7.50}$$

The general field equation (5.21) now becomes in curved space

$$\left(\gamma^{\mu}\nabla_{\mu} - i\gamma^{5}k_{5} + ik^{6}\right)\mathrm{X} = \Sigma\,. \tag{7.51}$$

With $\nabla_{\sigma}$ replacing $\partial_{\sigma}$ the Dirac-Clifford matrices in X behave like constant matrices, as if in flat space. Putting, Eqs. (5.31, 2), (5.45, 6) and (5.65, 6) into (7.51), we obtain in curved space, respectively the Klein-Gordon and Maxwell-Proca equations

$$\left.\begin{aligned} &\left(g^{\rho\sigma}\nabla_{\rho}\nabla_{\sigma} + k_{H}^{2}\right)\phi = s \\ &\left(g^{ab}\nabla_{a}\nabla_{b} + k_{Z}^{2}\right)Z^{a} = s^{a} \\ &\left(g^{ab}\nabla_{a}\nabla_{b} + k_{W}^{2}\right)W^{a} = l^{a} \end{aligned}\right\} \tag{7.52a}$$

with *a*, *b* = 0, 1, 2, 3, 4, 5. Because scalar and vector fields are spin-free, $\nabla_{\sigma}$ in these equations denotes the standard (semi-colon) covariant derivative. And so they may also be written

$$\left.\begin{aligned}&\left(\partial_\mu\partial^\mu+k_H^2\right)\phi=s\\&g^{\rho\sigma}Z^\mu{}_{;\rho;\sigma}+k_Z^2Z^\mu=s^\mu\\&\left(\partial_\mu\partial^\mu+k_Z^2\right)Z^5=s^5\\&g^{\rho\sigma}W^\mu{}_{;\rho;\sigma}+k_W^2W^\mu=l^\mu\\&\left(\partial_\mu\partial^\mu+k_W^2\right)W^5=l^5\end{aligned}\right\}. \tag{7.52b}$$

The scalar and pseudoscalar equations remain unchanged from their flat-space forms due to (7.16).

*7.2.3 Maxwell's Equations.* In flat space the differential form of operator $\hat{R}$ for the Maxwell field X is given by (5.27), with $k^6=0$. Hence from (7.50),

$$\hat{R}=\exp\left[\gamma^5\xi_5\left(-\gamma^\mu\nabla_\mu+\Sigma\mathrm{X}^{-1}\right)\right]. \tag{7.53}$$

The general field equation (6.1) now becomes in curved space

$$\gamma^\mu\nabla_\mu\mathrm{X}\left(x^\mu\right)=\Sigma\left(x^\mu\right). \tag{7.54}$$

Inserting (6.2) and (6.3) into (7.54) we find for Maxwell's equations in curved space:

Set I (8 component equations)

$$F^{\mu\nu}{}_{;\mu}+(g^{\mu\nu}\phi)_{;\mu}=\mu_0 j^\nu \tag{7.55}$$

$$\varepsilon^{\sigma\lambda\mu\nu}F_{\mu\nu;\lambda}=0 \tag{7.56}$$

Set II (7 component equations)

$$F^{5\mu}{}_{;\mu}=-\mu_5 j^5 \tag{7.57}$$

$$\varepsilon^{\mu\nu\rho\sigma}F_{5\nu;\mu}=0 \tag{7.58}$$

We notice first of all that both of the homogeneous equations, (7.56) and (7.58), are already covariant in their flat-space forms. In the case of (7.56), this is shown by substituting for the $F_{\alpha\beta;\gamma}$ using (7.18); we find that all terms involving the Christoffel symbols cancel, giving finally the flat-space form (6.8). For (7.58), substituting for the vector $F_{5\alpha}=-F_{\alpha 5}$ using (7.17) we once again find that the terms involving Christoffel symbols cancel, giving finally the flat-space form (6.10).

The second inhomogeneous equation, (7.57), has the expected covariant form and requires no special comment. We might, however, make two points about (7.55). First, the term involving the constant $\phi$ still vanishes in curved space. Consequently the first inhomogeneous Maxwell equation retains its usual covariant form. Second, with an eye to our work in the next chapter, we shall now show that charge is still conserved in curved space;[7] i.e., that $j^{\nu}_{\ ;\nu}=0$. Since $F^{\mu\nu}$ is antisymmetric we have from (7.55)

$$\begin{aligned}
2\mu_0 j^{\nu}_{\ ;\nu} &= F^{\mu\nu}_{\ \ ;\mu;\nu} - F^{\mu\nu}_{\ \ ;\nu;\mu} \\
&= F^{\nu\alpha} R^{\mu}_{\ \alpha\mu\nu} + F^{\mu\alpha} R^{\nu}_{\ \alpha\nu\mu} \\
&= 2F^{\mu\alpha} R^{\nu}_{\ \alpha\nu\mu} \qquad (7.59) \\
&= 2F^{\alpha\mu} R_{\alpha\mu} \\
&= 0
\end{aligned}$$

where we have made use of (7.40), (7.45) and the symmetry properties of the Riemann-Christoffel, Ricci and Maxwell tensors. Q. E. D. Needless to say, if charge had turned out *not* to be conserved, one would have reason to doubt the validity of the Maxwell theory, at least as it operates in curved space.

## Notes and references

[1] A. Einstein, The Foundation of the General Theory of Relativity, in *The Principle of Relativity* (Dover), pp. 111-164.

[2] Among the myriad textbook treatments of Riemannian geometry for physicists, the ones we consult most often [possibly because all three use the metrical signature (+−−−)] are: P .A. M.

Dirac, *General Theory of Relativity* (John Wiley & Sons, New York, 1975); Hans C. Ohanian, *Gravitation and Spacetime* (W. W. Norton & Co., New York, 1976); H. A. Atwater, *Introduction to General Relativity* (Pergamon Press, Oxford,1974).

[3] D. R. Brill and J. A. Wheeler, "Interaction of Neutrinos and Gravitational Fields," Rev. Mod. Phys. **29**, 465 (1957).

[4] D. A. McMahon, P. M. Alsing and P. Embid, "The Dirac equation in Rindler space: A pedagogical introduction," arXiv:gr-qc/0601010.

[5] P .A. M. Dirac, *ibid*, p. 21.

[6] P .A. M. Dirac, *ibid*, p. 20, 23.

[7] P .A. M. Dirac, *ibid*, p. 38, 39, 43

# Chapter 8

# Geometrodynamics in 4+2 dimensions

We have now studied in some detail the self-imaging properties of spinor, scalar, vector and rank-2 tensor fields. Moreover we have just upgraded their equations of motion, writing them in covariant form. As it happens, the intrinsic attributes of those self-imaging fields — their mass, charge and spin values—are independent of spacetime curvature. The fields and their associated particles are, one might say, actors on a stage—the background stage of spacetime, whether flat or curved. In this and the next chapter we take up the study of *geometrodynamics*— the interaction of matter and energy with spacetime itself. This means that we will no longer simply be describing the activites of actors on a stage. Rather we shall be describing the behavior of matter and spacetime considered as a dynamic whole. Specifically we shall be looking to uncover physical laws dependent on the properties of Riemanian spacetime, including curvature.

In the literature "geometrodynamics"—a term invented by J. A. Wheeler— generally signifies a theory of gravitation, one specifically emphasizing the geometric origin of gravitational attraction, and encompassing at the very least Einstein's general relativity. Here it means that and more. For the interactions we are concerned with are supposed to take place in 4+2 spacetime dimensions. And in that context whatever interactions there may be should proceed from the Law of Laws.

Now the Law of Laws is a machine, generating propagation laws in accordance with the form of the propagator $\hat{R}$ and behavior of the descriptor $\chi$ under spacetime transformations. Thus far it has given us laws of propagation for the series of descriptor types enumerated above, ending with the rank-2 tensor wave function. We propose now to ask the Law of Laws to give us the propagation laws for the next two descriptors in the series: rank-3 and rank-4 tensor wave functions. These new descriptors are formally similar to the one for bosons, as displayed in the opening paragraph of **Chapter 5**, the difference being that the expansion coefficients $\lambda$ now carry additional spacetime indices:

$$\text{Rank 3:} \qquad \mathrm{X}_a = I\lambda_a + \sum_{c,d=0,1,2,3,5,6} \lambda_a^{\ cd} M_{cd}, \qquad a = 0,1,2,3,5$$

$$\text{Rank 4:} \qquad \mathrm{X}_{ab} = I\lambda_{ab} + \sum_{c,d=0,1,2,3,5,6} \lambda_{ab}^{\ \ cd} M_{cd}, \qquad a,b = 0,1,2,3,5,$$

where the $M_{cd}$ are the Dirac matrices displayed in (4.8) and $I$ is the unit matrix. These are the wave functions required to describe fully the interaction between matter and spacetime. Of course one could not know this in advance. This is not model-building. Rather we are following the dictates of a formal construct, the Principle of True Representation. To gauge the relevance of these descriptors to the physical world one first needs to write down their equations of motion, then compare what the equations say with what we already know about the world. Taking that approach, we find as follows:

(1) The equations of motion for the rank-4 coefficients $\lambda^{cd}_{ab}$ reproduce the gravitational field equations of Einstein. But in addition they predict a repulsive *fifth* force, one that, 14 B years ago, may well have initiated the Big Bang and that, today, can give to the black-hole 'singularity' a finite radius.
(2) We do find a force-mediating particle for the fifth force—a particle of spin-1 we shall call the *riemann*—but no such particle for ordinary gravitation. According to the Principle of True Representation, there exists no such thing as a spin-2 graviton.
(3) The equations of motion for the rank-3 coefficients $\lambda^{cd}$ deal *not* with ordinary gravitating matter and energy, but wfth the non-gravitating, quantum-theoretic zero-point vacuum energy density, $u_{\mathrm{QFT}}$. These equations predict—or rather are consistent with—a bounded distribution of vacuum energy, i.e., a bounded universe, and in addition predict a repulsive *sixth* force, one that accounts for the accelerating expansion of the cosmos.
(4) There is no force-mediating particle for the sixth force.

It turns out that the interpretation of the rank-3 field depends on our findings for the rank-4 field. Therefore our study of geometrodynamics begins in the present chapter with a discussion of the rank-4 equation of motion. Following that, in **Chapter 9**, we apply the theory developed here to the rank-3 equation of motion and the problem of cosmic expansion.

## 8.1 Self-imaging spacetime

It is easy to imagine the propagation in six dimensions of self-imaging fields of spinor, scalar and vector form. By definition these fields propagate without dispersion in the direction of the $x^5$ axis; in other words, as depicted in **Fig. 3.5**, the field structure is the same in all hyperplanes $M^4$ perpendicular to $x^5$. But as we saw in the previous chapter, the phenomenon of self-imaging in $M^{4+2}$ occurs whether the background geometry is flat or curved. Evidently, then, it is not just the physical field $\chi$ that propagates; all departures from flatness in the background propagate too, and do so without dispersion. In this way one can see that spacetime, too, like the physical fields that propagate within it, is a self-imaging process

To show this formally, we simply add two spacetime indices to the field and source functions of massless Eq. (7.53). The Law of Laws (5.24) then reads

$$\hat{R}\mathrm{X}_{\alpha\beta} = r\mathrm{X}_{\alpha\beta}, \tag{8.1}$$

with, from (7.53), projection operator

$$\hat{R} = \exp\left[\gamma^5 \xi_5 \left(-\gamma^\mu \nabla_\mu + \Sigma_{\alpha\beta} \mathrm{X}_{\alpha\beta}^{-1}\right)\right]. \tag{8.2}$$

Note that the indices $\alpha$ and $\beta$ in this expression are *not* summed; there are as many such operators as there are component equations of (8.1). Operator $\hat{R}$ generates the field equations

$$\gamma^\mu \nabla_\mu \mathrm{X}_{\alpha\beta}\left(x^\mu\right) = \Sigma_{\alpha\beta}\left(x^\mu\right), \tag{8.3}$$

where, by adding two indices to the terms of (6.2) and (6.3), we have for *massless* fields,

$$\mathrm{X}_{\alpha\beta} = I\phi_{\alpha\beta} + \frac{i}{2}\sigma^{\mu\nu}F_{\alpha\beta\mu\nu} + i\gamma^5\gamma^\nu F_{\alpha\beta5\nu} \tag{8.4}$$

and

$$\Sigma_{\alpha\beta} = \tilde{\mu}\gamma^\mu j_{\alpha\beta\mu} + i\tilde{\mu}_5\gamma^5 j_{\alpha\beta5}\,. \tag{8.5}$$

and where $\tilde{\mu}$ and $\tilde{\mu}_5$ are positive constant analogs of magnetic permeability. Inserting these expansions into (8.3) we obtain the following two sets of field equations:

Set I
$$F_{\alpha\beta\ ;\mu}^{\ \ \mu\nu} + (g^{\mu\nu}\phi_{\alpha\beta})_{;\mu} = \tilde{\mu} j_{\alpha\beta}^{\ \ \nu} \tag{8.6}$$

$$\varepsilon^{\sigma\lambda\mu\nu}F_{\alpha\beta\mu\nu;\lambda} = 0 \tag{8.7}$$

Set II
$$F_{\alpha\beta\ ;\mu}^{\ \ 5\mu} = -\tilde{\mu}_5 j_{\alpha\beta}^{\ \ 5} \tag{8.8}$$

$$\varepsilon^{\mu\nu\rho\sigma}F_{\alpha\beta5\nu;\mu} = 0 \tag{8.9}$$

To make sense of these equations we first need to identify the rank-4 field tensor, $F_{\alpha\beta\mu\nu}$. The key here is to be found in the structure of the homogeneous equation (8.7), which can also be written

$$F_{\alpha\beta\mu\nu;\lambda} + F_{\alpha\beta\nu\lambda;\mu} + F_{\alpha\beta\lambda\mu;\nu} = 0\,. \tag{8.10}$$

Suppose we assume that $F_{\alpha\beta\mu\nu}$ is asymmetrical between $\alpha$ and $\beta$ , as it is between $\mu$ and $\nu$ , and further that $F_{\alpha\beta\mu\nu}$ is proportional the Riemann-Christoffel curvature tensor (7.36):

$$F_{\alpha\beta\mu\nu} = AR_{\alpha\beta\mu\nu}\,, \tag{8.11}$$

where $A$ is the constant of proportionality. With this assumption it appears that we may have come to a valid formulation of gravity. For the Bianchi relation (8.10) is known to lead directly to (or is at least consistent with) the Einstein field equations.[1]

But what about (8.6)? The main question is whether the rank-3 current $j_{\alpha\beta}^{\ \ \nu}$ is conserved. Thus we proceed to evaluate the covariant divergence, $j_{\alpha\beta\ ;\nu}^{\ \ \nu}$. Let us start with the first term on the left. Making use of the antisymmetry between $\mu$ and $\nu$, we have

$$\begin{aligned} 2F_{\alpha\beta\ \ ;\mu;\nu}^{\ \ \mu\nu} &= F_{\alpha\beta\ \ ;\mu;\nu}^{\ \ \mu\nu} - F_{\alpha\beta\ \ ;\nu;\mu}^{\ \ \mu\nu} \\ &= F_{\lambda\beta\mu\nu}R^{\lambda\ \mu\nu}_{\ \alpha} + F_{\alpha\lambda\mu\nu}R^{\lambda\ \mu\nu}_{\ \beta} + F_{\alpha\beta\lambda\nu}R^{\lambda\ \mu\nu}_{\ \mu} + F_{\alpha\beta\mu\lambda}R^{\lambda\ \mu\nu}_{\ \nu} \qquad (8.12) \\ &= A\left(R^{\lambda\ \mu\nu}_{\ \beta}R_{\lambda\alpha\mu\nu} + R_{\alpha\lambda\mu\nu}R^{\lambda\ \mu\nu}_{\ \beta} + 2R_{\alpha\beta\lambda\nu}R^{\lambda\nu}\right) \end{aligned}$$

where we have used (7.42) and (7.45). The first two terms within the parentheses of the last line cancel each other by antisymmetry between $\alpha$ and $\lambda$ ; and the last term vanishes due to the antisymmetry of the curvature tensor between $\lambda$ and $\nu$, combined with the *symmetry* of the Ricci tensor between those same indices. So the covariant divergence of the first term on the l. h. s. of (8.6) does indeed vanish. Consequently, to conserve the current one need only require that the divergence of the term involving $\phi_{\alpha\beta}$ should vanish as well. The divergence of that term is $g^{\mu\nu}\phi_{\alpha\beta;\mu;\nu}$, which can be shown to vanish if and only if $\phi_{\alpha\beta}$ itself vanishes. Now $\phi_{\alpha\beta}$ may reasonably be interpreted as the background Maxwell field of the vacuum, a field which as far as we know does not exist. Thus if we put

$$\phi_{\alpha\beta} = 0 \qquad (8.13)$$

then

$$j_{\alpha\beta\ ;\nu}^{\ \ \nu} = 0, \qquad (8.14)$$

and the current $j_{\alpha\beta}^{\ \ \nu}$ is conserved. This is a remarkable and perhaps unexpected result, given that covariant derivatives generally do not commute. It is a result entirely attributable to the algebraic (symmetry) properties of the curvature and Ricci tensors. It suggests that one now has in hand a plausible theory of geometrodynamics, one moreover that, in form, parallels closely Maxwell electrodynamics. Like the rank-2 Maxwell tensor, the rank-4 curvature tensor is self-imaging. The curvature tensor is, in other words, no mere mathematical tool; rather, it enjoys the same degree of physical reality as the Maxwell $F_{\mu\nu}$.

However, let us now look at the second set of field equations, (8.8) and (8.9). These are analogous to (7.57) and (7.58) of the Maxwell theory. They contain the curvature tensor $F_{\alpha\beta 5\mu}$, now extended to five dimensions. From (7.36)

$$R_{\alpha\beta 5\mu} = \frac{1}{2}(g_{\alpha\mu,\beta 5} - g_{\beta\mu,\alpha 5} - g_{\alpha 5,\beta\mu} + g_{\beta 5,\alpha\mu}) + \Gamma_{\lambda\alpha\mu}\Gamma^{\lambda}{}_{\beta 5} - \Gamma_{\lambda\alpha 5}\Gamma^{\lambda}{}_{\beta\mu} \quad . \tag{8.15}$$

But by (7.1), (7.2) and (7.15), each term on the r. h. s. of this expression vanishes identically. Then, since $F_{\alpha\beta\mu 5}$ vanishes, the current $j_{\alpha\beta 5}$ must also vanish. Consequently, in contrast to the Maxwell case, there is no second set of field equations. In the gravitational case one gets the first set only.

For the record, let us rewrite that first set now in a form that raises the first index of the field function:

$$F^{\mu}{}_{\nu\alpha\beta;\mu} = \tilde{\mu} j_{\nu\alpha\beta} \tag{8.16}$$

$$F^{\mu}{}_{\nu\beta\gamma;\alpha} + F^{\mu}{}_{\nu\gamma\alpha;\beta} + F^{\mu}{}_{\nu\alpha\beta;\gamma} = 0 \tag{8.17}$$

Note that in passing from (8.10) to (8.17) we have, for a later convenience, relabeled indices $\alpha,\beta \to \mu,\nu$ and $\lambda,\mu,\nu \to \alpha,\beta,\gamma$. These are the field equations of Riemann-Cristoffel (R-C).

## 8.2 The field equations of Einstein

The Einstein field equations are embedded in the Bianchi relation (8.17). To show this, first contract with respect to $\mu$ and $\gamma$. This gives, upon canceling the proportionality factor, $A$, a second Bianchi relation

$$R_{\nu\beta;\alpha} - R_{\nu\alpha;\beta} + R^{\mu}{}_{\nu\alpha\beta;\mu} = 0 . \tag{8.18}$$

where $R_{\mu\nu}$ is the Ricci tensor, (7.45). Contracting again, this time with respect to $\nu$ and $\beta$ , we obtain,

$$\left(R^{\mu}{}_{\alpha} - \frac{1}{2}g^{\mu}{}_{\alpha}R\right)_{;\mu} = 0\,, \tag{8.19}$$

where $R$ is the curvature scalar, (7.46). Integration of (8.19) yields the Einstein field equations

$$R_{\mu\nu} - \frac{1}{2}g_{\mu\nu}R = -\kappa T_{\mu\nu} - \Lambda g_{\mu\nu}\,, \tag{8.20}$$

where is $T_{\mu\nu}$ is the divergenceless energy-momentum tensor, $g_{\mu\nu}$ is the (divergenceless) fundamental metric tensor, and $\kappa$ and $\Lambda$ are constants of proportionality. Note that both $T_{\mu\nu}$ and $g_{\mu\nu}$ arise as if they were constants of integration; i.e., both have zero covariant divergence.

Now neither $\kappa$ nor $\Lambda$ on the r. h. s. of (8.20) are given theoretically. $\Lambda$ is of course the famous cosmological constant of Einstein and its value can only be determined by observation of features of the expanding universe. Constant $\kappa$ is far more accessible. To find it one makes use of an alternate form of the field equations. Contraction of $\mu$ and $\nu$ in (8.20) gives

$$R = \kappa T + 4\Lambda\,, \tag{8.21}$$

where $T = T^{\mu}{}_{\mu}$ is the *matter scalar*. Substituting for $R$ in (8.20) using (8.21) we obtain Einstein's equations in the form:

$$R_{\mu\nu} = -\kappa\left(T_{\mu\nu} - \frac{1}{2}g_{\mu\nu}T\right) + g_{\mu\nu}\Lambda\,. \tag{8.22}$$

Then, in the weak field approximation, and neglecting $\Lambda$, (8.22) reduces to the Poisson equation, from which one readily makes the identification[2]

$$\kappa = \frac{8\pi G}{c^4}\,, \tag{8.23}$$

where $G$ is the gravitational constant. In this notation $T_{\mu\nu}$ has the units of energy density and $R$ that of $[\,\text{Length}^{-2}\,]$.

From (8.22) we see that the cosmological term implies a background energy density

$$u_\Lambda = \frac{\Lambda}{\kappa}. \tag{8.24}$$

Now in Chapter 6 we found an expression for $u_\Lambda$ in terms of the universal constant, $\phi$, namely (6.75). Using that expression in (8.24) we obtain

$$\Lambda = \frac{\kappa}{2\mu_0}\phi^2. \tag{8.25}$$

The cosmological background energy of Einstein thus appears ultimately to derive from the Maxwell tensor's scalar companion, $\phi$, a product of the pseudotemporal dimension, $x^6$ (because $I = \gamma^6$; see *Sec. 4.2.2*). If that is so, then we seem to have found a deep and unexpected connection between Maxwell electrodynamics and the geometrodynamics of Einstein, a connection ultimately attributable to the existence of an infinite extra dimension.

## 8.3 The Weyl tensor

As shown above, the projection theory places the curvature tensor $R_{\mu\nu\alpha\beta}$ on the same plane of physical reality as that occupied by the Maxwell tensor, $F_{\mu\nu}$. The curvature tensor emerges as *the* fundamental entity—the starting point of geometrodynamics. Now the curvature tensor decomposes into two parts: one dependent on the Ricci tensor and its trace (the curvature scalar), the other a rank-4 tensor, $C_{\mu\nu\alpha\beta}$, known as the Weyl tensor.[3] Typically one says that the Ricci part of the curvature tensor measures the energy density of a volume of particles (cf. the Einstein field equations), whereas the Weyl part measures its tidal distortion. Thus freed of an energetic component, the Weyl tensor describes the unique gravitational distortion of spacetime.[4] On this ground the Weyl tensor might be considered more fundamental than the curvature tensor.[5] But to be truly fundamental—that is to say, ontologically real—the Weyl tensor must satisfy field equations of the form (8.16) and (8.17). Let us check to see if this is so.

The Weyl tensor is given by[3]

$$C_{\mu\nu\alpha\beta} = R_{\mu\nu\alpha\beta} - \frac{1}{2}\left(g_{\mu\beta}R_{\nu\alpha} - g_{\mu\alpha}R_{\nu\beta} + g_{\nu\alpha}R_{\mu\beta} - g_{\nu\beta}R_{\mu\alpha}\right) \\ - \frac{1}{6}\left(g_{\mu\alpha}g_{\nu\beta} - g_{\mu\beta}g_{\nu\alpha}\right)R \quad . \tag{8.26}$$

From its definition we see that the Weyl tensor exhibits the same algebraic symmetries as the curvature tensor. In addition, it vanishes on contraction of any two of its indices. For that reason the Weyl tensor is often described as curvature with Ricci removed. The covariant derivative of $C_{\mu\nu\alpha\beta}$—easily obtained through the use of (8.18) and (8.19)—can be expressed in the form

$$C^{\mu}{}_{\nu\alpha\beta;\mu} = J_{\nu\alpha\beta} \tag{8.27}$$

where the current

$$J_{\nu\alpha\beta} = \frac{1}{2}\left[R_{\nu\alpha;\beta} - R_{\nu\beta;\alpha} + \frac{1}{6}\left(g_{\nu\beta}R_{;\alpha} - g_{\nu\alpha}R_{;\beta}\right)\right]. \tag{8.28}$$

(The quantity in square brackets is known as the Cotton tensor.)

Now let us check to see whether Weyl satisfies the Bianci relation (8.17). We obtain, after much manipulation,

$$C^{\mu}{}_{\nu\beta\gamma;\alpha} + C^{\mu}{}_{\nu\gamma\alpha;\beta} + C^{\mu}{}_{\nu\alpha\beta;\gamma} = g_{\nu\beta}J^{\mu}{}_{\gamma\alpha} - g^{\mu}{}_{\beta}\,J_{\nu\gamma\alpha} \\ + g_{\nu\gamma}J^{\mu}{}_{\alpha\beta} - g^{\mu}{}_{\gamma}J_{\nu\alpha\beta} \quad . \\ + g_{\nu\alpha}J^{\mu}{}_{\beta\gamma} - g^{\mu}{}_{\alpha}J_{\nu\beta\gamma} \tag{8.29}$$

None of the terms on the r. h. s. vanishes, except in empty space, where Ricci itself vanishes. Thus, the Weyl tensor fails the first test of a fundamental theory of gravitation: the Bianci relation.

Now let us see about the current, $J_{\nu\alpha\beta}$. Is it conserved? Making use of the asymmetry between $\mu,\nu$ in (8.26), we may write (after a page of calculation)

$$\begin{aligned}
2J^{\nu}{}_{\alpha\beta;\nu} &= C^{\mu\nu}{}_{\alpha\beta;\mu;\nu} - C^{\mu\nu}{}_{\alpha\beta;\nu;\mu} \\
&= C_{\mu\nu\lambda\beta}R^{\lambda}{}_{\alpha}{}^{\mu\nu} + C_{\mu\nu\alpha\lambda}R^{\lambda}{}_{\beta}{}^{\mu\nu} + C_{\lambda\nu\alpha\beta}R^{\lambda}{}_{\mu}{}^{\mu\nu} + C_{\mu\lambda\alpha\beta}R^{\lambda}{}_{\nu}{}^{\mu\nu} \\
&= C_{\mu\nu\lambda\beta}R^{\lambda}{}_{\alpha}{}^{\mu\nu} - C_{\mu\nu\lambda\alpha}R^{\lambda}{}_{\beta}{}^{\mu\nu} + 2C_{\lambda\mu\alpha\beta}R^{\lambda\mu} \\
&= R^{\nu}{}_{\lambda}\left(R^{\lambda}{}_{\alpha\beta\nu} - R^{\lambda}{}_{\beta\alpha\nu}\right)
\end{aligned} \quad , \tag{8.30}$$

where once again we have made use of (7.42). Thus the current $J_{\nu\alpha\beta}$ is *not* conserved, except in empty space, where the curvature and Weyl tensors become equal. We conclude that the Weyl tensor cannot stand on its own as the basis of a fundamental theory of geometrodynamics. This does not mean that it is not useful—as a measure of tidal distortion, for example. It means only that the Weyl tensor does not enjoy the same level of physical reality as the tensors of Maxwell and Riemann-Cristoffel.

## 8.4 The Riemann-Cristoffel force and stress-energy tensor

The Einstein field equations, (8.20) or (8.22), describe (among other things) the gravitational force of attraction between massive bodies. In addition to the known force of attraction, our field equations (8.16) and (8.17) yield a second force—a fifth force for nature as a whole—one analogous to the Lorentz force of electrodynamics. We call it the Riemann-Christoffel (R-C) force and write it in the Lorenzian form

$$K^{\mu} = F^{\mu}{}_{\nu\alpha\beta}\, j^{\nu\alpha\beta} \,. \tag{8.31}$$

It may also be written, utilizing (8.16),

$$\begin{aligned}
K^{\mu} &= \frac{1}{\tilde{\mu}} F^{\mu}{}_{\nu\alpha\beta} F^{\lambda\nu\alpha\beta}{}_{;\lambda} \\
&= \tau^{\mu\nu}{}_{;\nu}
\end{aligned} \quad , \tag{8.32}$$

where $\tau^{\mu\nu}$ is the symmetrical stress-energy tensor:

$$\begin{aligned}\tau^{\mu\nu} &= \frac{1}{\tilde{\mu}}\left(-F^{\mu\lambda\alpha\beta}F^{\nu}{}_{\lambda\alpha\beta} + \frac{1}{4}g^{\mu\nu}F^{\sigma\tau\alpha\beta}F_{\sigma\tau\alpha\beta}\right) \\ &= \frac{A^2}{\tilde{\mu}}\left(-R^{\mu\lambda\alpha\beta}R^{\nu}{}_{\lambda\alpha\beta} + \frac{1}{4}g^{\mu\nu}R^{\sigma\tau\alpha\beta}R_{\sigma\tau\alpha\beta}\right)\end{aligned}, \tag{8.33}$$

which is in fact a contracted form (with raised indices) of the symmetric Bel-Robinson tensor $\tau_{\mu\nu\alpha\beta}$.[6] The similarity of $\tau^{\mu\nu}$ to the energy-momentum tensor (6.29) of electrodynamics is obvious. Now from the Einstein equation (8.20) we know that $\kappa\tau^{\mu\nu}$ has units of $[\text{Length}^{-2}]$. Thus to balance the units on the two sides of (8.33) we define

$$\frac{\kappa A^2}{\tilde{\mu}} \equiv \ell^2, \tag{8.34}$$

where $\ell$ is a new fundamental constant—on a par with $\hbar$, $c$ or $m_e$—with the dimensions of [Length]. If we further define

$$\tilde{\mu} = \kappa c^2 \tag{8.35}$$

then from (8.34)

$$A = \ell c. \tag{8.36}$$

Comparing (8.16) and (8.18) we may write for the conserved current

$$j_{\nu\alpha\beta} = \frac{A}{\tilde{\mu}}R^{\mu}{}_{\nu\alpha\beta;\mu} = \frac{A}{\tilde{\mu}}(R_{\nu\alpha;\beta} - R_{\nu\beta;\alpha}). \tag{8.37}$$

Then inserting into (8.37) the second form of the Einstein equations (8.22), and making use of (8.35) and (8.36), we obtain $j_{\nu\alpha\beta}$ in terms of gradients of the energy-momentum tensor and matter scalar:

$$j_{\nu\alpha\beta} = \frac{\ell}{c}\left[T_{\nu\beta;\alpha} - T_{\nu\alpha;\beta} - \frac{1}{2}\left(g_{\nu\beta}T_{;\alpha} - g_{\nu\alpha}T_{;\beta}\right)\right]. \tag{8.38}$$

As a result of definition (8.35), $j_{\nu\alpha\beta}$ has properly the units of mass current density: [mass × volume$^{-1}$× velocity]. According to (8.37) $j_{\nu\alpha\beta}$ depends functionally on the covariant gradients of the Ricci tensor. The R-C force (8.31), if it exists, is an edge effect.

Using (8.35) and (8.36) in (8.33) we have for the stress-energy tensor

$$\tau^{\mu\nu} = \frac{\ell^2}{\kappa}\left( -R^{\mu\lambda\alpha\beta} R^{\nu}{}_{\lambda\alpha\beta} + \frac{1}{4} g^{\mu\nu} R^{\sigma\tau\alpha\beta} R_{\sigma\tau\alpha\beta} \right). \tag{8.39}$$

As we shall soon see, unlike the Maxwell field $F^{\mu\nu}$, which is known to gravitate,[7] the Riemann-Cristoffel field $F^{\mu\nu}{}_{\alpha\beta}$ does not gravitate. It thus plays no part in general relativity as such; rather it is a separate effect. As we can see from (8.32) and (8.39), the strength of the R-C force is quadratically dependent on the constant length $\ell$. We do not yet know value of that constant and to determine it is going to require substantially more effort than is needed to determine the constant $\kappa$ appearing in the Einstein equations. See **Sec. 8.9** below.

## 8.5 Potential formulation of the field equations

To calculate the force density (8.31) we need to know the field strengths. These in turn are to be calculated from the field equation (8.16). Ideally one might hope to get the $F^{\mu\nu}{}_{\alpha\beta}$ by differentiating potential functions just as we do in electrodynamics. That is, we should like to define in terms of a rank-3 potential $A_{\alpha\beta\mu}$

$$F_{\mu\nu\alpha\beta} = F_{\alpha\beta\mu\nu} = A_{\alpha\beta\nu;\mu} - A_{\alpha\beta\mu;\nu}. \tag{8.40}$$

But this requires that (8.40) satisfy the Bianchi relation (8.17) identically. Unfortunately, it cannot do so, because the covariant derivatives do not commute. However, for weak fields—that is to say, in the Newtonian approximation—the covariant derivatives of (8.17) become ordinary ones, and the approximate form

$$F_{\alpha\beta\mu\nu} = A_{\alpha\beta\nu,\mu} - A_{\alpha\beta\mu,\nu} \tag{8.41}$$

then satisfies the Bianchi relation, just as in electrodynamics. The legitimacy of this approach to defining the curvature tensor is easily demonstrated. For weak gravitational fields the metric may be written in the form

$$g_{\mu\nu} = \eta_{\mu\nu} + h_{\mu\nu}, \tag{8.42}$$

where $h_{\mu\nu}$ is a small perturbation on the flat-space metric $\eta_{\mu\nu}$, and where only terms first order in $h_{\mu\nu}$ are retained in the analysis. Substituting this for the $g_{\mu\nu}$ in the curvature tensor (7.35) we obtain for weak fields

$$R_{\alpha\beta\mu\nu} = \frac{1}{2}\left(h_{\alpha\nu,\beta} - h_{\beta\nu,\alpha}\right)_{,\mu} - \frac{1}{2}\left(h_{\alpha\mu,\beta} - h_{\beta\mu,\alpha}\right)_{,\nu} \tag{8.43}$$

Thus, multiplying by the constant $A$ defined by (8.36) we have

$$A_{\alpha\beta\mu} = \frac{\ell c}{2}\left(h_{\alpha\mu,\beta} - h_{\beta\mu,\alpha}\right). \tag{8.44}$$

Just as the current (8.38) is given by (covariant) gradients of the Ricci tensor, the potential (8.44) is given by (ordinary) gradients of the (linearized) metric.

Now the purpose of displaying $A_{\alpha\beta\mu}$ in the form (8.44) is really two-fold. First, it shows that, for weak fields, the existence of a potential is already implicit in the curvature tensor. Second, it offers a straightforward way of finding the weak-field metric $\eta_{\mu\nu} + g_{\mu\nu}$ for any given energy-momentum distribution, $T^{\mu\nu}$; where by "straightforward" is meant not having to solve directly the Einstein field equations. And of course it offers a way of checking the validity of the potential method: the metric obtained from it should agree with that obtained by solving the equations of general relativity.

The potential $A_{\alpha\beta\mu}$ is calculated just as in electrodynamics. First substitute (8.41) into field equation (8.16), assuming ordinary derivatives. This gives

$$\partial_\mu \partial^\mu A_{\alpha\beta}^{\ \ \nu} - \partial^\nu \partial_\mu A_{\alpha\beta}^{\ \ \mu} = \tilde{\mu} j_{\alpha\beta}^{\ \ \nu}. \tag{8.45}$$

(Note that for consistency of notation we are now writing "$j_{\alpha\beta}^{\ \ \nu}$" rather than "$j^{\nu}_{\ \alpha\beta}$".) Next, we assume the Lorentz gauge condition

$$\partial_\mu A_{\alpha\beta}^{\ \ \mu} = 0 . \tag{8.46}$$

Eq. (8.45) then reduces to

$$\partial_\mu \partial^\mu A_{\alpha\beta}^{\ \ \nu} = \tilde{\mu} j_{\alpha\beta}^{\ \ \nu} , \tag{8.47}$$

the solution to which is the retarded potential

$$A_{\alpha\beta}^{\ \ \nu}(\mathbf{x},t) = \frac{\tilde{\mu}}{4\pi} \int \frac{j_{\alpha\beta}^{\ \ \nu}(\boldsymbol{\xi},t^*)}{r} d^3\boldsymbol{\xi} , \tag{8.48}$$

where current $j_{\alpha\beta}^{\ \ \nu}$ is given by (8.38), constant $\tilde{\mu}$ by (8.35), $t^* = t - r/c$ and $r = |\mathbf{x} - \boldsymbol{\xi}|$.

All of this looks perfectly familiar. To complete the analogy with electrodynamics [see (6.4)] we can define the rank-3 $E$ and $B$ fields:

$$\left.\begin{aligned} F_{\alpha\beta}^{\ \ 0j} &= -\frac{E_{\alpha\beta}^{\ \ j}}{c} \\ F_{\alpha\beta}^{\ \ ij} &= -\varepsilon_{ijk} B_{\alpha\beta}^{\ \ k} \end{aligned}\right\} \tag{8.49}$$

(We are now raising and lowering indices with $\eta_{\mu\nu}$, not $g_{\mu\nu}$.) In terms of the potential (8.48) we have from (8.49) and (8.41), in perfect analogy with (6.23),

$$\left.\begin{aligned} \frac{\mathbf{E}_{\alpha\beta}}{c} &= -\partial^0 \mathbf{A}_{\alpha\beta} - \nabla A_{\alpha\beta}^{\ \ 0} \\ \mathbf{B}_{\alpha\beta} &= \nabla \times \mathbf{A}_{\alpha\beta} \end{aligned}\right\} . \tag{8.50}$$

The R-C force (8.31) assumes the familiar Lorentzian form

$$\left.\begin{aligned} K^0 &= \mathbf{j}^{\alpha\beta} \cdot \frac{\mathbf{E}_{\alpha\beta}}{c} \\ \mathbf{K} &= j^{\alpha\beta}_{\ \ 0} \frac{\mathbf{E}_{\alpha\beta}}{c} + \mathbf{j}^{\alpha\beta} \times \mathbf{B}_{\alpha\beta} \end{aligned}\right\} . \tag{8.51}$$

## 8.6 Example of the potential method: spacetime metric and curvature induced by a spherical mass

*8.6.1 Potential components.* We now demonstrate the use of the potential method to find the metric and curvature tensor $R_{\alpha\beta}^{\ \ \mu\nu}$ interior and exterior to a spherical mass of constant mass density $\rho$. The case is an interesting one because it offers an easy route to the interior and exterior solutions of the Einstein equations.

First we get the current, (8.38). The energy-momentum tensor of a system of non-interacting mass particles of mass density $\rho(\mathbf{x})$ is given by

$$T^{\mu\nu} = \rho(\mathbf{x})\frac{dx^\mu}{d\tau}\frac{dx^\nu}{d\tau}, \tag{8.52}$$

where in a Lorentz frame with Lorentzian metric $\eta_{\mu\nu}$

$$d\tau = (1 - v^2 / c^2)^{1/2} dt. \tag{8.53}$$

The matter scalar $T^\mu_{\ \mu} \equiv T$ is

$$T = \rho(\mathbf{x})c^2. \tag{8.54}$$

For a stationary ($v = 0$) spherical mass distribution of radius $a$ and constant mass density $\rho$ centered at the origin, the only non-vanishing component of $T^{\mu\nu}$ is

$$T^{00} = T = \rho\sigma(\mathbf{x})c^2, \tag{8.55}$$

where $\sigma(\mathbf{x})$ is the spherical window function defined by

$$\sigma(\mathbf{x}) = \begin{cases} 1 & |\mathbf{x}| \le a \\ 0 & |\mathbf{x}| > a \end{cases} . \tag{8.56}$$

Putting (8.55) into (8.38) we find for the current components

$$\left.\begin{aligned} j_{0j}{}^{0} &= -\frac{\ell c\rho}{2}\partial_j\sigma \\ j_{ij}{}^{0} &= 0 \\ j_{0j}{}^{k} &= 0 \\ j_{ij}{}^{k} &= \frac{\ell c\rho}{2}\left(\eta^k_i\partial_j\sigma - \eta^k_j\partial_i\sigma\right) \end{aligned}\right\} \quad . \tag{8.57}$$

Inserting these currents into (8.48) and using (8.35) we obtain for the non-vanishing potential components

$$\left.\begin{aligned} A_{0j}{}^{0} &= -\frac{G\ell\rho}{c}I_j \\ A_{ij}{}^{k} &= \frac{G\ell\rho}{c}\left(\eta^k_i I_j - \eta^k_j I_i\right) \end{aligned}\right\}, \tag{8.58}$$

where $I_j$ is the integral with infinite limits

$$\begin{aligned} I_j &= \int \frac{\partial_j\sigma(\boldsymbol{\xi})}{|\mathbf{x}-\boldsymbol{\xi}|}d^3\boldsymbol{\xi} \\ &= -\int \frac{\partial}{\partial\xi^j}\left(\frac{1}{|\mathbf{x}-\boldsymbol{\xi}|}\right)\sigma(\boldsymbol{\xi})d^3\boldsymbol{\xi} \\ &= +\frac{\partial}{\partial x^j}\int \frac{\sigma(\boldsymbol{\xi})}{|\boldsymbol{x}-\boldsymbol{\xi}|}d^3\boldsymbol{\xi} \end{aligned} \tag{8.59}$$

and where we have integrated by parts. Carrying out the final integration we get

$$I_j = \begin{cases} -\dfrac{2\pi}{3}\partial_j r^2 = -\dfrac{4\pi}{3}x^j & r \le a \\ \dfrac{4\pi a^3}{3}\partial_j\left(\dfrac{1}{r}\right) = -\dfrac{4\pi}{3}\dfrac{a^3}{r^3}x^j & r > a \end{cases} \tag{8.60}$$

The potential (8.58) thus divides in two, one part interior, the other exterior to the spherical mass:

$$\left.\begin{aligned} A_{0j}^{\ 0} &= \frac{r_s \ell c}{2a^3} x^j = \frac{r_s \ell c}{4a^3} \partial_j r^2 \\ A_{ij}^{\ k} &= \frac{r_s \ell c}{2a^3}\left(\eta_j^k x^i - \eta_i^k x^j\right) = \frac{r_s \ell c}{4a^3}\left(\eta_j^k \partial_i - \eta_i^k \partial_j\right) r^2 \end{aligned}\right\} r \le a, \quad (8.61)$$

and

$$\left.\begin{aligned} A_{0j}^{\ 0} &= \frac{r_s \ell c}{2r^3} x^j = -\frac{r_s \ell c}{2} \partial_j \left(\frac{1}{r}\right) \\ A_{ij}^{\ k} &= \frac{r_s \ell c}{2r^3}\left(\eta_j^k x^i - \eta_i^k x^j\right) = -\frac{r_s \ell c}{2}\left[\eta_j^k \partial_i - \eta_i^k \partial_j\right]\left(\frac{1}{r}\right) \end{aligned}\right\} r > a, \quad (8.62)$$

where $r^2 = x^j x^j$, $r_s$ is the Schwarzschild radius

$$r_s = \frac{2mG}{c^2}, \qquad (8.63)$$

and $m$ is the mass of the sphere: $m = 4\pi a^3 \rho / 3$.

*8.6.2 The metric.* Comparing (8.61) and (8.62) with (8.44) we can easily pick off the weak-field metrical components, interior and exterior to the sphere, generated by the potential method. Their values are listed in **Table 8.1**. Appropriate constant terms have been added to the internal components to provide continuity with the external ones, which must reduce to flat-space values at infinity. Naturally we want to compare these components with the ones derived by Schwartzschild. The exact interior and exterior line elements as given by Schwartzschild are, respectively[8-10]:

$$ds^2 = \frac{1}{4}\left(3\sqrt{1-\frac{r_s}{a}} - \sqrt{1-\frac{r_s}{a}\cdot\frac{r^2}{a^2}}\right)^2 dx_0^{\,2} - \frac{dr^2}{1-\frac{r_s}{a}\cdot\frac{r^2}{a^2}} - r^2 d\theta^2 - r^2 \sin^2\theta \, d\phi^2 \tag{8.64}$$

and

$$ds^2 = \left(1-\frac{r_s}{r}\right)dx_0^{\,2} - \frac{dr^2}{1-\frac{r_s}{r}} - r^2 d\theta^2 - r^2 \sin^2\theta \, d\phi^2 . \tag{8.65}$$

The Schwartzschild metrical components, for small $r_s / a$ (weak field), are also listed in **Table 8.1.** For comparison the metrical components

**Table 8.1.** Potential method vs. Schwartzschild: Non-vanishing weak-field metrical components interior and exterior to spherical mass of radius $a$

| Solution | Component | Potential Method | Schwartzschild |
|---|---|---|---|
| Interior | $g_{00}$ | $1-\frac{r_s}{2a}\left(3-\frac{r^2}{a^2}\right)$ | $1-\frac{r_s}{2a}\left(3-\frac{r^2}{a^2}\right)$ |
| | $g_{ii}$ | $-1-\frac{r_s}{2a}\left(3-\frac{r^2}{a^2}\right)$ | $-1-\frac{r_s}{a}\cdot\frac{r^2}{a^2}$ |
| Exterior | $g_{00}$ | $1-\frac{r_s}{a}\cdot\frac{a}{r}$ | $1-\frac{r_s}{a}\cdot\frac{a}{r}$ |
| | $g_{ii}$ | $-1-\frac{r_s}{a}\cdot\frac{a}{r}$ | $-1-\frac{r_s}{a}\cdot\frac{a}{r}$ |

of Schwartzschild and those generated by the potential method are both plotted in **Fig. 8.1**. From the figure we at once see that the $g_{00}$ obtained by the potential method (P) and by Schwartzschild (S), interior and exterior, are identical. We see also that the exterior $g_{ii}$ for the potential method and Schwartzschild solution are identical as well. The interior $g_{ii}$, however, differ significantly. For although both curves are continuous at the boundary of the sphere, the slope of the component derived by the potential method is everywhere continuous, whereas that of the Schwartzschild component is discontinuous at the boundary. This discrepancy between the two methods will naturally affect the interior values of the curvature. Nevertheless, the broad agreement otherwise between the two methods gives reasonable confidence in the utility of the potential method.

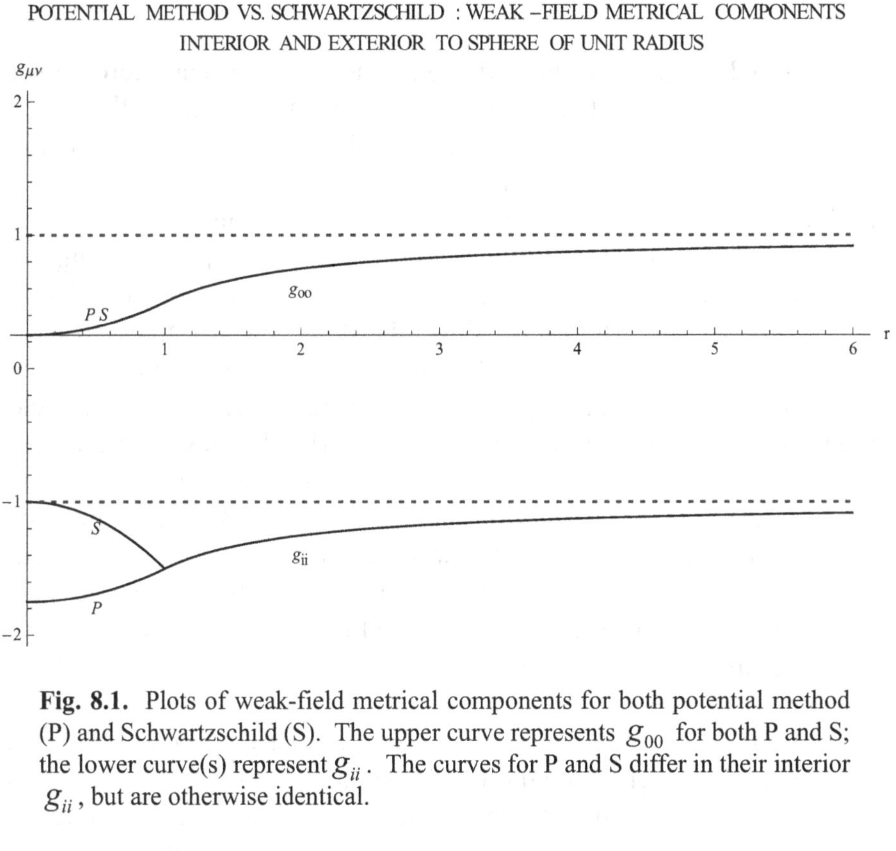


**Fig. 8.1.** Plots of weak-field metrical components for both potential method (P) and Schwartzschild (S). The upper curve represents $g_{00}$ for both P and S; the lower curve(s) represent $g_{ii}$. The curves for P and S differ in their interior $g_{ii}$, but are otherwise identical.

*8.6.3 Curvature.* The curvature tensor generated by the potential method is obtained by substituting (8.61) and (8.62) into (8.50), and then (8.50) into (8.49). After dividing by the constant $A$ of (8.36) we obtain

$$\left.\begin{aligned} R_{0j}^{\;\;0k} &= -\frac{1}{\ell c^2} E_{0j}^{\;\;k} = \frac{r_s}{2a^3}\delta_{jk} \\ R_{mn}^{\;\;ij} &= -\frac{1}{\ell c}\varepsilon_{ijk} B_{mn}^{\;\;k} = -\frac{r_s}{a^3}\varepsilon_{ijk}\varepsilon_{kmn} \end{aligned}\right\} \qquad r \le a \qquad (8.66)$$

$$\left.\begin{aligned} R_{0j}^{\ \ 0k} &= -\frac{1}{\ell c^2} E_{0j}^{\ \ k} = \frac{r_s}{2r^3}\left(\delta_{jk} - \frac{3x^j x^k}{r^2}\right) \\ R_{mn}^{\ \ ij} &= -\frac{1}{\ell c}\varepsilon_{ijk} B_{mn}^{\ \ k} = -\frac{r_s}{2r^3}\varepsilon_{ijk}\left[\frac{3x^\ell}{r^2}\left(\varepsilon_{k\ell m}x^n - \varepsilon_{k\ell n}x^m\right) + 2\varepsilon_{kmn}\right] \end{aligned}\right\} r > a \tag{8.67}$$

**Table 8.2** displays all *interior* components of the curvature tensor obtained by the potential method and compares them with values derived from the Schwartzschild metric. For the Schwartzschild case we have calculated the curvature in two ways: (1) Exactly, in spherical coordinates, using the metrical components exhibited in (8.64) in conjunction with definition (7.33); and approximately, in rectangular coordinates, using the weak-field metrical components given in **Table 8.1** in conjunction with (8.43). From **Table 8.2** we see that, for non-zero curvature components $R_{0j}^{\ \ 0j}$, the interior values obtained by the potential method are consistent with both the exact and weak-field values derived using the Schwartzschild metric. However, in the case of curvature components $R_{ij}^{\ \ ij}$, the highlighted exact and weak-field Schwartzschild values are grossly inconsistent with each other, perhaps throwing the form of the Schwartzschild interior line element (8.64) into doubt. We have little choice but to accept at face value the straightforwardly obtained values of the $R_{ij}^{\ \ ij}$ generated by the potential method, which in fact agree perfectly with the exact Schwartzschild values.

**Table 8.3** displays all *exterior* components of the curvature tensor obtained by the potential method and compares them with exact values derived from the Schwartzschild metric. The apparent difference between the values obtained by the potential method and the exact ones is due to the different coordinate systems employed. If we look on the $x$-axis, for example, where $y = z = 0$, we see that the values for the two methods of calculation are identical. Of course the values we are comparing are those of the mixed covariant-contravariant form $F_{\alpha\beta}^{\ \ \mu\nu}$. If we were to raise or lower any of the indices, the values obtained by the potential method would change at most by sign, whereas the exact values would change in accordance with the exact metric. Nevertheless, the point is made: the potential method offers a straightforward and practical way of calculating the curvature. Even more to the point: we

**Table 8.2** Riemann-Christoffel curvature $R_{\alpha\beta}^{\ \ \mu\nu}$ *interior* to spherical mass: potential method vs. values derived from Schwartzschild metric.[a]

| $R_{\alpha\beta}^{\ \ \mu\nu}$ | Potential method ( $\times r_s / a^3$ ) | From Schwartzschild metric ( $\times r_s / a^3$ ) | |
|---|---|---|---|
| | | Exact | Weak-field |
| $R_{01}^{\ \ 01} = -E_{01}^{\ 1} / \ell c^2$ | 1/2 | $\eta / 2$ | 1/2 |
| $R_{01}^{\ \ 02} = -E_{01}^{\ 2} / \ell c^2$ | 0 | 0 | 0 |
| $R_{01}^{\ \ 03} = -E_{01}^{\ 3} / \ell c^2$ | 0 | 0 | 0 |
| $R_{02}^{\ \ 01} = -E_{02}^{\ 1} / \ell c^2$ | 0 | 0 | 0 |
| $R_{02}^{\ \ 02} = -E_{02}^{\ 2} / \ell c^2$ | 1/2 | $\eta / 2$ | 1/2 |
| $R_{02}^{\ \ 03} = -E_{02}^{\ 3} / \ell c^2$ | 0 | 0 | 0 |
| $R_{03}^{\ \ 01} = -E_{03}^{\ 1} / \ell c^2$ | 0 | 0 | 0 |
| $R_{03}^{\ \ 02} = -E_{03}^{\ 2} / \ell c^2$ | 0 | 0 | 0 |
| $R_{03}^{\ \ 03} = -E_{03}^{\ 3} / \ell c^2$ | 1/2 | $\eta / 2$ | 1/2 |
| $R_{23}^{\ \ 23} = -B_{23}^{\ 1} / \ell c$ | −1 | −1 | +2 |
| $R_{23}^{\ \ 31} = -B_{23}^{\ 2} / \ell c$ | 0 | 0 | 0 |
| $R_{23}^{\ \ 12} = -B_{23}^{\ 3} / \ell c$ | 0 | 0 | 0 |
| $R_{31}^{\ \ 23} = -B_{31}^{\ 1} / \ell c$ | 0 | 0 | 0 |
| $R_{31}^{\ \ 31} = -B_{31}^{\ 2} / \ell c$ | −1 | −1 | +2 |
| $R_{31}^{\ \ 12} = -B_{31}^{\ 3} / \ell c$ | 0 | 0 | 0 |
| $R_{12}^{\ \ 23} = -B_{12}^{\ 1} / \ell c$ | 0 | 0 | 0 |
| $R_{12}^{\ \ 31} = -B_{12}^{\ 2} / \ell c$ | 0 | 0 | 0 |
| $R_{12}^{\ \ 12} = -B_{12}^{\ 3} / \ell c$ | −1 | −1 | +2 |

[a] Schwartzschild radius $r_s = 2mG / c^2$ and $a$ is the radius of the sphere.

Parameter $\eta = \dfrac{2f(r)}{3f(a) - f(r)}$, where $f(r) = (1 - r_s r^2 / a^3)^{1/2}$. A few values of $r_s / a$ are[11]: Proton $1.75 \times 10^{-39}$; Iron sphere (1 kg) $2.2 \times 10^{-26}$

Earth $1.4\times10^{-9}$; Neutron star 0.57. For ordinary object $\eta\approx 1+\frac{3}{4}\frac{r_s}{a}\left(1-\frac{r^2}{a^2}\right)\approx 1$.

**Table 8.3** Riemann-Christoffel curvature $R_{\alpha\beta}^{\ \ \mu\nu}$ *exterior* to spherical mass: potential method vs. exact value in Schwartzschild metric.[a]

| $R_{\alpha\beta}^{\ \ \mu\nu}$ | Potential method (rectangular coordinates) $\times r_s / r^3$ | Exact value[a] (polar coordinates) $\times r_s / r^3$ |
|---|---|---|
| $R_{01}^{\ \ 01}=-E_{01}^{\ 1}/\ell c^2$ | $(-2x^2+y^2+z^2)/2r^2$ | −1 |
| $R_{01}^{\ \ 02}=-E_{01}^{\ 2}/\ell c^2$ | $-3xy/2r^2$ | 0 |
| $R_{01}^{\ \ 03}=-E_{01}^{\ 3}/\ell c^2$ | $-3xz/2r^2$ | 0 |
| $R_{02}^{\ \ 01}=-E_{02}^{\ 1}/\ell c^2$ | $-3yx/2r^2$ | 0 |
| $R_{02}^{\ \ 02}=-E_{02}^{\ 2}/\ell c^2$ | $(-2y^2+z^2+x^2)/2r^2$ | 1/2 |
| $R_{02}^{\ \ 03}=-E_{02}^{\ 3}/\ell c^2$ | $-3yz/2r^2$ | 0 |
| $R_{03}^{\ \ 01}=-E_{03}^{\ 1}/\ell c^2$ | $-3zx/2r^2$ | 0 |
| $R_{03}^{\ \ 02}=-E_{03}^{\ 2}/\ell c^2$ | $-3zy/2r^2$ | 0 |
| $R_{03}^{\ \ 03}=-E_{03}^{\ 3}/\ell c^2$ | $(-2z^2+x^2+y^2)/2r^2$ | 1/2 |
| $R_{23}^{\ \ 23}=-B_{23}^{\ 1}/\ell c$ | $(-2x^2+y^2+z^2)/2r^2$ | −1 |
| $R_{23}^{\ \ 31}=-B_{23}^{\ 2}/\ell c$ | $-3xy/2r^2$ | 0 |
| $R_{23}^{\ \ 12}=-B_{23}^{\ 3}/\ell c$ | $-3xz/2r^2$ | 0 |
| $R_{31}^{\ \ 23}=-B_{31}^{\ 1}/\ell c$ | $-3xy/2r^2$ | 0 |
| $R_{31}^{\ \ 31}=-B_{31}^{\ 2}/\ell c$ | $(-2y^2+z^2+x^2)/2r^2$ | 1/2 |
| $R_{31}^{\ \ 12}=-B_{31}^{\ 3}/\ell c$ | $-3zy/2r^2$ | 0 |
| $R_{12}^{\ \ 23}=-B_{12}^{\ 1}/\ell c$ | $-3xz/2r^2$ | 0 |
| $R_{12}^{\ \ 31}=-B_{12}^{\ 2}/\ell c$ | $-3zy/2r^2$ | 0 |
| $R_{12}^{\ \ 12}=-B_{12}^{\ 3}/\ell c$ | $(-2z^2+x^2+y^2)/2r^2$ | 1/2 |

[a] Schwartzschild radius $r_s=\frac{2mG}{c^2}$, $r=(x^2+y^2+z^2)^{1/2}$ is the radial coordinate.

can be sure that the curvature tensor is a real physical field generated by a real conserved current. And that being the case, the curvature field should be capable of interacting with other such currents, producing a force of Lorentzian form, namely (8.31).

## 8.7 R-C self-force

Having found the potentials and curvatures interior and exterior to the sphere, we can read off from (8.66) and (8.67) the corresponding R-C $E$ and $B$ fields developed in those regions:

$$\left.\begin{aligned} \frac{1}{c}E_{0j}^{\ \ k} &= -\frac{r_s \ell c}{2a^3}\delta_{jk} \\ B_{mn}^{\ \ k} &= +\frac{r_s \ell c}{a^3}\varepsilon_{kmn} \end{aligned}\right\} \quad r \le a \tag{8.68}$$

and

$$\left.\begin{aligned} \frac{1}{c}E_{0j}^{\ \ k} &= \frac{r_s \ell c}{2r^3}\left(\frac{3x^j x^k}{r^2} - \delta_{jk}\right) \\ B_{mn}^{\ \ k} &= \frac{r_s \ell c}{2r^3}\left[\frac{3x^\ell}{r^2}\left(\varepsilon_{k\ell m}x^n - \varepsilon_{k\ell n}x^m\right) + 2\varepsilon_{kmn}\right] \end{aligned}\right\} \quad r > a\,. \tag{8.69}$$

Now there are two things one might do with these fields. The first and most obvious would be to find the R-C force on a test mass external to the source mass of radius $a$. To find the Lorentzian force density one would simply insert (8.69) into (8.51), together with the test mass's current density, $\left(j^0_{\ \alpha\beta}, \mathbf{j}_{\alpha\beta}\right)$. But the exercise is not an especially interesting one, because we do not know the value of constant $\ell$ or, consequently, what to look for experimentally. In fact, one might guess that if such a force were detectable, it would already have been observed.

But there is a second thing we can do, which is more interesting; namely, we can calculate the R-C force that the sphere exerts on itself. Suppose that the force is repulsive. We should then have a good candidate for the explosive force of the Big Bang, perhaps even of inflation. Moreover, if, in the central 'singularity' of a black hole, we assume a state of equilibrium between the opposing forces of gravitation and Riemann-Christoffel, then we may arrive at a finite radius and an estimate of the value of $\ell$.

To calculate the self-force we first substitute (8.68) into (8.51), the currents in this case being (8.57), the same currents that gave rise to the $E$ and $B$ fields (8.68). The time component of the force density, $K^0$, represents the power expended by the $E$ field acting on an element of the sphere. It is given by

$$K^0 = 2\mathbf{j}^{0j} \cdot \mathbf{E}_{0j} + \mathbf{j}^{ij} \cdot \mathbf{E}_{ij} / c = 0. \tag{8.70}$$

The power loss vanishes because both $\boldsymbol{j}^{0j} = 0$ and $\boldsymbol{E}_{ij} / c = 0$. Of course one expects vanishing power loss, because the sphere is stationary relative to itself: nothing moves.

Next we get the force-density three-vector, $\mathbf{K}$. Since $\mathbf{E}_{ij} / c = 0$ and $\boldsymbol{j}^{0j} = 0$ we have

$$\mathbf{K} = -2 j_{0j}{}^{0}\mathbf{E}_{0j} / c + \mathbf{j}_{ij} \times \mathbf{B}_{ij}. \tag{8.71}$$

In component form, after substituting for the currents and fields, this becomes

$$K^k = -C\delta_{jk}\partial_j \sigma(\mathbf{x}), \tag{8.72}$$

where $C$ is the positive constant

$$C = \frac{20\pi}{3} G\rho^2 \ell^2.$$

Now the self-force on an element $dV$ of the sphere is $K^k dV$. Hence the total force $F^k_{\text{R-C}}$ of the sphere on itself is given by the integral

$$\begin{aligned} F^k_{\text{R-C}} &= \int K^k dV \\ &= -C\int \delta_{jk}\partial_j \sigma(\mathbf{x}) dV \\ &= C\int (\partial_j \delta_{jk}) \sigma(\mathbf{x}) dV \\ &= C\oint \hat{n}^j \delta_{jk} dS \end{aligned}, \tag{8.73}$$

where we have applied both an integration by parts and the divergence theorem. The last integral is an integral over the surface of the sphere. The integral vanishes, as it must, because the sphere cannot move itself. However, the *pressure*, or force per unit area, $\mathbf{P}_{R\text{-}C}$, at the surface of the sphere does *not* vanish. It is given by differentiating the surface integral Eq. (8.73):

$$\mathbf{P}_{\text{R-C}} = \frac{d\mathbf{F}_{\text{R-C}}}{dS} = \frac{20\pi}{3} G\rho^2 \ell^2 \hat{\mathbf{n}} \tag{8.74}$$

and is directed parallel to the outer normal to the surface. Hence the R-C self-force is a *repulsive* force.

## 8.8 Gravitational self-force

We now obtain the self-force of the sphere comparable to (8.74) due to Newtonian gravitation. The gravitational self-force density, $\mathbf{f}$, at the surface of a sphere of radius $r$ concentric with and within the sphere of radius $a$ is

$$\begin{aligned} \mathbf{f} &= \left(\frac{4\pi}{3} r^3 \rho\right) \cdot \rho \cdot G\nabla\left(\frac{1}{r}\right) \\ &= -\frac{2\pi}{3} G\rho^2 \nabla r^2 \end{aligned} \tag{8.75}$$

The gravitational self-force on an element $dV = r^2 dr d\Omega$ of the sphere is $f dV$. Hence the total gravitational self-force is

$$\begin{aligned} \mathbf{F}_{\text{G}} &= \int \mathbf{f} dV \\ &= -\frac{2\pi}{3} G\rho^2 \int \left(\nabla r^2\right) dV \\ &= -\frac{2\pi}{3} G\rho^2 \oint a^2 \hat{\mathbf{n}} dS \end{aligned} \quad , \tag{8.76}$$

where to arrive at the last line we have used a well-known integral theorem. The pressure, $\mathbf{P}_{\text{G}}$, at the surface of the sphere due to

Newtonian gravitation is obtained by differentiating the surface intergral of (8.76):

$$\mathbf{P}_{\mathrm{G}} = \frac{d\mathbf{F}_{\mathrm{G}}}{ds} = -\frac{2\pi}{3} G\rho^2 a^2 \hat{\mathbf{n}} . \tag{8.77}$$

The gravitational pressure at the surface is directed oppositely to the outer normal, as expected.

## 8.9 An estimate of constant $\ell$

According to Einstein gravitation, at the center of every black hole there resides a singularity—a mass of zero diameter and infinite density. But such an object, with its unmeasurable attributes, seems physically implausible, particularly in the light of the quantum nature of the world. It is sometimes said that a quantum theory of gravity could give to this object a finite size and density.[12] But a finite ‘singularity’ might also be achieved by means of the repulsive R-C force, counteracting the force of gravity. Imagine a tiny volume held in equilibrium by the opposing forces of gravitation and Riemann-Christoffel. From (8.74) and (8.77) we have for equilibrium

$$a_{\mathrm{BH}} = \sqrt{10}\,\ell , \tag{8.78}$$

which prescribes a value for $\ell$ in terms of the black hole radius, $a_{\mathrm{BH}}$. What is $a_{\mathrm{BH}}$? A reasonable guess is that it is of the order of the Planck length , $\ell_{\mathrm{P}} = (\hbar G / c^3)^{1/2} \approx 1.6 \times 10^{-35}$ m , the smallest measurable distance.[12] So let us write

$$a_{\mathrm{BH}} = B\ell_{\mathrm{P}} , \tag{8.79}$$

where $B$ is a dimensionless constant of order unity. Then from (8.78) and (8.79)

$$\ell = \frac{B}{\sqrt{10}} \ell_P . \tag{8.80}$$

So $\ell$, by this estimate, is small indeed. Little wonder that the R-C force between separated massive objects—a force proportional to $\ell^2$ —is not observed.

Of course in making this argument from black hole geometry we have set aside our assumption of nearly-flat spacetime. If we were talking about conditions outside the 'singularity' that would be unacceptable. But we are talking about what happens *inside* the 'singularity', not outside. And inside, where the two opposing forces are in balance, one could easily argue, or at least imagine, that spacetime is more or less "flat". In any event, that is the assumption we have made here, leading to our estimate of $\ell$, Eq. (8.80).

## 8.10 The Big Bang

Suppose that we have a 'singularity' that is not in equilibrium, one whose radius $a$ is much smaller than the Planck length:

$$a^2 \ll \ell_P{}^2 . \tag{8.81}$$

Then from (8.74) and (8.77) we have

$$\mathbf{P}_{\text{R-C}} \gg B^2 \mathbf{P}_{\text{G}} . \tag{8.82}$$

So for radii $a$ much smaller than the Planck length, the repulsive R-C force greatly exceeds the gravitational force of attraction. According to some estimates, the radius of the initial singularity was perhaps $10^{-55}$ m, smaller than the Planck length by twenty orders of magnitude.[13] If so, then, at the surface of the initial singularity, the R-C force exceeded the gravitational force by some forty orders of magnitude, surely sufficient to initiate the Big Bang and inflation. In the next chapter we shall discover yet another repulsive force, a cosmic force, one that can continue the Universe's expansion long after its radius $a$ has exceeded the Planck length.

## 8.11 Energy density

It is of some interest to calculate the energy density $\tau^{00}$ interior to our spherical mass of radius $a$. Substituting definitions (8.49) into (8.33), we obtain in exact analogy with electrodynamics the general form

$$\tau^{00} = \frac{1}{2\tilde{\mu}}\left(\frac{1}{c^2}\mathbf{E}^{\alpha\beta}\cdot\mathbf{E}_{\alpha\beta} + \mathbf{B}^{\alpha\beta}\cdot\mathbf{B}_{\alpha\beta}\right). \tag{8.83}$$

Then substituting the $E$ and $B$ values of (8.68) we obtain

$$\tau^{00} = 2\pi G\rho^2\ell^2, \tag{8.84}$$

which, as expected, is a positive quantity. Now we are accustomed to the idea that all forms of energy gravitate. The density (8.84) is analogous to the electromagnetic energy density, which gravitates, and so we might expect the R-C density to gravitate too. But it does not, and it is not hard to see why. The $E$ and $B$ fields on which $\tau^{00}$ depends derive from the current density (8.38), which in turn depends on the gravitating stress-energy tensor $T^{\mu\nu}$ appearing on the r. h. s. of the Einstein field equation (8.22). But if the R-C $\tau^{\mu\nu}$ gravitates, then it should be added to the Einstein $T^{\mu\nu}$, leading to a new current density, new $E$ and $B$ fields, and a corrected value of $\tau^{\mu\nu}$. But this corrected value now contains energy that previously did not exist, an obvious violation of the conservation of energy. The energy density (8.84) therefore does not gravitate.

To drive this point home, let us rewrite (8.84) in the form

$$\tau^{00} = \frac{3}{4}\cdot\frac{r_s\ell^2}{a^3}\cdot\rho c^2, \tag{8.85}$$

where $r_s$ is the Schwartzschild radius (8.63). Here $\rho c^2$ is the energy density $T^{00}$ appearing in the Einstein equation and from which the R-C $E$ and $B$ fields ultimately derive. The constant of proportionality preceding it determines how much of that starting density goes into the R-C density $\tau^{00}$. For ordinary materials the constant of proportionality is exceedingly small; e.g., for a 1 kg iron sphere (see the bottom of **Table 8.2**),

$$\rho = 7.85\ \text{g/cm}^3,\ a = 3.12\ \text{cm},\ r_s/a = 2.2\times10^{-26},$$

we get

$$\frac{r_s \ell_P^{\ 2}}{a^3} = 0.6 \times 10^{-92}.$$

So for the ordinary object, whether $\tau^{00}$ gravitates is, as a practical matter, immaterial. But for a stellar-mass black hole, with

$$m \sim 10 m_\odot,\ m_\odot = 2 \times 10^{30}\ \text{kg},\ r_s \sim 30\ \text{km},\ a \sim \ell_P = 1.6 \times 10^{-35}\ \text{m}$$

we have

$$\frac{r_s \ell_P^{\ 2}}{a^3} = 2 \times 10^{38}.$$

With a multiplier of this magnitude, clearly the density (8.85) cannot be gravitating.

Note that, using (8.66), Eq. (8.85) can be written

$$\tau^{00} = \frac{3}{2} \cdot R_{0k}^{\ \ 0k} \ell^2 \cdot \rho c^2, \tag{8.86}$$

showing that the contribution to the energy density of the R-C field over and above the original $\rho c^2$ comes from the Riemann curvature, $R_{0k}^{\ \ 0k}$.

## 8. 12 A spin-1 *riemann* replaces the non-existent spin-2 graviton

We have argued that the projection operator $\hat{R}$ represents the particle belonging to the wave field $\chi$ on which it operates; see *Sec. 3.9.3* and the introductory paragraphs of **Ch. 4**. Thus in (5.24) (with $k_5 = k^6 = 0$) $\hat{R}$ operates on the Maxwell $F^{\mu\nu}$ and so represents the photon, the mediator of the Lorentz force of electromagnetism. In (8.1) above, $\hat{R}$ operates on the Riemann-Cristoffel $F^{\mu\nu}{}_{\alpha\beta}$ and so represents a new particle, the mediator of the R-C force defined by (8.31). We call this new particle the *riemann*. We shall later derive its internal structure.

Like the photon, the riemann is a spin-1 particle. We know this (and will later prove it) because the laws of propagation (5.24) and (8.1) are similar in structure. Unlike the Maxwell $F^{\mu\nu}$ and Riemann-Cristoffel $F_{\alpha\beta}^{\ \ \mu\nu}$, the Einstein tensor $G_{\mu\nu} = R_{\mu\nu} - g_{\mu\nu}R/2$ of (8.20) has no propagator associated with it. Indeed, the Einstein $G_{\mu\nu}$ are already two contractions and an integration removed from the curvature-bearing Bianchi relation (8.17). In the context of the present theory this can mean only one thing: there is no mediating boson in Einstein gravity. The massless spin-2 graviton, in other words, does not exist. From Nature's point of view this is no calamity, as the curvature of spacetime itself suffices to mediate the force of gravity. However, the implication for string theory is not insignificant. For this theory predicts the existence of a massless spin-2 particle—the graviton.[14] So if the graviton does not exist—as we claim here it does not—then string theory is simply wrong. And not only that. If there exists no field quantum belonging to the Einstein $G_{\mu\nu}$, then the very goal of unifying gravity and quantum mechanics—quantum gravity—is in principle an unachievable one. For the two domains operate differently: one via spacetime curvature, the other by exchange of field quanta. They are as different as night and day and, if the present theory is right, no marriage is possible. In retrospect none of this seems surprising. In the cases of the electromagnetic, weak and strong nuclear forces of nature, the force-mediating bosons are clearly distinct from the background stage of spacetime on which they move and operate. But if the gravitational force were to require a boson to mediate it, the clear distinction between mediating particle and stage of operation must then disappear, an arrangement not easy to visualize.

On the other hand, it might be possible to unify the forces of Lorentz and Riemann-Christoffel in a common formalism. The similarity of their governing equations of motion suggest the possibility. But this is far beyond the scope of the present volume and will not pursued here.

## Notes and references

[1] Hans C. Ohanian, *Gravitation and Spacetime* (W. W. Norton & Co., New York, 1976), p. 265; P .A. M. Dirac,*General Theory of Relativity* (John Wiley & Sons, New York, 1975), p. 44.

[2] H. A. Atwater, *Introduction to General Relativity* (Pergamon Press, Oxford,1974), pp. 85-8

[3] S. Weinberg, *Gravitation and Cosmology* ( John Wiley & Sons, New York, 1972), pp. 145-146.

[4] R. Penrose, *The Emperor's New Mind* (Penguin, London, 1989), pp. 210-211.

[5] R. Penrose, *The Road to Reality* (Alfred A. Knopf, New York, 2005). p. 464.

[6] C. W. Misner, K. S. Thorne and J. A. Wheeler, *Gravitation* (W. H. Freeman, San Francisco, 1973), p. 382, Eq. (15.30a).

[7] H. A. Atwater, *ibid*,pp. 89, 125.

[8] K. Schwartzschild, "Über das Gravitationsfeld eines Massenpunktes nach der Einsteinschen Theorie," *Sitzungsber. Preuss. Akad. Wiss.*, (1916),189-196; translated in *The Abraham Zelmonov Journal* **1** (2008),10-19.

[9] K. Schwartzschild, "Über das Gravitationsfeld einer Kugel aus incompressiebler Flüssigkeit nach der Einsteinschen Theorie," *Sitzungsber. Preuss. Akad. Wiss.*, (1916), 424-435; translated in *The Abraham Zelmonov Journal* **1** (2008), 20-32.

[10] H. G. Ellis, "Gravity Inside a Nonrotating, Homogeneous, Spherical Body," http://arXiv.org /1203.4750.

[11] Values taken from H. A. Atwater, *ibid*, p. 142.

[12] Luis J. Garay, "Quantum gravity and minimum length," International J. of Mod. Phys. A**10** (January, 1995), 145 ff.

[13] Alan H. Guth, *The Inflationary Universe* (Perseus Books, Reading Massachusetts, 1997), p. 185.

[14] For a comprehensive, nontechnical review of string theory, see Matthew Chalmers, "Stringscape," http://physicsworld.com/cws/article/indepth/30940 (2007).

# Chapter 9

# Cosmic expansion

## 9.1 Rank-3 tensor wave field

We come now to study the rank-3 tensor wave field, $F_{\alpha}^{\ \mu\nu}$, a physical field descriptor intermediate between the Maxwell $F^{\mu\nu}$ and Riemann-Christoffel $F_{\alpha\beta}^{\ \ \mu\nu}$. Like the latter two fields, the new tensor is antisymmetric in its two upper indices. Although 3-tensors are not unknown in theoretical physics—the Lanczos potential[1] comes to mind here—the one we are about to explore is unique in that it enjoys the same degree of physical reality as the Maxwell and Riemann tensors. This means that it can generate a force, complementary to but distinct from the Lorentz and Riemann-Christoffel forces studied above. Our main purpose here is to identify that force and describe its function in the universe.

The equations of motion of the $F_{\alpha}^{\ \mu\nu}$ are given freely to us by the Law of Laws. In a formal sense, the $F_{\alpha}^{\ \mu\nu}$ are just like the Riemann-Christoffel $F_{\alpha\beta}^{\ \ \mu\nu}$ but with one less spacetime index. Thus simply by removing a spacetime index from Eqs. (8.1) and (8.2) we obtain from the Law of Laws

$$\hat{R}\mathrm{X}_{\alpha} = r\mathrm{X}_{\alpha}, \tag{9.1}$$

where the projection operator

$$\hat{R} = \exp\left[\gamma^5 \xi_5 \left(-\gamma^{\mu}\nabla_{\mu} + \Sigma_{\alpha}\mathrm{X}_{\alpha}^{-1}\right)\right]. \tag{9.2}$$

Operator $\hat{R}$ generates the field equation

$$\gamma^{\mu}\nabla_{\mu}\mathrm{X}_{\alpha}\left(x^{\mu}\right) = \Sigma_{\alpha}\left(x^{\mu}\right), \tag{9.3}$$

where, upon removing an index from (8.4) and (8.5),

$$\mathrm{X}_{\alpha} = I\phi_{\alpha} + \frac{i}{2}\sigma^{\mu\nu}F_{\alpha\mu\nu} + i\gamma^{5}\gamma^{\nu}F_{\alpha5\nu} \tag{9.4}$$

and

$$\Sigma_{\alpha} = \tilde{\mu}\gamma^{\mu}j_{\alpha\mu} + \tilde{\mu}_{5}i\gamma^{5}j_{\alpha5}. \tag{9.5}$$

Note that the constants $\tilde{\mu}$ and $\tilde{\mu}_5$ and are unchanged from those defined in (8.5). We are still doing geometrodynamics: those constants defining the properties of spacetime remain the same throughout, independent of the wave field interacting with it. Inserting these expansions into (9.3) we obtain a new set of laws, again analogous to those of electromagnetism:

Set I
$$F_{\alpha\ ;\mu}^{\ \mu\nu} + (g^{\mu\nu}\phi_{\alpha})_{;\mu} = \tilde{\mu}j_{\alpha}^{\nu} \tag{9.6}$$

$$\varepsilon^{\lambda\mu\nu\sigma}F_{\alpha\mu\nu;\lambda} = 0 \tag{9.7}$$

Set II
$$F_{\alpha\ ;\mu}^{\ 5\mu} = -\tilde{\mu}_{5}j_{\alpha}^{5} \tag{9.8}$$

$$\varepsilon^{\mu\nu\rho\sigma}F_{\alpha5\nu;\mu} = 0 \tag{9.9}$$

Ideally, to facilitate our understanding of these equations, we should like to be able to identify the rank-3 field tensor, $F_{\alpha}^{\ \mu\nu}$, with some known mathematical object, in the same way that we could identify $F_{\alpha\beta}^{\ \ \mu\nu}$ with the curvature tensor. Unfortunately, there is nothing in the known portfolio of geometric objects to compare it to or identify it with. In particular, we know of no 3-tensor that would cause the Bianchi-like sum of terms in (9.7) to vanish. Apart from its asymmetry in the indices $\mu$ and $\nu$, the properties of $F_{\alpha}^{\ \mu\nu}$ are at this point unknown. To get at the meaning of these equations we shall have to proceed step by step.

Let us start with inhomogeneous equation (9.6) and see whether the current density $j_{\alpha}^{\nu}$ is conserved. The covariant divergence of the term involving vector $\phi_{\nu}$ can be shown to vanish if and only if $\phi_{\nu}$ itself vanishes. Thus if (9.6) is to be physically relevant it must be the case that

$$\phi_\alpha = 0, \tag{9.10}$$

corresponding to (8.13). Just as there is no background Maxwell field of the vacuum, there is no vector field of the vacuum either. Then utilizing the antisymmetry in $\mu$ and $\nu$, we can write (9.6) in the form

$$\begin{aligned} 2\tilde{\mu} j_{\alpha\ ;\nu}^{\ \nu} = 2F_{\alpha\ ;\mu;\nu}^{\ \mu\nu} &= F_{\alpha\ ;\mu;\nu}^{\ \mu\nu} - F_{\alpha\ ;\nu;\mu}^{\ \mu\nu} \\ &= F_{\lambda}^{\ \mu\nu} R^{\lambda}_{\ \alpha\mu\nu} - F_{\alpha}^{\ \lambda\nu} R^{\mu}_{\ \lambda\mu\nu} + F_{\alpha}^{\ \lambda\mu} R^{\nu}_{\ \lambda\mu\nu}, \\ &= F_{\lambda}^{\ \mu\nu} R^{\lambda}_{\ \alpha\mu\nu} + 2F_{\alpha}^{\ \lambda\nu} R_{\lambda\nu} \end{aligned} \tag{9.11}$$

where we have used (7.41) and (7.45). Now because $F_{\alpha}^{\ \lambda\nu}$ is antisymmetric in $\lambda$ and $\nu$ , whereas $R_{\lambda\nu}$ is symmetric in those indices, the second term on the r.h.s. of this expression vanishes for all $\alpha$. However, the first term, which entails the curvature tensor $R^{\lambda}_{\ \alpha\mu\nu}$, vanishes in flat space, but not in curved space, where $R^{\lambda}_{\ \alpha\mu\nu} \neq 0$. In other words, current $j_{\alpha}^{\ \nu}$ is conserved in flat spacetime but not in curved spacetime. Now what are we to make of this? Our first inclination is to dismiss the new equations on the ground that, even though *form*-invariant against general coordinate transformations, they are distinctly *content*-variant and hence bear no relation to the real world.[2] But this is not necessarily so. For if the world *were* flat, not curved, then our new equations might well have a place in its description: in a flat world there would be no need for general coordinate transformations, and no concern for variation in physical content.

Now in fact there is such a world, namely, the universe as a whole. Historically one of the great questions of cosmology has been whether the universe is flat, spherical or saddle-shaped.[3] The Principle of True Representation yields an immediate answer: it must be flat. This is easily seen by inspection of **Fig. 3.5**. The $M^4$ hyperplane is everywhere perpendicular to the flat and infinite fifth coordinate axis, $x^5$. Therefore $M^4$ —and the universe as a whole—is flat. Recent WMAP measurements of features of the microwave background fluctuations confirm that the universe is indeed flat, with a 0.4% margin of error.[4]

So, rather than discard our new equations as being unphysical, we shall—at least tentatively—assume them to apply to the universe as a whole. This means that all covariant derivatives of both Set I and Set II can be replaced by ordinary derivatives. The equations of Set I now read

$$\partial_\mu F^{\mu\nu}_{\alpha} = \tilde{\mu} j_\alpha^{\ \nu} \tag{9.12}$$

$$\partial_\lambda F_{\alpha\mu\nu} + \partial_\mu F_{\alpha\nu\lambda} + \partial_\nu F_{\alpha\lambda\mu} = 0 . \tag{9.13}$$

Once again we have field equations analogous to those of electrodynamics. And like (8.16) and (8.17), they read directly on the field of geometrodynamics.

There remains one undefined aspect of this first set of equations, namely, the meaning and structure of the current density $j_\alpha^{\ \nu}$. In the case of (8.16), we could define $j_{\alpha\beta}^{\ \ \mu}$ by relating it to the Einstein equations (8.22), which in turn came from a contraction of the Bianchi identity (8.17). But if we try to duplicate that procedure here, by performing a contraction of (9.13), say between $\lambda$ and $\mu$, we obtain merely the identity

$$\partial^\lambda F_{\alpha\lambda\nu} + \partial^\lambda F_{\alpha\nu\lambda} = 0 , \tag{9.14}$$

which tells nothing about the current $j_\alpha^{\ \nu}$. To define $j_\alpha^{\ \nu}$ it appears that we shall have to take an indirect approach. To do this, go to (8.38) and from both sides remove the index $\beta$. Then, omitting the two terms involving the derivative with respect to the now missing $\beta$, we are left with the schematic form

$$j_\alpha^{\ \nu} = \frac{\ell}{c}\left(T^\nu - \frac{1}{2} g^\nu T\right)_{;\alpha} , \tag{9.15}$$

which defines the basic structure of the current. To give it physical content we make the identifications

$$\left.\begin{aligned} T^\nu &\rightarrow j^\nu c \\ T &\rightarrow \rho c^2 \\ g^\nu &\rightarrow v^\nu / c \end{aligned}\right\} , \tag{9.16}$$

where $j^\nu$ is the convection current density

$$j^{\nu} = \rho \upsilon^{\nu}, \tag{9.17}$$

$\upsilon^{\nu}$ is the four velocity

$$\upsilon^{\nu} = \frac{dx^{\nu}}{d\tau}, \tag{9.18}$$

$d\tau = \sqrt{1 - \upsilon^2 / c^2}\, dt$ is an interval of proper time and $\rho$ is mass density. Inserting these into (9.15) we obtain the rank-2 current density

$$j_{\alpha}^{\ \nu} = \frac{\ell}{2} j^{\nu}_{\ ;\alpha} . \tag{9.19}$$

Note that the fundamental length $\ell$, which lends to $j_{\alpha}^{\ \nu}$ units of mass current density, is the same $\ell$ as that defined in (8.80); our discussion of geometrodynamics requires only the one new fundamental length, $\ell$. Note also that the current defined by (9.19) is *not* generally conserved, but *is* conserved in flat space: utilizing (7.39) and (7.45) we have

$$j^{\nu}_{\ ;\alpha;\nu} = j^{\nu}_{\ ;\nu;\alpha} - j^{\beta} R_{\alpha\beta} . \tag{9.20}$$

Here the first term vanishes generally by conservation of mass:

$$j^{\nu}_{\ ;\nu} = 0 , \tag{9.21}$$

but second term is assured to vanish in flat space only, thanks to the presence here of the Ricci tensor.

We turn now to Set II, Eqs. (9.8) and (9.9). The key here is to define the current $j_{\alpha}^{\ 5}$. We can do this using the same approach taken above to define $j_{\alpha}^{\ \nu}$. In the discussion following (8.15) it was noted that, since the curvature $F_{\alpha\beta}^{\ \ \nu 5}$ vanishes, so must the current $j_{\alpha\beta}^{\ \ 5}$ vanish. So, by removing index $\beta$ from $j_{\alpha\beta}^{\ \ 5}$ we have

$$j_{\alpha}^{\ 5} = 0 , \tag{9.22}$$

from which it follows that the fields $F_{\alpha}^{\ \nu 5} = 0$. Thus there is no second set of field equations; we get the first set only.

## 9.2 Large-scale structure of the physical universe: flat and bounded

We now have in hand a new set of field equations, Eqs. (9.12) and (9.13), defining the behavior of the 3-tensor $F_\alpha^{\ \mu\nu}$. These equations are structurally similar to those of Maxwell and Riemann-Christoffel, and so there will be no difficulty deriving from them a new force density analogous to the Lorentz and R-C force densities. The question is, what is the meaning of this new force; how does it arise, and what is it being applied to?

The answer is supplied by the current density, (9.19). It tells us two things. First, it contains a new mass density, $\rho$. But this mass density cannot refer to ordinary, gravitating matter; for the physical effects of ordinary, gravitating matter are already fully described by the Riemann-Christoffel equations in general, and the Einstein field equations in particular. The mass density $\rho$ must therefore refer to a different kind of matter entirely, namely, matter that does not gravitate. But we know of only one form of non-gravitating mass/energy [aside from the R-C energy density (8.83)!] : the quantum-field-theoretic vacuum energy[5] $u_{\text{QFT}} = \rho_{\text{QFT}} c^2$. Given its enormous magnitude,[6,7] we know that this vacuum energy—assuming it exists—cannot gravitate, as otherwise the universe would have collapsed soon after its inception. The widespread assumption that $\rho_{\text{QFT}}$ must gravitate—as do all (or rather, *most*) other forms of energy—has led some to say that its calculated value represents the worst-ever prediction of theoretical physics. We now know that that calculation need not be wrong. For the vacuum energy finds its natural place in the theory of the 3-tensor $F_\alpha^{\ \mu\nu}$ developed here. To avoid any confusion as to the meaning of the symbol $\rho$, we will rename it to reflect its non-gravitating character:

$$\rho \to \rho_{\text{NG}}, \tag{9.23}$$

where $\rho_{\text{NG}} \propto \rho_{\text{QFT}}$, with the constant of proportionality to be determined below.

The second thing we see from current (9.19) is that all fields and forces derive from the *gradient* of the convection current, $j^\nu$. In other words, the fields and forces are essentially edge effects. So if $\rho_{\text{NG}}$ is spatially uniform, and the fields and forces are to be non-vanishing, *the vacuum energy distribution must be bounded.* This requirement presents a definite picture of the universe. We know that the universe is flat, and this means that, in total, it can be assumed infinite and mostly empty of anything at all. But at least one finite region of it— the one we live in and believe originated in the Big Bang— is occupied by a great, expanding ball of vacuum energy, $\rho_{\mathrm{QFT}}c^2$. It is at the outer edge of this great ball that fields and forces are created and felt. If the force created there produces positive acceleration (away from the center of the universe), it can account for the continuing and accelerating expansion of the universe. Of course to uphold the cosmological principle the diameter of the ball must be something larger than that of the visible universe. **Figure 9.1** summarizes the universal geometry implied by the current (9.19).

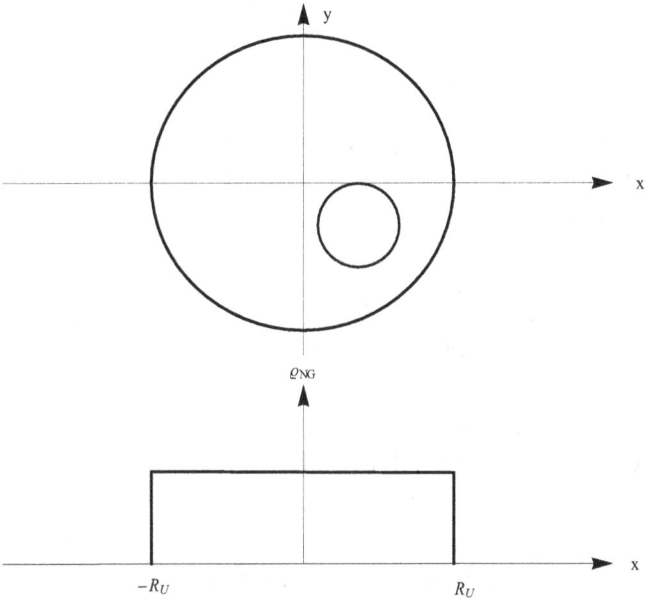


**Fig. 9.1** Depicting the universe in 2-D (Flatland) form. The universe in its entirety is infinite in all directions and is mostly empty. However, there exists within it a bounded disk of spatially-uniform energy density $\rho_{\mathrm{NG}}c^2$ and radius $R_{\mathrm{U}}$, comprising the *physical* universe. The smaller circle bounds the *observable* universe, which expands along with the physical universe of which it is a part.

## 9.3 Cosmic Casimir force and stress-energy tensor

As intimated above, our new field equations (9.12) and (9.13) imply the existence of a *sixth* force of nature, one we shall call the *cosmic Casimir force*, since it arises, like the Casimir force between metal plates,[5] from a step in the vacuum energy distribution. It is *cosmic*, because it occurs at the boundary of the physical universe. In analogy with the Lorentz force of electromagnetism, it is given by

$$\kappa^{\mu} = F_{\alpha}^{\ \mu\nu} j^{\alpha}_{\ \nu} . \tag{9.24}$$

It may also be written, utilizing (9.12)

$$\begin{aligned} \kappa^{\mu} &= \frac{1}{\tilde{\mu}} F_{\alpha}^{\ \mu\nu} F^{\alpha\lambda}_{\ \ \ \nu;\lambda} , \\ &= -\sigma^{\mu\nu}_{\ \ ;\nu} \end{aligned} \tag{9.25}$$

where $\sigma^{\mu\nu}$ is the stress-energy tensor arising from the field $F_{\alpha}^{\ \mu\nu}$. In perfect analogy with electrodynamics and Riemann-Christoffel geometrodynamics,

$$\sigma^{\mu\nu} = \frac{1}{\tilde{\mu}} \left( -F^{\alpha\mu\lambda} F_{\alpha\ \lambda}^{\ \nu} + \frac{1}{4} \eta^{\mu\nu} F^{\alpha\sigma\tau} F_{\alpha\sigma\tau} \right). \tag{9.26}$$

Note that we are here using the flat-space metric $\eta^{\mu\nu} = (+---)$. Note also that constant $\tilde{\mu}$ is still defined by (8.23) and (8.35): we are bound to geometrodynamics through constant *G*.

## 9.4 Potential formulation of the field equations

We propose to calculate the cosmic Casimir force density $\kappa^{\mu}$ at the edge of the physical universe. But to do this we first need to find the field $F_{\alpha}^{\ \mu\nu}$, and this is most easily accomplished by way of a potential. Thus we define

$$F_{\alpha\mu\nu} = \partial_{\mu} A_{\alpha\nu} - \partial_{\nu} A_{\alpha\mu} , \tag{9.27}$$

where $A_{\alpha\mu}$ is a rank-2 tensor potential, and which definition satisfies identically the Bianchi-like relation (9.13). Putting this into (9.12) and assuming the Lorentz gauge condition

$$\partial_\mu A_\alpha^{\ \mu} = 0 \tag{9.28}$$

we obtain the wave equation

$$\partial_\mu \partial^\mu A_\alpha^{\ \nu} = \tilde{\mu} j_\alpha^{\ \nu}, \tag{9.29}$$

the solution to which is the retarded potential

$$A_\alpha^{\ \nu}(\mathbf{x},t) = \frac{\tilde{\mu}}{4\pi}\int \frac{j_\alpha^{\ \nu}(\xi,t^*)}{r}\, d^3\xi, \tag{9.30}$$

where current $j_\alpha^{\ \nu}$ is given by (9.19), constant $\tilde{\mu}$ by (8.23) and (8.35), $t^* = t - r/c$ and $r = |\mathbf{x} - \xi|$.

To complete the analogy with electrodynamics (and Riemann-Christoffel geometrodynamics) we can define the rank-2 $E$ and $B$ fields:

$$\left.\begin{aligned} F_\alpha^{\ 0j} &= -\frac{E_\alpha^{\ j}}{c} \\ F_\alpha^{\ ij} &= -\varepsilon_{ijk} B_\alpha^{\ k} \end{aligned}\right\}. \tag{9.31}$$

In terms of the potential we have,

$$\left.\begin{aligned} \frac{\mathbf{E}_\alpha}{c} &= -\partial^0 \mathbf{A}_\alpha - \nabla A_\alpha^{\ 0} \\ \mathbf{B}_\alpha &= \nabla \times \mathbf{A}_\alpha \end{aligned}\right\}. \tag{9.32}$$

The force density $\kappa^\mu$ then takes the Lorentzian form

$$\left.\begin{aligned} \kappa^0 &= \mathbf{j}^\alpha \cdot \frac{\mathbf{E}_\alpha}{c} \\ \boldsymbol{\kappa} &= j^{\alpha 0}\frac{\mathbf{E}_\alpha}{c} + \mathbf{j}^\alpha \times \mathbf{B}_\alpha \end{aligned}\right\}. \tag{9.33}$$

## 9.5 The Casimir self-force of the universe

Let us start with a static universe and compute the cosmic Casimir self-force associated with it. The components of current (9.19) are

$$\left.\begin{aligned} j_0{}^0 &= 0 \\ j_0{}^k &= 0 \\ j_j{}^0 &= \frac{\ell \rho_{\mathrm{NG}} c}{2} \partial_j \sigma(\mathbf{x}) \\ j_j{}^k &= 0 \end{aligned}\right\}, \tag{9.34}$$

where $\sigma(\mathbf{x})$ is the spherical window function

$$\sigma(\mathbf{x}) = \begin{cases} 1 & |\mathbf{x}| \le R_{\mathrm{U}} \\ 0 & |\mathbf{x}| > R_{\mathrm{U}} \end{cases}, \tag{9.35}$$

with $R_{\mathrm{U}}$ the radius of the static universe. Then from (9.30) and definition (8.35) we obtain for the non-vanishing potential components

$$A_j{}^0 = \frac{G \ell \rho_{\mathrm{NG}}}{c} I_j, \tag{9.36}$$

where $I_j$ is the integral (8.59), and whose value is displayed in (8.60). In the present instance we have

$$I_j = \begin{cases} -\dfrac{2\pi}{3} \partial_j r^2 = -\dfrac{4\pi}{3} x^j, & r \le R_{\mathrm{U}} \\ \dfrac{4\pi R_{\mathrm{U}}{}^3}{3} \partial_j \left(\dfrac{1}{r}\right) = -\dfrac{4\pi}{3} \dfrac{R_{\mathrm{U}}{}^3}{r^3} x^j, & r > R_{\mathrm{U}} \end{cases}, \tag{9.37}$$

giving the potential components both inside and outside the physical universe:

$$A_j^{\ 0} = \begin{cases} -\dfrac{2\pi G\ell\rho_{\text{NG}}}{3c}\partial_j r^2 = -\dfrac{4\pi G\ell\rho_{\text{NG}}}{3c}x^j, & r \le R_{\text{U}} \\ \dfrac{G\ell m_{\text{NG}}}{c}\partial_j\left(\dfrac{1}{r}\right) = -\dfrac{G\ell m_{\text{NG}}}{c}\dfrac{x^j}{r^3}, & r > R_{\text{U}} \end{cases} \tag{9.38}$$

where $m_{\text{NG}}$ is the total (non-gravitating) mass of the vacuum energy of the physical universe. Knowing the potential we can now compute the $E$ and $B$ fields from (9.32). Since $A_0^{\ k} = A_j^{\ k} = 0$, we see that there is no $B$-field contribution to the force. But there is an $E$-field, given by

$$\begin{aligned} E_j^{\ k}/c &= -\partial_k A_j^{\ 0} \\ &= \begin{cases} \dfrac{4\pi G\ell\rho_{\text{NG}}}{3c}\delta_{jk}, & r \le R_{\text{U}} \\ \dfrac{G\ell m_{\text{NG}}}{c}\partial_k\left(\dfrac{x^j}{r^3}\right), & r > R_{\text{U}} \end{cases} \end{aligned} \tag{9.39}$$

Of immediate interest is the $E$-field internal to the universe, i.e., where $r \le R_{\text{U}}$, which enables us to calculate the Casimir self-force of the universe. From (9.33) we have for the self-force density

$$\begin{aligned} \kappa^k &= j^{j0}E_j^{\ k} \\ &= -j_j^{\ 0}E_j^{\ k} \\ &= -\frac{2\pi}{3}G\ell^2\rho_{\text{NG}}^{\ \ 2}\partial_k\sigma(\mathbf{x}) \end{aligned} \tag{9.40}$$

The total self-force $\mathrm{K}^k$ is given by an integral over the volume of the universe

$$\begin{aligned} \mathrm{K}^k &= \int \kappa^k dV \\ &= +\frac{2\pi}{3}G\ell^2\rho_{\text{NG}}^{\ \ 2}\oint n^k dS \end{aligned} \tag{9.41}$$

where to arrive at the surface integral we have followed the procedure outlined for (8.73). And so the force per unit area, or pressure, on the boundary of the universe is, from (9.41),

$$\frac{d\mathbf{K}}{dS} = +\frac{2\pi}{3} G\ell^2 \rho_{\text{NG}}{}^2 \hat{\mathbf{n}} \tag{9.42}$$

The pressure is indeed positive, pushing out on the surface of the universe. Thus we have in hand a plausible and readily-understood explanation for the expansion of the universe—one entirely dependent on the existence of a non-gravitating vacuum energy density, $\rho_{\text{QFT}} c^2$.

## 9.6 The dynamics of cosmic expansion

*9.6.1 Robertson-Walker metric.* We now go from the static universe to a dynamic one. To do this we are going to assume a Robertson-Walker metric[3,8] appropriate to a flat universe. It may be written in the form

$$ds^2 = c^2 dt^2 - a(t)^2 \left(dr^2 + r^2 d\Omega^2\right), \tag{9.43}$$

where:

*ds* is the four-dimensional line element;

*dt* is an interval of universal (Newtonian) time, $t$, the time since the Big Bang, the same everywhere in the universe;

$r$ is a time-independent spatial coordinate, called *comoving*, with the dimensions of length;

$d\Omega^2 = d\theta^2 + \sin^2\theta d\phi^2$; and

$a(t)$ is a dimensionless scale factor; at the present time, $t_0$, $a(t_0) \equiv a_0 = 1$.

Along a radius, where $d\Omega = 0$, and at a fixed time $t$, integration of $ds \equiv dD(t) = a(t)dr$ yields the *proper distance*

$$D(t) = a(t)r. \tag{9.44}$$

Generally one says that distances between objects within a galaxy are more or less constant in time; hence their relative coordinates are said to be comoving. However, the distances between galaxies are ever increasing; those distances are proper distances.[3]

*9.6.2 Force of acceleration.* Let us now consider an element of mass, $dm_{\mathrm{NG}} = \rho_{\mathrm{NG}} dV$, located a proper distance $D(t)$ from the center of the universe, as depicted in **Fig. 9.2**. Because the universe is expanding under the force (9.42), the element of mass experiences an acceleration, $\ddot{\mathbf{D}}$, and a corresponding Newtonian force, $\boldsymbol{\kappa} dV = \rho_{\mathrm{NG}} \ddot{\mathbf{D}} dV$, where $\boldsymbol{\kappa}$ is the Casimir force density, (9.40). But from (9.44)

$$\frac{\ddot{D}}{D} = \frac{\ddot{a}}{a} = \text{constant in time}. \tag{9.45}$$

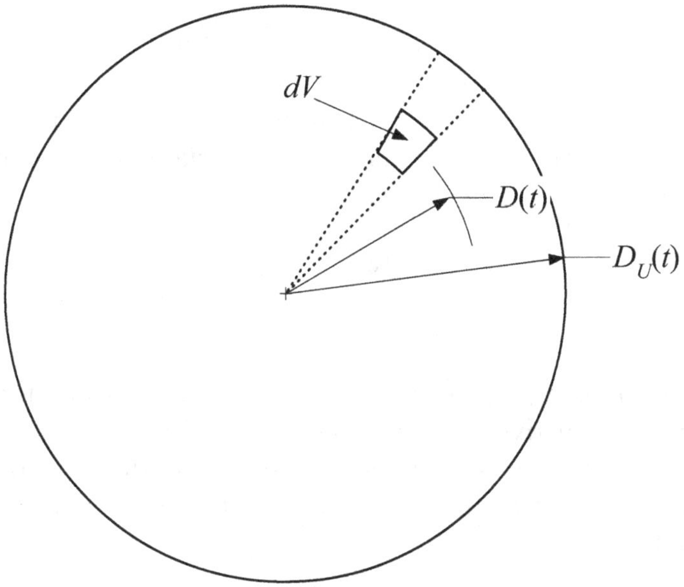


**Fig. 9.2.** An element of non-gravitating mass $dm_{\mathrm{NG}} = \rho_{\mathrm{NG}} dV$, occupying volume element $dV$ of an expanding universe of radius $D_U(t)$, experiences radial acceleration $\ddot{\boldsymbol{D}}(t)$ and corresponding force $dm_{\mathrm{NG}} \ddot{\boldsymbol{D}}(t)$.

Hence the total force on the universe (which sums to zero vectorially) is expressed by the integral

$$
\begin{aligned}
\mathbf{K} &= \int \boldsymbol{\kappa} dV \\
&= \frac{\ddot{a}}{a} \int \rho_{NG} \mathbf{D} dV \\
&= \frac{\ddot{a}\rho_{NG}}{2a} \int \nabla D^2 dV \\
&= \frac{\ddot{a}\rho_{NG}}{2a} \oint D_U{}^2 \hat{\mathbf{n}} dS
\end{aligned} \qquad (9.46)
$$

where $D_U(t)$ is the radius of the universe at time $t$. Although $\rho_{NG}$ may vary with time, it is assumed to be spatially uniform and may be brought outside the integral, allowing the total force on the universe to be expressed as an integral over its surface. Let us define[9]

$$
D_U(t) = a(t) D_{U0}, \qquad (9.47)
$$

where $D_{U0}$ is the *current* radius of the universe. Then, from (9.46), the force per unit area (pressure) at the surface of the universe due to accelerated expansion is given by

$$
\frac{d\mathbf{K}}{dS} = \frac{1}{2} \ddot{a} a \rho_{NG} D_{U0}{}^2 \hat{\mathbf{n}}. \qquad (9.48)
$$

*9.6.3 Law of expansion.* Thus we have two expressions for the pressure: one, (9.42), derived from the Casimir self-force density of the universe, the other, (9.48), from an application of Newton's second law to the mass of the expanding universe. Equating the two expressions we obtain

$$
\ddot{a} = \frac{4\pi}{3} \frac{G\ell^2 \rho_{NG}}{a D_{U0}{}^2}, \qquad (9.49)
$$

which can also be written

$$
\frac{d}{dt} \dot{a}^2 = \frac{8\pi}{3} \frac{G\ell^2}{D_{U0}{}^2} \frac{\rho_{NG}}{a} \dot{a}. \qquad (9.50)
$$

Thus we arrive at an equation for the rate of increase of the scale factor, $a$:

$$\dot{a}^2 = \frac{8\pi G}{3} \frac{\ell^2}{D_{U0}{}^2} \int \frac{\rho_{NG}}{a} da . \tag{9.51}$$

*9.6.4 Non-gravitating matter density of the universe.* We cannot evaluate the integral of (9.51) as it stands because we do not know the dependence of $\rho_{NG}$ on scale factor $a$. To get this dependence we shall have to see how (9.51) relates to conventional cosmology based on the Einstein field equations. Solving those equations in the Robertson-Walker metric, one obtains, for a *flat* universe[10]

$$\frac{\ddot{a}}{a} + \frac{2\dot{a}^2}{a^2} = 4\pi G\left(\rho_{TOT} - \frac{p_{TOT}}{c^2}\right) \tag{9.52}$$

and

$$\frac{3\ddot{a}}{a} = -4\pi G\left(\rho_{TOT} + \frac{3p_{TOT}}{c^2}\right), \tag{9.53}$$

where we have written the mass density and pressure with a subscript TOT to indicate that these are *total* quantities, meaning that they include the vacuum energy $\rho_\Lambda$, i.e., energy deriving from the cosmological constant. The total mass density here may be expressed in the form[11]

$$\rho_{TOT} = \left(\Omega_\Lambda + \frac{\Omega_B + \Omega_C}{a^3} + \frac{\Omega_R}{a^4}\right)\rho_{CRIT} \tag{9.54}$$

$$\text{critical density } \rho_{CRIT} \cong 10^{-29} \text{ g/cm}^3 \tag{9.55}$$

and with fractional components[12, 13]

$$\left.\begin{array}{ll} \text{dark energy} & \Omega_\Lambda = 0.683 \\ \text{baryonic matter} & \Omega_B = 0.049 \\ \text{cold dark matter} & \Omega_C = 0.268 \\ \text{radiation} & \Omega_R = 0.0000485 \end{array}\right\}. \tag{9.56}$$

Note that $\Omega_\Lambda + \Omega_B + \Omega_C + \Omega_R = 1$.

Eliminating $\ddot{a}$ between (9.52) and (9.53) we obtain the Friedmann equation for the time evolution of the scale factor $a$:

$$\dot{a}^2 = \frac{8\pi G}{3} \rho_{\text{TOT}} a^2 . \tag{9.57}$$

Equating the right-hand sides of (9.51) and (9.57) we have

$$\int \frac{\rho_{\text{NG}}}{a} da = \frac{D_{U0}^{\ 2}}{\ell^2} a^2 \rho_{\text{TOT}} , \tag{9.58}$$

which yields for the non-gravitating mass density of the universe

$$\rho_{NG} = -2qa^2 \cdot \frac{D_{U0}^{\ 2}}{\ell^2} \rho_{\text{CRIT}} , \tag{9.59}$$

where $q$ is the *deceleration parameter*:[15]

$$q = -\Omega_\Lambda + \frac{\Omega_B + \Omega_C}{2a^3} + \frac{\Omega_R}{a^4} . \tag{9.60}$$

This is the non-gravitating mass density required to reproduce the Friedmann equation (9.57) governing the expansion of the universe.

Present day, using the above Planck data, we have $q_0 = -0.524$, giving

$$\rho_{\text{NG},0} = 1.048 \frac{D_{U0}^{\ 2}}{\ell^2} \rho_{\text{CRIT}} , \tag{9.61}$$

which is positive and, from (9.49), thus properly indicates an expanding universe.

*9.6.5 Connection between $\rho_{\text{QFT}}$ and $\rho_{\text{CRIT}}$.* In **Sec. 9.2** above, it was proposed that $\rho_{\text{NG}} \propto \rho_{\text{QFT}}$, i.e., that the expansion of the cosmos was in fact driven by the huge vacuum energy density given by quantum field theory, that density acting as the source of a cosmic Casimir force at edge of the universe. In the light of (9.61), we see our proposal can be met by taking

$$\frac{D_{U0}^{\;2}}{\ell^2}\rho_{CRIT} = C\rho_{QFT}\,, \tag{9.62}$$

where $C$ is a constant of proportionality of the order unity. And so we finally see the reason for the vast discrepancy between the observed $\rho_{CRIT}$ and the calculated $\rho_{QFT}$: it is due to the enormous value of the proportionality constant $D_{U0}^{\;2}/\ell^2$ --a constant equal to the square of the ratio of the largest and smallest measurable dimensions in the physical universe.

## 9.7 Size of the physical universe

An interesting side benefit of relation (9.62) is that it yields an estimate of the current radius $D_{U0}$ of the physical universe (as distinct from the observable one). If for illustration we take constant $C=1$, and substitute for $\ell$ using (8.80) with constant $B = \sqrt{10}$, we obtain

$$D_{U0} = \left(\frac{\rho_{QFT}}{\rho_{CRIT}}\right)^{1/2} \ell_P\,. \tag{9.63}$$

With the following values of the constants [see (9.55), fn. 6 below and (8.79)] :

$$\left.\begin{aligned} \rho_{CRIT} &= 10^{-29}\ \text{g/cm}^3 \\ \rho_{QFT} &= 1.017\times10^{95}\ \text{g/cm}^3 \\ \ell_P &= 1.6\times10^{-33}\ \text{cm} \end{aligned}\right\} \tag{9.64}$$

we obtain

$$D_{U0} = 1.63\times10^{29}\ \text{cm}\,. \tag{9.65}$$

The radius $D_{OBS}$ of the observable universe is[16]

$$D_{OBS} = 4.35\times10^{28}\ \text{cm}\,. \tag{9.66}$$

Hence

$$\frac{D_{\mathrm{U0}}}{D_{\mathrm{OBS}}} = 3.75 . \tag{9.67}$$

(The two circles of **Fig. 9.1** are drawn to this ratio.) This is a surprisingly small number. In contrast, Alan Guth, arguing from inflation theory, suggests a ratio of the order $\sim 10^{23}$.[17] One might as well call it infinite. What makes the value (9.67) interesting is that, should it be correct, the observable universe stands a good chance of its being intersected by the edge of the physical universe. The area of intersection presumably would then reveal itself to us as a kind of localized anisotropy, in violation of the cosmological principle. The probability $P$ that the observable universe is intersected by or is tangent to the boundary of the physical universe is

$$P = 1 - \left(1 - \frac{D_{\mathrm{OBS}}}{D_{\mathrm{U0}}}\right)^3 = 0.61 , \tag{9.68}$$

where we have inserted the ratio (9.67). In other words, on the present theory of cosmic expansion caused by the Casimir force, there is roughly a 60% chance of detecting a break in isotropy somewhere in the observable universe. The 2008 claim for observation of 'dark flow' is suggestive in this regard.[18]

## 9.8 No force-mediating particle for the cosmic Casimir force

In terms of the Casimir $E$ and $B$ fields, the 00 component of the stress-energy tensor (9.29) takes the familiar form [cf. (6.30) and (8.83)]

$$\begin{aligned} \sigma^{00} &= \frac{1}{2\tilde{\mu}}\left(\frac{1}{c^2}\mathbf{E}_\alpha \cdot \mathbf{E}^\alpha + \mathbf{B}_\alpha \cdot \mathbf{B}^\alpha\right) \\ &= -\frac{1}{2\tilde{\mu}c^2} E_j{}^k E_j{}^k \end{aligned} . \tag{9.69}$$

where for the second line we have used the fact that there is no $B$ field. Then from (9.39), and after inserting the constant $\tilde{\mu} = \kappa c^2 = 8\pi G / c^2$, we obtain the mass equivalent

$$\begin{aligned}\frac{\sigma^{00}}{c^2} &= -\frac{\pi G}{3c^2}\ell^2 \rho_{\mathrm{NG}}{}^2 \\ &\approx -\frac{\pi G}{3c^2}\ell_P{}^2 \rho_{\mathrm{QFT}}{}^2 \\ &= -2\times 10^{96}\ \mathrm{g/cm^3}\end{aligned} \quad . \tag{9.70}$$

This is the energy density of the rank-3 tensor field. In magnitude it is just one-half the pressure formed at the surface of the universe and only a little larger than $\rho_{\mathrm{QFT}}$. Remarkably, it is algebraically negative, which fact by itself does not necessarily prevent it from gravitating. Nevertheless, it obviously does not gravitate, just as $\rho_{\mathrm{QFT}}$, though positive, does not gravitate.

Of interest also is the Poynting vector $S^k = c\sigma^{0k}$, or

$$\begin{aligned}\mathbf{S} &= \frac{1}{\tilde{\mu}}\mathbf{E}_\alpha \times \mathbf{B}^\alpha \\ &= 0\end{aligned} \quad . \tag{9.71}$$

The Poynting vector vanishes because $\mathbf{B}^\alpha$ vanishes.

By now we are familiar with the idea that operator $\hat{R}$ in equations such as (5.24), (8.1) and (9.1) represents the quantum of the field on which it operates. Indeed in the cases of the Maxwell and Riemann-Christoffel fields $\hat{R}$ does represent the particle, viz., the photon and riemann, respectively. However, in the present case—that of the rank-3 Casimir field— $\hat{R}$ does not represent the particle, because there *is* no force-mediating particle. There is no such particle because no particle can have zero momentum and still mediate force. The cosmic Casimir force is similar to the force of gravitation, in that it acts without benefit of an associated field quantum.

## 9.9 Two alternative models of expansion

We know that Friedmann equation (9.57) describes the expansion of the universe, and this is true whether the expansion is accelerating or decelerating. What is not necessarily known is the nature of the force causing the expansion. The present chapter has argued for a cosmic Casimir force as being the cause of the expansion. The Casimir force is unique in that its source is the non-gravitating vacuum energy, $\rho_{\text{QFT}}$, and requires for its production a bounded universe. For comparison with the theory just developed, we want to discuss two alternative models of expansion, both driven by a gravitating source of vacuum energy. The first of these is the currently-accepted standard model based on the Einstein field equations and which assumes an infinite universe. The second relies solely on Newtonian gravitation operating in a finite universe.

*9.9.1 The standard model of cosmic expansion.*[19] This model is based on the Einstein/Friedmann equation, (9.53). Respecting isotropy and homogeneity out to infinity, the model is boundary-free. According to (9.53), for accelerating expansion, one evidently must have $\rho_{\text{TOT}}c^2 + 3p_{\text{TOT}} < 0$, or $p_{\text{TOT}} < -\rho_{\text{TOT}}c^2/3$. Since $\rho_{\text{TOT}}$ is positive, we see that for accelerating expansion, $p_{\text{TOT}}$ must be negative. Let us see how this comes about. We write (9.53) in the form

$$\frac{d}{dt}\dot{a}^2 = -\frac{8\pi G}{3}\left(\rho_{\text{TOT}} + \frac{3p_{\text{TOT}}}{c^2}\right)a\dot{a}\,. \tag{9.72}$$

Multiplying by *dt* and integrating, we get

$$\dot{a}^2 = -\frac{8\pi G}{3}\int\left(\rho_{\text{TOT}} + \frac{3p_{\text{TOT}}}{c^2}\right)ada\,. \tag{9.73}$$

On comparing this with (9.57) we see that we must have

$$\int\left(\rho_{\text{TOT}} + \frac{3p_{\text{TOT}}}{c^2}\right)ada = -\rho_{\text{TOT}}a^2, \tag{9.74}$$

from which it follows that

$$\rho_{\text{TOT}} + \frac{3p_{\text{TOT}}}{c^2} = 2q\rho_{\text{CRIT}}, \tag{9.75}$$

where $q$ is the deceleration parameter, (9.60). Replacing $\rho_{\text{TOT}}$ and $q$ by means of (9.54) and (9.60) we obtain

$$\begin{aligned}\frac{p_{\text{TOT}}}{c^2} &= \left(-\Omega_\Lambda + \frac{\Omega_R}{3a^4}\right)\rho_{\text{CRIT}} \\ &\simeq -\rho_\Lambda \\ &= -\frac{\Lambda c^2}{8\pi G}\end{aligned} \tag{9.76}$$

where $\Lambda$ is the cosmological constant and where in the second and third lines we have neglected the tiny radiation term. This is the pressure predicted by the standard theory of cosmic expansion. It is attributable to the presence of the vacuum energy $\rho_\Lambda$, which in turn comes from the presence in the Einstein equations of a cosmological constant. The pressure is indeed negative and, as stated above, is generally said to be the cause of the accelerating expansion of the universe.

*9.9.2 Cosmic expansion by negative gravitating mass.* The concept of negative pressure is (for this writer) difficult to visualize, because it is not pushing against anything. The universe in the standard model is, after all, infinite. If there were nothing in the infinite universe but constant dark energy $\rho_\Lambda$, what exactly would be expanding? Conceptually an expanding *finite* universe is perhaps more plausible than an infinite one, because one can then explain expansion by way of a force acting on its boundary. Let us try to derive such a force assuming a finite universe of radius $D_{\text{U}}$, filled with *gravitating* matter of uniform density, $\rho_{\text{G}}$. The pressure $d\mathbf{F}_{\text{G}}/dS$ at the surface of this finite universe due to Newtonian gravitation—the gravitational self-force of the universe—is given by (8.77) (with the replacements $\mathbf{F} \to \mathbf{F}_{\text{G}}$, $a \to D_{\text{U}}$ and $\rho \to \rho_{\text{G}}$):

$$\frac{d\mathbf{F}_{\text{G}}}{dS} = -\frac{2\pi G}{3}\rho_{\text{G}}{}^2 D_{\text{U}}{}^2\hat{\mathbf{n}} \tag{9.77}$$

But as a result of this self-force each element of the universe experiences radial acceleration, leading to a pressure at the surface of the universe. This pressure from acceleration must, by Newton's second law, exactly match the pressure $d\mathbf{F}_{\text{G}}/dS$ due to the gravitational self-force. It is obtained by differentiating (9.46) (with the replacements $\mathbf{K} \to \mathbf{F}_G$, $\rho_{NG} \to \rho_G$):

$$\frac{d\mathbf{F}_G}{dS} = \frac{\ddot{a}\rho_G}{2a} D_U^{\ 2}\hat{\mathbf{n}}. \qquad (9.78)$$

Thus combining (9.77) and (9.78), we have

$$\ddot{a} = -\frac{4\pi G}{3}\rho_G a, \qquad (9.79)$$

which is an equation of the Friedmann form (9.53), with $\rho_G$ replacing $\rho_{TOT} + 3p_{TOT}/c^2$. The energy density $\rho_G$ driving the expansion is thus given by [see (9.75)]

$$\rho_G = 2q\rho_{CRIT}. \qquad (9.80)$$

Now at the present time, $q$ is negative [see (9.60)]. Consequently $\rho_G$ is now negative and (9.79) indeed describes accelerating expansion.

## 9.10 Discussion

Thus we appear to have in hand three viable theories of the accelerating expansion of the cosmos. **Table 9.1** summarizes for comparison the features of these competing theories.

Of the three theories, the standard model perhaps inspires the most confidence because it is entirely self-contained; it uses only the field equations of Einstein. The other two theories, in constrast, begin with non-Ricci field quantities, but ultimately must turn to Friedmann (9.57) to determine the mass densities $\rho_{NG}$ and $\rho_G$. If the standard model has a weakness, it is that it involves the *ad hoc* introduction of a cosmological constant into the Einstein equations. But that constant is central to the model, resulting in the negative pressure widely assumed responsible for the accelerating expansion of the universe.

The theory we have labeled 'Newtonian' enjoys an advantage of conceptual simplicity, the mechanism of expansion being that of ordinary Newtonian gravitation. It is also very peculiar. For according to (9.77) the force producing accelerated expansion is pressing *inward* on the boundary of the universe. Of course this is to be expected: when the mass is negative, as it is in (9.79), the force and acceleration in Newton's $F = ma$ are oppositely directed. (Similarly, in

**Table 9.1** Comparison of three theories of cosmic expansion.

| Feature | Law of Laws (Casimir) | Standard Model (Einstein) | Newtonian |
|---|---|---|---|
| Predicts flat universe | Yes | No. Must be assumed. | No. Must be assumed. |
| Predicts bounded universe | Yes | No. Universe is infinite. | No, but boundary is assumed. |
| Field | Rank-3 tensor $F_{\alpha}^{\ \mu\nu}$ | Rank-2 Ricci $R_{\mu\nu}$ | Newtonian gravitational |
| Field source | Non-gravitating rank-2 mass current density $j_{\alpha}^{\ \nu} = \frac{\ell \rho_{\mathrm{NG}}}{2} \partial[v^{\nu} \sigma(\mathbf{x})]$ | Stress-energy $S_{\mu\nu} = T_{\mu\nu} - \frac{1}{2} g_{\mu\nu} T^{\mu}_{\ \mu}$ | Gravitating mass density $\rho_{\mathrm{G}}$ |
| Force per unit area acting on boundary | $\frac{d\mathbf{K}}{dS} = +\frac{2\pi G}{3} \ell^2 \rho_{\mathrm{NG}}^{\ 2} \hat{\mathbf{n}}$ | None: no boundary | $\frac{d\mathbf{F}_{\mathrm{G}}}{dS} = -\frac{2\pi G}{3} \rho_{\mathrm{G}}^{\ 2} D_{\mathrm{U}}^{\ 2} \hat{\mathbf{r}}$ |
| Dynamics from | Newton's $\mathbf{F} = m\ddot{\mathbf{x}}$ | Einstein's $R_{\mu\nu} = -\kappa S_{\mu\nu}$ | Newton's $\mathbf{F} = m\ddot{\mathbf{x}}$ |
| Resulting equation for $\ddot{a}$ | $\ddot{a} = \frac{4\pi}{3} \frac{G \ell^2 \rho_{\mathrm{NG}}}{a D_{\mathrm{U0}}^{\ 2}}$ | $\ddot{a} = -\frac{4\pi G}{3} (\rho_{\mathrm{TOT}} + \frac{3 p_{\mathrm{TOT}}}{c^2}) a$ | $\ddot{a} = -\frac{4\pi G}{3} \rho_{\mathrm{G}} a$ |
| Friedmann equation for $\dot{a}$ | $\dot{a}^2 = \frac{8\pi G}{3} \rho_{\mathrm{TOT}} a^2$ | $\dot{a}^2 = \frac{8\pi G}{3} \rho_{\mathrm{TOT}} a^2$ | $\dot{a}^2 = \frac{8\pi G}{3} \rho_{\mathrm{TOT}} a^2$ |
| From Friedmann: source of expansion | Positive non-gravitating mass density $\rho_{NG} = -2qa^2 \cdot \frac{D_{U0}^{\ 2}}{\ell^2} \rho_{\mathrm{CRIT}}$ | Negative pressure $\frac{p_{\mathrm{TOT}}}{c^2} \simeq -\rho_{\Lambda}$ | Negative gravitating mass density $\rho_{\mathrm{G}} = 2q \rho_{\mathrm{CRIT}}$ |
| Mechanism of expansion | Force on boundary | Negative pressure | Force on boundary |
| Constants of the theory | $G$, $c$, $D_{\mathrm{U0}}$, $\ell$ | $G$, $c$ | $G$ |

| | | | |
|---|---|---|---|
| Connection between $\rho_{CRIT}$ and $\rho_{QFT}$ | $\frac{D_{U0}^{\ 2}}{\ell^2}\rho_{CRIT} = C\rho_{QFT}$ | None | None |

the hole theory of the electron, the momentum and velocity of the negative energy electron are oppositely directed.) The question is, conceptual simplicity aside, does the Newtonian theory offer an acceptable explanation of the accelerating expansion of the universe? The answer is that it probably does. In the first place, as pointed out by Bondi,[20] nothing in gravitation theory prohibits negative mass. Secondly, while negative mass may seem exotic, it is perhaps no more so than negative pressure. Interestingly, although the Newtonian theory begins by assuming a bounded universe, the radius $D$ of the universe ultimately cancels out, yielding an equation (9.79) for $\ddot{a}$ independent of radius and structurally similar to equation (9.53) for $\ddot{a}$ produced by the standard model. Suppose in the Newtonian model we let $D$ go to infinity. Then we just reproduce the infinite geometry of the standard model. But this does not mean that the two models then become identical. On the contrary, each retains its own mechanism for expansion: one, a force on the boundary of infinite radius; the other, negative pressure.

At present there seems no obvious way of distinguishing theoretically between the standard and Newtonian models, or, for that matter, between those models and the one generated by the Law of Laws. All three theories appear capable of describing correctly the accelerating expansion of the universe. There is, however, one practical distinction to be made: of the three theories, the one deriving from the Law of Laws is more predictive. In particular, it predicts a flat universe and requires that it be bounded. Through (9.61) it *requires* a huge ratio $\rho_{NG,0} / \rho_{CRIT}$ to produce the observed expansion of the physical universe. It predicts, too, that the physical universe may not be much larger than the observable one, in which case the very edge of the physical universe might be observable.

## Notes and references


[1] Hyôitirô Takeno, "On the spintensor of Lanczos," Tensor **15** (1964), 103-119.

[2] Hans C. Ohanian, *Gravitation and Spacetime* (W. W. Norton & Co., New York, 1976), pp. 253-254.

[3] S. Weinberg, *Cosmology* (Oxford University Press, Oxford, UK, 2008), pp.2-9.

[4] http://map.gsfc.nasa.gov/universe/uni_shape.html.

[5] Steve K. Lamoreaux, "Casimir forces: Still surprising after 60 years," *Physics Today*, February 2007, 40-45.

[6] To calculate $u_{\text{QFT}}$ (for a quantized electromagnetic field) one assumes that the vacuum is filled with simple harmonic oscillators, all fluctuating at zero-point energy $E_0 = h\sigma c/2$, where wave number $\sigma = 1/\lambda$. All normal modes oscillating at reciprocal wavelengths $<\sigma$ occupy the positive octant of a spherical volume of radius $\sigma$ in wave-number space. Hence the density of modes is $dn = d(4\pi\sigma^3/3) = 4\pi\sigma^2 d\sigma$. Then, taking account of the two polarization states of each mode, we get for the energy density of the vacuum,

$$
\begin{aligned}
u_{\text{QFT}} &= 2\int_0^{\sigma_{\max}} E_0 dn \\
&= \pi hc\sigma_{\max}^4
\end{aligned}
$$

where $\sigma_{\max}$ is the cut-off value. Typically one takes for $\sigma_{\max}$ the reciprocal of the Planck length $\ell_{\text{P}} = (\hbar G/c^3)^{1/2} = 1.616\times10^{-33}$ cm This value yields $u_{\text{QFT}}/c^2 \;= 1.017\times10^{95}$ g/cm$^3$, and hence the infamous ratio

$$
\frac{u_{\text{QFT}}}{u_\Lambda} \sim 10^{124},
$$

where $u_\Lambda/c^2 \sim 10^{-29}$ g/cm$^3$ is the dark energy density (see **Sec. 6.7**). Now $u_\Lambda$ is an observed value, whereas $u_{\text{QFT}}$ is a calculated one. The task of understanding why the two values should differ at all—let alone so outrageously—is known as the cosmological constant problem. See Ref. [7].

[7] S. Weinberg, "The cosmological constant problem," Rev. Mod. Phys. **61**, 1 (1989).

[8] S. Weinberg, *Cosmology* (Oxford University Press, Oxford, UK, 2008), pp. 34-45.

[9] S. Weinberg, ibid, p. 5, Eq. (1.1.15).

[10] S. Weinberg, *ibid*, p. 36.

[11] S. Weinberg, *ibid*, p. 41.

[12] Planck Collaboration, "Planck 2013 results. I. Overview of products and scientific results," arXiv:1303.5062 [astro-ph.CO].

[13] www.int.washington.Edu/PHYS554/2011/Chapter2_11.pdf

[14] S. Weinberg, *ibid*, p. 37.

[15] S. Weinberg, *ibid*, p. 13, Eq. (1.5.48).

[16] T. M. Davis and C. H. Lineweaver, "Expanding Confusion: Common Misconceptions of Cosmological Horizons and the Superluminal Expansion of the Universe," *Astron. Soc. of Australia* **21** (2004), 97-109.

[17] Alan H. Guth, *The Inflationary Universe*, (Perseus Books, Reading Massachusetts, 1997), p. 186.

[18] A. Kashlinsky *et al.*, "A measurement of large-scale peculiar velocities of clusters of galaxies: results and cosmological implications," arXiv:0809.3734 [astro-ph] 22 Sep 2008.

[19] S. Perlmutter, "Supernovae, Dark Energy, and the Accelerating Universe," *Physics Today* (April 2003), 53-60.

[20] H. Bondi, "Negative Mass in General Relativity," Rev. Mod. Phys. **29** (1957), 423.

# PART IV

# INTERNAL STRUCTURE OF THE FIELD QUANTA

# Chapter 10

# Internal structure of the leptons, internal spacetime and baryon asymmetry

For more than a century we have known that at a fundamental level the physical world comprises a duality of wave and particle. As we saw in **Chapter 3**, this dualistic behavior can be attributed to a condition of self-consistency present in the world of nature—a condition we have called the Principle of True Representation—expressed formally by the Law of Laws, Eq. (2.6). The Law of Laws is law of propagation, a wave equation describing the propagation of a non-diffracting wave $\chi_n$ along the $x^5$ axis of 6-D spacetime. As mentioned already in *Sec. 3.9.3*, the wave propagator $\hat{R}$ in its convolution form represents the particle, or quantum, of the propagating wave field. In this and the next chapter we are going to show in detail the internal spatial structure of the known quanta. Specifically, in the present chapter we shall study the leptons, in both their charged ($e,\ \mu,\ \tau$) and electrically neutral ($\nu_e,\ \nu_\mu,\ \nu_\tau$) forms. In the next chapter we shall study internal structures of the bosons, including those of the photon and riemann particle discussed in **Sec. 8.12.**

We take on the leptons first, not because they are the simplest of the quanta—which they are not—but because for a proper description they demand that we address a full range of particle issues: internal mass and charge currents, spin density and family structure. As it happens, some of the features of the particle picture can be treated classically, and it is at this level we begin our discussion. But to get at the true nature of the field quanta we shall ultimately need to make use of the Lagrangian variational calculus. From there, one can derive the internal details of the charged leptons and their associated neutrinos. This seems straightforward, but a surprise awaits. For the internal structures of these particles entail negative probability densities. As such densities cannot be understood in ordinary Euclidean $E^3$, one is led to account for them by way of a new, 2-*D* internal Minkowskian spacetime, $N^2$. The existence of $N^2$ at once leads to a possible solution to the problem of the missing world of antimatter.

## 10.1 Special relativity and the *Zitterbewegung*

By the 'internal structure' of a particle we mean the locus of oscillatory motion, or *Zitterbewegung*,[1] of the point mass comprising that particle. The Zitterbewegung of the electron may in fact be viewed as the source of that particle's intrinsic spin and magnetic moment.[2] Its potential significance for describing the intrinsic properties of leptons in general would thus seem obvious. Nevertheless, in the literature there appear to be no detailed accounts of the Zitterbewegung in mu and tau,[3] perhaps owing to the dictum that these objects are simply more massive versions of the electron. Nor to our knowledge does the literature refer except in passing to the Zitterbewegung of the neutrino.[3] Accordingly our immediate aim is to present a generalized account of the relativistic kinematics of the Zitterbewegung, one applicable to all leptons, not just the electron.

Although the Zitterbewegung is in the last analysis a quantum phenomenon, its main kinematical features are already present in the formalism of special relativity. To show this, let us take a free lepton from family $\ell$ and write down the classical 4-vector energy-momentum relation for it:

$$p_0^{\,2} - \mathbf{p}^2 = m_{\ell n}^{\,2} c^2 , \tag{10.1}$$

where, repeating (4.92),

$$m_{\ell n} = n m_\ell , \tag{10.2}$$

and where $n = 0$ for massless neutrino $\nu_\ell$, and $n = 1$ for the charged lepton of mass $m_\ell$. But we know that $M^4$ is embedded in six-dimensional $M^4 \times \bar{M}^2$, and that in six dimensions lepton mass is given in terms of two invariants, $p_{5\ell n}$ and $p^6 = G_\ell$ [see (4.38)]:

$$m_{n\ell}^{\,2} c^2 = G_\ell^{\,2} - p_{5\ell n}^{\,2} , \tag{10.3}$$

where, repeating (4.89)

$$p_{\ell n5} = G_\ell - n m_e c \tag{10.4}$$

and (4.91)

$$G_\ell = \frac{m_\ell^{\,2} + m_e^{\,2}}{2m_e} c . \tag{10.5}$$

Eliminating mass $m_{\ell n}$ between (10.1) and (10.3) we obtain

$$p_0^{\ 2} - \mathbf{p}^2 - G_\ell^{\ 2} = -p_{5\ell n}^{\ \ 2}, \tag{10.6}$$

which is the energy momentum relation for our free lepton in six dimensions. Written this way, (10.6) reveals the dynamics of a point $P$ of *negative* mass squared $-p_{5\ell n}^{\ \ 2}/c^2$ and *total* mechanical momentum $p_T$, given by

$$p_T^{\ 2} = \mathbf{p}^2 + G_\ell^{\ 2}. \tag{10.7}$$

So the total momentum of the generic lepton, whether charged or neutral, has two components—the constant external linear momentum $\mathbf{p}$ and a time-varying *internal* momentum of constant magnitude $p^6 = G_\ell$. The latter is to be interpreted as the momentum of the Zitterbewegung, a phenomenon now seen to be dependent on the existence of the pseudotemporal dimension, $x^6$. The precise internal geometry implied by (10.2) depends on whether the lepton under consideration is electrically neutral or charged.

*10.1.1 Classical internal geometry of the neutrino.* For the massless neutrino ($n=0$) we have from (10.1), $p_0 = |\mathbf{p}|$. Thus for the neutrino we have from (10.7)

$$p_T^{\ 2} = p_0^{\ 2} + G_\ell^{\ 2} \tag{10.8}$$

The point mass $P$ comprising the neutrino thus describes a helical path about the average direction of motion with longitudinal momentum $p_0$ and transverse momentum $G_\ell$; see **Fig. 10.1.** The helical motion of point $P$ accounts for the intrinsic spin of the neutrino. Let $b_\ell$ denote the radius of the helix. The neutrino's spin angular momentum can then be written $b_\ell G_\ell = \hbar/2$, giving for the radius

$$b_\ell = \frac{\hbar}{2G_\ell} = \frac{m_e}{m_\ell^{\ 2} + m_e^{\ 2}} \frac{\hbar}{c}, \tag{10.9}$$

where for $G_\ell$ we have substituted (10.5). For the electron neutrino ($\ell = e$) the radius is $\hbar/2m_e c$, or one-half the reduced Compton wavelength of the electron. For neutrinos $\nu_\mu$ and $\nu_\tau$, the helix radius $b_\ell$ is $\approx m_e / m_\ell$ times smaller than the reduced Compton wavelength of the associated charged lepton.

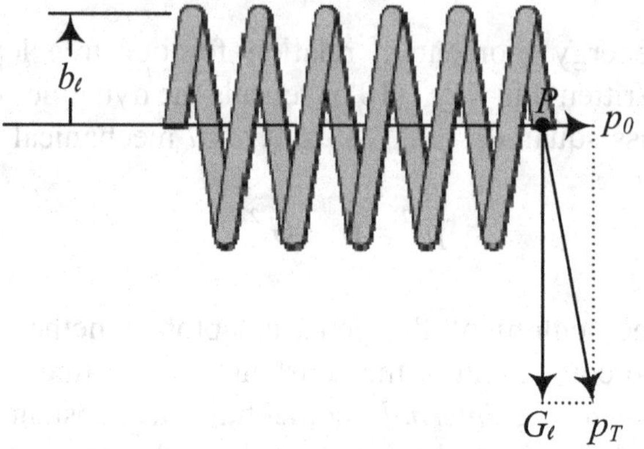


**Fig. 10.1** Classical picture of a left-handed neutrino $\nu_\ell$ executing Zitterbewegung. Mass point $P$ of negative mass-squared $-p_5^{\,2} / c^2$ traces a helical path with longitudinal momentum $p_0$, transverse momentum $G_\ell$ and total momentum $p_T = (p_0^{\,2} + G_\ell^{\,2})^{1/2}$. Radius of the helix is $b_\ell$, which in the case of neutrino $\nu_e$ is one-half the reduced Compton wavelength of the electron, $e^-$.

The massless neutrino moves as a whole with velocity $c$. Thus internal point $P$, taking a helical path about the centerline of motion, is superluminal. Multiplying both sides of (10.4) by $c^2 / p_0^{\,2}$ we find for $P$'s total velocity $\upsilon_T$,

$$\upsilon_T = \sqrt{c^2 + \upsilon_5^{\,2}}\,, \tag{10.10}$$

where for neutrino $\nu_\ell$

$$\upsilon_5 = \frac{G_\ell c}{p_0}\,. \tag{10.11}$$

In the case of the electron neutrino the $\upsilon_5$ component of the velocity reduces to $m_e c^2 / p_0$.

*10.1.2 Classical internal geometry of the charged leptons.* We now describe the Zitterbewegung of the charged lepton, $\ell^-$. From (10.7) and (10.3) we have for the total momentum of internal point $P$

$$p_T^{\ 2} = \mathbf{p}^2 + m_\ell^{\ 2}c^2 + p_5^{\ 2}, \tag{10.12}$$

where from (10.4) and (10.5)

$$\begin{aligned} p_5 &= G_\ell - m_e c \\ &= \frac{m_\ell^{\ 2} - m_e^{\ 2}}{2m_e} c \end{aligned}. \tag{10.13}$$

To interpret this geometrically it proves helpful to consider first the case of the electron, $e^-$. For this particle, $p_5 = 0$ and (10.2) reduces to

$$p_T^{\ 2} = \mathbf{p}^2 + m_e^{\ 2}c^2 . \tag{10.14}$$

The massless point $P$ comprising the electron thus moves with longitudinal momentum $\mathbf{p}$ and tangential momentum $m_e c$, describing in general a helical path about the line of external motion. For an electron at rest, the helical path degenerates to a circle. The internal geometry of the electron at rest, shown in **Fig. 10.2**, reproduces the

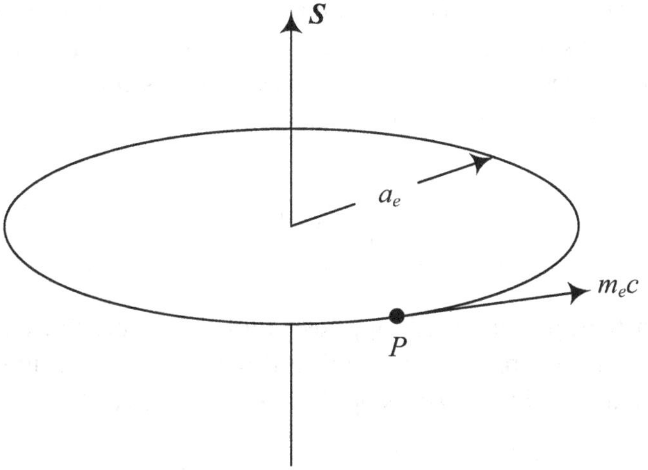


**Fig. 10.2** Classical picture of the electron at rest. Massless point $P$ orbits its mean position in a circle of radius $a_e$ with speed $c$ and tangential momentum $m_e c$, accounting for the electron's intrinsic angular momentum $\mathbf{S}$ and its rest mass $m_e c^2$.

classical model of the electron proposed by Hönl[4] and others[5]. The tangential momentum $m_e c$ accounts for the electron's spin angular momentum. If $a_e$ is the radius of the helix, then the spin can be written $a_a m_e c = \hbar/2$, giving for the radius of of the electron's Zitterbewegung

$$a_e = \frac{\hbar}{2m_e c} \tag{10.15}$$

This agrees with Schrödinger's estimate of the radius,[1] and also with radius of the electron neutrino's Zitterbewegung. Note that the tangential momentum $m_e c$ accounts not only for the electron's spin angular momentum but for its rest mass $m_e c^2$ as well.

Let us now consider the heavy leptons $\mu^-$ and $\tau^-$. If they are at rest, with $\mathbf{p} = 0$ Eq. (10.12) looks very much like (10.8). The point *P* comprising either heavy lepton thus behaves much like the one comprising the neutrino: it moves in a helical path about its average direction of motion with longitudinal momentum $m_\ell c$ and transverse momentum $p_5$, with $p_5$ given by (10.13). Its average direction of motion is, however, not the neutrino's straight line, but a circular one similar to that of the electron, but smaller in radius, yielding a compact charged heavy lepton. This peculiar internal geometry is depicted in **Fig. 10.3**. As in the case of the electron, point *P*'s longitudinal momentum is responsible for the heavy lepton's intrinsic spin and its rest mass $m_\ell c^2$ as well. The radius of the internal orbit is therefore

$$a_\ell = \frac{\hbar}{2m_\ell c}, \quad \ell = e, \mu, \tau \tag{10.16}$$

a formula that, as indicated, applies to all three charged leptons. We state here without proof the defining dimensions of the internal helix (where the approximate forms apply to mu and tau only):

$$\text{Radius} = \frac{\hbar p_5}{2G_\ell^2} \approx \frac{\hbar}{2p_5} = \frac{1}{2k_5} \tag{10.17}$$

$$\text{Pitch} = \frac{\pi\hbar m_\ell c}{G_\ell^2} \approx \frac{4\pi\hbar m_e^2}{m_\ell^3 c} \tag{10.18}$$

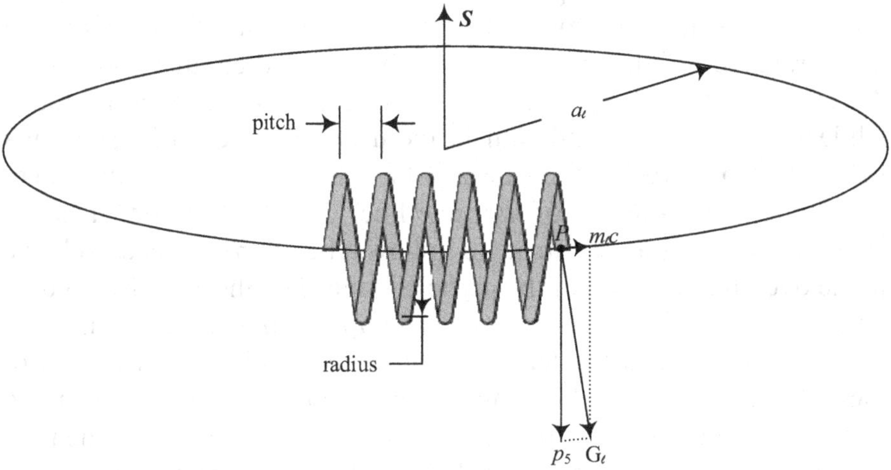


**Fig. 10.3** Classical picture of the heavy lepton $\ell^-$ ($\ell = \mu, \tau$) at rest. Point *P* of negative mass- squared $-p_5^2/c^2$ traces a helical path centered on a circle of radius $a_\ell$ with tangential speed $c$ and momentum $m_\ell c$, accounting for the heavy lepton's intrinsic angular momentum, **S** and rest mass $m_\ell c^2$. Radius of helix $= \hbar p_5 / 2G_\ell^2$. Pitch of helix $= \pi \hbar m_\ell c / G_\ell^2$. These formulas apply to the electron as well: radius of helix $= 0$; pitch of helix $= 2\pi a_e$.

Note that these formulas reduce properly to the circular orbit of massless point *P* of the electron: Radius = 0, Pitch $= 2\pi a_e$. **Table 10.1** shows

**Table 10.1** Features of internal geometry of charged lepton $\ell^-$

| Lepton $\ell^-$ | Mass $\mu_\ell \equiv \dfrac{m_\ell}{m_e}$ | # of turns of helix $= \dfrac{\text{Circumference } 2\pi a_\ell}{\text{Pitch of helix}} = \left(\dfrac{\mu_\ell^2 + 1}{2\mu_\ell}\right)^2$ | $\dfrac{\text{Radius of helix}}{\text{Radius } a_\ell} = 2\mu_\ell \dfrac{\mu_\ell^2 - 1}{\left(\mu_\ell^2 + 1\right)^2}$ |
|---|---|---|---|
| $e^-$ | 1 | 1 | 0 |
| $\mu^-$ | 206.8 | $1.07 \times 10^4$ | $0.94 \times 10^{-2}$ |
| $\tau^-$ | 3477.0 | $3.02 \times 10^6$ | $0.6 \times 10^{-3}$ |

for each of the three charged leptons, the number of turns in the helix for one complete orbit around the center of mass; and the radius of the helix relative to the radius $a_\ell$ of the helix's circular centerline. According to the table, the muon's helix contains ten-thousand turns, while its radius is one-hundredth the radius $a_\mu$ of its centerline; the tau's helix contains three million turns while its radius is less than one-thousandth the radius $a_\tau$ of *its* centerline. The helix of either heavy lepton thus closely approximates a hollow tube of tiny diameter relative to the overall dimension of the particle induced by the Zitterbewegung. And so, as far as appearance goes, and apart from scale, the internal geometry of either heavy lepton is hardly distinguishable from that of the electron. The crucial distinguishing feature of these three charged leptons is their widely differing values of the internal momentum $p^6 = G_\ell$, a feature which, as we shall see later, is responsible for their widely differing mass values.

The heavy lepton's internal mass point $P$ is, like that of the neutrino, superluminal. Multiplying both sides of (10.12) by $c^2 / p_0{}^2$, we obtain for its total velocity

$$\begin{aligned} \upsilon_T &= \sqrt{\mathbf{v}^2 + (1 - \frac{\mathbf{v}^2}{c^2})c^2 + \upsilon_5{}^2}, \\ &= \sqrt{c^2 + \upsilon_5{}^2} \end{aligned} \tag{10.19}$$

where $\mathbf{v} = \mathbf{p} / p_0$ and

$$\upsilon_5 = \frac{p_5 c}{p_0}. \tag{10.20}$$

Of course, as the electron's internal point $P$ is massless, its total velocity $\upsilon_T = c$. Tachyons do indeed exist but are confined unobservably to the volumes created by the Zitterbewegung of both the neutrino and the heavy leptons.

The image presented by **Fig. 10.3** is suggestive. It appears that the neutrinos into which the heavy lepton decays are already present inside it, ready to be released. Take the muon, for example. The transverse momentum of its internal mass point can be written $p_5(\mu^-) = G_\mu - G_e$, which expresses the conservation of $p_5$ in the muon's principal decay mode $\mu^- \to \nu_\mu + \bar{\nu}_e + e^-$. But $G_\mu$ and $-G_e$ represent the transverse momenta internal to $\nu_\mu$ and $\bar{\nu}_e$, respectively. Thus the helix internal to the muon encodes by way of its transverse momentum $p_5$ those very leptonic objects into which the muon must decay.

The tau lepton is even odder. The transverse momentum of its internal mass point $P$ can be written in either of two ways:

$$p_5(\tau^-) = G_\tau - G_e$$

or

$$p_5(\tau^-) = G_\tau - G_\mu + G_\mu - G_e.$$

These relations express the conservation of $p_5$ in alternate decay modes $\tau^- \to \nu_\tau + \bar{\nu}_e + e^-$ and $\tau^- \to \nu_\tau + \bar{\nu}_\mu + \mu^-$, respectively. Thus even at the classical level one sees a foreshadowing of the superposition of states predicted by quantum field theory. Of course this classical preview of QFT is possible only in 6-dimensional spacetime.

Before leaving this classical exposition of the Zitterbewegung, let us look briefly at the magnetic moment predicted for the charged leptons depicted in **Figs. 10.2** and **10.3**. In each case the orbital motion of charged point $P$ produces an electrical current $i = -ec/2\pi a_\ell$, where $2\pi a_\ell / c$ is the time for $P$ to orbit the center of mass. The orbit surrounds an area $A = \pi a_\ell^{\;2}$, and so the magnetic moment $\mu_\ell$ predicted for lepton $\ell^-$ is

$$\begin{aligned} \mu_\ell &= i \cdot A \\ &= -\frac{ec}{2} a_\ell , \qquad (10.21) \\ &= -\frac{e\hbar}{4m_\ell} \end{aligned}$$

which is (to a close approximation) one-half the experimentally-confirmed value given by the Dirac wave equation. This is typical of classical calculations of the magnetic moment of the electron: they generally give the wrong value. Actually one can get the correct value (or any value you please) by assuming that the distributions of mass and charge differ slightly from each other in just the right way. But this is obviously contrived, and anyway the electron is known from scattering experiments to be point-like, as assumed in the above picture of the Zitterbewegung. Something more fundamental is needed to obtain the correct magnetic moment from the particle picture, and one of the tasks of this chapter is to discover what it is.

The classical depictions presented above are not, of course, to be construed as literal representations of leptons. They do, however, enable one to visualize internal momentum and its connection with spin and magnetic moment, and to see the connection between the heavy charged leptons and their decay products.

## 10.2 Charged leptons: extracting the spatial part of propagator $\hat{R}$

We propose now to develop in quantum-mechanical terms the internal structure of the charged leptons. We have shown that the probability and charge amplitude wave functions for these particles are non-diffracting fields in 6-dimensional spacetime [see Eqs. (3.104), (3.105), (4.31) and (4.42)]. As such they obey propagation law (3.106), which we write in the form

$$\ell(x) = \Delta(x) * \ell(x)\,, \tag{10.22}$$

where $\ell(x)$ stands for either the probability or charge amplitude of charged lepton $\ell^-$,

$$\Delta(x) = e^{ik_5\xi^5} f(x)\,, \tag{10.23}$$

$f(x)$ is the wave propagator (3.81), and $*$ denotes the 4-dimensional convolution operation. The function $\Delta(x)$, as mentioned already in *Sec. 3.9.3*, represents the invariant distribution of the field quantum and evidently acts like a Dirac $\delta$-function operating on $\ell(x)$. Let us consider a particle at rest, with

$$\ell(x) = Ae^{-ik_C x^0} \tag{10.24}$$

where $A$ is a constant spinor, $k_C = 2\pi / \lambda_C$ and $\lambda_C = h / m_\ell c$ is the Compton wavelength of lepton $\ell^-$. We can then integrate out the temporal part of the convolution (10.22), revealing the purely spatial part of the distribution $\Delta(x)$. In detail, for the particle at rest, we have from (10.22),

$$\ell(x) = e^{ik_5\xi^5} \int f(\xi)\ell(x-\xi)d^4\xi$$

$$\begin{aligned} &= e^{ik_5\xi^5}\int f(\xi)e^{ik_C\xi^0}\,d\xi^0 d^3\boldsymbol{\xi}\cdot\ell(x) \\ &= \int D(\boldsymbol{\xi},\xi^5)d^3\boldsymbol{\xi}\cdot\ell(x) \end{aligned}, \tag{10.25}$$

where the purely spatial distribution $D(\boldsymbol{\xi},\xi^5)$ is given by the temporal inverse Fourier transform of the propagator $f$:

$$D(\boldsymbol{\xi},\xi^5) = e^{ik_5\xi^5}\int f(\xi^0,\boldsymbol{\xi})\; e^{ik_C\xi^0}\,d\xi^0\,. \tag{10.26}$$

To evaluate this integral, we replace $f$ by its Fourier integral representation, the first line of (3.81). After integrating over $\xi^0$ and $k_0$, and then over the polar coordinate angles $\theta$ and $\varphi$, we obtain

$$D(\boldsymbol{\xi},\xi^5) = -\frac{e^{ik_5\xi^5}}{2\pi^2}\frac{1}{\rho}\frac{\partial}{\partial\rho}I(\rho,\xi^5), \tag{10.27}$$

where

$$I(\rho,\xi^5) = \int_0^\infty e^{-i\xi^5\sqrt{k_5^2+k^2}}\cos k\rho\; dk\,, \tag{10.28}$$

$k = |\mathbf{k}|$ and $\rho = |\boldsymbol{\xi}|$. In terms of the real and imaginary parts of $D$ and $I$ we have

$$D_r = -\frac{1}{2\pi^2}\frac{1}{\rho}\frac{\partial}{\partial\rho}\left(I_r\cos k_5\xi^5 - I_i\sin k_5\xi^5\right) \tag{10.29}$$

$$D_i = -\frac{1}{2\pi^2}\frac{1}{\rho}\frac{\partial}{\partial\rho}\left(I_r\sin k_5\xi^5 + I_i\cos k_5\xi^5\right). \tag{10.30}$$

Together, the two equations form a rotation transformation, but this is a formality and of no apparent consequence. In fact, according to (10.25), the integral over $D_i$ vanishes. And so, in terms of internal structure, our interest focusses entirely on $D_r$, over which, according to (10.25), the integral is unity. The integrals $I_r$ and $I_i$ may be evaluated with the aid of a table of Fourier cosine transforms. We have[6]

$$I_r = \frac{\partial}{\partial \xi^5}\int_0^\infty \frac{\sin\left(\xi^5\sqrt{k_5^{\,2}+k^2}\right)}{\sqrt{k_5^{\,2}+k^2}}\cos k\rho \, dk = \frac{\partial}{\partial \xi^5}\left[\frac{\pi}{2}J_0(k_5 z_1)U\left(\xi^5-\rho\right)\right]$$
$$= \frac{\pi}{2}\left[-k_5\xi^5\frac{J_1(k_5 z_1)}{z_1}U\left(\xi^5-\rho\right)+\delta\left(\rho-\xi^5\right)\right] \tag{10.31}$$

$$I_i = \frac{\partial}{\partial \xi^5}\int_0^\infty \frac{\cos\left(\xi^5\sqrt{k_5^{\,2}+k^2}\right)}{\sqrt{k_5^{\,2}+k^2}}\cos k\rho \, dk = \frac{\partial}{\partial \xi^5}\left[-\frac{\pi}{2}Y_0(k_5 z_1)U\left(\xi^5-\rho\right)\right.$$
$$\left.+K_0(k_5 z_2)U\left(\rho-\xi^5\right)\right]$$
$$= k_5\xi^5\left[\frac{\pi}{2}\frac{Y_1(k_5 z_1)}{z_1}U\left(\xi^5-\rho\right)+\frac{K_1(k_5 z_2)}{z_2}U\left(\rho-\xi^5\right)\right] \tag{10.32}$$

Where he $J_n$, $Y_n$ and $K_n$ are, respectively, the Bessel, Neumann and modified Bessel functions of order $n$, and $U$ is the unit step function:

$$U(x) = \begin{cases} 0 \text{ for } x<0 \\ 1 \text{ for } x\geq 0 \end{cases}. \tag{10.33}$$

For brevity we have defined $z_1 = \sqrt{\xi_5^{\,2}-\rho^2}$ and $z_2 = \sqrt{\rho^2-\xi_5^{\,2}}$. In carrying out the derivatives with respect to $\xi^5$ we have used the relations[7]

$$U'(x) = \delta(x) \tag{10.34a}$$

$$J_0'(x) = -J_1(x)\,,\; Y_0'(x) = -Y_1(x)\,,\; K_0'(x) = -K_1(x) \tag{10.34b}$$

$$J_0(0) = 1 \tag{10.34c}$$

$$\frac{\pi}{2}Y_0(0)+K_0(0) = 0 \quad \text{(cancelling infinities)} \tag{10.34d}$$

Expressions (10.31) and (10.32) are now to be inserted into (10.29).

After taking the derivative with respect to $\rho$, we obtain for the radial distribution $4\pi\rho^2 D_r$

$$4\pi\rho^2 D_r = \left\{ k_5\xi^5 \rho \frac{\partial}{\partial\rho}\left[\frac{J_1(k_5 z_1)}{z_1}\right] U\left(\xi^5-\rho\right) - \frac{1}{2}\left(k_5\xi^5\right)^2 \delta\left(\rho-\xi^5\right) - \rho\delta'\left(\rho-\xi^5\right) \right\} \cos k_5\xi^5$$
$$+\left\{ \rho\frac{\partial}{\partial\rho}\left[\frac{Y_1(k_5 z_1)}{z_1}\right] U\left(\xi^5-\rho\right) - \xi^5 \frac{Y_1(k_5 z_1)}{z_1}\delta\left(\rho-\xi^5\right) \right.$$
$$\left. +\frac{2}{\pi}\rho\frac{\partial}{\partial\rho}\left[\frac{K_1(k_5 z_2)}{z_2}\right] U\left(\rho-\xi^5\right) + \xi^5 \frac{2}{\pi}\frac{K_1(k_5 z_2)}{z_2}\delta\left(\rho-\xi^5\right) \right\} k_5\xi^5 \sin k_5\xi^5 \tag{10.35}$$

Here we see at once that the radius $a_\ell$ of the particle is to be identified with the propagation distance $\xi^5$: particle size as generated by the Zitterbewegung comes from the fifth dimension. Now according to (10.17), the reciprocal quantity $1/k_5$ represents the diameter of the classical helix internal to either $\mu$ or $\tau$. Hence the parameter $1/k_5\xi^5 \equiv \omega^{-1}$ (<<1) represents the diameter of the classical helix relative to the particle radius $\xi^5$.

To make this ungainly expression easier to read, let us rewrite it in terms of the dimensionless variable $x = \rho/\xi^5$ and the constant $\omega = k_5\xi^5$. Then with $t_1 = (1-x^2)^{1/2}$, $t_2 = (x^2-1)^{1/2}$, (10.35) takes the slightly more intelligible form

$$4\pi\rho^2 D_r \frac{d\rho}{dx} \equiv R(x) = \left\{ \omega x \frac{\partial}{\partial x}\left[\frac{J_1(\omega t_1)}{t_1}\right] U(1-x) - \frac{1}{2}\omega^2\delta(x-1) - x\delta'(x-1) \right\} \cos\omega$$
$$+\left\{ x\frac{\partial}{\partial x}\left[\frac{Y_1(\omega t_1)}{t_1}\right] U(1-x) - \frac{Y_1(\omega t_1)}{t_1}\delta(1-x) \right.$$
$$\left. +\frac{2}{\pi}x\frac{\partial}{\partial x}\left[\frac{K_1(\omega t_2)}{t_2}\right] U(x-1) + \frac{2}{\pi}\frac{K_1(\omega t_2)}{t_2}\delta(x-1) \right\} \omega \sin\omega \tag{10.36}$$

It is this form we choose to plot, rather than that of (10.35).

Let us first consider the case of the electron. For this particle $k_5 = p_5/\hbar = \omega/\xi^5 = 0$, and (10.35) reduces to

$$4\pi\rho^2 D_r = -\rho\delta'\left(\rho-\xi^5\right) \tag{10.37a}$$

while (10.36) becomes

$$R(x) = -x\delta'(x-1). \tag{10.37b}$$

The integral over either distribution, the one with respect to $\rho$, the other with respect to $x$, is indeed unity, as it must be. The electron's radial probability distribution, the result of its Zitterbewegung, is depicted in **Fig. 10.4**, where to render the distribution visible we have approximated it with the derivative of a Gaussian distribution. This image of the internal structure of the electron is consistent with the classical picture presented in **Fig. 10.2**, in that the distribution's amplitude is concentrated in a shell of zero thickness at the radius $\xi^5 = a_e$ of the

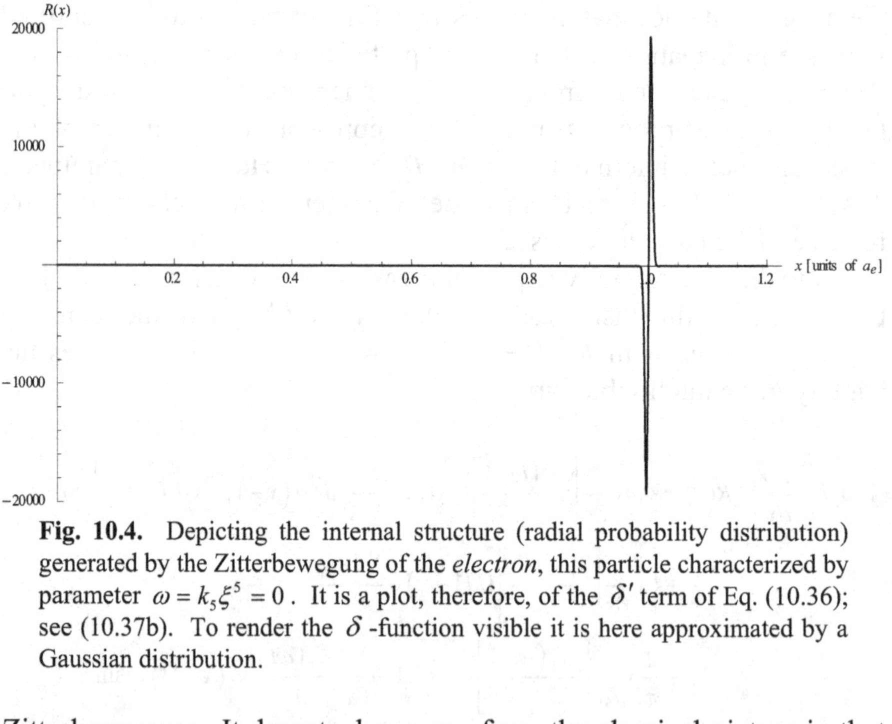


**Fig. 10.4.** Depicting the internal structure (radial probability distribution) generated by the Zitterbewegung of the *electron*, this particle characterized by parameter $\omega = k_5\xi^5 = 0$. It is a plot, therefore, of the $\delta'$ term of Eq. (10.36); see (10.37b). To render the $\delta$-function visible it is here approximated by a Gaussian distribution.

Zitterbewegung. It departs, however, from the classical picture in that it goes algebraically negative. Note that $D_r$ describes a material particle, not a wave, and as such is not to be squared. The meaning of this occurrence of negative probability will be discussed later in **Sec. 10.7.** Meanwhile, because $D_r$ does go negative, it is clear that it has no empirical significance outside the integral sign of (10.25). This is another way of saying that the Zitterbewegung is unobservable.[5]

Next we turn to the study of (10.36) for non-zero $\omega$. For illustrative purposes, so that we can readily see the basic internal structure common to all charged leptons heavier than the electron, we will pick values of helix diameter $\omega^{-1}$ that are much larger than the actual values associated with leptons $\mu$ and $\tau$ (see **Table 10.1**, right-hand column). Moreover, to avoid a confusion of curves we intend, by appropriate choices of $\omega$, to plot separately the $\cos\omega$ and $\sin\omega$ terms of (10.36). It should be mentioned that, whatever the choice of $\omega$, integration over the $\cos\omega$ term always yields $\cos^2\omega$ and integration over the $\sin\omega$ term always yields $\sin^2\omega$, thus satisfying (10.25).

**Fig. 10.5** depicts the radial probability density distribution for $\omega = 4\pi$, forcing to zero the $\sin\omega$ term of (10.36). The non-vanishing $\cos\omega$ term consists of two main parts, labeled **a** and **b**: curve **a** involves the derivative of the special function $J_1(\omega t_1)/t_1$, while curve **b** plots the

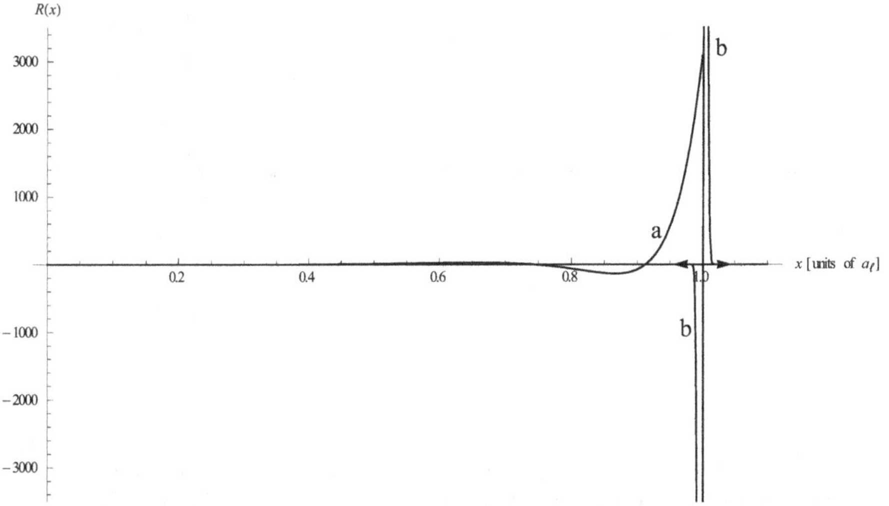


**Fig. 10.5.** This plot depicts the internal structure (radial probability distribution) generated by the Zitterbewegung of a fictitious charged lepton characterized by parameter $\omega = k_5\xi^5 = 4\pi$. It is a plot, therefore, of the $\cos\omega$ term of Eq. (10.36), the $\sin\omega$ term having vanished owing to the value of $\omega$. Curve **a** plots the term involving Bessel function $J_1$, while curve **b** plots the *sum* of the $\delta$ and $\delta'$ terms. To render the $\delta$-function visible it is here approximated by a Gaussian distribution. The arrowheads indicate the location and diameter $1/k_5 = 1/4\pi$ of the classical helix depicted in **Fig. 10.3**.

*sum* of the $\delta$-function and its derivative; for purposes of visualization we again approximate the $\delta$-function with a Gaussian distribution. Curve **b** is concentrated in a shell of zero thickness at the radius $\xi^5 = a_\ell$ of the Zitterbewegung, revealing the kinship of this fictitious lepton to the electron shown in **Fig. 10.4**. Curve **a**, however, is of finite width and serves to distinguish this particle from the electron. In fact, judging by the dimension indicated by the arrowheads in **Fig. 10.5**, curve **a** appears to correspond to the classical helical path depicted in **Fig. 10.3**.

**Fig. 10.6** depicts the radial probability density distribution for $\omega = 9\pi/2$, this value forcing to zero the $\cos\omega$ term of (10.36). The non-vanishing $\sin\omega$ consists of four terms, shown here as curves **a-d**. Curves **a** and **b** are plots, respectively, of the special functions $Y_1(\omega t_1)/t_1$ and $K_1(\omega t_2)/t_2$, each multiplied by $\delta(x-1)$; while curves **c**

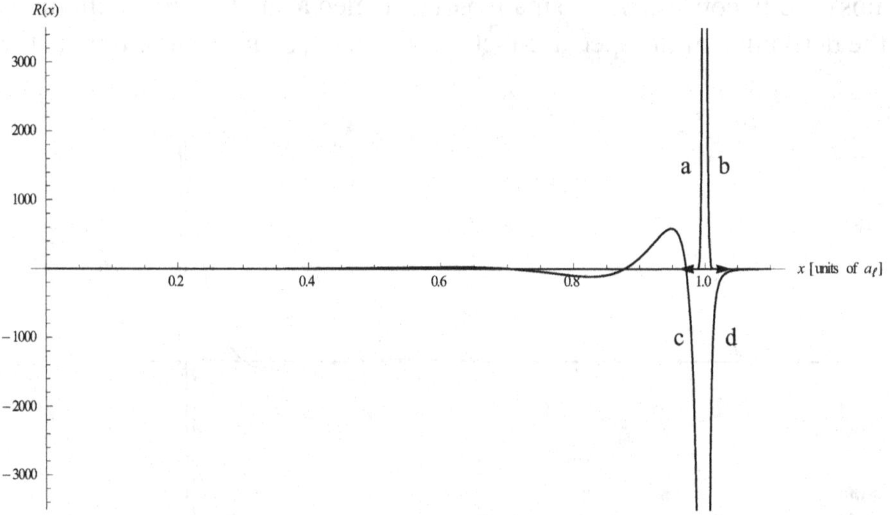


**Fig. 10.6.** This plot depicts the internal structure (radial probability distribution) generated by the Zitterbewegung of a fictitious charged lepton characterized by parameter $\omega = k_5\xi^5 = 9\pi/2$. It is a plot, therefore, of the $\sin\omega$ term of Eq. (10.36), the $\cos\omega$ term having vanished owing to the value of $\omega$. Curves **a** and **b** plot, respectively, the special functions $Y_1(\omega t_1)/t_1$ and $K_1(\omega t_2)/t_2$, each multiplied by $\delta(x-1)$; while curves **c** and **d** are plots of the derivatives of the aforementioned special functions, each multiplied by $x$. To render the $\delta$-function visible it is here approximated by a Gaussian distribution. The arrowheads indicate the location and diameter $1/k_5 = 1/(9\pi/2)$ of the classical helix depicted in **Fig. 10.3**.

and **d** are plots of the derivatives of the aforementioned special functions, each multiplied by $x$. In general appearance the distribution depicted here differs markedly from the one shown in **Fig. 10.5**. Nevertheless, key similarities remain. In particular, a first piece of the present distribution, bounded by curves **a** and **b**, is concentrated in a shell of zero thickness at the radius $\xi^5 = a_\ell$ of the Zitterbewegung, while a second piece, bounded by curves **c** and **d,** comprises a shell of finite thickness, a thickness comparable, as indicated by the arrowheads in **Fig. 10.5**, to the diameter of the classical helical path depicted in **Fig. 10.3**.

For purposes of illustration we have divided the distribution (10.36) into two fictitious types, the $\cos\omega$ type, illustrated by **Fig. 10.5**; and the sin $\omega$ type, illustrated by **Fig. 10.6**. Presumably real charged leptons would consist of a superposition of both types of distribution. **Table 10.2** provides the $\sin^2\omega$ and $\cos^2\omega$ values associated with the real charged leptons e, mu and tau. According to

**Table 10.2** $\cos^2\omega$ and $\sin^2\omega$ values for charged Leptons $\ell^-$

| Lepton $\ell^-$ | Mass $\mu_\ell \equiv \dfrac{m_\ell}{m_e}$ | $\omega = k_5 a_\ell = \dfrac{1}{4}\left(\mu_\ell - \dfrac{1}{\mu_\ell}\right)$ | $\cos^2\omega$ | $\sin^2\omega$ |
|---|---|---|---|---|
| $e^-$ | 1 | 0 | 1 | 0 |
| $\mu^-$ | 206.77 | 51.69 | 0.02 | 0.98 |
| $\tau^-$ | 3477.50 | 869.38 | 0.44 | 0.56 |

the table, 98% of the muon's internal distribution is of the $\sin\omega$ type and only 2% of the $\cos\omega$ type. In contrast, 56% of tau's internal distribution is of the $\sin\omega$ type and 44% of the $\cos\omega$ type. Clearly, based on internal structure, neither mu nor tau can be considered simply a more massive version of the electron, the electron being entirely of the $\cos\omega$ type.

## 10.3 Charged lepton internal five-current density

The probability current density components $s^\mu, s^5$ for the Dirac particle in $M^4 \times \bar{M}^2$ were derived in *Sec. 4.6.1*; see Eqs. (4.55) and (4.56). Multipying these by electric charge $-e$ and denoting the lepton probability amplitude by $\ell$, we obtain the corresponding charge current density components

$$j^\mu = -ec\tilde{\ell}\gamma^\mu\eta_-\ell \tag{10.38a}$$

$$j^5 = -ecp_5\tilde{\ell}\ell\,. \tag{10.38b}$$

in which $\tilde{\ell} = \ell^\dagger\gamma^0$, and where we write for the metric operator

$$\eta_\mp = G_\ell \mp \gamma^5 p_5\,. \tag{10.39}$$

We will now employ these external densities in conjunction with (10.25) to deduce the internal charge-current densities of the charged lepton at rest.

The equation of motion of the charge amplitude $\ell$ is given by (4.37):

$$\eta_+\left(\gamma^\mu \hat{p}_\mu - \eta_-\right)\ell = 0 \tag{10.40}$$

In the chiral representation of the Dirac matrices,[8] i.e.,

$$\gamma^0 = \begin{pmatrix} 0 & -I \\ -I & 0 \end{pmatrix},\ \boldsymbol{\gamma} = \begin{pmatrix} 0 & +\boldsymbol{\sigma} \\ -\boldsymbol{\sigma} & 0 \end{pmatrix},\ \gamma^5 = \begin{pmatrix} I & 0 \\ 0 & -I \end{pmatrix}, \tag{10.41}$$

where $I$ is the $2\times 2$ identity matrix and $\boldsymbol{\sigma}$ denotes the Pauli matrices (3.42), a normalized solution of (10.40) for a particle at rest is

$$\ell = \left(\frac{G_\ell + p_5}{2V}\right)^{1/2} \begin{pmatrix} \dfrac{1}{m_\ell c}u \\ \dfrac{1}{G_\ell + p_5}u \end{pmatrix} e^{-ik_C x^0}, \tag{10.42}$$

where $u^\dagger u = 1$ and $V$ is the volume of the box normalization. The normalization is such that, in accordance with (10.38a), $\ell^\dagger\eta_-\ell = V^{-1}$. To find the internal charge density $J_0$, in (10.38a) we replace $\ell$ (but not

$\tilde{\ell}$) with the right-hand side of (10.25), giving

$$\begin{aligned} j^0 &= -ec\ell^\dagger \eta_- \ell \int D_r(\boldsymbol{\xi}) d^3\boldsymbol{\xi} \\ &= -ecV^{-1} \int D_r(\boldsymbol{\xi}) d^3\boldsymbol{\xi} \end{aligned} . \qquad (10.43)$$

This expresses the external charge density $j^0$ as the spatial average of an internal distribution $J^0$, where

$$J^0(\boldsymbol{\xi}) = -ecD_r(\boldsymbol{\xi}), \qquad (10.44)$$

and $D_r$ is the distribution (10.35). (Note for simplicity we are omitting constant $\xi^5$ from the argument of $D_r$.) We identify $J^0$ with the spread-out, time-averaged distribution of charge associated with the Zitterbewegung.

Having found $J^0$, we now seek the internal charge current $\mathbf{J}$ associated with intrinsic spin. For this we shall require the Gordon decomposition of the external current, $j^\mu$. Employing (10.40) and its conjugate in conjunction with (10.38a) we obtain the decomposition

$$j^\mu = -\frac{e\hbar c}{2}\left[i\left(\tilde{\ell}\cdot\partial^\mu \ell - \partial^\mu \tilde{\ell}\cdot\ell\right) + \partial_\nu\left(\tilde{\ell}\sigma^{\mu\nu}\ell\right)\right], \qquad (10.45)$$

where

$$\sigma^{\mu\nu} = \frac{i}{2}\left(\gamma^\mu\gamma^\nu - \gamma^\nu\gamma^\mu\right). \qquad (10.46)$$

The spatial part of (10.45) can be written

$$\mathbf{j} = -\frac{e\hbar c}{2i}\left(\tilde{\ell}\cdot\nabla\ell - \nabla\tilde{\ell}\cdot\ell\right) + \partial_0\mathbf{P} + \nabla\times\mathbf{m}, \qquad (10.47)$$

where $\mathbf{P} = (ie\hbar c/2)\ell^\dagger\boldsymbol{\gamma}\ell$ is the electric moment density (which vanishes for a free particle, at rest or otherwise),

$$\mathbf{m} = -\frac{e\hbar c}{2}\tilde{\ell}\boldsymbol{\Sigma}\ell \qquad (10.48)$$

is the *magnetic moment density* and $\boldsymbol{\Sigma} = \mathrm{diag}(\boldsymbol{\sigma}, \boldsymbol{\sigma})$. The term $\nabla \times \mathbf{m}$ represents the contribution to the total current coming from the spin:

$$\mathbf{j}_{\text{spin}} = \nabla \times \mathbf{m}. \tag{10.49}$$

Just as was done in the derivation of $J^0$, in (10.48) we replace $\ell$ (but not $\tilde{\ell}$) with the right-hand side of (10.25), resulting in

$$\mathbf{m} = -\frac{e\hbar}{2m_\ell V}\hat{\mathbf{s}}\int D_r(\boldsymbol{\xi})d^3\xi, \tag{10.50}$$

where $\hat{\mathbf{s}}$ is a unit vector pointing in the direction of spin, i.e., $\boldsymbol{\Sigma}\ell = \hat{\mathbf{s}}\ell$. This expresses **m** as the spatial average of an internal magnetic-moment density $\mathbf{M}$, where

$$\mathbf{M} = -\frac{e\hbar}{2m_\ell}D_r(\boldsymbol{\xi})\hat{\mathbf{s}}. \tag{10.51}$$

Then, in parallel with (10.49), the internal current $\mathbf{J}$ generated by this moment density is

$$\begin{aligned} \mathbf{J} &= \nabla \times \mathbf{M} \\ &= -\frac{e\hbar}{2m_\ell}\nabla D_r \times \hat{\mathbf{s}} \\ &= -\frac{e\hbar}{2m_\ell}D_r'(\rho)\hat{\boldsymbol{\rho}} \times \hat{\mathbf{s}} \end{aligned} \tag{10.52}$$

where $\hat{\boldsymbol{\rho}} = \boldsymbol{\xi}/\rho$ and the factor $e\hbar/2m_\ell$ is the Bohr magneton in rationalized MKS units. We have a definite picture of current circulating in a thin shell about the axis of spin; see **Fig. 10.7**. As emphasized by Huang,[2] both the spin and resulting magnetic moment may be said to originate in the Zitterbewegung.

To find the fifth internal current density $J^5$ from (10.39) we use the same method as was used to find $J^0$. The result is

$$J_5(\boldsymbol{\xi}) = -\frac{ep_5}{m_\ell}D_r(\boldsymbol{\xi}). \tag{10.53}$$

Note that $p_5 / m_\ell = v_5$, the fifth component of velocity.

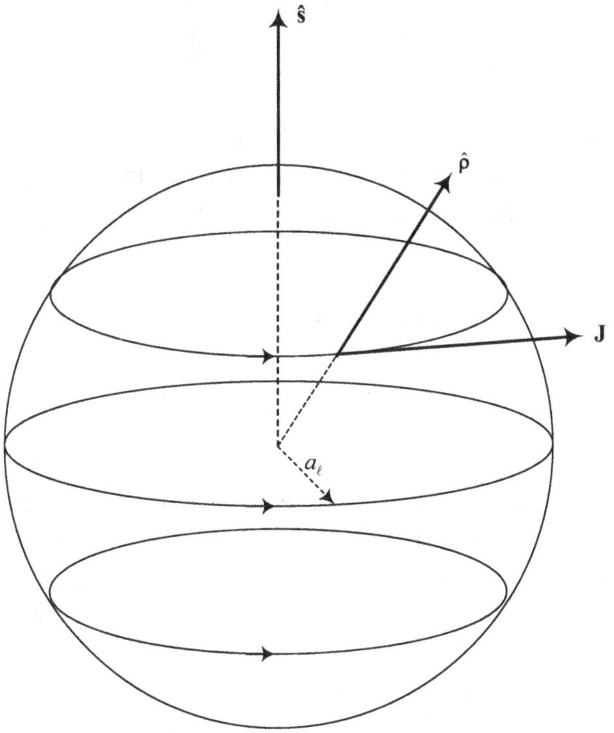


**Fig. 10.7** Electric currents in the charged lepton produced by Zitterbewegung. Internal current density $\mathbf{J} \propto \hat{\mathbf{s}} \times \hat{\mathbf{p}}$ is confined to a thin spherical shell of radius $a_\ell$ and circulates about the spin axis denoted by unit spin vector $\hat{\mathbf{s}}$ .

## 10.4 Charged lepton internal five-momentum density

In this section we derive the internal five-momentum density $\Pi^a$ of the charged lepton. This is the momentum analog of the internal charge current density $J^a$ just obtained. The *external* five-momentum density $\varpi^a$ is just the $0a$ component of the energy-momentum tensor $T^{ab}$ (divided by $c$, to give it units of momentum density): $\varpi^a \equiv T^{0a} / c$, $a = 0,1,2,3,5$ . To be entirely accurate, only the $4 \times 4$ components $T^{\mu\nu}$ form a true tensor; the 4 components $T^{\mu 5}$ form a vector; and $T^{55}$ is a scalar. This mixed object $T^{ab}$ is nevertheless derivable from a single

Lagrangian density $\mathcal{L}\left(\ell,\ell_{,a};\tilde{\ell},\tilde{\ell}_{,a}\right)$, where $\ell_{,a}$ denotes the derivative $\partial_a \ell = \partial\ell/\partial x^a$ :

$$T^{ab} = \frac{\partial\mathcal{L}}{\partial\left(\partial_b \ell\right)}\partial^a \ell + \partial^a \tilde{\ell}\frac{\partial\mathcal{L}}{\partial\left(\partial_b \tilde{\ell}\right)} - g^{ab}\mathcal{L}. \tag{10.54}$$

The form of $\mathcal{L}$ is to be such that variations of $\tilde{\ell}$ and $\ell$ yield, respectively, the equations of motion of $\ell$ and $\tilde{\ell}$, i.e., (10.40) and its conjugate. An appropriate form for this density is

$$\begin{aligned}\mathcal{L} = &\frac{c}{2}\tilde{\ell}\left(G_\ell - \gamma^5 \hat{p}_5\right)_{\text{op}}\left(\gamma^\mu \hat{p}_\mu + \gamma^5 \hat{p}_5 - G_\ell\right)\ell \\ &-\frac{c}{2}\tilde{\ell}\left(\gamma^\mu \hat{p}_\mu - \gamma^5 \hat{p}_5 + G_\ell\right)_{\text{op}}\left(G_\ell - \gamma^5 \hat{p}_5\right)\ell\end{aligned} , \tag{10.55}$$

which, given the normalization cited for (10.42), has the units of energy density. Operators $\left(\hat{p}_5\right)_{\text{op}}$ and $\left(\hat{p}_\mu\right)_{\text{op}}$ act backwards on $\tilde{\ell}$. In fact, $\mathcal{L}=0$ owing to the equations of motion. Thus the final term in (10.54) vanishes. Putting (10.55) into (10.54) we get for the relevant terms of $T^{ab}$

$$\begin{aligned}T^{00}/c = \varpi^0 &= \frac{i\hbar}{2}\left(\tilde{\ell}\hat{\eta}_+\gamma^0\partial^0\ell - \partial^0\tilde{\ell}\gamma^0\hat{\eta}_-\ell\right) \\ &= p^0\ell^\dagger\hat{\eta}_-\ell\end{aligned} \tag{10.56}$$

$$\begin{aligned}T^{0k}/c = \varpi^k &= \frac{i\hbar}{4}\left(\tilde{\ell}\hat{\eta}_+\gamma^k\partial^0\ell - \partial^0\tilde{\ell}\gamma^k\hat{\eta}_-\ell\right) \\ &+\frac{i\hbar}{4}\left(\tilde{\ell}\hat{\eta}_+\gamma^0\partial^k\ell - \partial^k\tilde{\ell}\gamma^0\hat{\eta}_-\ell\right)\end{aligned} \tag{10.57}$$

and

$$\begin{aligned}T^{05}/c = \varpi^5 &= \frac{i\hbar}{2}\left(\tilde{\ell}\hat{\eta}_+\gamma^5\partial^0\ell + \partial^0\tilde{\ell}\gamma^5\hat{\eta}_-\ell\right) \\ &= p^0 p^5\tilde{\ell}\ell\end{aligned} \tag{10.58}$$

Note that tensor $T^{\mu\nu}$ must be symmetrical;[9] accordingly we have symmetrized the result given by (10.54) for the off-diagonal components $T^{0k}$; $T^{05}$ need not be symmetrized because it is the $\mu = 0$ component of a vector.

To find the internal energy density $\Pi^0$ we use the same method as was used to find the internal charge density (10.44). From (10.56) and (10.25) we have for the particle at rest

$$\begin{aligned}\varpi^0 &= m_\ell c \ell^\dagger \hat{\eta}_- \ell \int D_r(\xi) d^3\xi \\ &= m_\ell c V^{-1} \int D_r(\xi) d^3\xi\end{aligned}, \qquad (10.59)$$

which expresses the external energy density $\varpi^0$ as the spatial average over an internal energy density

$$\Pi^0(\xi) = m_\ell c D_r(\xi). \qquad (10.60)$$

In the same way, from (10.58) and (10.25) we find, since $\tilde{\ell}\ell = 1/m_\ell c$,

$$\begin{aligned}\varpi^5 &= m_\ell c p^5 \tilde{\ell}\ell \int D_r(\xi) d^3\xi \\ &= V^{-1} p^5 \int D_r(\xi) d^3\xi\end{aligned} \qquad (10.61)$$

which gives for the fifth component of the internal momentum density vector

$$\Pi^5(\xi) = p^5 D_r(\xi). \qquad (10.62)$$

We aim now to obtain the internal momentum density $\mathbf{\Pi}$ responsible for intrinsic spin angular momentum. For this we will need a Gordon-type decomposition of the external density $\varpi^k$. From the equation of motion and its adjoint we obtain

$$i\hbar\partial^0 \ell = -i\hbar\gamma^0\gamma^k\partial_k \ell + \gamma^0(G_\ell - \gamma^5 p_5)\ell \qquad (10.63)$$

and

$$i\hbar\partial^0 \tilde{\ell} = +i\hbar\partial_k \tilde{\ell}\gamma^0\gamma^k - \tilde{\ell}\gamma^0(G_\ell - \gamma^5 p_5). \qquad (10.64)$$

Putting these values of $\partial^0\ell$ and $\partial^0\tilde{\ell}$ into (10.57) we obtain

$$\varpi^k = \frac{i\hbar}{2}\left(\ell^\dagger \hat{\eta}_- \partial^k \ell - \partial^k \ell^\dagger \hat{\eta}_- \ell\right) + \frac{\hbar}{4}\partial_m\left(\ell^\dagger \hat{\eta}_- \sigma^{km} \ell\right). \qquad (10.65)$$

In vector form this becomes

$$\boldsymbol{\varpi} = \frac{\hbar}{2i}\left(\ell^{\dagger}\hat{\eta}_{-}\nabla\ell - \nabla\ell^{\dagger}\hat{\eta}_{-}\ell\right) + \nabla\times\mathbf{w}\,, \tag{10.66}$$

where **w** is the *spin moment density*, analogous to the magnetic moment density **m** of (10.47), and

$$\mathbf{w} = \frac{\hbar}{4}\ell^{\dagger}\hat{\eta}_{-}\boldsymbol{\Sigma}\ell\,. \tag{10.67}$$

The term $\nabla\times\mathbf{w}$ represents the contribution to the total momentum density coming from the spin:

$$\boldsymbol{\varpi}_{\text{spin}} = \nabla\times\mathbf{w}\,. \tag{10.68}$$

Using (10.25) we can put (10.67) in the form

$$\mathbf{w} = \frac{\hbar}{4V}\hat{\mathbf{s}}\int D_r(\boldsymbol{\xi})d^3\boldsymbol{\xi} \tag{10.69}$$

which expresses **w** as the spatial average of an *internal* spin moment density

$$\mathbf{W} = \frac{\hbar}{4}D_r(\boldsymbol{\xi})\hat{\mathbf{s}}\,. \tag{10.70}$$

Thus the internal momentum density giving rise to spin angular momentum is

$$\begin{aligned}\boldsymbol{\Pi} &= \nabla\times\mathbf{W}\\ &= \frac{\hbar}{4}\nabla D_r\times\hat{\mathbf{s}}\\ &= \frac{\hbar}{4}D_r'(\rho)\hat{\boldsymbol{\rho}}\times\hat{\mathbf{s}}\end{aligned} \tag{10.71}$$

And so we have a definite picture of momentum streaming about the unit spin vector $\hat{\mathbf{s}}$.

## 10.5 Calculating with internal density functions

We now have in hand a complete picture of what is going on inside the volume of the charged lepton's Zitterbewegung. It bears repeating: the internal probability, charge and momentum densities obtained above describe not a wave packet, but a material particle, the corpuscular quantum companion of the lepton wave function. And again: these densities are the exclusive products of the law of self-imaging in 6-dimensional spacetime, Eq. (10.22). They are quite unavailable to QFT in any of its present forms. (The fact that the radius $a_\ell = \xi^5$ comes from the fifth dimension proves this.) Also it should be understood that the internal distributions are not directly observable. For they are hidden behind integral signs. Nevertheless, they are perfectly real, with real observable consequences. The internal distributions $J^a$ and $\Pi^a$ obtained above can in fact be considered examples of Kantian noumena: they are unobservable, but knowable, facts of the external world, giving rise to observable phenomena in the world of appearance (magnetic moment and spin).

But how do we know that the distributions obtained above are correct? Could the Principle of True Representation be wrong, giving distributions of the wrong form? The only way to know is to perform calculations with them, to see whether they reproduce correctly what is observed in nature. Here follow some examples:

*10.5.1 Spin angular momentum.* The charged lepton's spin vector $\mathbf{S}$ should be calculable from the volume integral

$$\mathbf{S} = \int \boldsymbol{\xi} \times \boldsymbol{\Pi} \, d\boldsymbol{\xi}^3 , \tag{10.72}$$

with internal momentum density $\boldsymbol{\Pi}$ given by (10.71). For simplicity, let the unit spin vector $\hat{\mathbf{s}}$ point along the z-axis, i.e. $\hat{\mathbf{s}} = \mathbf{k}$. After integrating over polar angles $\theta$ and $\varphi$, the r.h.s. of (10.72) reduces to the radial integral

$$\mathbf{S} = -\frac{2\pi\hbar}{3} \hat{\mathbf{s}} \int_0^\infty D_r'(\rho) \rho^3 d\rho . \tag{10.73}$$

Then, integrating by parts and making use of the normalization $\langle 1 \rangle = \int D_r(\boldsymbol{\xi}) d\boldsymbol{\xi}^3 = 1$ we obtain finally

$$\mathbf{S} = \frac{\hbar}{2}\hat{\mathbf{s}}, \tag{10.74}$$

which is of course the correct value of spin angular momentum. We should not be too surprised at getting the right value, since the Dirac wave equation, to which our particle is companion, is a spin one-half equation. Nevertheless, it does show that the structure of the momentum density given by (10.71) is correct.

*10.5.2 Spin magnetic moment.* "By definition the magnetic dipole moment of a volume distribution of current with respect to the origin *O* is"[10] (in our notation)

$$\mathbf{m} = \frac{1}{2}\int \boldsymbol{\xi} \times \mathbf{J} d\boldsymbol{\xi}^3 . \tag{10.75}$$

Inserting for current $\mathbf{J}$ the internal density (10.52) and integrating in the same way as we did the integral of (10.72), we obtain

$$\mathbf{m} = -\frac{e\hbar}{2m_\ell}\hat{\mathbf{s}} , \tag{10.76}$$

which is indeed the correct value of the charged lepton's magnetic moment due to spin.

*10.5.3 Amplitude of the Zitterbewegung.* The Zitterbewegung is a jitter phenomenon which upon time averaging generates the internal densities $J^a$ and $\Pi^a$. The *amplitude* of the electron's jitter was defined by Schrödinger[1] to be the average normal distance $\hat{\mathbf{n}}\cdot\boldsymbol{\rho}$ from a reference plane defined by normal $\hat{\mathbf{n}}$ and containing the centroid of the motion. Using operator methods he found an amplitude $\sim ƛ_C/2 = \hbar/2m_e c$, where $ƛ_C$ is the reduced Compton wavelength of the electron. For comparison with Schrödinger's result, let us calculate the average normal distance from the $\xi$-$\eta$ plane, say, using for the probability distribution the function $D_r(\rho)$, Eq. (10.29) with $I_r$ and $I_i$ given by (10.31) and (10.32). We are then called to evaluate

$$\begin{aligned}\langle|\hat{\mathbf{n}}\cdot\boldsymbol{\rho}|\rangle &= \int |\rho\cos\theta| D_r(\rho)\rho^2 d\rho \sin\theta d\theta d\phi \\ &= 2\pi\int D_r(\rho)\rho^3 d\rho\end{aligned} \tag{10.77}$$

Thus the amplitude has two main parts, corresponding to the $\cos k_5\xi^5$ and $\sin k_5\xi^5$ terms of (10.29). Let us designate these $X_1$ and $X_2$, so that $\langle|\hat{\mathbf{n}}\cdot\boldsymbol{\rho}|\rangle = X_1 + X_2$. After a straightforward if cumbersome process of integration we find

$$X_1 = \xi^5 J_0(\omega)\cos\omega \tag{10.78}$$

and

$$X_2 = \xi^5 Y_0(\omega)\sin\omega, \tag{10.79}$$

where we have written $k_5\xi^5 = \omega$, $k_5 = p_5/\hbar$ and $\xi_5 = a_\ell$ is the particle radius depicted in **Figs. 10.5, 6.** Hence on the present theory the amplitude of the Zitterbewegung of charged lepton $\ell$ is given by

$$\langle|\hat{\mathbf{n}}\cdot\boldsymbol{\rho}|\rangle = \xi^5\left[J_0(\omega)\cos\omega + Y_0(\omega)\sin\omega\right]. \tag{10.80}$$

The amplitude as a function of parameter $\omega$ is plotted in **Fig. 10.8**. For the electron $\omega = 0$ and (10.80) gives

$$\langle|\hat{\mathbf{n}}\cdot\boldsymbol{\rho}|\rangle_{\text{electron}} = a_e. \tag{10.81a}$$

So, for the electron, Schrödinger's amplitude $\hbar/2m_e c$ is equivalent to our radius $\xi_5 = a_e$, giving confidence in the radius (10.15) found already in our classical analysis of the Zitterbewegung. In a moment we will confirm this value by showing that it yields exactly the Darwin correction in the spectrum of the hydrogen atom. But first we must draw attention to an unexpected result, namely, the precipitous drop in amplitude as $\omega$ increases from zero. Using the values of $\omega$ given in **Table 10.2** we find for the amplitudes of the other two charged leptons

$$\langle|\hat{\mathbf{n}}\cdot\boldsymbol{\rho}|\rangle_{\text{muon}} = 0.078 a_\mu \tag{10.81b}$$

$$\langle|\hat{\mathbf{n}}\cdot\boldsymbol{\rho}|\rangle_{\text{tau}} = 0.019 a_\tau \tag{10.81c}$$

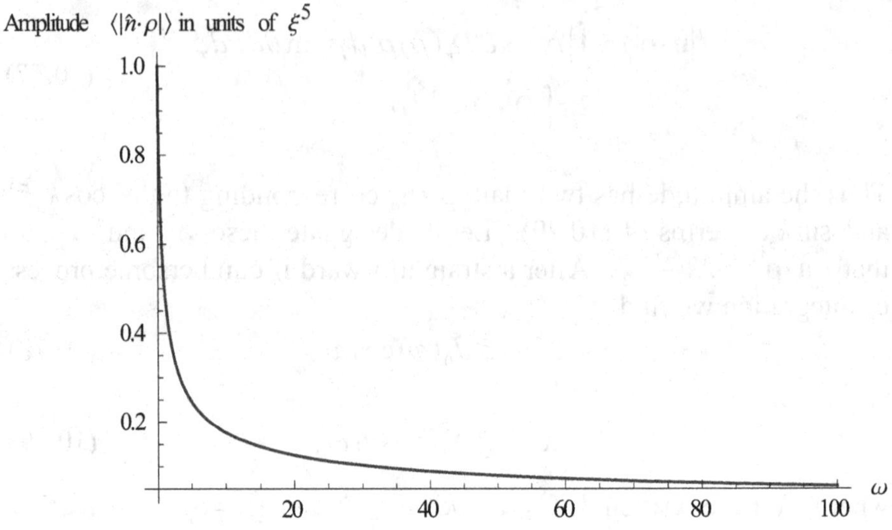


**Fig. 10.8.** Plot of the amplitude of the Zitterbewegung against parameter $k_5\xi^5 = \omega$, where $\xi_5 = a_\ell$ is the particle radius depicted in **Figs. 10.5, 6.** *Amplitude* is defined to be the average normal distance $\hat{\mathbf{n}}\cdot\boldsymbol{\rho}$ from a reference plane defined by normal $\hat{\mathbf{n}}$ and containing the centroid of the motion.

i.e. the muon's amplitude (relative to its Compton wavelength) is just 8% that of the electron's, and tau's, a mere 2%. How can this be? The answer is of course the peculiar nature of the probability distribution. Given the presence in $D_r$ of negative probability one should perhaps be not too surprised to obtain from it peculiar average values.

*10.5.4. Mean radius of the Zitterbewegung.* Equally strange is the average radius, defined by

$$\begin{aligned}\langle\rho\rangle &= \int \rho D_r(\rho)\rho^2 d\rho d\Omega \\ &= 4\pi\int D_r(\rho)\rho^3 d\rho\end{aligned} \quad . \tag{10.82}$$

Comparing this with (10.77) we find $\langle\rho\rangle = 2\langle|\hat{\mathbf{n}}\cdot\boldsymbol{\rho}|\rangle$, so that

$$\langle\rho\rangle_{\text{electron}} = 2a_e \tag{10.83a}$$

$$\langle\rho\rangle_{\text{muon}} = 0.156a_\mu \tag{10.83b}$$

$$\langle \rho \rangle_{\text{tau}} = 0.038 a_\tau \tag{10.83c}$$

While these results run against intuition, they would appear to have no particular observational significance. But that is not so in the case of our next calculation of average value.

*10.5.5. Mean square radius of the Zitterbewegung.* The mean square radius $\langle \rho^2 \rangle$ figures prominently in the Darwin correction mentioned above. It is defined by

$$\begin{aligned} \langle \rho^2 \rangle &= \int \rho^2 D_r(\rho) \rho^2 d\rho d\Omega \\ &= 4\pi \int D_r(\rho) \rho^4 d\rho \end{aligned} \tag{10.84}$$

For the electron $\omega = 0$ and $D_r(\rho)$ is given by (10.37a). Using this in (10.84) we obtain

$$\langle \rho^2 \rangle_{\text{electron}} = 3 a_e^{\ 2} = \frac{3}{4} \lambda_C^{\ 2}, \tag{10.85}$$

yet another unpredictable result.

Now to get comparable results for mu and tau, we need to integrate (10.84) for $\omega > 0$. As in (10.77) the integral has two main parts, corresponding to the $\cos k_5 \xi^5$ and $\sin k_5 \xi^5$ terms of (10.29). Let us as before designate these $X_1$ and $X_2$, so that $\langle |\rho^2| \rangle = X_1 + X_2$. We then find, for $\omega > 0$,

$$X_1 = \frac{3 \sin\omega \cos\omega}{\omega} \xi_5^{\ 2} \tag{10.86a}$$

$$X_2 = -\frac{3 \sin\omega \cos\omega}{\omega} \xi_5^{\ 2} \tag{10.86b}$$

Hence

$$\langle \rho^2 \rangle_{\text{mu, tau}} = 0\,, \tag{10.87}$$

a most remarkable result. And yet, judging by the shape of the plot of **Fig. 10.8**, one could almost see it coming.

*10.5.6. Zitterbewegung explains the form of the Dirac Hamiltonian of the hydrogen atom.* To confirm the value $\hbar/2m_e c$ obtained above for the radius $a_e$ of the electron's Zitterbewegung, we will calculate a physical effect generally attributed to the Zitterbewegung, namely, the Darwin correction to the s states of hydrogen.[11] To second order in $v/c$ the Dirac Hamiltonian for this system is given by[12]

$$H = \frac{\mathbf{p}^2}{2m_e} - e\phi - \frac{\mathbf{p}^4}{8m_e^3c^2} + \frac{e\hbar\boldsymbol{\sigma}\cdot\mathbf{E}\times\mathbf{p}}{4m_e^2c^2} + \frac{e\hbar^2}{8m_e^2c^2}\nabla\cdot\mathbf{E}. \quad (10.88)$$

The first two terms are equivalent to the non-relativistic Hamiltonian for a spinless point charge interacting with the electric potential $\phi(\mathbf{x}) = e/4\pi\varepsilon_0|\mathbf{x}|$, while the third term is a relativistic correction to the leading kinetic term. As for the fourth term, here is Sakurai's description of it:

> The fourth term represents the spin interaction of the moving electron with the electric field. Crudely speaking, we say this arises because the moving electron "sees" an apparent magnetic field given by $\mathbf{E}\times(\mathbf{v}/c)$. Naively, we expect in this way [a term] which is just twice the fourth term of [(10.82)]. That this argument is incomplete was shown within the framework of classical electrodynamics two years before the advent of the Dirac theory by L. H. Thomas, who argued that a more careful treatment which would take into account the energy associated with the precession of the electron spin would result in reduction by a factor of two, in agreement with the fourth term of [(10.82)].

The last term in (10.88) is the Darwin term, whose presence in the Hamiltonian was shown by C. G. Darwin (grandson of Charles Darwin) to be due to the Zitterbewegung. Now as it turns out, the fourth term—the spin-interaction term—too, can be explained in terms of the Zitterbewegung. In fact all three terms involving the electric charge can be obtained by integrating over an appropriate interaction density. Let us first try the standard and obvious one, $\mathcal{H} = J^\mu A_\mu$:

$$\begin{aligned} H_{\text{inter}}(\mathbf{x}) &= \int J^\mu(\boldsymbol{\xi}-\mathbf{x})A_\mu(\boldsymbol{\xi})d^3\boldsymbol{\xi} \\ &= \int J^\mu(\boldsymbol{\xi})A_\mu(\boldsymbol{\xi}+\mathbf{x})d^3\boldsymbol{\xi} \end{aligned},$$

$$=\int\left[J^0(\boldsymbol{\xi})A^0(\boldsymbol{\xi}+\mathbf{x})-\mathbf{J}(\boldsymbol{\xi})\cdot\mathbf{A}(\boldsymbol{\xi}+\mathbf{x})\right]d^3\boldsymbol{\xi}, \qquad (10.89)$$

where $J^\mu=(J^0,\mathbf{J})$ is the electron's internal four-current defined by (10.44) and (10.52); and $A^\mu=(A^0,\mathbf{A})$ is the electromagnetic four-potential seen by the moving electron. From (6.39) we obtain for the potentials

$$A^0(\mathbf{x})=\frac{\mu_0}{4\pi}\int\frac{ec\delta^3(\boldsymbol{\xi})}{r}d^3\boldsymbol{\xi}=\frac{1}{c}\phi(\mathbf{x}) \qquad (10.90)$$

$$\mathbf{A}(\mathbf{x})=\frac{\mu_0}{4\pi}\int\frac{e(-\mathbf{v})\delta^3(\boldsymbol{\xi})}{r}d^3\boldsymbol{\xi}=-\frac{\mathbf{p}}{m_ec^2}\phi(\mathbf{x}) \qquad (10.91)$$

where $r=|\mathbf{x}-\boldsymbol{\xi}|$ and the Coulomb potential

$$\phi(\mathbf{x})=\frac{e}{4\pi\varepsilon_0|\mathbf{x}|}. \qquad (10.92)$$

Note that $\mathbf{v}=\mathbf{p}/m_e$ is the electron's velocity as seen from the nucleus, and so $-\mathbf{v}$ is the nucleus's velocity as seen by the electron.

Now we have two integrals to consider in (10.89), both of which, according to (10.90) and (10.91), depend on the instantaneous Coulomb potential $\phi(\mathbf{x}+\boldsymbol{\xi})$, where vector $\boldsymbol{\xi}$ defines the fluctuating point electron's current position relative to its mean position $\mathbf{x}$. To evaluate the integrals we expand $\phi(\mathbf{x}+\boldsymbol{\xi})$ in powers of $\boldsymbol{\xi}$, giving the approximate form

$$\phi(\mathbf{x}+\boldsymbol{\xi})=\phi(\mathbf{x})+\boldsymbol{\xi}\cdot\nabla\phi+\frac{1}{2}(\boldsymbol{\xi}\cdot\nabla)^2\phi+\cdots \qquad (10.93)$$

Putting this into the first of the two integrals of (10.89) we obtain

$$\begin{aligned}H^{(1)}_{\text{inter}}&=-e\int D_r(\boldsymbol{\xi})\phi(\boldsymbol{\xi}+\mathbf{x})d^3\boldsymbol{\xi}\\&=-e\phi(\mathbf{x})-\frac{e}{6}\left[\int D_r(\boldsymbol{\xi})\rho^2d^3\boldsymbol{\xi}\right]\cdot\nabla^2\phi(\mathbf{x})\end{aligned}. \qquad (10.94)$$

The remaining integral in square brackets here is just the mean square radius, which for the electron has the value given by (10.85). Thus we have finally

$$H^{(1)}_{\text{inter}} = -e\phi(\mathbf{x}) + \frac{e\hbar^2}{8m_e^{\,2}c^2}\nabla\cdot\mathbf{E}, \tag{10.95}$$

where we have introduced the definition $\mathbf{E} = -\nabla(cA^0) = -\nabla\phi$ [see (6.23)]. So we have indeed accounted for the Darwin term by way of the Zitterbewegung. Even the factor 1/8 has been reproduced. Typically in demonstrations of this kind one uses for $D_r$ the simple $\delta$-function, $\delta(\rho - \lambda\!\!\!^-_C)$, this yielding a factor 1/6 rather than 1/8. The internal electron distribution (10.37a) would thus appear to be the correct one. Note that if one were to replace the orbital electron with a muon or tauon, thereby creating a muonic or tauonic atom, the integral of the second line of (10.94), in the light of (10.87), must vanish. *For the muonic or tauonic atom, there is no Darwin term.*

Let us now consider the second integral of (10.89). This is

$$\begin{aligned} H^{(2)}_{\text{inter}}(\mathbf{x}) &= -\int \mathbf{J}(\boldsymbol{\xi})\cdot\mathbf{A}(\boldsymbol{\xi}+\mathbf{x})d^3\boldsymbol{\xi} \\ &= -\frac{e\hbar}{2m_e^{\,2}c^2}\mathbf{p}\cdot\int \nabla D_r(\boldsymbol{\xi})\times\hat{\mathbf{s}}\ \phi(\mathbf{x}+\boldsymbol{\xi})d^3\boldsymbol{\xi} \end{aligned} \tag{10.96}$$

where to obtain the second line we have used both (10.52) and (10.91). We now insert (10.93), retaining only the first two terms of that expression. The integral involving the first of those terms vanishes, leaving

$$\begin{aligned} H^{(2)}_{\text{inter}}(\mathbf{x}) &= \frac{e\hbar}{2m_e^{\,2}c^2}\mathbf{p}\cdot\left[\int \nabla D_r(\boldsymbol{\xi})\ \boldsymbol{\xi}\cdot\mathbf{E}\ d^3\boldsymbol{\xi}\times\hat{\mathbf{s}}\right] \\ &= -\frac{e\hbar}{2m_e^{\,2}c^2}\int D_r(\boldsymbol{\xi})\ d^3\boldsymbol{\xi}\ \mathbf{p}\cdot(\mathbf{E}\times\hat{\mathbf{s}}) \\ &= \frac{e\hbar}{2m_e^{\,2}c^2}\ \boldsymbol{\sigma}\cdot(\mathbf{E}\times\mathbf{p}) \end{aligned} \tag{10.97}$$

where we have made the operator replacement $\hat{\mathbf{s}} \to \boldsymbol{\sigma}$. Comparing this with (10.88) we see that the Zitterbewegung has given us a spin interaction term of the right form, including algebraic sign, but in magnitude twice the correct value. Our result thus corresponds to the naïve physics alluded to in Sakurai's commentary on the fourth term of (10.88). Justifiably one could say that the Zitterbebegung has provided a partial explanation of the fourth term, but one that must be corrected by adding to it the (negative) contribution of the Thomas precession.

But there may still be a way of getting the correct spin interaction without invoking the Thomas correction. For suppose it is not the internal charge current density $J^{\mu}$, but rather the momentum density $\Pi^{\mu}$, as defined by (10.60) and 10.71), that interacts with the vector potential $A^{\mu}$. Because the electron is charged, this circulation of momentum, which is responsible for the electron's spin, creates in addition an effective internal charge current density

$$\tilde{J}^{\mu} = -\frac{e}{m_e}\Pi^{\mu}. \tag{10.98}$$

The effective interaction density would then be $\tilde{J}^{\mu}A_{\mu}$ rather than $J^{\mu}A_{\mu}$. On comparing $\tilde{J}^{\mu}$ with $J^{\mu}$ as defined by (10.44) and (10.52), we see that

$$\tilde{J}^{0} = J^{0} \tag{10.99a}$$

$$\tilde{\mathbf{J}} = \frac{1}{2}\mathbf{J}. \tag{10.99b}$$

Consequently, with $\tilde{J}^{\mu}$ replacing $J^{\mu}$, the Darwin term remains unchanged, but the spin interaction term stemming from (10.97) is reduced by a factor of two, and now reads correctly

$$H^{(2)}_{\text{inter}}(\mathbf{x}) = \frac{e\hbar}{4m_e^{2}c^{2}}\,\boldsymbol{\sigma}\cdot(\mathbf{E}\times\mathbf{p}). \tag{10.100}$$

From the middle line of (10.97) we see that this result depends on $D_r$ through the simple normalization integral $\langle 1\rangle = 1$. Thus our hypothetical muonic and tauonic atoms exhibit the same spin interaction as normal hydrogen, but with $m_{\mu}$ or $m_{\tau}$ replacing $m_e$.

In summary, the Zitterbewegung appears to explain exactly both the Darwin and spin interaction terms of (10.88). It does this in unified fashion, namely by way of an interaction density $\tilde{J}^{\mu}A_{\mu}$, where the effective charge current density $\tilde{J}^{\mu}$ yields correctly the spin interaction term without appeal to the Thomas correction.

## 10.6 Internal structure of the massless flavor neutrino

Neutrinos are massless spinning objects and, like their massive, charged companions, possess internal probability and momentum current densities traceable to the Zitterbewegung. Our aim here is to display those currents. Little can be done with them in the way of calculations, apart from demonstrating the source of the neutrino's spin. They are nevertheless interesting in their own right, a further demonstration the Principle of True Representation's capacity to reveal hidden features of the external world.

*10.6.1 Neutrino internal probability distribution.* The starting point is the same as for our study of the internal structure of the charged lepton, namely, the law of propagation in the form (10.22), which we rewrite here in terms of the flavor neutrino wave function $\nu(x)$ :

$$\nu(x) = \Delta(x) * \nu(x) \tag{10.101}$$

with $\Delta(x)$ again given by (10.23). Now we know that neutrinos when propagating oscillate between flavors. To avoid the formal complications of oscillation, we shall study the properties of the neutrino of flavor $\ell$ in the form it has immediately after its creation by charged-current interaction, before it has changed into anything else. As we showed in **Sec. 4.7**, the neutrino is at that point in a definite state of mass—namely, zero—and its wave function is a solution Dirac equation (4.37) with $p^6 \equiv G = p_5$ and $m^2 = 0$:

$$\left(1+\gamma^5\right)\gamma^\mu \hat{p}_\mu \nu(x) = 0\,. \tag{10.102}$$

This massless equation is actually reducible to the two-component Weyl equation, but for purposes of computing internal probability and momentum currents we want to leave it in four-component form. Let our neutrino be born at the origin $(x^0, \mathbf{x}) = 0$, subsequent to which it travels at light speed along the positive $x^3$ axis. Its guiding plane wave may be written [see (10.24) for comparison with the wave function of a massive particle at rest]:

$$\nu(x) = Ae^{-iK_\mu x^\mu} = Ae^{-iK(x^0 - x^3)}, \tag{10.103}$$

where $A$ is a constant spinor, and for the massless particle, we have written $K^0 = K^3 \equiv K$.

Let us denote by $\mathrm{N}(k)$ the four-dimensional Fourier transform of $\nu(x)$ and, making use of the convolution theorem, rewrite (10.101) as a Fourier integral:

$$\nu(x) = \frac{e^{ik_5\xi^5}}{(2\pi)^4} \int F(k)\mathrm{N}(k)e^{-ik\cdot x} d^4k, \tag{10.104}$$

where $F(k)$ is given by (3.79) with $k^6 = G/\hbar = k_5$. Now

$$\begin{aligned} \mathrm{N}(k) &= A\int e^{-iK(x^0 - x^3)} e^{i(k_0 x^0 - \mathbf{k}\cdot\mathbf{x})} d^4x \\ &= A(4\pi)^4 \delta(k^0 - K)\delta(k^1)\delta(k^2)\delta(k^3 - K) \end{aligned}. \tag{10.105}$$

Notice that $\mathrm{N}(k)$ vanishes unless $k^0 = k^3$. Putting (10.105) and (3.79) into (10.104) and setting $x^1 = x^2 = 0$ we obtain

$$\nu(x) = Ae^{ik_5\xi^5} \int e^{-i\xi^5\sqrt{k_5^2 + k_1^2 + k_2^2}} \delta(k^0 - K)\delta(k^1)\delta(k^2)\delta(k^3 - K)e^{ik^3(x^3 - x^0)} d^4k \tag{10.106}$$

Now make the substitutions

$$\left.\begin{aligned} k^0 - K &= k'^0 \\ k^3 - K &= k'^3 \end{aligned}\right\}. \tag{10.107}$$

We then get

$$\nu(x) = e^{ik_5\xi^5} \int e^{-i\xi^5\sqrt{k_5^2 + k_1^2 + k_2^2}} \delta(k^1)\delta(k^2)dk^1dk^2 \cdot \int \delta(k'^0)\delta(k'^3)dk'^0dk'^3 \cdot \nu(x) \tag{10.108}$$

Next, we deal with the $\delta$-functions. For the first two we simply substitute their integral definitions:

$$\delta(k^1)\delta(k^2) = \frac{1}{(2\pi)^2} \int_{-\infty}^{\infty} e^{-i(k^1\xi^1 + k^2\xi^2)} d\xi^1 d\xi^2. \tag{10.109}$$

For the second two we write (after omitting the primes on $k'^0, k'^3$)

$$\begin{aligned}\int \delta(k^0)\delta(k^3)dk^0dk^3 &= \frac{1}{(2\pi)^2}\iint e^{i(k^0\xi^0 - k^3\xi^3)} d\xi^0 d\xi^3 dk^0 dk^3 \\ &= \frac{1}{(2\pi)^2}\iint e^{ik^3(\xi^0-\xi^3)} d\xi^0 d\xi^3 dk^0 dk^3 \qquad (1.110) \\ &= \frac{1}{2\pi}\int \delta(\xi^0 - \xi^3) d\xi^3 \cdot d\xi^0 dk^0\end{aligned}$$

Now by the time/frequency uncertainty relation we have $d\xi^0 dk^0 = 2\pi$. And so, putting (10.109) and (10.110) into (10.108) we obtain, in analogy with the third line of (10.25) (but with a different *D*!),

$$\nu(x) = \int D(\xi^0 - \xi^3, \boldsymbol{\rho}, \xi^5)\, d^2\boldsymbol{\rho}\, d\xi^3 \cdot \nu(x), \qquad (10.111)$$

where

$$D(\xi^0 - \xi^3, \boldsymbol{\rho}, \xi^5) = Q(\boldsymbol{\rho}, \xi^5)\delta(\xi^0 - \xi^3) \qquad (10.112)$$

and

$$\begin{aligned}Q(\boldsymbol{\rho}, \xi^5) &= \frac{e^{ik_5\xi^5}}{(2\pi)^2}\int e^{-i\xi^5\sqrt{k_5{}^2 + \boldsymbol{\kappa}^2}} e^{i\boldsymbol{\kappa}\cdot\boldsymbol{\rho}} d^2\boldsymbol{\kappa} \\ &= \frac{e^{ik_5\xi^5}}{2\pi}\int_0^\infty e^{-i\xi^5\sqrt{k_5{}^2+\boldsymbol{\kappa}^2}} J_0(\kappa\rho)\kappa d\kappa\end{aligned} \qquad (10.113)$$

where the two-dimensional vectors $\boldsymbol{\rho} = (\xi^1, \xi^2)$ and $\boldsymbol{\kappa} = (k^1, k^2)$. Function *D*—or rather its real part, $D_r$—defines the internal structure of the neutrino. From (10.112) we see that this particle is a disk of zero thickness, flattened as if by Lorentz contraction, and moving at the speed of light. The real part of rotationally symmetric function *Q* defines its cross sectional probability structure—a structure due to the Zitterbewegung.

To see what this structure looks like we need to evaluate the integral (10.113). This may be accomplished with the aid of a table of Hankel transforms. In analogy with (10.29) we write for the real part of *Q*

$$Q_r = \frac{1}{2\pi}\left(I_r \cos k_5\xi^5 - I_i \sin k_5\xi^5\right) \tag{10.114}$$

where[13]

$$\begin{aligned} I_r &= \frac{\partial}{\partial \xi^5}\int_0^\infty \frac{\sin\left(\xi^5\sqrt{k_5^{\,2}+\kappa^2}\right)}{\sqrt{k_5^{\,2}+\kappa^2}} J_0(\kappa\rho)\kappa d\kappa \\ &= \frac{\partial}{\partial \xi^5}\left[\frac{\cos(k_5 z_1)}{z_1}U(\xi^5-\rho)\right] \\ &= \frac{\partial}{\partial \xi^5}\left[\frac{\cos(k_5 z_1)}{z_1}\right]U(\xi^5-\rho) + \frac{1}{z_1}\delta(\xi^5-\rho) \end{aligned} \tag{10.115}$$

and

$$\begin{aligned} I_i &= \frac{\partial}{\partial \xi^5}\int_0^\infty \frac{\cos\left(\xi^5\sqrt{k_5^{\,2}+\kappa^2}\right)}{\sqrt{k_5^{\,2}+\kappa^2}} J_0(\kappa\rho)\kappa d\kappa \\ &= \frac{\partial}{\partial \xi^5}\left[-\frac{\sin(k_5 z_1)}{z_1}U(\xi^5-\rho) + \frac{\exp(-k_5 z_2)}{z_2}U(\rho-\xi^5)\right] \\ &= -\frac{\partial}{\partial \xi^5}\left[\frac{\sin(k_5 z_1)}{z_1}\right]U(\xi^5-\rho) - k_5\delta(\xi^5-\rho) \\ &\quad + \frac{\partial}{\partial \xi^5}\left[\frac{\exp(-k_5 z_2)}{z_2}\right]U(\rho-\xi^5) - \frac{1}{z_2}\delta(\rho-\xi^5) \end{aligned} \tag{10.116}$$

In these expressions $U$, $z_1$ and $z_2$ have the same meanings as in (10.31) and (10.32). Distribution $Q_r$ is now completely determined. However, we propose to display it as we did the distribution of (10.36), namely, in terms of the dimensionless variable $x = \rho/\xi^5$ and the constant $\omega = k_5\xi^5$. Then, with $t_1 = (1-x^2)^{1/2}$, $t_2 = (x^2-1)^{1/2}$, we obtain the complete neutrino radial distribution

$$
\begin{aligned}
2\pi\rho Q_r \frac{d\rho}{dx} &\equiv R(x) \\
&= \left\{ \frac{\omega \sin[\omega(1-t_1)]}{t_1^{\,2}} - \frac{\cos[\omega(1-t_1)]}{t_1^{\,3}} \right\} xU(1-x) \\
&\quad - \sin\omega \left( \frac{\omega}{t_2^{\,2}} + \frac{1}{t_2^{\,3}} \right) \exp(-\omega t_2) xU(x-1) \\
&\quad + \left( \frac{\cos\omega}{t_1} + \frac{\sin\omega}{t_2} + \omega \sin\omega \right) \delta(x-1)
\end{aligned}
\tag{10.117}
$$

Now in our classical treatment of the neutrino we found for the radius of the neutrino's helical path $\xi^5 = b_\ell = \hbar/2G_\ell$ [see (10.9)]. But for the neutrino, $\hbar k_5 = p_5 = G_\ell$. Hence for all flavor neutrinos, $\omega = 1/2$. Consequently the dimensionless radial distribution we have called $R(x)$ has exactly the same form for all flavor neutrinos, whether $\nu_e, \nu_\mu$ or $\nu_\tau$. This is in contradistinction to the charged lepton distributions, which differ from each other owing to differing values of $\omega$. The actual neutrino distributions do of course differ hugely in scale, due to their widely differing values of radius, $b_\ell \propto 1/G_\ell$ (see **Table 4.1**). Distribution $R(x)$ is plotted in **Fig. 10.9.** The two terms of (10.117) involving unit step functions $U$ give rise to the negative peak at $x = 1$. The positive peak at $x = 1$ is due to the $\delta$-function term. Note that the $\delta$ function is multiplied by the infinite factors $1/t_1$ and $1/t_2$. In the figure we have represented that doubly infinite term by a vertical straight line. This image of the neutrino's inner structure corresponds closely to the classical picture presented in **Fig. 10.1**, but differs from the classical one in that it goes algebraically negative, as do the charged leptons' inner distributions.

*10.6.2 Neutrino internal probability current density.* Electrically neutral, the neutrino has no magnetic moment. It nevertheless has an internal probability current density, and for completeness we want to derive it and show that it exists. In form the external five-vector $s^a$ is similar to that of the charge-current density (10.38), but with the charge $-e$ removed [see also (4.55-57)]:

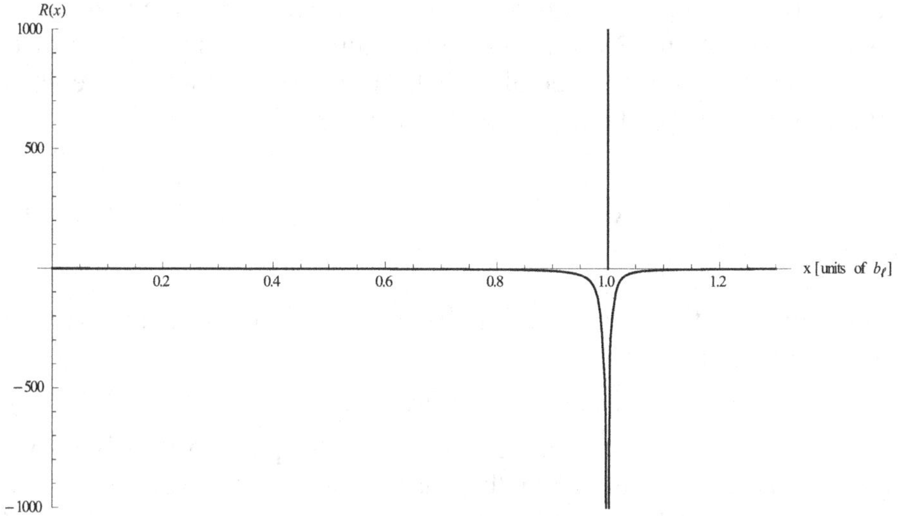


**Fig. 10.9.** Plot of the neutrino's internal probability distribution $R(x)$, defined by Eq. (10.117). This dimensionless distribution, depicting the time-averaged Zitterbewegung of the neutrino, has the same form for all three flavors of neutrino, although the three distributions differ in scale owing to the differences in radius, $b_\ell$, as defined by (10.9). The negative peak at $x=1$ is produced by the first and second terms of the r. h. s. of (10.117), whereas the (doubly infinite) positive peak comes from the last line of the equation.

$$s^\mu = c\tilde{\nu}\gamma^\mu \eta_{-}\nu \tag{10.118a}$$

$$s^5 = cp_5\tilde{\nu}\nu . \tag{10.118b}$$

in which $\tilde{\nu} = \nu^\dagger \gamma^0$, and where we write for the neutrino's metric operator (since for the neutrino $p_5 = G_\ell$)

$$\eta_\mp = G_\ell(1 \mp \gamma^5) . \tag{10.119}$$

In the chiral representation (10.41), and assuming negative helicity (left-handed neutrino), a normalized solution of (10.102) is

$$\nu = \sqrt{\frac{G_\ell}{2V}} \begin{bmatrix} \dfrac{u}{p_0} \\ \dfrac{-u}{G_\ell} \end{bmatrix} e^{-iK(x^0 - x^3)}, \text{ with } u = \begin{bmatrix} 0 \\ 1 \end{bmatrix} . \tag{10.120}$$

The normalization is such that $\nu^{\dagger}\eta\_\nu = V^{-1}$, where $V$ is the volume of box normalization. Now using a procedure that we have performed many times, to get the internal probability density $S^0$ we first replace $\nu$ (but not $\tilde{\nu}$) in (10.118a) by the r. h. s. of (10.111) to get

$$s^0 = c\frac{1}{V}\int Q(\boldsymbol{\rho})d^2\boldsymbol{\rho}\,\delta(\xi^0 - \xi^3). \qquad (10.121)$$

This expresses the external probability density $s^0$ as the spatial average of an internal density

$$\mathcal{S}^0 = cQ(\boldsymbol{\rho})\delta(\xi^0 - \xi^3). \qquad (10.122)$$

The external probability current density due to spin is obtainable from (10.47) merely by removing the charge factor $-e$. This yields by analogy with (10.48) a *probability moment density* $\mathbf{l}$, where

$$\mathbf{l} = \frac{\hbar c}{2}\tilde{\nu}\,\boldsymbol{\Sigma}\,\nu \qquad (10.123)$$

and a total probability current density $\mathbf{s}_{\text{spin}}$ due to spin:

$$\mathbf{s}_{\text{spin}} = \nabla \times \mathbf{l} \qquad (10.124)$$

Now using (10.111) we can write (10.123) in the form

$$\mathbf{l} = -\frac{\hbar c}{2p_0 V}\hat{\mathbf{k}}\int Q(\boldsymbol{\rho})d^2\boldsymbol{\rho}\,\delta(\xi^0 - \xi^3), \qquad (10.125)$$

where we have used the fact that, for a left-handed neutrino moving in the $+z$ direction, $\boldsymbol{\Sigma}\nu = \hat{\mathbf{s}}\nu = -\hat{\mathbf{k}}\nu$, and also that $\tilde{\nu}\nu = G_\ell / p_0 V$. From (10.125) we extract the *internal probability moment density* $\mathbf{L}$, where

$$\mathbf{L} = -\frac{\hbar c}{2p_0}\hat{\mathbf{k}}Q(\boldsymbol{\rho})\delta(\xi^0 - \xi^3). \qquad (10.126)$$

The internal probability current density $\boldsymbol{\mathcal{S}}$ is thus given by

$$\begin{aligned}\mathcal{S} &= \nabla \times \mathbf{L} \\ &= \frac{\hbar c}{2p_0}\hat{\mathbf{k}} \times \nabla Q \cdot \delta(\xi^0 - \xi^3)\end{aligned} \quad (10.127)$$

In the same way as we obtained density $\mathcal{S}^0$ we obtain from (10.118b) and (10.111) the fifth probability current density $\mathcal{S}^5$:

$$\mathcal{S}^5 = \frac{cG_\ell}{p_0} Q(\rho)\delta(\xi^0 - \xi^3). \quad (10.128)$$

Note that $cG_\ell / p_0 = v_5$, the fifth component of the neutrino's velocity along $x^5$.

As shown in **Fig. 10.10** we have a definite picture of a current of probability circulating in a thin ring about the neutrino's axis of spin.

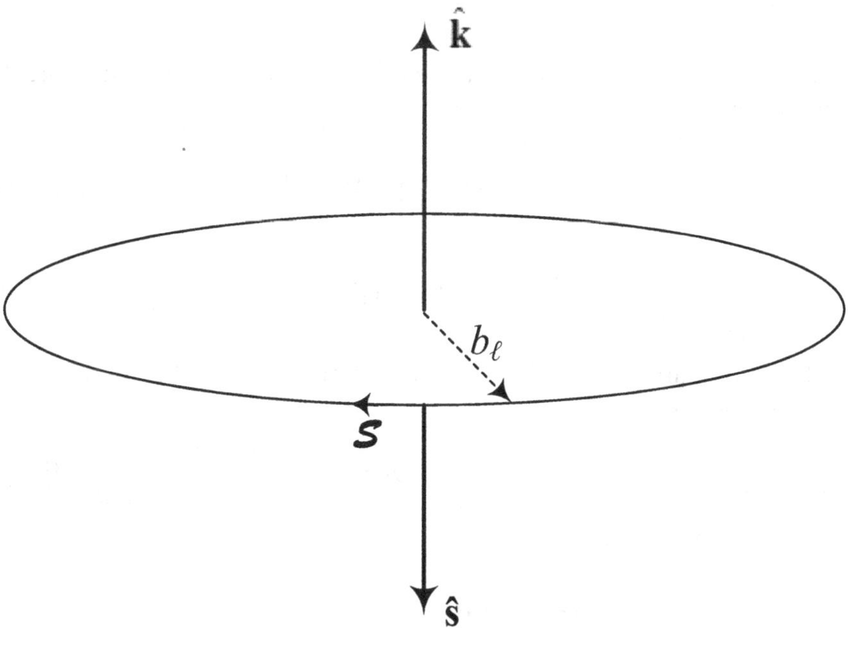


**Fig. 10.10** Probability current in left-handed flavor neutrino produced by Zitterbewegung. Neutrino propagates in direction $\hat{\mathbf{k}}$ opposite to its unit spin vector $\hat{\mathbf{s}}$. Internal probability current density $\mathcal{S}$ is confined to a thin ring of radius $b_\ell$ in the plane of the neutrino and circulates about $\hat{\mathbf{s}}$.

*10.6.3 Neutrino internal five-momentum density.* To obtain the neutrino's internal momentum density, in formulas (10.56), (10.66) and (10.58), one need only replace the charged lepton's wave function $\ell(x)$ with the neutrino's $\nu_\ell(x)$. This gives for the neutrino's external momentum density components

$$\varpi^0 = p^0 \nu^\dagger \eta_- \nu \tag{10.129}$$

$$\varpi^5 = p^0 G_\ell \tilde{\nu}\nu \tag{10.130}$$

$$\boldsymbol{\varpi} = \frac{\hbar}{2i}\left(\nu^\dagger \hat{\eta}_- \nabla \nu - \nabla \nu^\dagger \hat{\eta}_- \nu\right) + \nabla \times \mathbf{w} \tag{10.131}$$

where the neutrino's spin moment density

$$\mathbf{w} = \frac{\hbar}{4} \nu^\dagger \hat{\eta}_- \boldsymbol{\Sigma} \nu\,. \tag{10.132}$$

From these external momentum densities we readily obtain the internal momentum densities

$$\Pi^0 = p^0 Q(\boldsymbol{\rho})\delta(\xi^0 - \xi^3) \tag{10.133}$$

and

$$\Pi^5 = p^5 Q(\boldsymbol{\rho})\delta(\xi^0 - \xi^3)\,. \tag{10.134}$$

As for the spatial part $\boldsymbol{\Pi}$, we have from (10.132) the internal spin moment density

$$\mathbf{W} = \frac{\hbar}{4} Q(\boldsymbol{\rho})\delta(\xi^0 - \xi^3)\hat{\mathbf{s}} \tag{10.135}$$

from which we obtain

$$\begin{aligned} \boldsymbol{\Pi} &= \nabla \times \mathbf{W} \\ &= \frac{\hbar}{4} \nabla Q(\boldsymbol{\rho}) \times \hat{\mathbf{s}}\,\delta(\xi^0 - \xi^3) \end{aligned} \tag{10.136}$$

From this we calculate the neutrino's spin vector

$$\begin{aligned} S &= \int \xi \times \Pi \, d\xi^3 \\ &= \frac{\hbar}{2}\hat{\mathbf{s}} \\ &= -\frac{\hbar}{2}\hat{\mathbf{k}} \end{aligned} \qquad (10.137)$$

This is for the negative helicity neutrino. The sign is reversed for the positive helicity antineutrino. Thus we have shown that the Zitterbewegung accounts for the spin of both the charged lepton $\ell^-$ and its associated neutrino $\nu_\ell$.

## 10.7 Implications of negative probability

*10.7.1 Inner spacetime* $N^2$**.** We close this long chapter with an inquiry into the meaning of the negative probability densities encountered in our study of the lepton's Zitterbewegung. Now while negative probability is decidedly strange, there is something else strange about leptons, and in fact about all particles of spin one-half: upon rotation through $2\pi$, the lepton wave function changes sign, $+ \rightarrow -$.[14] So to bring the wave function back to its original state it must undergo an additional $2\pi$ rotation. Could these two oddities be related? The answer is yes, they are, as we shall now demonstrate.

First we show that the wave function is not alone in exhibiting $4\pi$ rotational symmetry: the point particle $P$ executing Zitterbewegung shows this same symmetry. Let us take an electron, whose Zitterbewegung is illustrated classically in **Fig. 10.2**, and compute its phase integral. We have

$$\oint p \cdot dl = \oint m_e c \cdot \frac{\lambda\!\!\!^{-}_C}{2} d\phi = \frac{h}{4\pi} \oint d\phi \, . \qquad (10.138)$$

So to obtain integral $h$, thus ensuring reëntrance of $P$'s associated de Broglie wave, point $P$ must circulate through $4\pi$ radians. Note also that, if in (10.21) area $A$ is counted twice, corresponding to two complete circuits around that area, then we get the correct value of the electron's magnetic moment.

Next, let us look at the electron's radial probability distribution, illustrated in **Fig. 10.4**. There are two peaks: one positive and one negative. The connection between $4\pi$ rotational symmetry and probability is now obvious. With the positive peak we are encouraged to correlate one of the two successive orbits of $2\pi$ radians; and with the negative peak, the other of the two orbits. The problem remaining is to locate the geometric origin of these correlations.

Can this be accomplished in ordinary Euclidian $E^3$? It has been suggested that, topologically, the electron is a toroidal knot in $E^3$, such as the one depicted in **Fig. 10.11**.[15] The knot replaces the simple ring

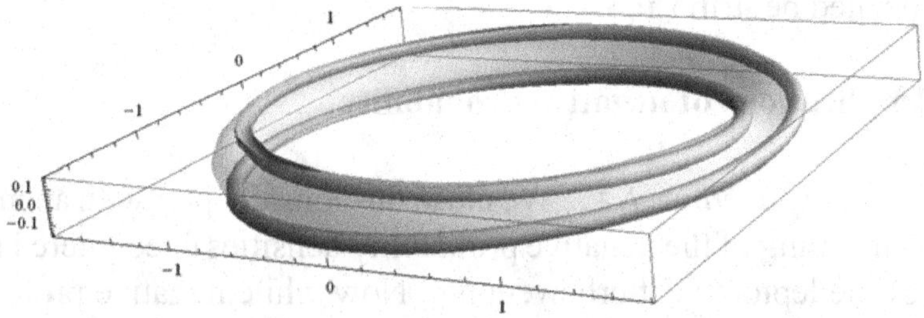


**Fig. 10.11** A torus knot model of the electron. The torus around which the knot wraps is faintly shown. To return to its starting point, a massless point $P$ traveling round the knot passes through $4\pi$ radians of arc.

structure of **Fig. 10.2**. Such models are satisfactory to the extent that point $P$, by way of its Zitterbewegung, passes through $4\pi$ radians of rotation before returning to its original position. But the knot model carries no hint of negative probability and on that ground fails to represent correctly the hidden structure of the electron. We conclude that one cannot properly describe all features of the electron in ordinary 3-space.

And so once again we infer the presence of higher dimensions. But the extra dimensions called for here are very unlike the infinite external ones $x^5$ and $x^6$ demanded by the Principle of True Representation. Our new dimensions instead form a two-dimensional, Minkowskian *inner* spacetime, one we shall designate $N^2$. This inner spacetime is not entirely new to us: it was already mentioned at the end of **Sec. 4.4** as one in which the electron may become trapped, generating the masses of leptons mu and tau. It consists of one temporal dimension, $u^0 = c\tau$

and one spatial dimension, $u$. The defining feature of $N^2$ is that $u$ is bounded, creating a slab geometry of thickness W, equal (as we shall see) to one-half the Compton wavelength of the electron. Total spacetime is now the 8-dimensional product space $M^4 \times \bar{M}^2 \times N^2$. This total world geometry is depicted in **Fig. 10.12** and contrasted with the

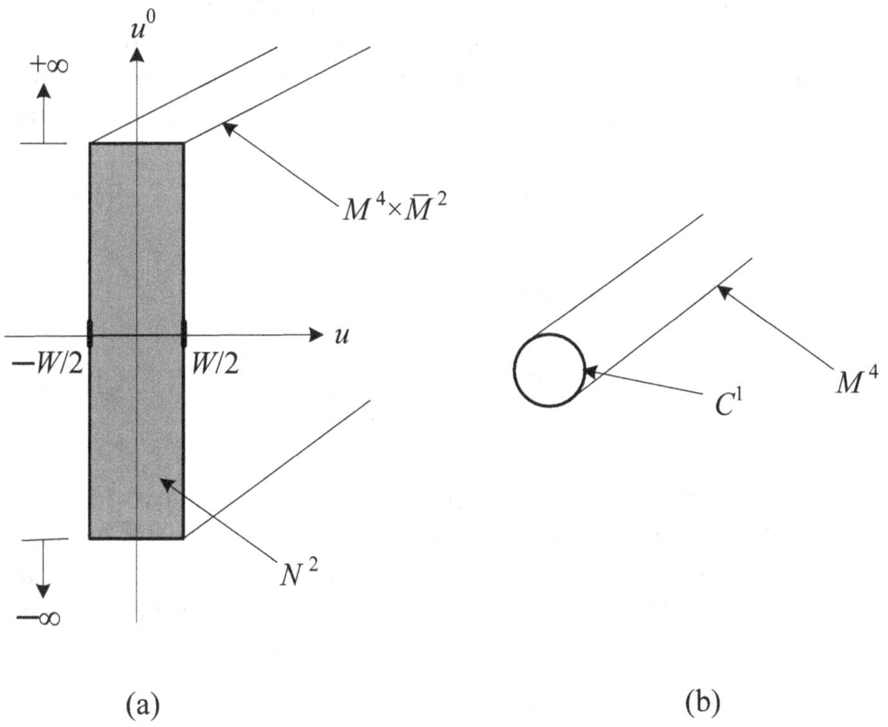


**Fig. 10.12** (a) Depicting world geometry $M^4 \times \bar{M}^2 \times N^2$, where $N^2$, shown lightly shaded, represents a 2-D internal Minkowski space with time coordinate $u^0 = c\tau$ and space coordinate $u$. $N^2$ is bounded, lying interior to lines $u = \pm W/2$, where width $W$ equals one-half the Compton wavelength of the electron. The product of $N^2$ and 6-D $M^4 \times \bar{M}^2$, shown here as a line, creates the infinite 8-D slab $M^4 \times \bar{M}^2 \times N^2$. (b) Depicting the Kaluza-Klein tubular world geometry $M^4 \times C^1$,[16] where $C^1$ is a circle whose circumference is of the order of, or a few orders larger than, a Planck length.

5-dimensional tubular world geometry of Kaluza-Klein.[16] $N^2$ is, like the Zitterbewegung itself, hidden from view. Presumably we live, not *within* $N^2$, but on one of the two 6-dimensional $M^4 \times \bar{M}^2$ hyperplanes bounding it. What happens on the hyperplane we are *not* living on is open to question; see *Sec. 10.7.3* below.

We are now in a position to explain the occurrence of negative probability in the Zitterbewegung. Let us again take for simplicity the case of the electron. The orbital motion of massless point $P$ continues as above in Euclidian $E^3$. However, as it does so, point $P$ executes, with the same linear momentum $m_e c$, a back-and-forth, oscillatory motion within $N^2$. Now we must demand reëntrance of the de Broglie wave, not only for the orbital motion, but for the linear motion within $N^2$ as well. For one complete cycle of the linear motion we have the phase integral

$$\oint p \cdot du = m_e c \cdot 2W \,. \tag{10.139}$$

So for reëntrance, the well width

$$W = \frac{h}{2m_e c} = \frac{\lambda_C}{2}. \tag{10.140}$$

The two reëntrance conditions effectively coordinate the two forms of motion. These motions are, moreover, linked parametrically by inner time $\tau$. Suppose that $u$ is confined to the interval $-W/2 \le u \le W/2$ as shown in **Fig. 10.12** and that the motion of point $P$ begins on the plane $u = +W/2$ at time $\tau = 0$. Then, defining the parameter

$$\theta = \frac{4\pi c\tau}{\lambda_C} = \frac{4\pi\tau}{T},$$

where $T$ is the period $\lambda_C / c$ of either of the two periodic motions, we can describe the path of counterclockwise motion in the subspace $E^3 \times N^2$ by the set of parametric equations

$$\left.\begin{aligned} \xi^1 &= \frac{\lambda_C}{4\pi}\cos\theta \\ \xi^2 &= \frac{\lambda_C}{4\pi}\sin\theta \\ u &= \frac{\lambda_C}{4\pi}[(\theta - 2\pi)\,\mathrm{sgn}(\theta - 2\pi) - \pi] \end{aligned}\right\}. \tag{10.141}$$

Note that dimension $u$ is perpendicular to *all* planes in $E^3$; the $\xi^1$-$\xi^2$ plane is just one of an infinite set of such planes in $E^3$. As shown in **Fig. 10.13**, the path of motion consists of two helices of one turn each,

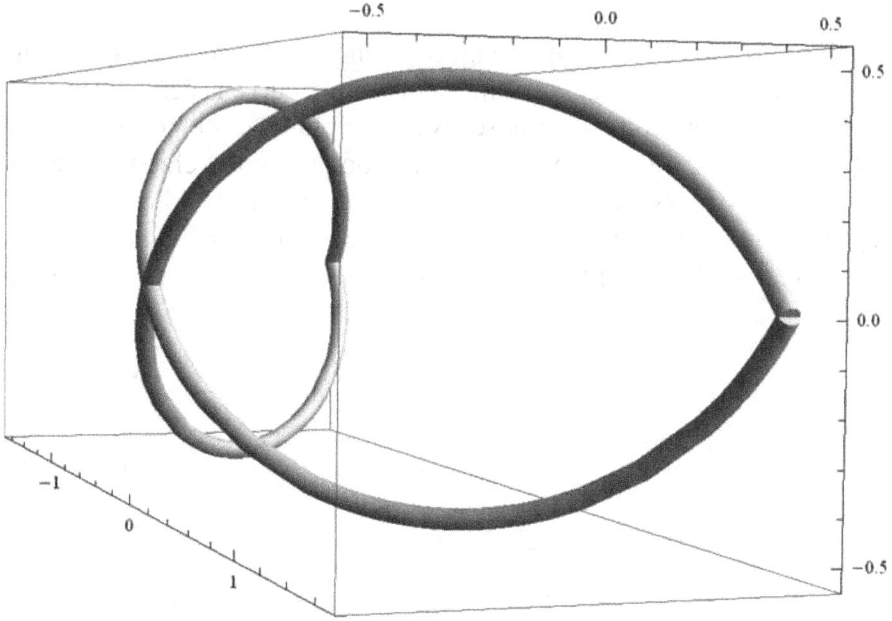


**Fig. 10.13** Path of classical motion of point $P$ in $E^3 \times N^2$, described parametrically by Eqs. (10.135). For plotting purposes, $\lambda_C = 2\pi$. Hence motion in $E^3$ alone describes a circle of radius $\lambda_C / 4\pi = 0.5$, whereas motion in $N^2$ describes two straight lines, an outgoing one of length $W = \lambda_C / 2 = \pi$ directed *against* positive $u$, and an incoming one of the same length directed *with* positive $u$. In combination, and assuming counterclockwise motion in $E^3$, the two types of motion yield a pair of helices of one turn each, the outgoing one being of negative helicity, the returning one of positive helicity.

the outgoing one (moving in the negative $u$ direction) being of negative helicity, the returning one (moving in the positive $u$ direction) of positive helicity. (The helicities are reversed by changing the sign of $\xi^2$.) For plotting purposes we have put $\lambda_C = 2\pi$. This path of motion in $E^3 \times N^2$ replaces the torus knot model of the electron shown in **Fig. 10.11**.

Let us now calculate the momentum $m_e \dot{u}$ of point $P$ as it moves within $E^3 \times N^2$. Taking the derivative $\partial / \partial \tau$ of Eqs. (10.141) we get the parametric representation

$$\left.\begin{aligned} m_e\dot{\xi}^1 &= -m_e c\sin\theta \\ m_e\dot{\xi}^2 &= m_e c\cos\theta \\ m_e\dot{u} &= m_e c\,\mathrm{sgn}(\theta-2\pi) \end{aligned}\right\} \tag{10.142}$$

This gives (correctly) the outgoing momentum as $-m_e c$ and returning momentum as $+m_e c$. The (disjoint) momentum curve is depicted in **Fig. 10.14**, where for plotting purposes we have put $m_e c = 1$. (Note that in $N^2$ there is no relativistic upper bound on velocity; hence $\text{momentum} = \text{mass} \times \text{velocity}$.) The plot consists of two rings, separated in momentum space by a distance $\Delta(m_e\dot{u}) = 2m_e c = 2$. The ring at $m_e\dot{u} = -1$ corresponds to *P*'s motion *against* positive *u*, while the

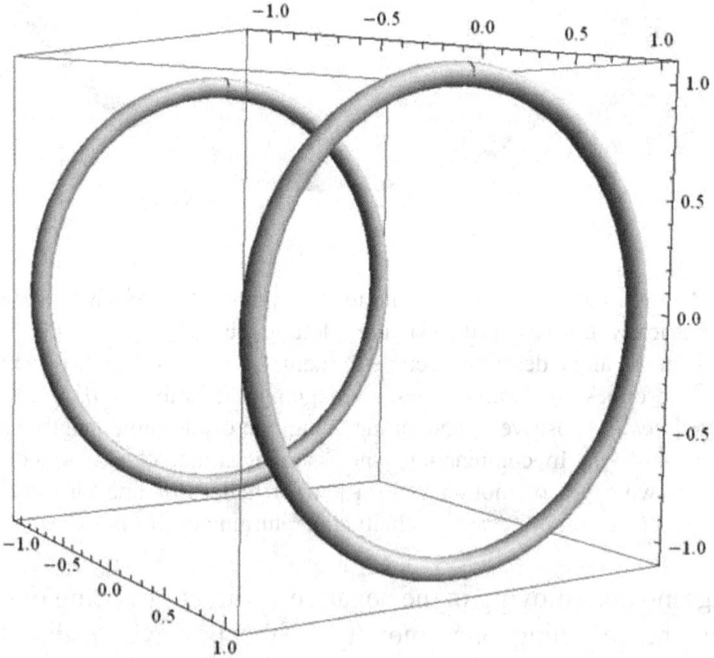


**Fig. 10.14** Plot of the momentum of point *P* whose path of motion in $E^3 \times N^2$ is depicted in **Fig. 10.13**. The plot is giving parametrically by Eqs. (10.136), where for plotting purposes, $m_e c = 1$. The plot consists of two rings, separated in momentum space by a distance $\Delta(m_e\dot{u}) = 2m_e c = 2$. The ring at $m_e\dot{u} = -1$ corresponds *P*'s motion *against* positive *u*, while the ring at $m_e\dot{u} = +1$ corresponds to its motion *with* positive *u*. In both rings the spin vector points out of the page.

ring at $m_e\dot{u} = +1$ corresponds to its motion *with* positive $u$. In both rings the spin vector points out of the page. (To reverse the direction of spin, change the sign of $\dot{\xi}_2$.)

Next we write down the momentum density $\boldsymbol{\pi}(\rho, p)$ corresponding to the momentum plot of **Fig. 10.14**. This is an object in four dimensions, a function of 3-radius $\rho$ and 1-momentum $p = m_e\dot{u}$, a function whose integral over $p$—providing an effective dimensional reduction—yields the three dimensional momentum density $\boldsymbol{\Pi}$ given by (10.71):

$$\boldsymbol{\Pi}(\rho) = \int \boldsymbol{\pi}(\rho, p)dp\,, \qquad (10.143)$$

where

$$\boldsymbol{\pi}(\rho, p) = \frac{\hbar}{4}\nabla\left[d_1(\rho)\delta(p - m_e c) - d_2(\rho)\delta(p + m_e c)\right]\times\hat{\mathbf{s}}\,. \qquad (10.144)$$

Here $d_1$ and $d_2$ are algebraically positive, spherically symmetric functions of $\rho$, shells of the form $A(\rho)\delta(\rho - \rho_0)$, which contain the Zitterbewegung in $E^3$. Two such functions are required: one ($d_1$) corresponding to motion in the direction of positive $u$ in $N^2$, the other ($d_2$) in the direction of negative $u$. The spherical symmetry is the result of averaging over the rotational motions in all planes passing through the centroid of motion in $E^3$ that give spin pointing in the direction of positive $u$. (Recall that all such planes are perpendicular to dimension $u$.) Carrying out the integration in (10.143) we obtain

$$\boldsymbol{\Pi}(\rho) = \frac{\hbar}{4}\nabla\left[d_1(\rho) - d_2(\rho)\right]\times\hat{\mathbf{s}}\,. \qquad (10.145)$$

On comparing this with (10.72) and (10.37) we see that we must have

$$\begin{aligned} d_1(\rho) - d_2(\rho) &= D_r(\rho) \\ &= -\frac{1}{4\pi\rho}\delta'(\rho - a_e) \\ &= -\frac{1}{4\pi\rho}\lim_{\varepsilon\to 0}\left[\frac{\delta(\rho - a_e + \varepsilon) - \delta(\rho - a_e - \varepsilon)}{2\varepsilon}\right] \end{aligned} \qquad . (10.146)$$

Hence

$$\left.\begin{aligned} d_1(\rho) &= \frac{1}{4\pi\rho}\lim_{\varepsilon\to 0}\frac{\delta(\rho-a_e-\varepsilon)}{2\varepsilon} \\ d_2(\rho) &= \frac{1}{4\pi\rho}\lim_{\varepsilon\to 0}\frac{\delta(\rho-a_e+\varepsilon)}{2\varepsilon} \end{aligned}\right\} \tag{10.147}$$

Finally we can understand the presence of positive and negative peaks in the probability distribution $D_r$ depicted in **Fig. 10.4**. These peaks encode the positive and negative momenta associated with back and forth motion of massless point $P$ in inner spacetime, $N^2$. Interestingly, according to (10.147), the radius of the spherical distribution $d_1$ in momentum plane $p = m_e c$ is infinitesimally larger than the radius of distribution $d_2$ in momentum plane $p = -m_e c$. Mathematically this difference is needed to prevent the two peaks from cancelling each other upon integration of (10.143); but nothing in the geometry as such tells us that there should be such a difference. The presence of an infinitesimal $\varepsilon$ in the denominators of Eqs. (10.147) is not too surprising; we saw the double infinity as well in the neutrino's radial distribution, (10.117).

*10.7.2 The missing world of antimatter.* By itself negative probability density is not excessively interesting, as it is unobservable, being always under an integral sign. On the other hand, inner spacetime $N^2$ *is* interesting, because (as we shall see later) it is central to an understanding of the masses of the charged leptons $\mu$ and $\tau$. But there is another reason to be interested in $N^2$: it may hold the solution to the puzzle of the missing world of antimatter.[17]

According to the logic of the Big Bang, in the very early universe matter and antimatter should have been produced in equal amounts. Such is obviously required for charge neutrality of the universe.[18] Today, however, what we mostly observe is matter, and very little antimatter to go with it. So the question arises, where is—or what happened to—the antimatter? Since most matter is baryonic, this perplexing problem is often called the problem of *baryon asymmetry*.

One possible answer has to do with *baryogenesis*, the process by which baryons are created by weak interaction. The idea is that, with a violation of CP (charge-parity) symmetry, there could result an excess of baryons over antibaryons, even if at the beginning of the universe their numbers were perfectly balanced. However, Andrei Sakharov showed[17] that for this process to work (1) the universe must be in a state of disequilibrium and (2) baryon number cannot be a conserved quantum number. Since baryon number is by all accounts conserved, this particular approach to the asymmetry problem would appear unpromising.

Another approach is to assume that the unobserved antimatter exists, but is sequestered in a region of the universe where it cannot interact with the matter we *can* observe; or is so distant that radiation arising from its interaction with matter has not yet reached us. These explanations are possible, of course, but not especially satisfying. The question remains, how could these (unlikely) spatial distributions of matter and antimatter come about?

All of which leads us to look into the symmetry properties of our hidden inner spacetime $N^2$. To help visualize these properties, **Fig. 10.15** reproduces in simplified form the world geometry in **Fig. 10.12 (a)**. This new figure omits the internal time $u^0$ and full $M^4 \times \bar{M}^2$ geometry and replaces them with the simple flatland geometry $E^2 \in (\xi^1, \xi^2)$. We now have only three spatial dimensions to visualize, a number sufficient for our purposes. Also, the planar surfaces $u = W/2$ and $u = -W/2$ bounding $N^2$ are now labeled $U_1$ and $U_2$, respectively, with their outward normal vectors $\mathbf{n}_1$ and $\mathbf{n}_2$ shown as well. The surfaces $U_1$ and $U_2$ represent parallel flatland universes, separated by a distance $\Delta u = W = \lambda_C / 2$. It is assumed that the inhabitants of $U_1$ and $U_2$ (shown in profile) face the interior of $N^2$, and hence each other, to make their observations; for it is interior to $N^2$ that the back-and-forth motion of the Zitterbewegung takes place, and this motion is part of their worlds.

Now let us look again at the motion of point $P$ described parametrically by Eqs. (10.141). We recall that this motion was set up in $U_1$ and thus judged to be counterclockwise from that point of view. Now suppose we set up this same motion from the point of view of $U_2$. To do this we use exactly the same set of parametric equations (10.141) as before. However from our new point of view we now judge the motion to be clockwise, not counterclockwise. So we see that the motions in $U_1$ and $U_2$ are mirror images of each other. In fact, even our two observers are mirror images of one another. More broadly, one can say that the two universes $U_1$ and $U_2$ are related by reflection in the

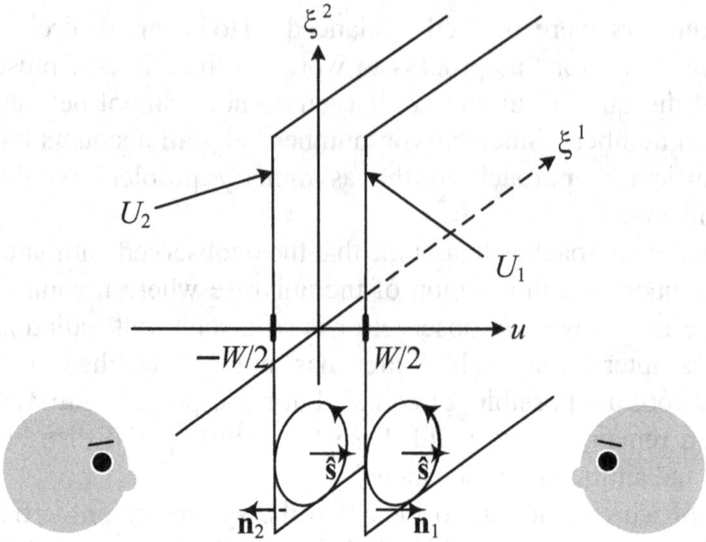


**Fig. 10.15** This figure presents a modified version of the world geometry depicted in **Fig. 10.12 (a)**. It replaces internal time $u^0$ and full $M^4 \times \bar{M}^2$ geometry with the simple flatland geometry $E^2 \in (\xi^1, \xi^2)$. Surfaces $U_1$ and $U_2$, with outward normal vectors $\mathbf{n}_1$ and $\mathbf{n}_2$ and separated by a distance $\Delta u = W = \lambda_C / 2$, bound the internal spacetime $N^2$. These two surfaces represent parallel flatland universes. Their respective inhabitants, shown in profile, face the interior of $N^2$ to make their observations.

$\xi^1$-$\xi^2$ plane .

We can go a step further. From **Fig. 10.15** we see that the unit spin vector $\hat{\mathbf{s}}$ corresponding to the counterclockwise motion in $U_1$ points in the direction of the outward normal $\mathbf{n}_1$, whereas in $U_2$ vector $\hat{\mathbf{s}}$ points *against* the direction of the outward normal $\mathbf{n}_2$. This suggests than we can identify electric charge with the scalar product of unit spin vector and surface normal. Then, since $\mathbf{n}_2 \cdot \hat{\mathbf{s}} = -\mathbf{n}_1 \cdot \hat{\mathbf{s}}$, we conclude that electrons on either side of the $\xi^1$-$\xi^2$ plane are of opposite electric charge.

In summary, overall, electrons on opposite sides of the $\xi^1$-$\xi^2$ plane are mirror images of each other and also of opposite charge. They are, in short, each other's antiparticle. Note that one can easily include the electron neutrino in this picture. For the radius of its Zitterbewegung is, according to (10.9) and (10.15), equal to that of the electron. Because the neutrino is electrically neutral, our definition of electric charge will have to be modified to read [see Eq. (4.68)]:

$$q/e = -\frac{m}{m_e}\mathbf{n}\cdot\hat{\mathbf{s}}, \tag{10.148}$$

where for the positive or negative electron, $m = m_e$, and for the massless neutrino or antineutrino, $m = m_\nu = 0$.

Now what are we to make of all this? Does it solve the problem of the missing antimatter? The answer is, maybe. If the world consisted only of electrons and electron neutrinos, then the answer would lean towards the affirmative. But as we have not shown that *all* matter in our universe appears as antimatter in a parallel universe, we can only assume that it is so. Suppose we do this. What is the upshot?

First we should emphasize that the matter/antimatter symmetry under discussion here does *not* arise from *CP* violation in weak interactions. Rather, it is a product of geometry, the presence of a hidden spacetime $N^2$ whose two boundaries $U_1$ and $U_2$ constitute space- and charge-conjugate universes.

Second, it is possible that the particle in $U_1$ and its antiparticle image in $U_2$ are in reality one and the same particle, seen from two different perspectives. They do after all share the same unit spin vector, $\hat{\mathbf{s}}$, as may be seen from **Fig. 10.15**; what causes them to be in a conjugate relation is the fact that universes $U_1$ and $U_2$ are *oriented*: their normals point in opposite directions. This possibility, that we are dealing here merely with single particles seen from two different perspectives, has two significant consequences: (a) The electrical neutrality of the universe as whole ($U_1 + U_2$) is ensured; and (b) there is (fortunately) no chance of this one-and-the-same particle annihilating itself.

However, it seems more likely that the particles in $U_1$ and antiparticles in $U_2$ are distinct and separate from each other. Suppose that is the case. Then, for example, the point *P* comprising an electron observable in $U_1$ would, in the course of its back-and-forth motion in $N^2$, reverse its direction at $U_2$ as shown in **Fig. 10.13**, but would not be observable at that point in $U_2$. Similarly, the point *P* comprising a positron observable in $U_2$ would reverse it motion in $N^2$ at $U_1$, but would not be observable there in $U_1$. Conceivably, it is the infinitesimal difference in diameters of the two momentum rings of **Fig. 10.14** that reminds oscillating point *P* which of the two planes, $U_1$ or $U_2$, it belongs to. This point of view—that objects in $U_1$ and $U_2$ are separate from and act independently of each other is reinforced when we take account of the behavior of the leptons $\mu$ and $\tau$. We know from **Fig. 10.5**, **Fig. 10.6** and **Table 10.2** that the internal geometries of these objects entail negative probabilities. Thus the Zitterbewegung of the heavy leptons, like that of the electron, involves back-and-forth motion in $N^2$. Although the exact paths of motion of these particles in $N^2$ are not known, and may not even be calculable, we do know that these paths penetrate $N^2$ a distance $(m_e / m_\ell)W$ before returning to home position. The relatively tiny volumes of motion within $N^2$ associated with the heavy leptons are indicated schematically in **Fig. 10.16.** Not only can such particles $U_1$ and antiparticles in $U_2$ be considered separate objects, they are sufficiently separated to preclude annihilation.

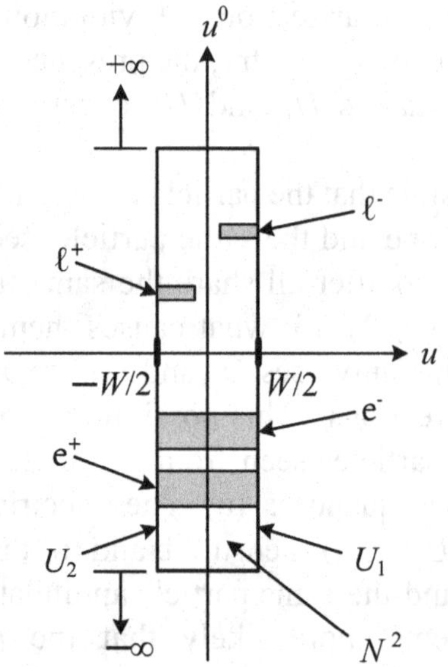


**Fig. 10.16**. Inner spacetime $N^2 \in (u^0, u)$, bounded by parallel universes $U_1$ and $U_2$, separated a distance $\Delta u = W = \lambda_C / 2$, where $\lambda_C$ is the Compton wavelength of the electron. Leptons occupy $U_1$, whereas antileptons occupy $U_2$, thus resolving the problem of matter/antimatter asymmetry. Shaded areas denote volumes within $N^2$ occupied by the Zitterbewegung of the various leptons and antileptons. The volumes belonging to the electron and positron extend across the entire width of $N^2$, raising the question whether they are separate particles or one particle seen from two different perspectives. Here they are depicted as separate particles. The volumes belonging to the leptons $\ell^- = \mu^-, \tau^-$ and antileptons $\ell^+ = \mu^+, \tau^+$ extend a short distance $(m_e / m_\ell)W$ into $N^2$ and thus unmistakenly represent individual particles.

And so, if this scenario is correct, there is no missing antimatter. It resides in the universe $U_2$ conjugate to the one we live in. What is missing is the ability to observe this conjugate universe. Unfortunately, barring interaction between the electrons of $U_1$ and positrons of $U_2$, that could mean that our solution to the problem of baryon asymmetry is untestable.

## Notes and references


[1] E. Schrödinger, Sitzber. Preuss. Akad. Wiss., Physik-Math. Kl. **24**, 418 (1930).

[2] K. Huang, Am. J. Phys. **20**, 479 (1952).

[3] D. Hestenes, Found. Phys. **15**, 63 (1985).

[4] H. Hönl, Ann. Physik (V) **33**, 565 (1938).

[5] A. O. Barut and A. J. Bracken, Phys. Rev. D **23**, 2454 (1981), and references therein.

[6] *Tables of Integral Transforms*, Vol. I, A. Erdélyi, Ed. (McGraw-Hill, New York, 1954), p. 26, transform pairs (30) and (34).

[7] E. Jahnke, F. Emde, F. Lösch, *Tables of Higher Functions* (McGraw-Hill, New York, 1960), Ch. IX.

[8] C. Itzykson and J. B. Zuber, *Quantum Field Theory* (McGraw-Hill, New York, 1980), p. 694.

[9] For guidance on symmetry and symmetrization see W. Heitler, *The Quantum Theory of Radiation* (Oxford, London, 1954), p. 418; W. Pauli, Handbuch der Physik, 2$^{nd}$ Ed., **24/I** (J. Springer, Berlin, 1933), p. 235; J. J. Sakurai, *Advanced Quantum Mechanics* (Addison-Wesley, Reading, MA 1967), p. 107.

[10] J. A. Stratton, *Electromagnetic Theory* (McGraw-Hill, New York, 1941), p. 235.

[11] J. J. Sakurai, *op. cit.*, p. 119.

[12] J. J. Sakurai, *ibid*, p.87. Note that Sakurai takes charge $e$ to be negative (see his footnote on p. 15), whereas we take it to be positive and write for the charge of the electron $-e$.

[13] *Tables of Integral Transforms*, Vol. II, A. Erdélyi, Ed. (McGraw-Hill, New York, 1954), p. 11, transform pair (41) and p. 12, transform pair (50).

[14] S. Schweber, *An Introduction to Relativistic Quantum Field Theory* (Row, Peterson, Evanston, IL, 1961), pp. 77, 78. Y. Aharonov and L. Susskind, Phys. Rev. **158**, 1237 (1967); H. J. Bernstein, Phys. Rev. Lett. **18,** 1102 (1967); A. G. Klein and G. I. Opat, Phys. Rev. Lett. **37**, 238 (1976).
[15] See J. Duffield, *Relativity + the Theory of Everything* (Corella Ltd., Poole, UK. 2009), p. 41, and references therein.
[16] O. Klein, Z. Phys. **37**, 875 (1926); Nature **118**, 516 (1926); J. –M. Souriau, Nuovo Cimento **30**, 565 (1963); A. Einstein and P. Bergman, Ann. Math. **39**, 683 (1938).
[17] Peter Rogers, "Where did all the antimatter go?" http://physicsweb.org/article/world/14/8/9/1.
[18] U. Sarkar, *Particle and Astroparticle Physics* (Taylor & Francis, New York, 2007).

# Chapter 11

# Internal structure of the bosons

Here we seek to describe the internal structure of the bosons. Specifically we are after the internal momentum densities of these fundamental particles. Knowing the momentum density we can calculate the spin. The method employed to get at these structures is same as that used in the previous chapter for the leptons. One starts with the external momentum density $\varpi^{\nu} = T^{0\nu}/c$ and, making use of the fact that the boson wave fields are self-imaging, extracts from $\varpi^{\nu}$ the internal momentum density $\Pi^{\nu} = (\Pi^0, \mathbf{\Pi})$.

## 11.1 Scalar boson

Let us take a scalar boson of mass $m_H$ at rest and write for its wave function

$$\phi = A\sin(k_H x^0) . \qquad (11.1)$$

Putting this into (5.43) and (5.44) we get for the wave's momentum density $\varpi^{\nu} = (\varpi^0, \boldsymbol{\varpi})$

$$\varpi^0 = \frac{1}{2c}\left[\left(\partial_0\phi\right)^2 + \left(k_H\phi\right)^2\right] = \frac{1}{2c}A^2 k_H{}^2 \qquad (11.2)$$

$$\boldsymbol{\varpi} = 0 . \qquad (11.3)$$

We note two things. First, from (11.2), since $\varpi^0 = m_H c / V$, where $V$ is the volume of box normalization, coefficient $A$ is given by

$$A^2 = \frac{2m_H c^2}{k_H{}^2 V} . \qquad (11.4)$$

Second, from (11.3), the momentum density of the rest particle shows no evidence of internal spin momentum; the scalar boson is a particle of spin 0. Energy density $\Pi^0$ is its only internal attribute.

To find the internal energy density we observe that scalar boson wave function $\phi$, like the lepton wave functions $\ell$ and $\nu$, is self-imaging. This means that it obeys a propagation law identical to those exhibited in (10.22) and (10.101):

$$\phi(x) = \Delta(x) * \phi(x), \tag{11.5}$$

where $\Delta(x)$ is given by (10.23) with $k_5 = 0$. Replacing *one* of the $\phi$s in each of the two terms of (11.2) with the r.h.s. of (11.5), then carrying out the integration, we obtain

$$\begin{aligned} \varpi^0 &= \frac{1}{2c}\left[(\partial_0\phi)^2 + (k_H\phi)^2\right]\int D_r(\boldsymbol{\xi})d^3\boldsymbol{\xi} \\ &= \frac{m_H c}{V}\int D_r(\boldsymbol{\xi})d^3\boldsymbol{\xi} \end{aligned} \tag{11.6}$$

where $D_r$ is given by (10.35) with $k_5 = 0$:

$$D_r(\boldsymbol{\xi}) = -\frac{\delta'(\rho - \xi^5)}{4\pi\rho} \tag{11.7}$$

and $\xi^5$ is the radius of the Zitterbewegung. Equation (11.6) expresses the external energy density $\varpi^0$ as the spatial average of an internal energy density $\Pi^0$, where

$$\Pi^0 = m_H c D_r(\boldsymbol{\xi}) \tag{11.8}$$

This has the same form as the energy density of the charged lepton, Eq. (10.60); the rest energies of the scalar boson and charged lepton are both attributable to the Zitterbewegung. Moreover, the probability distribution $\Pi^0 / m_H c$ is, apart from radius, identical with that of the electron, shown in **Fig. 10.4**. This is not surprising. The propagator $\Delta(x)$ is common to all wave fields, spinor and scalar in particular. Since it is the propagator that becomes the particle, result (11.8) is to be expected. Still, there are two major differences:

(1) Intrinsic spin in the case of the leptons can be attributed to the Zitterbewegung. Now the lepton is a spin-1/2 particle. Thus the radius $\xi^5$ of the lepton's Zitterbewegung is $\lambda_C / 4\pi$, where $\lambda_C$ is the Compton

wavelength of the particle, ensuring reëntrance of the de Broglie wave after $4\pi$ radians of internal orbital motion. Bosons, however, are integer-spin spin particles, and so their de Broglie waves (if indeed such waves exist) are reëntrant after $2\pi$ --not $4\pi$ --radians of internal motion. In units of Compton wavelength, the radius of the boson's Zitterbewegung is thus twice that of the lepton's. In the case of the scalar boson it is

$$\xi^5 = \frac{\lambda_H}{2\pi} = \frac{\hbar}{m_H c} \tag{11.9}$$

Nevertheless, as can be seen from **Fig. 10.4**, the scalar boson's Zitterbewegung still generates negative probability and thus penetrates internal spacetime $N^2$ in a manner similar to that of the electron. The difference is one of scale: in units of Compton wavelength, the radius and depth of penetration of the boson's Zitterbewegung are twice that of the electron's. Consequently, whereas in the case of the electron one complete circuit within $N^2$ corresponds to a single reëntrance of the de Broglie wave, in the case of the boson it yields *two* such reëntrances—one of positive probability, the other of negative probability.

(2) If, apart from scale, the electron and scalar boson have the same internal probability distribution, how is it that the electron has finite spin but the scalar boson spin 0? The answer is that, while the spin of a lepton is indeed directly attributable to its Zitterbewegung, the spin of a boson cannot be so attributed. Rather, the spin of a boson is a product of its *polarization*, not its Zitterbewegung. And since the scalar boson, being scalar, has no polarization, its spin is 0. We will have more to say about this at the end of **Sec. 11.3**.

## 11.2 Vector bosons $Z^0$, $W^{\pm}$.

*11.2.1 Momentum density.* The study of the internal structure of the vector bosons begins as it did for the scalar boson: with the momentum density $\varpi^{\nu} = T^{0\nu}/c$. We shall focus on getting internal formulas for the $W^{\pm}$ at rest, formulas that reduce to those for $Z^0$ by putting $k_5 = 0$. In addition, we shall limit our study to transverse fields, as it is these that give rise to the boson's intrinsic spin. Thus rewriting (5.89) and (5.90) to make the conversion from $Z^0$ to $W^{\pm}$, we have for the momentum density:

$$\varpi^0 = \frac{1}{2c}\left(\frac{\mathbf{E}_W{}^2}{c^2} + \mathbf{B}_W{}^2 + k_W{}^2\mathbf{W}^2\right) \tag{11.10}$$

and

$$\boldsymbol{\varpi} = \frac{\mathbf{E}_W}{c^2} \times \mathbf{B}_W, \tag{11.11}$$

where in these expressions $\mathbf{W}$ is the Maxwell-Proca wave function satisfying (5.78), functions $\mathbf{E}_W$ and $\mathbf{B}_W$ are defined exactly as in (5.47), but with Z replaced by W everywhere it occurs, and $k_W = m_W c/\hbar$.

We are not quite done setting up the momentum density: Eq. (11.11) needs to be rewritten to bring out the presence of spin. From (5.72) and the second of Eqs. (5.47) we have, exactly as in electromagnetism,

$$B_W{}^\ell = (\nabla \times \mathbf{W})^\ell = \varepsilon_{\ell mn}\partial_m W^n. \tag{11.12}$$

Using this for $\mathbf{B}_W$ in (11.11) we obtain for the momentum 3-density

$$\varpi^k = \frac{1}{c^2}\left(\mathbf{E}_W \cdot \partial_k \mathbf{W} - \mathbf{E}_W \cdot \nabla W^k\right), \tag{11.13}$$

an expression comparable in form to the Gordon decomposition (10.65). In either expression the first term refers to the particle's translational motion and the second term to its spin momentum. The second term in (11.13) is in fact 0. Nevertheless, as we shall see, we can extract from it the internal spin momentum density, which is non-vanishing.

*11.2.2 Transverse wave function for the particle at rest.* To start let us take a transversely polarized plane wave traveling in the positive $x^3$ direction and whose polarization vectors therefore run perpendicular to the $x^3$ axis. We can then write for the components of wave vector $\mathbf{W}$,

$$W^i = w^i \cos\left(\phi + \alpha^i\right),\ i = 1,\ 2 \tag{11.14}$$

where the $w^i$ are amplitudes, $\alpha^i$ phases, $\phi = k_0 x^0 + k_3 x^3$ and from wave equation (5.78), $k_0{}^2 = k_3{}^2 + k_W{}^2$. Then utilizing standard definitions of the form (6.23) for $\mathbf{E}_W$ and $\mathbf{B}_W$ we have

$$\left.\begin{aligned}
\frac{E_W{}^1}{c} &= -\partial_0 W^1 = k_0 w^1 \sin\left(\phi+\alpha^1\right) \\
\frac{E_W{}^2}{c} &= -\partial_0 W^2 = k_0 w^2 \sin\left(\phi+\alpha^2\right) \\
B_W{}^1 &= -\partial_3 W^2 = -k^3 w^2 \sin\left(\phi+\alpha^2\right) \\
B_W{}^2 &= \partial_3 W^1 = k^3 w^1 \sin\left(\phi+\alpha^1\right)
\end{aligned}\right\}. \tag{11.15}$$

Putting (11.14) and (11.15) into (11.10), after time averaging we get for the energy density

$$\overline{\varpi^0} = \frac{1}{2c} k_0{}^2 \mathbf{w}^2 \tag{11.16}$$

Since for the particle in motion $\overline{\varpi^0} = \hbar k_0 / V$, where $V$ is the volume of box normalization, we have

$$\mathbf{w}^2 = \frac{2\hbar c}{k_0 V}. \tag{11.17}$$

Similarly, for the translational momentum we have from (11.11), after time averaging,

$$\overline{\varpi^3} = \frac{1}{2c} k_0 k^3 \mathbf{w}^2. \tag{11.18}$$

For the particle in motion, $\overline{\varpi^3} = \hbar k^3 / V$, and so we reproduce the result (11.17).

Specializing now to a particle at rest, we have from (11.10), (11.13), (11.14), (11.15) and (11.17):

$$\varpi^0 = \frac{1}{2c}\left(\frac{\mathbf{E}_W{}^2}{c^2} + k_W{}^2 \mathbf{W}^2\right) \tag{11.19}$$

$$\varpi^k = -\frac{1}{c^2} \mathbf{E}_W \cdot \nabla W^k \tag{11.20}$$

$$W^i = w^i \cos\left(\phi+\alpha^i\right),\ i=1,\ 2 \tag{11.21}$$

$$\frac{E_W{}^i}{c} = k_W w^i \sin\left(\phi + \alpha^i\right) \tag{11.22}$$

$$\mathbf{w}^2 = \frac{2\hbar c}{k_W V}, \tag{11.23}$$

in which $\phi = k_0 x^0$. Coefficients $w^i$ are adjustable, subject to constraint (11.23).

*11.2.3 Internal energy density.* Vectors $\mathbf{W}$ and $\mathbf{E}_W$ appearing in (11.19) are subject to the same self-imaging law as charged-lepton spinor $\ell$ in (10.22) and scalar $\phi$ in (11.5). Hence we have immediately, as in (11.6),

$$\begin{aligned} \varpi^0 &= \frac{1}{2c}\left(\frac{\mathbf{E}_W{}^2}{c^2} + k_W{}^2\mathbf{W}^2\right)\int D_r(\boldsymbol{\xi})d^3\boldsymbol{\xi} \\ &= \frac{m_W c}{V}\int D_r(\boldsymbol{\xi})d^3\boldsymbol{\xi} \end{aligned}, \tag{11.24}$$

where $D_r$ is given by (10.35), with

$$\xi^5 = \frac{\lambda_W}{2\pi} = \frac{\hbar}{m_W c} \tag{11.25}$$

and $k_5 = \pm m_e c/\hbar$ [see Eqs. (5.9) and (5.10)]. In parallel with the case of the scalar boson, (11.24) expresses the external energy density $\varpi^0$ as the spatial average of an internal energy density $\Pi^0$, where

$$\Pi^0 = m_W c D_r(\boldsymbol{\xi}). \tag{11.26}$$

Note that, in (10.35), the parameter $\omega = k_5\xi^5 = m_e / m_W \ll 1$. Thus $D_r$ has essentially the simple form (11.7), with $\xi^5$ given by (11.25).

*11.2.4 Internal momentum density and derivation of spin.* To find the internal momentum density $\boldsymbol{\Pi}$, in (11.20) we first replace $W^k$ by its convolution equivalent $W^k * \Delta$ exactly as we did in (11.24). This gives

$$\begin{aligned}
\varpi^k c^2 &= -E_W{}^j \partial_j W^k \\
&= -E_W{}^j \int \partial_j \Delta(x-\xi) W^k(\xi) d^4\xi \\
&= -E_W{}^j \int \partial_j \Delta(\xi) W^k(x-\xi) d^4\xi \\
&= -E_W{}^j \int \partial_j D_r(\boldsymbol{\xi}) d^3\boldsymbol{\xi} \cdot W^k \\
&= -E_W{}^j W^k X^j
\end{aligned} \tag{11.27}$$

where we have defined

$$X^j = \int \partial_j D_r(\boldsymbol{\xi}) d^3\boldsymbol{\xi} \tag{11.28}$$

and again $D_r$ is given by (10.35). Next we perform the time average of (11.27), giving

$$\overline{\varpi^k} = -\frac{k_W}{2c} w^j w^k \sin(\alpha^j - \alpha^k) X^j , \tag{11.29}$$

the r.h.s. of which vanishes for $j = k$. Writing out the components,

$$\left.\begin{aligned}
\overline{\varpi^1} &= -\frac{\hbar w^1 w^2}{\mathbf{w}^2 V} \sin\varphi \int \partial_2 D_r(\boldsymbol{\xi}) d^3\boldsymbol{\xi} \\
\overline{\varpi^2} &= +\frac{\hbar w^1 w^2}{\mathbf{w}^2 V} \sin\varphi \int \partial_1 D_r(\boldsymbol{\xi}) d^3\boldsymbol{\xi}
\end{aligned}\right\} \tag{11.30}$$

where in the coefficient we have used (11.17) and for the phase difference we write

$$\alpha^2 - \alpha^1 \equiv \varphi . \tag{11.31}$$

Now in vectorial form (11.30) can be generalized to

$$\overline{\boldsymbol{\varpi}} = \frac{\hbar w^1 w^2}{\mathbf{w}^2 V} \sin\varphi \int \hat{\mathbf{k}} \times \nabla D_r(\boldsymbol{\xi}) d^3\boldsymbol{\xi} , \tag{11.32}$$

where vector $\hat{\mathbf{k}}$ denotes an arbitrarily chosen direction in space (in our case, the direction of the positive $x^3$ axis). Equation (11.32) expresses the external spin momentum density (which equals 0!) as the spatial average of an internal spin momentum density

$$\boldsymbol{\Pi} = \frac{\hbar w^1 w^2}{\mathbf{w}^2} \sin\varphi \, \hat{\mathbf{k}} \times \nabla D_r \tag{11.33}$$

This formula gives the internal momentum for the general case of elliptical polarization. It vanishes for plane polarization, i.e., when $w^1$, $w^2$ or $\varphi$ equals 0. It achieves its extreme values for circular polarization, i.e., when $w^1 = w^2$ and $\varphi = \pm\pi/2$:

$$\boldsymbol{\Pi}_{\text{circ}} = \pm\frac{\hbar}{2} \hat{\mathbf{k}} \times \nabla D_r \tag{11.34}$$

where the upper (lower) sign represents right (left) circular polarization. (In right circular polarization, the **E** vector turns clockwise when viewed *against* direction $\hat{\mathbf{k}}$.) The spin angular momentum **S** resulting from this internal motion is given by

$$\mathbf{S}_{\text{circ}} = \int \boldsymbol{\xi} \times \boldsymbol{\Pi}_{\text{circ}} d^3\xi = \mp\hbar\hat{\mathbf{k}}\,, \tag{11.35}$$

where we have used the same method of integration as was used to get (10.74). Thus we have shown that $W^{\pm}$ and $Z^0$ are indeed particles of spin 1. We have also shown that the direction of spin is opposite to the direction of polarization as defined by vector $\hat{\mathbf{k}}$, a fact well known in vector wave optics, where $\hat{\mathbf{k}}$ denotes the direction of propagation.[1]

## 11.3 Tensor boson $\gamma$ (photon)

*11.3.1 Momentum density.* Let us rewrite the electromagnetic momentum density (6.30) and (6.31), omitting the contribution of dark radiation:

$$\varpi^0 = \frac{1}{2\mu_0 c}\left(\frac{1}{c^2}\mathbf{E}^2 + \mathbf{B}^2\right) \tag{11.36}$$

$$\boldsymbol{\varpi} = \frac{1}{\mu_0 c^2}\mathbf{E}\times\mathbf{B}\,. \tag{11.37}$$

The similarity of these equations to (11.10) and (11.11) suggests that we can carry over much of the work done in our study of the internal dynamics of the massive vector bosons to a comparable study of the massless photon. And that is true, although differences will arise owing to the fact that the photon cannot be brought to rest. In particular, it means that for a free electromagnetic wave, one cannot put $\mathbf{B}=0$ as was done to arrive at (11.19). With the definition $\mathbf{B}=\nabla\times\mathbf{A}$, where $\mathbf{A}$ is the vector potential, we can put (11.37) in a form comparable to (11.13):

$$\varpi^k = \frac{1}{\mu_0 c^2}\left(\mathbf{E}\cdot\partial_k\mathbf{A} - \mathbf{E}\cdot\nabla A^k\right). \tag{11.38}$$

Here the first term describes the wave's translational momentum density and the second, its spin momentum density (which is 0). The first term never vanishes.

*11.3.2 Photon wave functions.* We take a plane wave traveling in the positive $x^3$ direction and write for the components of vector potential $\mathbf{A}$,

$$A^i = a^i \cos\left(\phi + \alpha^i\right),\ i = 1,\ 2 \tag{11.39}$$

where the $a^i$ are amplitudes, $\alpha^i$ phases, $\phi = k_0 x^0 + k_3 x^3$ and from wave equation (6.45), $k_0 = -k_3 \equiv k$. Then utilizing standard definitions (6.23) for $\mathbf{E}$ and $\mathbf{B}$ we have

$$\left.\begin{aligned}
\frac{E^1}{c} &= -\partial_0 A^1 = ka^1 \sin\left(\phi+\alpha^1\right)\\
\frac{E^2}{c} &= -\partial_0 A^2 = ka^2 \sin\left(\phi+\alpha^2\right)\\
B^1 &= -\partial_3 A^2 = -ka^2 \sin\left(\phi+\alpha^2\right)\\
B^2 &= \partial_3 A^1 = ka^1 \sin\left(\phi+\alpha^1\right)
\end{aligned}\right\}. \tag{11.40}$$

Putting these into (11.36), after time averaging we get for the energy density

$$\overline{\varpi^0} = \frac{1}{2\mu_0 c} k^2 \mathbf{a}^2 \tag{11.41}$$

Since $\overline{\varpi^0} = \hbar k / V$, where $V$ is the volume of box normalization, we have

$$\mathbf{a}^2 = \frac{2\mu_0 c\hbar}{kV}. \tag{11.42}$$

Similarly, for the translational momentum we have from (11.38), after time averaging,

$$\overline{\varpi^3} = \frac{1}{2\mu_0 c} k^2 \mathbf{a}^2 \tag{11.43}$$

which is the same as (11.41). For the photon, $\overline{\varpi^3} = \hbar k / V$, and so we reproduce the result (11.42).

*11.3.3 Photon internal probability distribution and energy density.* Vectors $\mathbf{E}$ and $\mathbf{B}$ appearing in (11.19) are subject to the same self-imaging law as the massless neutrino spinor $\nu$ in (10.101). Hence we have, as in (10.111),

$$\begin{pmatrix} \mathbf{E}(x) \\ \mathbf{B}(x) \end{pmatrix} = \int D(\xi^0 - \xi^3, \boldsymbol{\rho}, \xi^5)\, d^2\boldsymbol{\rho}\, d\xi^3 \cdot \begin{pmatrix} \mathbf{E}(x) \\ \mathbf{B}(x) \end{pmatrix} \tag{11.44}$$

where

$$D(\xi^0 - \xi^3, \boldsymbol{\rho}, \xi^5) = Q(\boldsymbol{\rho}, \xi^5)\delta(\xi^0 - \xi^3) \tag{11.45}$$

and, since for the photon, $k_5 = 0$ [see (5.6)], function $Q$ as given by (10.113) simplifies to

$$\begin{aligned} Q(\boldsymbol{\rho}, \xi^5) &= \frac{1}{(2\pi)^2} \int e^{-i\xi^5\sqrt{\boldsymbol{\kappa}^2}} e^{i\boldsymbol{\kappa}\cdot\boldsymbol{\rho}} d^2\boldsymbol{\kappa} \\ &= \frac{1}{2\pi} \int_0^\infty e^{-i\xi^5\kappa} J_0(\kappa\rho)\kappa d\kappa \end{aligned} \tag{11.46}$$

where the two-dimensional vectors $\boldsymbol{\rho} = (\xi^1, \xi^2)$ and $\boldsymbol{\kappa} = (k^1, k^2)$. The real part $D_r$ of function $D$ defines the internal structure of the photon (the imaginary part vanishes upon integration). From (11.45) we see that the photon is a disk of zero thickness, flattened as if by Lorentz contraction, and of course moving at speed c as shown by the $\delta$-function. The real part of rotationally symmetric function $Q$ defines its cross sectional probability structure—a structure due to the Zitterbewegung. For the real part of $Q$ we have from (10.114) and (10.115)

$$\begin{aligned} 2\pi Q_r &= \frac{\partial}{\partial \xi^5} \int_0^\infty \frac{\sin\left(\kappa \xi^5\right)}{\kappa} J_0(\kappa\rho)\kappa d\kappa \\ &= \frac{\partial}{\partial \xi^5}\left[\frac{1}{\sqrt{\xi_5^{\,2}-\rho^2}} U(\xi^5-\rho)\right] \qquad (11.47) \\ &= \frac{\partial}{\partial \xi^5}\left(\frac{1}{\sqrt{\xi_5^{\,2}-\rho^2}}\right) U(\xi^5-\rho) + \frac{1}{\sqrt{\xi_5^{\,2}-\rho^2}} \delta(\xi^5-\rho) \end{aligned}$$

where $U$ is the unit step function (10.33) and where the radius $\xi^5$ of the Zitterbewegung is given by the reduced radiation wavelength:

$$\xi^5 = \frac{\lambda}{2\pi}. \qquad (11.48)$$

We plot this distribution in terms of the dimensionless radial variable $x = \rho / \xi^5$. Then, with $t_1 = (1-x^2)^{1/2}$, we obtain the photon radial probability distribution

$$\begin{aligned} 2\pi\rho Q_r \frac{d\rho}{dx} &\equiv R(x) \\ &= -\frac{1}{t_1^3} x U(1-x) + \frac{1}{t_1}\delta(x-1) \end{aligned} \qquad (11.49)$$

Distribution $R(x)$ is plotted in **Fig. 11.1**. Like those of the scalar and vector bosons, the photon's internal probability distribution contains both positive and negative values. Thus the photon's Zitterbewegung penetrates inner spacetime $N^2$ in the same way as described for the scalar boson. For each round trip within $N^2$, the photon experiences *two* reëntrances of the de Broglie wave in Euclidian $E^3$, one of positive probability, the other of negative probability.

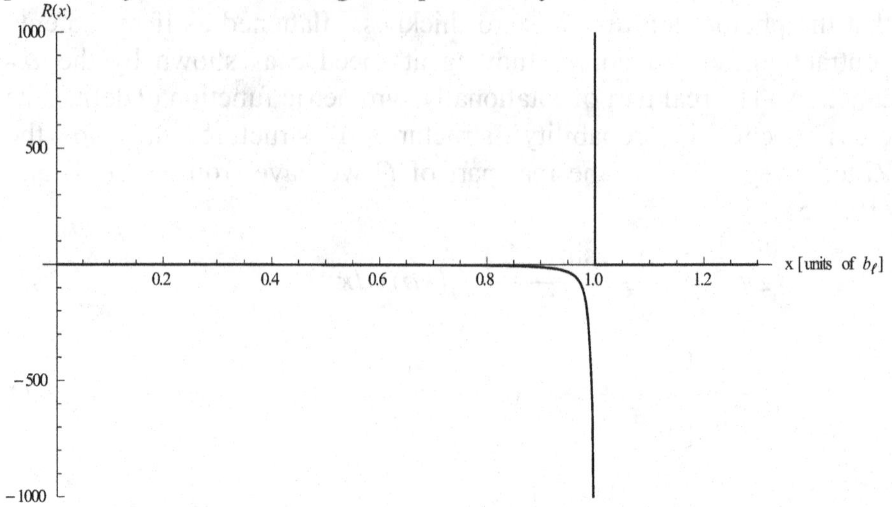


**Fig. 11.1.** Plot of the photon's internal probability distribution $R(x)$, defined by Eq. (11.49). This dimensionless distribution, depicting the time-averaged Zitterbewegung of the photon, has the same form for all photons. Its scale changes with photon radius, $\xi^5 = \lambda/2\pi$ . The negative peak at $x=1$ is produced by the first term of the r. h. s. of (11.49), whereas the (doubly infinite) positive peak comes from the last term of the equation.

Now in consequence of (11.44) we have from (11.36) the time averaged energy density

$$\begin{aligned}\overline{\varpi^0} &= \frac{1}{2\mu_0 c}\left(\frac{1}{c^2}\mathbf{E}^2 + \mathbf{B}^2\right)\int D_r(\xi^0 - \xi^3, \boldsymbol{\rho}, \xi^5)\, d^2\boldsymbol{\rho}\, d\xi^3 \\ &= \frac{\hbar k}{V}\int D_r(\xi^0 - \xi^3, \boldsymbol{\rho}, \xi^5)\, d^2\boldsymbol{\rho}\, d\xi^3\end{aligned} \quad , \quad (11.50)$$

in exact analogy to (11.24). Thus the photon's internal energy density distribution is

$$\Pi^0 = \hbar k Q_r(\boldsymbol{\rho}, \xi^5)\delta(\xi^0 - \xi^3), \quad (11.51)$$

which is analogous to the vector boson's energy density (11.26).

*11.3.4 Internal momentum density and derivation of spin.* To find the photon's internal momentum density $\mathbf{\Pi}$, we proceed as we did for the vector boson. In the spin piece of (11.38) we first replace $A^k$ by its convolution equivalent $A^k * \Delta$. This gives, as in (11.27),

$$\begin{aligned}\varpi^k \mu_0 c^2 &= -E^j \partial_j A^k \\ &= -E^j A^k X^j\end{aligned} \tag{11.52}$$

where we have defined

$$X^j = \int \partial_j D_r(\boldsymbol{\xi}) d^3\boldsymbol{\xi} \tag{11.53}$$

and $D_r$ is given by (11.45) (we are now taking real parts). Next we perform the time average of (11.52), giving

$$\overline{\varpi^k} = -\frac{k}{2\mu_0 c} a^j a^k \sin(\alpha^j - \alpha^k) X^j, \tag{11.54}$$

Writing out the components as in (11.30)

$$\left.\begin{aligned}\overline{\varpi^1} &= -\frac{\hbar a^1 a^2}{\mathbf{a}^2 V} \sin\varphi \int \partial_2 D_r(\boldsymbol{\xi}) d^3\boldsymbol{\xi} \\ \overline{\varpi^2} &= +\frac{\hbar a^1 a^2}{\mathbf{a}^2 V} \sin\varphi \int \partial_1 D_r(\boldsymbol{\xi}) d^3\boldsymbol{\xi}\end{aligned}\right\} \tag{11.55}$$

where in rewriting the coefficients we have used (11.42) and phase difference $\varphi$ is still given by (11.31). In vectorial form (11.55) can be generalized to

$$\overline{\boldsymbol{\varpi}} = \frac{\hbar a^1 a^2}{\mathbf{a}^2 V} \sin\varphi \int \hat{\mathbf{k}} \times \nabla D_r(\boldsymbol{\xi}) d^3\boldsymbol{\xi}, \tag{11.56}$$

where vector $\hat{\mathbf{k}}$ denotes an arbitrarily chosen direction of propagation (in our case, the direction of the positive $x^3$ axis). Equation (11.56) expresses the external spin momentum density as the spatial average of an internal spin momentum density

$$\mathbf{\Pi} = \frac{\hbar a^1 a^2}{\mathbf{a}^2} \sin\varphi\, \hat{\mathbf{k}} \times \nabla Q_r \cdot \delta(\xi^0 - \xi^3) \tag{11.57}$$

This is for the general case of elliptical polarization. It vanishes for plane polarization, i.e., when $a^1$, $a^2$ or $\varphi$ equals 0. It attains its extreme values for circular polarization, i.e., when $a^1 = a^2$ and $\varphi = \pm\pi/2$:

$$\mathbf{\Pi}_{\text{circ}} = \pm\frac{\hbar}{2} \hat{\mathbf{k}} \times \nabla Q_r \cdot \delta(\xi^0 - \xi^3) \tag{11.58}$$

where the upper (lower) sign represents right (left) circular polarization. We note again that in right hand polarization vector $\mathbf{E}$ rotates clockwise when viewed against the direction $\hat{\mathbf{k}}$ of propagation. The spin angular momentum $\mathbf{S}$ resulting from this internal motion is given by

$$\mathbf{S}_{\text{circ}} = \int \boldsymbol{\xi} \times \mathbf{\Pi}_{\text{circ}} d^3\boldsymbol{\xi} = \mp\hbar\hat{\mathbf{k}}\,, \tag{11.59}$$

where we have used the same method of integration as was used to get (10.137). Thus we have shown that the photon is a particle of spin 1. We have also shown that for right circular polarization the spin is antiparallel to the direction of motion $\hat{\mathbf{k}}$, and for left circular polarization it is parallel to it.[1]

This has been a lot of work to confirm what is already known about the photon. And yet it is helpful to know what constitutes a photon internally. The photon's unalterable foundation is its internal probability distribution, the planar ring depicted in **Fig. 11.1**. Multiplied by $\hbar k$, the probability distribution becomes the photon's internal energy density distribution $\Pi^0$, also unalterable. The internal spin momentum density $\mathbf{\Pi}$ is, however, another matter. Its form and value changes according to the values assigned to the polarization constants $a^1$, $a^2$ and $\varphi$ (also showing that $\Pi^\nu$ is not actually a 4-vector). In sum, a photon is a disk of energy density $\Pi^0$, on which is impressed a spin momentum density $\mathbf{\Pi}$ whose form and value are determined by the polarized vector potential $\mathbf{A}$. That is what it means to speak of a polarized photon.

It is of interest to compare the photon and the neutrino, both massless particles endowed with spin. In the case of the neutrino, we could attribute its spin directly to the Zitterbewegung, i.e., to the helical motion of a tachyonic point $P$ about the direction of propagation. We could even depict this origin of spin classically, as in **Fig. 10.1**, owing to the fact that the neutrino possesses non-vanishing fifth and sixth momenta, $p_5 = p^6 \equiv G$ [see (4.38) and **Fig. 4.2**]. However, in the case of the photon, the value of both of these extra momenta is 0! [See (5.6) and (5.7).] Consequently, photon spin cannot be attributed to, or pictured classically as, helical motion of an internal point $P$. Instead, the spin arises as a turning of probability ring $Q_r$, a turning produced by the operator $\hat{\mathbf{k}} \times \nabla$ and regulated in magnitude by the polarization parameters $a^1$, $a^2$ and $\varphi$ of vector potential $\mathbf{A}$.

There have been attempts to describe the photon as a composite particle, one made up of a neutrino and antineutrino.[2] Our work here shows that the photon is not, and cannot be, formed in that way. Of course one could not know this without going into the internal dynamics of the neutrino and photon as we have done in this and the previous chapter.

## 11.4 Curvature tensor boson: the riemann

In **Chapter 8** we described the action of a fifth force of nature, one we have called the R-C force, as the field giving rise to it is the curvature tensor $R^{\mu\nu}{}_{\alpha\beta}$ of Riemann-Cristoffel. The R-C force is Lorentzian in form, is repulsive, and can exist as a force between separated gravitating bodies or as a self-force. In either case, because the curvature tensor is self-imaging, there must exist a force-mediating boson to go with it. We call this particle the *riemann*.

The energy-momentum tensor for the free tensor field is given by (8.33). In vector form the energy and momentum densities for this field are, utilizing definitions (8.49)

$$\varpi^0 = \frac{1}{2\tilde{\mu}c}\left(\frac{1}{c^2}\mathbf{E}^{\alpha\beta}\cdot\mathbf{E}_{\alpha\beta} + \mathbf{B}^{\alpha\beta}\cdot\mathbf{B}_{\alpha\beta}\right) \tag{11.60}$$

$$\boldsymbol{\varpi} = \frac{1}{\tilde{\mu}c^2}\mathbf{E}^{\alpha\beta}\times\mathbf{B}_{\alpha\beta}\,. \tag{11.61}$$

Now these equations, apart from the summations over spacetime indices, are identical with relations (11.36) and (11.37) attaching to the radiation field of Maxwell. Our work with electromagnetic radiation therefore carries over directly to radiation generated by the curvataure tensor. The riemann is a spin-1 boson having formally the same internal energy and spin momentum densities as the photon, namely, (11.51) and (11.57). We cannot be sure of the values of the reciprocal reduced wavelength $k = 2\pi / \lambda$. For the R-C force between objects of ordinary matter as discussed in **Sec. 8.9**, *k* is no doubt essentially 0. But for the self-force of the Big Bang discussed in **Sec. 8.10**, it may have been at the beginning unimaginably large, but now tiny, owing to expansion of the cosmos.

**Notes and references**


[1] G. R. Fowles, *Introduction to Modern Optics* (Holt, Rinehart and Winston, New York, 1968), p. 217.

[2] W. A. Perkins, "Neutrino Theory of Photons," Phys. Rev. **137**, B1291 (1965).

# PART V

# THE FAMILY PROBLEM

# Chapter 12

# Neutrino mixing and oscillation

The present chapter follows directly from **Chapter 4**, where it was argued that fermion flavor generation and family replication derive not from gauge symmetry but from higher dimensions. Here we apply the general theory of flavor developed in that earlier chapter to some unresolved problems in neutrino mixing and oscillation. In particular we want to address the following questions:

- Where do the mass-squared eigenstates come from?
- Is the neutrino a Dirac or Majorana particle?
- Can massless neutrinos oscillate?
- Why must neutrinos oscillate at all?
- Why are there no right-handed neutrinos?
- Can neutrinos oscillate and still conserve lepton number?
- Where does the sterile neutrino come from and why does it get involved in oscillation?
- What explains the nearly maximal mixing of $\nu_\mu$ and $\nu_\tau$?

Neutrino mixing and oscillation offer an ideal proving ground for our theory of higher dimensions and the theory of flavor proceeding from it.

## 12.1 What we know so far

Let us begin by reviewing the description of neutrinos of definite mass in the infinite product space $M^4 \times \bar{M}^2$ with signature $(+---)\times(+-)$. In this six-dimensional world, each lepton family $\ell$ is represented by a self-imaging wave field $\chi_\ell(x^\mu, x^5, x^6) = \psi(x^\mu, x^5)\exp(iG_\ell x^6/\hbar)$, a field whose amplitude $\psi(x^\mu, x^5)$ repeats periodically, up to a constant phase factor, in planes perpendicular to the time-like $x^5$ axis. The subscript $\ell$ indicates that $\chi_\ell$ is an eigenstate of the pseudoenergy operator $\hat{p}^6 = -\hat{p}_6 = -i\hbar\partial_6$ with eigenvalue $p^6 = G_\ell$; in other words $\chi_\ell$ is monochromatic in the $x^6$ pseudotemporal domain and thus belongs to the five-momentum sphere of radius $G_\ell$ in $M^4 \times \bar{M}^2$. The charged and neutral flavors belonging to family $\ell$ are generated by an expansion of the amplitude $\psi(x^\mu, x^5)$ in eigenstates of the fifth momentum operator, $\hat{p}_5$ [see (4.32) and (4.33)]:

$$\chi_\ell(x^\mu, x^5, x^6) = e^{iG_\ell x^6/\hbar} \sum_n \psi_{\ell n}(x^\mu) e^{-ip_{5\ell n} x^5/\hbar}, \qquad (12.1)$$

where [see (4.85)]

$$p_{5\ell n} = L_\ell G_\ell - nm_e c, \qquad (12.2)$$

$p_{5\ell n}$ is the invariant fifth component of momentum (pseudomomentum) of the $n$th member of lepton family $\ell$, $L_\ell$ is the lepton family number and $m_e$ is the mass of the electron. Thus *flavor* can be considered a physical attribute jointly specified by 5-radius $G_\ell$ and expansion index $-n = Q/e$ (electric charge).

Flavor neutrino $\nu_\ell$ arises as the zero-order mode of expansion (12.1). Let us write for the neutrino term of the expansion $(n = 0,\ L_\ell = +1)$

$$\nu_\ell(x^\mu, x^5, x^6) = e^{iG_\ell x^6/\hbar} \psi_{\ell 0}(x^\mu) e^{-iG_\ell x^5/\hbar} \qquad (12.3)$$

Neutrino wave function $\nu_\ell$ is an eigenstate of the mass-squared operator defined by (4.36): $\widehat{m^2} = (\hat{p}_6{}^2 - \hat{p}_5{}^2)/c^2$. This yields for the mass-squared value of the flavor neutrino, $m_{\nu_\ell}^2 = 0$. This must be so, for according to (4.68), the flavor neutrino's mass-squared value is proportional to its electric charge.

## 12.2 Neutrino mixing and the spacetime origin of mass-squared eigenstates

Now the flavor neutrinos described by (12.3) are objects of definite mass. As such they cannot oscillate and in fact cannot even exist, as we showed in **Sec. 4.7**. To describe real neutrinos, which *do* oscillate and do exist, we must replace representation (12.3) with one involving a superposition of non-degenerate mass-squared eigenstates, rendering $\nu_\ell$ no longer an eigenstate of $m^2$ and making possible the transition from flavor $\ell$ to other flavors $\ell'$. This is readily accomplished. One need only replace the monochromatic (pseudotemporal) vibration $\exp(iG_\ell x^6/\hbar)$ of (12.3) with a polychromatic one whose spectrum consists of $N$ distinct pseudotemporal frequencies $p_{\ell j}^6$, $(j = 1,\ 2, \ldots,\ N)$

differing only slightly from frequency $G_\ell$. This results in the modified wave function

$$\nu_\ell(x^\mu, x^5, x^6) = e^{-iG_\ell x^5/\hbar} \sum_{j=1}^{N} U_{\ell j} e^{+i(G_\ell + g_{\ell j})x^6/\hbar} \psi_{\ell j}(x^\mu), \qquad (12.4)$$

where the $g_{\ell j}$ represent small departures from the main frequency $G_\ell$, the $U_{\ell j}$ are the mixing parameters and the $\psi_{\ell j}(x^\mu)$ are plane-wave Dirac spinors—plane-wave solutions of the Dirac equation (4.37). The latter are normalized in accordance to (4.58):

$$\left\langle \psi_{\ell j} \middle| \left( p^6_{\ell j} - \gamma^5 G_\ell \right) \middle| \psi_{\ell j} \right\rangle = 1. \qquad (12.5)$$

Interference between the $N$ terms of expansion (12.4) leads to neutrino oscillation.

We can put representation (12.4) in a more compact and familiar form by defining, respectively, eigenstates of $\hat{p}_5$ and $\hat{p}^6$:

$$|\ell\rangle = \frac{1}{\sqrt{L}} e^{-iG_\ell x^5/\hbar} \qquad (12.6)$$

and

$$|\ell j\rangle = \frac{1}{\sqrt{L}} e^{i(G_\ell + g_{\ell j})x^6/\hbar}, \qquad (12.7)$$

where $\hat{p}_5 |\ell\rangle = G_\ell |\ell\rangle$, $\hat{p}^6 |\ell j\rangle = (G_\ell + g_{\ell j}) |\ell j\rangle$ and the factors $1/\sqrt{L}$ provide for box normalization:

$$\begin{aligned} \langle \ell' k | \langle \ell' | \ell \rangle | \ell j \rangle &= \frac{1}{L^2} \int_{-L/2}^{L/2} dx^5 e^{-i(G_\ell - G_{\ell'})x^5/\hbar} \int_{-L/2}^{L/2} dx^6 e^{i(G_\ell + g_{\ell j} - G_{\ell'} - g_{\ell' k})x^6/\hbar} \\ &= \delta_{\ell\ell'} \delta_{jk} \end{aligned} \qquad (12.8)$$

Wave function (12.4) then reads

$$|\nu_\ell(x^\mu, x^5, x^6)\rangle = \sum_{j=1}^{N} U_{\ell j} |\ell\rangle |\ell j\rangle |\psi_{\ell j}(x^\mu)\rangle. \qquad (12.9)$$

The product state $|\ell\rangle|\ell j\rangle$ appearing here is the $j$th eigenstate of $\widehat{m^2c^2}$, with eigenvalue

$$m_j^2c^2 = (G_\ell + g_{\ell j})^2 - G_\ell^2 \\ = 2G_\ell g_{\ell j} + g_{\ell j}^2 \tag{12.10}$$

Then, provided $|g_{\ell j}|/G_\ell \ll 1$, we have

$$g_{\ell j} = \frac{m_j^2c^2}{2G_\ell}. \tag{12.11}$$

Since the $G_\ell$ are known from **Table 4.1** and the mass-squared differences $\Delta m_{ij}^2 = m_i^2 - m_j^2$ are known experimentally, the frequency differences $\Delta g_{\ell ij} = g_{\ell i} - g_{\ell j}$ giving rise to oscillation are completely defined:

$$\Delta g_{\ell ij} = \frac{\Delta m_{ij}^2c^2}{2G_\ell}. \tag{12.12a}$$

Here we note a remarkable feature of the formalism: Since the $\Delta m_{ij}^2$ are independent of flavor $\ell$, so too are the products $2G_\ell\Delta g_{\ell ij}$ independent of $\ell$; i.e.,

$$2G_e\Delta g_{eij} = 2G_\mu\Delta g_{\mu ij} = 2G_\tau\Delta g_{\tau ij} = \Delta m_{ij}^2c^2. \tag{12.12b}$$

The system of neutrino flavors forms a self-communicating whole: it provides the frequency difference $\Delta g_{\ell ij}$ required to produce for each $G_\ell$ the same constant $\Delta m_{ij}^2$. Algebraically, as shown in **Fig. 12.1**, the three datum pairs $\left(G_\ell, \Delta g_{\ell ij}\right)$, $\ell = e, \mu, \tau$, represent points on a hyperbola of semitransverse axis $(\Delta m_{ij}^2)^{1/2}c$. There are two such curves, corresponding to $(i, j) = (2,1)$ and $(3,1)$.

Expansion (12.9) is to be compared with the conventional expression for the propagating neutrino wave function[1]:

$$\left|\nu_\ell(x^\mu)\right\rangle = \sum_{j=1}^{N} U_{\ell j}\left|j\right\rangle\left|\psi_j(x^\mu)\right\rangle. \tag{12.13}$$

Here $\left|j\right\rangle$ is the $j$th mass-squared eigenstate of the expansion. In this conventional theory the vectors $\left|j\right\rangle$ and corresponding $m_j^2$ come from diagonalizing the weak interaction Lagrangian.[2] As the $m_j^2$ are

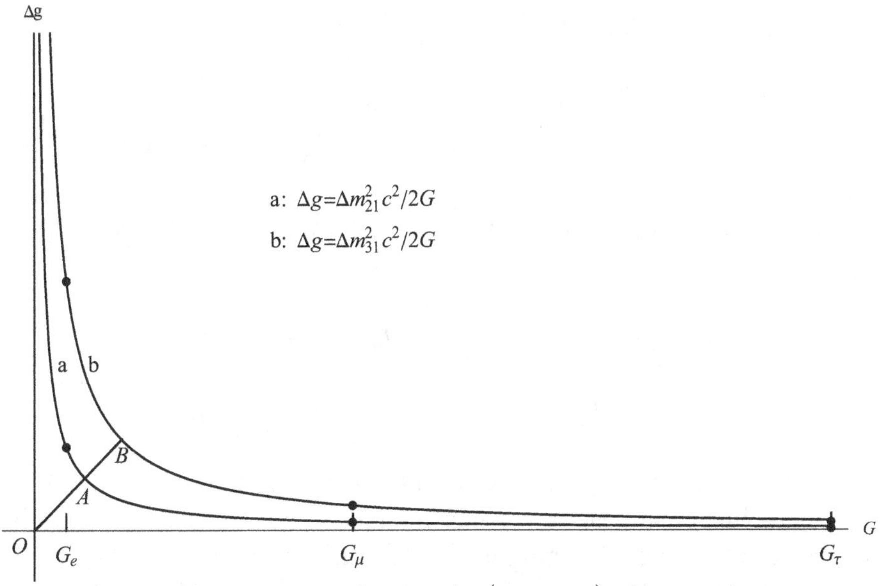


**Fig. 12.1** Momentum pairs $(G_e, \Delta g_{e21})$, $(G_\mu, \Delta g_{\mu 21})$, $(G_\tau, \Delta g_{\tau 21})$ are here depicted as three points on hyperbola "a" of semitransverse axis $OA = (\Delta m_{21}^2)^{1/2} c$; and momentum pairs $(G_e, \Delta g_{e31})$, $(G_\mu, \Delta g_{\mu 31})$, $(G_\tau, \Delta g_{\tau 31})$ as three points on hyperbola "b" of semitransverse axis $OB = (\Delta m_{31}^2)^{1/2} c$. The three 6-momentum shells of radii $G_\ell$ are thus algebraically in communication with each other: the product of frequency difference $\Delta g_{\ell ij}$ and diameter $2G_\ell$ invariably yields $\Delta m_{ij}^2 c^2$, independent of flavor $\ell$.

defined in terms of the unknown flavor neutrino masses $m_{\nu_\ell}$, their values are neither known nor calculable. In contrast, the mass-squared eigenstates $|\ell\rangle|\ell j\rangle$ of (12.9) arise not from a weak interaction Lagrangian, but as complex waveforms defined within the extra-dimensional spacetime $\bar{M}^2$. Consequently we know well the origin and structure of the corresponding $m_j^2$; they are defined by Eq. (12.10). In addition, the vector $|\ell\rangle$, which can be brought outside the summation sign, protects the wave function against change of lepton family number. We shall see a little later in **Sec. 12.8** how one can have flavor oscillation while conserving $L_\ell$.

## 12.3 On the Dirac nature of the flavor neutrino

According to (12.9) flavor neutrino $\nu_\ell$ depends on $x^5$ only through the exponential eigenstate $|\ell\rangle \propto \exp(-iG_\ell x^5/\hbar)$. As a consequence of this exponential structure, invariance under translations in $x^5$ is equivalent to global gauge invariance under the substitution

$$\begin{aligned} \nu_\ell(x^\mu, x^5, x^6) &\to \nu_\ell(x^\mu, x^5 + a, x^6) \\ &\to \nu_\ell(x^\mu, x^5, x^6)\exp(-iG_\ell a/\hbar) \end{aligned}, \tag{12.14}$$

where $a$ is independent of $x^5$. Thus the Lagrangian [see (10.55)]

$$\begin{aligned} \mathcal{L} = &\frac{c}{2}\tilde{\nu}_\ell\left(\hat{p}^6 - \gamma^5\hat{p}_5\right)_{\text{op}}\left(\gamma^\mu\hat{p}_\mu + \gamma^5\hat{p}_5 - \hat{p}^6\right)\nu_\ell \\ &-\frac{c}{2}\tilde{\nu}_\ell\left(\gamma^\mu\hat{p}_\mu - \gamma^5\hat{p}_5 + \hat{p}^6\right)_{\text{op}}\left(\hat{p}^6 - \gamma^5\hat{p}_5\right)\nu_\ell \end{aligned} \tag{12.15}$$

and all physical currents derived from it, are necessarily independent of $x^5$. It is thus clear that flavor neutrinos carry a $U(1)$ charge, namely $p_5 = G_\ell$, and the Lagrangian is invariant under those $U(1)$ transformations. It is well-known, however, that a $U(1)$ symmetry forbids Majorana mass terms in the Lagrangian.[2] Thus, according to this theory, *neutrinos are Dirac particles, not Majorana particles.* The observation of neutrinoless double beta decay $(0\nu\beta\beta)$,[3] a process that requires that neutrinos be Majorana particles, would automatically invalidate the present theory.

## 12.4 Mass of the flavor neutrino

In **Sec. 4.7** it was noted that by definition flavor neutrino $\nu_\ell$ belongs to family $\ell$ if and only if its five-momentum vector terminates on the momentum shell of radius $G_\ell$, i.e., $p_\mu p^\mu + {p_{5\nu_\ell}}^2 = {G_\ell}^2$. However, the propagating vector $|\nu_\ell\rangle$ now consists of a series of terms, the squared length of the five-momentum vector of the $j$th term being $(G_\ell + g_{\ell j})^2$. Hence, in order that $\nu_\ell$ belong to family $\ell$, we demand that the average value

$$\left\langle \hat{p}_6{}^2 \right\rangle = G_\ell{}^2 . \tag{12.16}$$

But $G_\ell{}^2 = \left\langle \hat{p}_5{}^2 \right\rangle$. Hence (12.16) may be written

$$\left\langle \widehat{m^2} \right\rangle = m_{\nu_\ell}{}^2 = 0 \tag{12.17}$$

That is, if flavor neutrino $\nu_\ell$ is to belong to family $\ell$, it must have vanishing mass. We can easily show that this is so.

We first write down an expression for the electric charge $q_{\nu_\ell}$ of neutrino $\nu_\ell$. Our previously derived Eq. (4.68) will not do, because it refers to a neutrino of definite mass. An expression good for neutrino states consisting of a superposition of mass-squared eigenstates is available from (4.51). To transform this expression for probability current to one for charge current we need only make the replacement $\gamma^5 \to -\gamma^5$ [cf. (4.28) and (4.44)]. This gives

$$q_{\nu_\ell} = \frac{-e}{2}\left[\langle \nu_{Q\ell}|(\hat{p}^6 + \gamma^5 \hat{p}^5)|\nu_{Q\ell}\rangle - \langle \nu_{Q\ell}|(\hat{p}^6 + \gamma^5 \hat{p}^5)_{\rm op}|\nu_{Q\ell}\rangle\right], \qquad (12.18)$$

where $|\nu_{Q\ell}\rangle$ is the charge wave function of neutrino $\nu_\ell$. According to (4.43) the charge and probability wave functions are related by

$$|\nu_{Q\ell}\rangle = \frac{\hat{p}^6 - \gamma^5 \hat{p}^5}{m_\ell c}|\nu_\ell\rangle, \quad \langle \nu_{Q\ell}| = -\langle \nu_\ell|\frac{(\hat{p}^6 - \gamma^5 \hat{p}^5)_{\rm op}}{m_\ell c}, \qquad (12.19)$$

where $m_\ell$ is the mass of the charged lepton of family $\ell$. Inserting Eqs. (12.19) into (12.18) we obtain

$$\begin{aligned} q_{\nu_\ell} &= \frac{-e}{2m_\ell^{\,2}}\left[-\langle \nu_\ell|(\hat{p}^6 - \gamma^5 \hat{p}^5)_{\rm op}\widehat{m^2}|\nu_\ell\rangle + \langle \nu_\ell|(\widehat{m^2})_{\rm op}(\hat{p}^6 - \gamma^5 \hat{p}^5)|\nu_\ell\rangle\right] \\ &= \frac{-e}{m_\ell^{\,2}}\langle \nu_\ell|\widehat{m^2}(\hat{p}^6 - \gamma^5 \hat{p}^5)|\nu_\ell\rangle \\ &= \frac{-e}{m_\ell^{\,2}} m_{\nu_\ell}^{\;2} \end{aligned} \qquad (12.20)$$

where, in accordance to (4.59), $m_{\nu_\ell}^{\;2}$ denotes the expectation value of the mass squared operator:

$$m_{\nu_\ell}^{\;2} = \left\langle \widehat{m^2} \right\rangle = \sum_{j=1}^{N} |U_{\ell j}|^2 m_j^{\;2}, \qquad (12.21)$$

and where we have used expansion (12.9) and normalization relations

(12.5) and (12.8). But the neutrino has vanishing electric charge. Hence

$$m_{\nu_\ell}{}^2 = 0\,, \tag{12.22}$$

as was to be proved. Note that (12.22) holds in both vacuum and in matter. We shall see in **Sec. 12.6** how massless neutrinos can be massless and still oscillate.

It should be noted that the term "expectation value" used in connection with (12.21) does not mean that $m_{\nu_\ell}{}^2$ vanishes on average only. In ordinary quantum mechanics as formulated in $M^4$, (12.21) would be interpreted to mean that, in an ensemble of measurements of neutrino mass, neutrino $\nu_\ell$ would be found to have the mass-squared value $m_j{}^2$ a fraction $|U_{\ell j}|^2$ of the time, with $j = 1,2,3,...$ .[4] But it cannot mean that here. For charge $q_{\nu_\ell}$ is an invariant, and consequently $m_{\nu_\ell}{}^2$ is also an invariant, with the value 0. The summation in (12.21) takes place in $\bar{M}^2$, not $M^4$, and has a purely classical interpretation. It merely adds up the weighted mass-squared values of the terms of the expansion, resulting in *the* mass-squared value of the flavor neutrino, one that obtains in every direct measurement of neutrino mass. This means, of course, that the neutrinos created or detected in charge-current interactions are flavor states, not individual mass-squared eigenstates.[5]

## 12.5 The oscillation imperative (again)

Why must neutrinos oscillate? At first sight it seems that Nature has made things more complicated than they need be. Indeed, if spacetime were limited to the four dimensions of $M^4$ there would be no need for oscillation. But, as we showed in **Sec. 4.7**, in six-dimensional $M^4 \times \bar{M}^2$ neutrinos must oscillate to exist. Given the peculiarity of this result, we are going to prove it again here, this time from the perspective of our expansion in mass-squared eigenstates, (12.9).

There are two steps. First we show that if the *j*th mixing parameter $U_{\ell j}$ is non-vanishing, then the corresponding frequency increment $g_{\ell j}$ must also be non-vanishing. Inserting (12.9) into Dirac equation (4.28) written in the form

$$\left(\hat{p}^6 + \gamma^5 \hat{p}_5\right)\left(\gamma^\mu \hat{p}_\mu + \gamma^5 \hat{p}_5 - \hat{p}^6\right)|\nu_\ell\rangle = 0\,, \tag{12.23}$$

we find that the $j$th spinor $\left|\psi_{\ell j}\right\rangle$ satisfies the four-dimensional wave equation

$$\left[\left(G_\ell+g_{\ell j}\right)+\gamma^5 G_\ell\right]\left[\gamma^\mu \hat{p}_\mu+\gamma^5 G_\ell-\left(G_\ell+g_{\ell j}\right)\right]\left|\psi_{\ell j}\right\rangle=0, \quad (12.24)$$

provided that $U_{\ell j}\neq 0$. Let us assume a plane-wave solution of the form

$$\left|\psi_{\ell j}\right\rangle=\begin{pmatrix}\phi_{\ell j}\\ \chi_{\ell j}\end{pmatrix}e^{-i\left(p_{0j}x^0-\mathbf{p}.\mathbf{x}\right)/\hbar}, \quad (12.25)$$

where $\phi_{\ell j}$ and $\chi_{\ell j}$ are two-component spinors and

$$p_{0j}=\left(\mathbf{p}^2+m_j^2c^2\right)^{1/2}. \quad (12.26)$$

In the chiral representation of the Dirac matrices [see Eqs. (10.41)], Eq. (12.24) can be written as two coupled equations

$$g_{\ell j}\left[\left(p_{0j}-\boldsymbol{\sigma}\cdot\mathbf{p}\right)\phi_{\ell j}+\left(2G_\ell+g_{\ell j}\right)\chi_{\ell j}\right]=0 \quad (12.27a)$$

$$\left(2G_\ell+g_{\ell j}\right)\left[g_{\ell j}\phi_{\ell j}+\left(p_{0j}+\boldsymbol{\sigma}\cdot\mathbf{p}\right)\chi_{\ell j}\right]=0, \quad (12.27b)$$

where the components of $\boldsymbol{\sigma}$ are the Pauli matrices (3.42). Now suppose that $g_{\ell j}=0$. Equation (12.27a) is annihilated while (12.27b) reduces to

$$\left(p_{0j}+\boldsymbol{\sigma}\cdot\mathbf{p}\right)\chi_{\ell j}=0. \quad (12.28)$$

This is the standard Weyl two-component description of massless neutrinos. Note, however, that the spinor $\phi_{\ell j}$ remains undefined in such a description. Consequently, the fifth and sixth components of the probability current density, Eqs. (4.56) and (4.57), which components depend on $\phi_{\ell j}$, are also undefined. But then the neutrino itself is undefined. Thus, if $U_{\ell j}\neq 0$, then the definition and existence of the neutrino requires that the $j$th frequency increment $g_{\ell j}$ be non-vanishing.

Second, and finally, suppose that for every $\ell$, we have $U_{\ell j}=\delta_{\ell j}$. The mixing matrix is then diagonal and oscillation cannot occur. The flavor state $|\nu_\ell\rangle$ is in this case an eigenstate of the mass-squared operator $m^2$ with eigenvalue $m_{\nu_\ell}{}^2$, where from (12.10) and (12.17)

$$m_{\nu_\ell}{}^2c^2 = m_\ell{}^2c^2 = 2G_\ell g_{\ell\ell} + g_{\ell\ell}{}^2 = 0, \qquad (12.29)$$

implying that $g_{\ell\ell} = 0$. But this contradicts the requirement that $g_{\ell j} \neq 0$ when $U_{\ell j} \neq 0$. The diagonal matrix is thus impermissible: each row $\ell$ of the unitary mixing matrix must contain at least two non-vanishing elements, proving that flavor neutrinos must oscillate to exist.

## 12.6 Tardyonic and tachyonic mass-squared eigenstates

Since flavor neutrinos have no choice but to oscillate, we know that, for any $\ell$, the summation (12.21) contains at least two non-vanishing terms. Therefore, because $m_{\nu_\ell}{}^2 = 0$, at least one of the mass-squared eigenstates representing $\nu_\ell$ must be tachyonic.

According to (12.10), $m_j{}^2$ can indeed have either algebraic sign, depending on the sign of the corresponding $g_{\ell j}$. This freedom of sign at once allows the expectation value (12.21) to vanish and provides the finite values of $\Delta m^2$ required for oscillation. To illustrate, suppose that neutrino state $|\nu_\ell\rangle$ can be represented by a superposition of two eigenstates of the pseudoenergy operator $\hat{p}^6$: $|\ell i\rangle$ and $|\ell j\rangle$. If the mixing is maximal, i.e., $|U_{\ell i}|^2 = |U_{\ell j}|^2 = 1/2$, then from (10.22), $m_i{}^2 = -m_j{}^2$, and

$$\Delta m_{ji}{}^2 = m_j{}^2 - m_i{}^2 = 2m_j{}^2 = 2\left(2G_\ell g_{\ell j} + g_{\ell j}{}^2\right). \qquad (12.30)$$

If it is assumed that $m_j{}^2 > 0$, then $\Delta m_{ji}{}^2 > 0$. The speed $\Delta v$ of the tachyonic (tardyonic) mass-squared eigenstate above (below) $c$ is given by

$$\Delta v = \Delta\left(\frac{pc}{p_{0j}}\right) = pc\Delta\left(p^2 + m_j{}^2c^2\right)^{-1/2} \approx \frac{\Delta m_{ji}{}^2c^2}{4p^2}c, \qquad (12.31)$$

where $p = p_\nu$ is the momentum of the neutrino. In the solar case,[6, 7] $\ell = e$ and

$$\Delta m_{\text{SOL}}{}^2 = \Delta m_{21}{}^2 = (7.6 \pm 0.2)\times 10^{-5}\ \text{eV}^2/c^4\ ,^{[8]} \qquad (12.32)$$

$$p_\nu(^8B \text{ energy average}) = 8.5 \text{ MeV}/c \qquad ,^{[9]} \tag{12.33}$$

giving $\Delta v_{\text{SOL}} = 2.6\times10^{-19}c$. Similarly, in the atmospheric case,[10] $\ell = \mu$

$$\Delta m_{\text{ATM}}^2 = \left|\Delta m_{31}^2\right| = 2.32^{+0.12}_{-0.08}\times10^{-3}\text{ eV}^2/c^4 \qquad ,^{[11]} \tag{12.34}$$

$$p_\nu(\text{ATM } \nu_\mu \text{ energy range}) \sim 1 \text{ GeV}/c \qquad ,^{[12]} \tag{12.35}$$

which gives $\Delta v_{\text{ATM}} = 5.8\times10^{-22}c$. These departures $\Delta v$ from $c$ are small indeed. (A human fingernail grows at a rate $\sim 10^{-18}c$.) In any case the mass-squared eigenstates occur only in superposition and, to conserve charge, are not individually detectable. On the present theory, tachyons do exist but are unobservably confined to the propagating flavor neutrino state.

Let us now estimate the magnitudes of the $g_{\ell j}$ occurring in the solar and atmospheric anomalies, assuming as above two-neutrino mixing. In the solar case, from (12.12) and **Sec. 4.8, Table 4.1**

$$\frac{g_{e2}}{G_e} = \frac{\Delta g_{e21}}{2G_e} = \frac{\Delta m_{\text{SOL}}^2 c^2}{4G_e^2} \sim 0.7\times10^{-16}, \tag{12.36}$$

and in the atmospheric case

$$\frac{g_{\mu3}}{G_\mu} = \frac{\Delta g_{\mu31}}{2G_\mu} = \frac{\Delta m_{\text{ATM}}^2 c^2}{4G_\mu^2} \sim 4.7\times10^{-24}. \tag{12.37}$$

The solar and atmospheric neutrino oscillations are thus driven by extremely small fractional departures from the shell radii, $G_e$ and $G_\mu$. The quadratic term in the second line of (12.10) is entirely negligible and the sign of $g_{\ell j}$ determines the sign of $m_j^2$.

## 12.7 Why there are no right-handed neutrinos

Let us take a closer look at the propagating neutrino state (12.9). We know from **Sec. 12.5** that for each non-vanishing term there exists a corresponding non-vanishing $g_{\ell j}$ and mass-squared value $m_j^2$. We can then solve (12.27a) for the lower component $\chi_{\ell j}$. Putting this into (12.25) we have for the plane-wave solution

$$\left|\psi_{\ell j}\right\rangle = C_{\ell j}\begin{pmatrix} \phi_{\ell j} \\ \dfrac{p_{0j} - \boldsymbol{\sigma}\cdot\mathbf{p}}{2G_\ell + g_{\ell j}}\phi_{\ell j} \end{pmatrix} e^{-i\left(p_{0j}x^0 - \mathbf{p}.\mathbf{x}\right)/\hbar}, \tag{12.38}$$

where from (12.26)

$$p_{0j} \cong |\mathbf{p}| + \frac{m_j^2 c^2}{2|\mathbf{p}|}, \tag{12.39}$$

and we have now included a constant of normalization, $C_{\ell j}$. From (12.5) the $\left|\psi_{\ell j}\right\rangle$ are normalized in accordance to

$$\left\langle\psi_{\ell j}\right|\left[G_\ell\left(1-\gamma^5\right)+g_{\ell j}\right]\left|\psi_{\ell j}\right\rangle = 1. \tag{12.40}$$

This yields for the squared magnitude of the normalization constant

$$\left|C_{\ell j}^{\pm}\right|^2 = \frac{2G_\ell + g_{\ell j}}{2p_{0j}\left(p_{0j} \mp |\mathbf{p}|\right)} \cong \frac{G_\ell}{\mathbf{p}^2 + \dfrac{m_j^2 c^2}{2} \mp \mathbf{p}^2}. \tag{12.41}$$

where the upper sign refers to positive helicity states $\left|\psi_{\ell j}^{+}\right\rangle$ and the lower sign to negative helicity states $\left|\psi_{\ell j}^{-}\right\rangle$. Now according to (12.41) the squared magnitudes $\left|C_{\ell j}^{+}\right|^2$ corresponding to tachyonic right-handed mass-squared eigenstates $\left(m_j^2 < 0\right)$ are algebraically negative, a manifest absurdity. Tachyonic right-handed states $\left|\psi_{\ell j}^{+}\right\rangle$ possess no positive norm and consequently do not exist. But we know from **Sec. 12.6** that at least one of the states $\left|\psi_{\ell j}^{+}\right\rangle$ comprising a right-handed flavor neutrino must be tachyonic. As such states are non-existent, we conclude that *right-handed flavor neutrinos do not exist.* The constants $\left|C_{\ell j}^{-}\right|^2$, on the other hand, are positive definite and accordingly support the existence of left-handed flavor neutrinos.

We should note that the handedness of the neutrino depends on the direction of propagation along $x^5$. Neutrinos have arisen here left-handed because we chose (with foresight) to have neutrinos propagate in the positive $x^5$ direction [see the exponential structure of (12.4)], and antineutrinos in the negative $x^5$ direction. Had we chosen the opposite convention, neutrinos would have come out right-handed and antineutrinos left-handed.

Since right-handed neutrinos do not exist, there can be no neutrino mass terms generated in the SM Lagrangian as a result of spontaneous symmetry breaking. The masslessness of flavor neutrinos deduced from self-imaging theory is thus correctly reflected in the structure of the Lagrangian. From the present viewpoint, the minimal SM,[13] which assumes massless neutrinos and conserves lepton number by virtue of $U(1)$ symmetry in the Lagrangian, is complete as it stands. Neutrino oscillation is, therefore, not a prediction of the SM and must be accounted for by other means. It is this we now pursue.

## 12.8 Neutrino oscillation and conservation of lepton family number

In **Sec. 4.13** we proved that, as a consequence of the conservation of charge $Q$ and pseudomomentum $p_5$, in all processes lepton number is strictly conserved. This result does not, however, preclude neutrino oscillation. For in 6-D $M^4 \times \bar{M}^2$, lepton family number ($L_\ell$) and flavor ($G_\ell$, $Q$) are separate quantum numbers; and while lepton number is invariably conserved, flavor need not be.

To see how neutrino flavor oscillation can occur without violating lepton-number conservation, it suffices to consider a world having just two neutrino flavors $\ell$— $\alpha$ and $\beta$, say. With each of these one may associate a flavor eigenstate $|\nu_\ell\rangle = |\nu_\ell(x^5, x^6)\rangle$ represented by an expansion in mass squared eigenstates $|\ell\rangle|\ell j\rangle$, with $j = 1,2$. To find these expansions, we integrate out the $x^\mu$ dependence of (12.9) using normalization relation (12.5). This gives for our two flavor eigenstates the expansions

$$|\nu_\alpha\rangle = \sum_{j=1}^{2} U_{\alpha j} |\alpha\rangle |\alpha j\rangle . \tag{12.42}$$

and

$$|\nu_\beta\rangle = \sum_{j=1}^{2} U_{\beta j} |\beta\rangle |\beta j\rangle , \tag{12.43}$$

where the eigenstates $|\ell\rangle$ and $|\ell j\rangle$ are defined by (12.6) and (12.7). Now the states $|\alpha\rangle$ and $|\beta\rangle$ are eigenstates of $\hat{p}_5$, with eigenvalues $G_\alpha$ and $G_\beta$, respectively. Since $p_5$ is conserved in any process and $G_\alpha \neq G_\beta$, it is clear that "oscillation" cannot mean transition from $|\nu_\alpha\rangle$ to $|\nu_\beta\rangle$ or *vice versa.* So what, then, oscillates? The answer is clear. There must exist two additional states, one with the lepton number of $|\nu_\alpha\rangle$ and the direction in flavor space of $|\nu_\beta\rangle$; the other with the lepton number of $|\nu_\beta\rangle$ and direction in flavor space of $|\nu_\alpha\rangle$. We define the state capable of entering into oscillation with $|\nu_\alpha\rangle$ as $|\nu_{\alpha\to\beta}\rangle$, where

$$\Big|\nu_{\underbrace{\alpha}_{\text{Lepton number}} \to \underbrace{\beta}_{\text{Flavor}}}\Big\rangle = \sum_{j=1}^{2} \underbrace{U_{\beta j}|\alpha\rangle}_{\substack{\text{Flavor}\\ \beta}}\underbrace{|\alpha j\rangle}_{\substack{\text{Lepton number}\\ L_\alpha}} \tag{12.44}$$

and that capable of entering into oscillation with $|\nu_\beta\rangle$ as $|\nu_{\beta\to\alpha}\rangle$, where

$$\Big|\nu_{\underbrace{\beta}_{\text{Lepton number}} \to \underbrace{\alpha}_{\text{Flavor}}}\Big\rangle = \sum_{j=1}^{2} \underbrace{U_{\alpha j}|\beta\rangle}_{\substack{\text{Flavor}\\ \alpha}}\underbrace{|\beta j\rangle}_{\substack{\text{Lepton number}\\ L_\beta}}. \tag{12.45}$$

The oscillation-capable pair (12.42) and (12.44) can be combined to form a single matrix equation:

$$\begin{pmatrix} |\nu_{\alpha\to\alpha}\rangle \\ |\nu_{\alpha\to\beta}\rangle \end{pmatrix} = U \begin{pmatrix} |\alpha\rangle|\alpha 1\rangle \\ |\alpha\rangle|\alpha 2\rangle \end{pmatrix} \tag{12.46}$$

and similarly, for the pair (12.43) and (12.45) we can write

$$\begin{pmatrix} |\nu_{\beta\to\alpha}\rangle \\ |\nu_{\beta\to\beta}\rangle \end{pmatrix} = U \begin{pmatrix} |\beta\rangle|\beta 1\rangle \\ |\beta\rangle|\beta 2\rangle \end{pmatrix}, \tag{12.47}$$

where in both matrix equations the unitary mixing matrix

$$U = \begin{pmatrix} U_{\alpha 1} & U_{\alpha 2} \\ U_{\alpha 1} & U_{\alpha 2} \end{pmatrix} = \begin{pmatrix} \cos\theta & \sin\theta e^{-i\delta} \\ -\sin\theta e^{i\delta} & \cos\theta \end{pmatrix}. \qquad (12.48)$$

Note that in (12.46) and (12.47) for consistency of notation we have defined $|\nu_\alpha\rangle \equiv |\nu_{\alpha\to\alpha}\rangle$ and $|\nu_\beta\rangle \equiv |\nu_{\beta\to\beta}\rangle$. Also, in the definition of matrix $U$ we have included a possible charge-parity (CP)-violating phase $\delta$.

We can illustrate the process of oscillation as follows. Suppose that at time $x^0 = 0$ flavor neutrino $\nu_\alpha$ is created by charged current (CC) weak interaction. The time development of the propagating neutrino state is given by (12.9) with $\ell = \alpha$ and $N = 2$:

$$\left|\nu_\alpha(x^\mu, x^5, x^6)\right\rangle = \sum_{j=1}^{2} U_{\alpha j} \left|\alpha\right\rangle \left|\alpha j\right\rangle \left|\psi_{\alpha j}(x^\mu)\right\rangle, \qquad (12.49)$$

where the Dirac wave function $\left|\psi_{\alpha j}(x^\mu)\right\rangle$ is given by (12.38) and the energy $p_{0j}$ by (12.39). Now because the mixing matrix $U$ is unitary, (12.46) can be inverted, expressing the two mass-eigenstates in terms of the flavor eigenstates:

$$\begin{pmatrix} \left|\alpha\right\rangle\left|\alpha 1\right\rangle \\ \left|\alpha\right\rangle\left|\alpha 2\right\rangle \end{pmatrix} = U^\dagger \begin{pmatrix} \left|\nu_{\alpha\to\alpha}\right\rangle \\ \left|\nu_{\alpha\to\beta}\right\rangle \end{pmatrix}, \qquad (12.50)$$

where the dagger denotes Hermitan conjugate: $U^\dagger = \left(U^T\right)^*$. In series form the expansion of the mass-squared eigenstates in flavor eigenstates reads

$$\left|\alpha\right\rangle \left|\alpha j\right\rangle = \sum_{\ell=\alpha,\beta} U^*_{\ell j} \left|\nu_{\alpha\to\ell}\right\rangle, \quad j = 1, 2. \qquad (12.51)$$

Putting (12.51) into (12.49) we obtain an expansion of the neutrino wave function in flavor eigenstates:

$$\left|\nu_\alpha(x^\mu, x^5, x^6)\right\rangle = \sum_{\ell=\alpha,\beta} \left|C_{\alpha\to\ell}(x^\mu)\right\rangle \left|\nu_{\alpha\to\ell}\right\rangle, \qquad (12.52)$$

where the spacetime-dependent coefficients $\left|C_{\alpha\to\ell}\right\rangle$ are given by

$$\left|C_{\alpha\to\ell}(x^{\mu})\right\rangle = \sum_{j=1}^{2} U_{\ell j}^{*} U_{\alpha j} \left|\psi_{\alpha j}(x^{\mu})\right\rangle, \quad \ell = \mu, \tau \tag{12.53}$$

Equation (12.52) describes oscillation, not between flavor eigenstates $\nu_\alpha$ and $\nu_\beta$, but between $\nu_\alpha$ and the new flavor eigenstate $\nu_{\alpha\to\beta}$. The coefficients $\left|C_{\alpha\to\alpha}\right\rangle$ and $\left|C_{\alpha\to\beta}\right\rangle$ represent, respectively, the probability amplitudes of survival $\nu_\alpha \to \nu_\alpha$ and of the transition $\nu_\alpha \to \nu_{\alpha\to\beta}$. The probabilities themselves are given by

$$\begin{aligned} P\left(\nu_\alpha \to \nu_\alpha\right) &= \left\langle C_{\alpha\to\alpha}\right| G_\alpha\left(1-\gamma^5\right)\left|C_{\alpha\to\alpha}\right\rangle \\ &= 1 - 4\left|U_{\alpha 1}\right|^2 \left|U_{\alpha 2}\right|^2 \sin^2\left(\frac{\Delta m_{21}^2 c^2 x^0}{4\hbar p}\right) \end{aligned} \tag{12.54}$$

and

$$\begin{aligned} P\left(\nu_\alpha \to \nu_{\alpha\to\beta}\right) &= \left\langle C_{\alpha\to\beta}\right| G_\alpha\left(1-\gamma^5\right)\left|C_{\alpha\to\beta}\right\rangle \\ &= 4\left|U_{\alpha 1}\right|^2 \left|U_{\alpha 2}\right|^2 \sin^2\left(\frac{\Delta m_{21}^2 c^2 x^0}{4\hbar p}\right), \end{aligned} \tag{12.55}$$

where in carrying out the operations $\langle C|\hat{O}|C\rangle$ we have neglected the tiny $g_{\alpha j}$ in relation to $G_\alpha$. Naturally the sum of the two probabilities is unity.

As stated above, the state $\nu_{\alpha\to\beta}$ has the same direction in flavor space as $\nu_\beta$; namely, the direction defined by coefficients $(U_{\beta 1}, U_{\beta 2})$. Now, as in standard oscillation theory, it is the flavor neutrino's orientation in flavor space that determines the flavor of the charged lepton created in neutrino reactions such as: $\nu_\tau + \text{nucleon} \to \tau^- + \text{hadrons}$. Thus, as indicated in (12.44), we say that $\nu_{\alpha\to\beta}$ carries "flavor $\beta$", because in the detection reaction it results in the creation of a charged lepton of flavor $\beta$ ($Q=-1$, $p^6 = G_\beta$). However, the presence of vector $|\alpha\rangle$ in (12.44) shows that $\nu_{\alpha\to\beta}$ carries the lepton number of $\nu_\alpha$, not $\nu_\beta$ [lepton number being related to $p_5$ by (12.2)]. The process of oscillation between flavors $\alpha$ and $\beta$ described by (12.52) thus conserves the original lepton family number $L_\alpha$.

The oscillation scenario just described was initiated with the CC production of neutrino $\nu_\alpha$. One could just as easily start with $\nu_\beta$. In that case the relevant survival and transition probabilities $P(\nu_\beta \to \nu_\beta)$ and $P(\nu_\beta \to \nu_{\beta\to\alpha})$ are given by (12.54) and (12.55) with subscripts $\alpha$ and $\beta$ interchanged. Moreover, the oscillation takes place while conserving lepton number $L_\beta$. The form of the argument of the $\sin^2$ function remains unchanged.

In the literature one normally sees parametrization of the oscillation probabilities with energy $E$ rather than momentum $p$, i.e., one puts $p = E/c$; and for the distance traveled in vacuum from the production point, $x^0 \equiv L$. The argument of the $\sin^2$ function may then be expressed in various ways:

$$\frac{\Delta m^2 c^3 L}{4\hbar E} = \pi \frac{L}{l} = 1.27 \frac{\Delta m^2 L}{E} \ . \qquad (12.56)$$

In the first variation, $l = 4\pi\hbar E / \Delta m^2 c^3$, where $l$ denotes the *oscillation length*. This is the distance between minima of the $\sin^2$ function; or, referring to probability (12.54), the distance it takes to oscillate from a beam of pure $\nu_\alpha$ back to one of pure $\nu_\alpha$. The second variation emphasizes the sensitivity ratio $L/E$; for example, different experimental configurations can give the same sensitivity to $\Delta m^2$, provided they have the same $L/E$. In this second variation, $\Delta m^2$ is expressed in $[\mathrm{eV}^2]$, $L$ in [m (km)], and $E$ in $[\mathrm{MeV}\ (\mathrm{GeV})]$.

To get the oscillation probabilities for antineutrinos, one simply replaces the matrix elements $U_{\ell j}$ with their complex conjugates. We see from (12.54) and (12.55) that in a two-flavor world the oscillation probabilities for neutrinos and antineutrinos are identical. In such a world oscillation yields no evidence of CP violation in the neutrino sector.

In summary, in an $M^4 \times \bar{M}^2$ world with two lepton families and two neutrino flavors, $\alpha$ and $\beta$, we have a total of $2^2$ flavor states to work with, as each neutrino flavor may carry lepton number $L_\alpha$ or $L_\beta$. We call this the $2 \times 2$ configuration. The connection between the four flavor states and four mass-squared eigenstates may be summarized in matrix form:

$$\begin{pmatrix} |\nu_{\alpha\to\alpha}\rangle \\ |\nu_{\alpha\to\beta}\rangle \\ |\nu_{\beta\to\alpha}\rangle \\ |\nu_{\beta\to\beta}\rangle \end{pmatrix} = \begin{pmatrix} \cos\theta & \sin\theta & 0 & 0 \\ -\sin\theta & \cos\theta & 0 & 0 \\ 0 & 0 & \cos\theta & \sin\theta \\ 0 & 0 & -\sin\theta & \cos\theta \end{pmatrix} \begin{pmatrix} |\alpha\rangle|\alpha 1\rangle \\ |\alpha\rangle|\alpha 2\rangle \\ |\beta\rangle|\beta 1\rangle \\ |\beta\rangle|\beta 2\rangle \end{pmatrix}. \quad (12.57)$$

The block diagonal unitary mixing matrix effectively divides the mixing into two sectors, one carrying lepton number $L_\alpha$, the other $L_\beta$.

In a world with three lepton families the total number of neutrino states would be at a minimum $3^2$—a $3\times 3$ configuration. This does not contradict the finding, based on the decay width of $Z^0$, that there exist three and only three families of leptons.[14] For neutrino states of the form $\nu_{\alpha\to\beta}$, where $\alpha \neq \beta$, represent neutrinos of family $\alpha$, i.e., neutrinos bearing lepton family number $L_\alpha$, evolved by propagation from flavor $\alpha$ to flavor $\beta$ (the arrow in the subscript helps to remind us of this). Such states do not appear in the SM and do not contribute to the partial width for the decay of $Z^0$ into neutrino-antineutrino pairs.

The critical reader will have noticed that in the above we have used the term "neutrino flavor" in two different senses. These two senses relate to the two representations of the propagating neutrino state: one in terms of mass-squared eigenstates, the other in terms of flavor eigenstates. In the first sense, neutrino flavor is considered to be a physical attribute defined jointly by the neutrino's charge $Q = 0$ and expected squared length $G_\ell^2$ of the energy-momentum 5-vector as defined by (12.16). By that definition the propagating neutrino represented by wave function (12.49) retains its original flavor $\alpha$ right up to (but not including) the moment of detection. In the second sense of the term, neutrino flavor is defined probabilistically in terms of a weighted superposition of flavor eigenstates, as in (12.52). In this latter sense the neutrino's flavor is declared only at the moment of detection. There is no conflict between our two uses of "neutrino flavor."

## 12.9 Charged lepton partners of the $2\times 2$ neutrino states

Once manifested, any one of the four neutrino types in the $2\times 2$ configuration may enter into CC interaction with a nucleus, resulting in a charged lepton of the same lepton number and flavor as the interacting neutrino. Thus in our hypothetical two-family world there are four types of charged lepton. They are conveniently displayed in a column matrix, paralleling the four neutrino flavor states of (12.57):

$$\begin{pmatrix} l_{\alpha\to\alpha} \\ l_{\alpha\to\beta} \\ l_{\beta\to\alpha} \\ l_{\beta\to\beta} \end{pmatrix}, \tag{12.58}$$

where, for consistency of notation, for the charged lepton states normally denoted $\ell^-$ ($\ell^+$) we write $l_{\ell\to\ell}$ ($\overline{l}_{\ell\to\ell}$), $\ell = \alpha,\ \beta$. For example, charged lepton $l_{\alpha\to\beta}$, the weak interaction partner of $\nu_{\alpha\to\beta}$, carries lepton number $L_\alpha$ and flavor $\beta$ (defined by shell radius $G_\beta$ and charge $Q = -1$). The wave function for this particle is

$$\left|l_{\alpha\to\beta}\right\rangle = \left|\alpha\right\rangle\left\langle\beta\middle|l_{\beta\to\beta}\right\rangle, \tag{12.59}$$

where the vectors $|\alpha\rangle$ and $|\beta\rangle$ are eigenstates of $\hat{p}_5$; see (12.6). The operator $|\alpha\rangle\langle\beta|$ converts the $p_5$ value of $\left|l_{\beta\to\beta}\right\rangle$ to that of $\left|l_{\alpha\to\alpha}\right\rangle$, i.e., from $G_\beta - m_e c$ to $G_\alpha - m_e c$ [see (12.2)]. This means that $\hat{p}_5\left|l_{\alpha\to\beta}\right\rangle = (L_\alpha G_\alpha - m_e c)\left|l_{\alpha\to\beta}\right\rangle$ and $\hat{p}^6\left|l_{\alpha\to\beta}\right\rangle = G_\beta\left|l_{\alpha\to\beta}\right\rangle$. Note that $\left|l_{\alpha\to\beta}\right\rangle$ satisfies wave equation (4.28) (with $G_\ell = G_\beta$) and yields $\left\langle\widehat{m^2}\right\rangle = m_\beta^{\ 2}$, provided one makes the replacement

$$\hat{p}_5 \to \hat{p}_5 - L_\alpha B_5 \tag{12.60}$$

where

$$B_5 = G_\alpha - G_\beta. \tag{12.61}$$

Here $B_5$ formally resembles a constant external electromagnetic potential and $L_\alpha$ plays the role of charge. Of course $B_5$ is not a gauge field and no force derives from it.

To illustrate the above, let us consider neutrino detection in the LSND experiment.[15] Ostensibly LSND detects the appearance by oscillation of $\overline{\nu}_e$ in a beam of $\overline{\nu}_\mu$ by means of the reaction (in conventional notation)

$$\begin{array}{c}\bar{\nu}_\mu \\ \downarrow \\ \bar{\nu}_e + p \to e^+ + n\end{array} \tag{12.62}$$

However, according to the present theory, LSND actually detects (in our new notation) $\bar{\nu}_{\mu\to e}$, not $\bar{\nu}_e$; and the lepton number of $\bar{\nu}_{\mu\to e}$ is $L_\mu$, not $L_e$. Thus the LSND reaction is not (12.62) but

$$\begin{array}{c}\bar{\nu}_{\mu\to\mu} \\ \downarrow \\ \bar{\nu}_{\mu\to e} + p \to \bar{l}_{\mu\to e} + n\end{array} \tag{12.63}$$

where $\bar{l}_{\mu\to e}$ is a positron carrying lepton number $L_\mu$. The $p_5$ balance of reaction (12.63) reads

$$(-G_\mu) + (R + 2m_e c) \to (-G_\mu + m_e c) + (R + m_e c), \tag{12.64}$$

where the $p_5$ values of the leptons are derived from (12.2), and those for the nucleons from (4.101). Pseudomomentum $p_5$ is conserved. Experimentally $\bar{l}_{\mu\to e}$ is indistinguishable from $e^+$; it behaves as if it were a normal positron, with rest mass $m_e$ and charge $+e$. The main point is that $\bar{\nu}_\mu$'s lepton number $L_\mu = -1$ is conserved in the LSND detection process, just as it is in the oscillation process $\bar{\nu}_{\mu\to\mu} \leftrightarrow \bar{\nu}_{\mu\to e}$.

A more interesting example is this. In 1998 the Super-Kamiokande Collaboration reported a strong deficit of atmospheric muon neutrinos and showed the data to be consistent with two-flavor oscillation $\nu_\mu \leftrightarrow \nu_\tau$ at nearly maximal mixing.[16] The Super-K result has since been confirmed by the K2K Collaboration[17] and MINOS Collaboration.[18] Ideally, to provide "smoking gun" evidence of the process $\nu_\mu \leftrightarrow \nu_\tau$, one would like to see the *appearance* of $\nu_\tau$ in a beam of $\nu_\mu$ derived, most conveniently, from a neutrino factory. In our new notation, this means looking for the appearance of $\nu_{\mu\to\tau}$ —a tau neutrino carrying lepton number $L_\mu$—in a beam of $\nu_{\mu\to\mu}$. To do this one observes the CC interaction of $\nu_{\mu\to\tau}$ by identifying the $l_{\mu\to\tau}$ lepton created at the interaction vertex. Such an experiment has in fact been performed by the OPERA Collaboration, which as of July, 2015 has identified five candidate events.[19] The five identifying decay channels and corresponding balance of $p_5$ are displayed in **Table 12.1.** Here we see that $\nu_{\mu\to\mu}$'s lepton number $L_\mu = +1$ is conserved in the OPERA detection process, just as is $\bar{\nu}_{\mu\to\mu}$'s in the LSND detection process. Of particular interest, though, are the third and fourth channels listed in the **Table**. Since $l_{\mu\to\tau}$ and $\mu^- = l_{\mu\to\mu}$ already have the same lepton number, the neutrino sum $\bar{\nu}_{\mu\to\mu} + \nu_{\mu\to\tau}$ in the third channel is superfluous, i.e., it is not needed to conserve anything. Similarly, the neutrino sum $\bar{\nu}_{\mu\to e} + \nu_{\mu\to\tau}$ in the fourth channel is superfluous. Thus there is allowed the possibility of neutrinoless lepton decays

$$l_{\mu\to\tau} \to l_{\mu\to\ell} \quad , \ \ell = e, \mu \tag{12.65}$$

**Table 12.1** Observations of lepton $l_{\mu\to\tau}$ produced in CC $\nu_{\mu\to\tau}$ interactions at OPERA ("$h^\pm$" stands for $\pi^\pm$ or $K^\pm$.)

| Channel and $p_5$ balance | Observed |
|---|---|
| $l_{\mu\to\tau} \to h^- + \nu_{\mu\to\tau}$<br>$(G_\mu - m_e c) \to (-m_e c) + (G_\mu)$ | 3 |
| $l_{\mu\to\tau} \to h^- h^- h^+ + \nu_{\mu\to\tau}$<br>$(G_\mu - m_e c) \to (-m_e c - m_e c + m_e c) + (G_\mu)$ | 1 |
| $l_{\mu\to\tau} \to l_{\mu\to\mu} + \bar{\nu}_{\mu\to\mu} + \nu_{\mu\to\tau}$<br>$(G_\mu - m_e c) \to (G_\mu - m_e c) + (-G_\mu) + (G_\mu)$ | 1 |
| $l_{\mu\to\tau} \to l_{\mu\to e} + \bar{\nu}_{\mu\to e} + \nu_{\mu\to\tau}$<br>$(G_\mu - m_e c) \to (G_\mu - m_e c) + (-G_\mu) + (G_\mu)$ | 0 |

Reactions (12.65) thus provide a unique and—in principle—testable prediction of the present theory of flavor.

## 12.10 Review: three-neutrino mixing parameters

To apply our lepton-number-conserving theory of oscillation to problems in neutrino mixing, we shall need the values of the mixing parameters already determined by way of the standard or conventional

$3\nu$ theory of oscillation. In this standard theory—the basics of which were proposed long ago by Pontecorvo[20, 21] and by Maki, Nakagawa and Sakata[22]—transitions between the three flavor eigenstates $|\nu_\ell\rangle$ ($\ell = e, \mu, \tau$) follow from their representation in terms of three non-degenerate mass eigenstates, $|\nu_j\rangle$ ($j = 1, 2, 3$). In contrast to the $2\times 2$ and $3\times 3$ configurations described above, we might call this the $3\times 1$ model, because flavor $\ell$ can carry only one lepton number, namely $L_\ell$. Thus flavor transitions in this model necessarily violate conservation of lepton number. Nevertheless, all of what we now know of the oscillation phenomenon comes from exploring experimentally the structure of the $3\times 1$ model.[23] In this model the propagation of a neutrino born with flavor $\ell$ is expressed by (12.13), with $N = 3$. The connection between flavor and mass eigenstates is then found by integrating out the $x^\mu$ dependence of (12.13) using the normalization condition $\langle \psi_j | \psi_j \rangle = 1$:

$$|\nu_\ell\rangle = \sum_{j=1}^{3} U_{\ell j} |j\rangle \tag{12.66}$$

which is invertible to

$$|j\rangle = \sum_{\ell = e,\mu,\tau} U^*_{j\ell} |\ell\rangle \tag{12.67}$$

The unitary PMNS mixing matrix $U$ (so-named in recognition of the above authors) is customarily parametrized in the form

$$\begin{aligned} U = \left[U_{\ell j}\right] &= \begin{pmatrix} 1 & 0 & 0 \\ 0 & c_{23} & s_{23} \\ 0 & -s_{23} & c_{23} \end{pmatrix} \begin{pmatrix} c_{13} & 0 & s_{13}e^{-i\delta} \\ 0 & 1 & 0 \\ -s_{13}e^{i\delta} & 0 & c_{13} \end{pmatrix} \begin{pmatrix} c_{12} & s_{12} & 0 \\ -s_{12} & c_{12} & 0 \\ 0 & 0 & 1 \end{pmatrix} \\ &= \begin{pmatrix} c_{13}c_{12} & c_{13}s_{12} & s_{13}e^{-i\delta} \\ -c_{23}s_{12} - s_{13}s_{23}c_{12}e^{-i\delta} & c_{23}c_{12} - s_{13}s_{23}s_{12}e^{i\delta} & c_{13}s_{23} \\ s_{23}s_{12} - s_{13}c_{23}c_{12}e^{i\delta} & -s_{23}c_{12} - s_{13}c_{23}s_{12}e^{i\delta} & c_{13}c_{23} \end{pmatrix} \end{aligned} \tag{12.68}$$

where $s_{ij} = \sin\theta_{ij}$, $c_{ij} = \cos\theta_{ij}$, the $\theta_{ij}$ are the rotation angles in the $ij$ planes, and $\delta$ denotes a possible CP-violating phase. There should be included two additional phases $\alpha$ if neutrinos were Majorana particles. But as we saw in **Sec. 12.3** neutrinos are Dirac not Majorana particles, and so we can omit the $\alpha$. Inserting (12.67) into (12.13) we obtain an expansion of the propagating wave function in flavor eigenstates:

$$\left|\nu_\ell(x^\mu)\right\rangle = \sum_{\ell'=e,\mu,\tau} \left|C_{\ell\to\ell'}(x^\mu)\right\rangle \left|\nu_{\ell'}\right\rangle \tag{12.69}$$

where the spacetime-dependent coefficients—probability amplitudes—are given by

$$\left|C_{\ell\to\ell'}(x^\mu)\right\rangle = \sum_{j=1}^{3} U^*_{\ell' j} U_{\ell j} \left|\psi_j(x^\mu)\right\rangle . \tag{12.70}$$

The survival probabilities are thus given by[24, 25]

$$\begin{aligned} P(\nu_\ell \to \nu_\ell) &= \langle C_{\ell\to\ell} | C_{\ell\to\ell} \rangle \\ &= 1 - 4\sum_{j>i} \left|U_{\ell i}\right|^2 \left|U_{\ell j}\right|^2 \sin^2 \Delta_{ji} \end{aligned} \tag{12.71}$$

and the transition probabilities by

$$\begin{aligned} P(\nu_\ell \to \nu_{\ell'}) &= \langle C_{\ell\to\ell'} | C_{\ell\to\ell'} \rangle \\ &= -4\sum_{j>i} \mathrm{Re}\left[U^*_{\ell i} U_{\ell' i} U_{\ell j} U^*_{\ell' j}\right] \sin^2 \Delta_{ji} \\ &\quad + 2\sum_{j>i} \mathrm{Im}\left[U^*_{\ell i} U_{\ell' i} U_{\ell j} U^*_{\ell' j}\right] \sin 2\Delta_{ji} \end{aligned} \tag{12.72}$$

where, as in (12.56), we have defined $\Delta_{ji} = \Delta m^2_{ji} c^3 L / 4\hbar E$ and $\Delta m^2_{ji} = m_j^2 - m_i^2$. The probabilities for antineutrinos $\bar{\nu}_\ell$ are obtained by replacing the matrix elements $U_{\ell j}$ by their complex conjugates. The second sum here obviously vanishes for real-valued $U_{\ell j}$ ($\delta = 0$ or $\pi$). For complex-valued $U_{\ell j}$ the difference $P(\nu_\ell \to \nu_{\ell'}) - P(\bar{\nu}_\ell \to \bar{\nu}_{\ell'})$ measures the degree of CP-violation in the neutrino sector.[25]

The oscillation problem has seven unknowns: three mass values $m_j$ ($j = 1,2,3$) coming from the exponentials in the Dirac spinors $\psi_j$ of (12.13); three Euler (mixing) angles ($\theta_{12}$, $\theta_{13}$, $\theta_{23}$); and one CP-violating phase $\delta$. However, experiment detects not masses but mass-squared differences, and only two of the three $\Delta m^2_{ji}$ are independent. So the effective number of unknowns is six, not seven. Moreover, neglecting the CP-dependent sum in (12.72), we see that the probabilities depend not on the $\Delta m_{ji}^2$ as such but on their absolute values $\left|\Delta m_{ji}^2\right|$. So if by definition $\Delta m_{21}^2 \equiv \Delta m_{\text{SOL}}^2 > 0$, then in defining $\Delta m_{\text{ATM}}^2$ we have two mass-ordering patterns to take into account:

$$\begin{aligned}
&\text{Normal Ordering (NO): } m_1 < m_2 < m_3 \Rightarrow \; \Delta m_{\text{ATM}}^2 \equiv (\Delta m_{31}^2)_{\text{NH}} \\
&\text{Inverted Ordering (IO): } m_3 < m_1 < m_2 \Rightarrow \; \Delta m_{\text{ATM}}^2 \equiv (\Delta m_{23}^2)_{\text{IH}} \\
&\qquad\qquad\qquad\qquad\qquad\qquad\qquad\qquad = (\Delta m_{13}^2)_{\text{IH}} + \Delta m_{21}^2
\end{aligned} \tag{12.73}$$

As indicated in **Table 12.2**, oscillation parameters $\theta_{12}$ and $\Delta m_{21}^2$ are accessible mainly through solar and long-baseline reactor data; $\theta_{23}$ and $\Delta m_{31}^2 \approx \Delta m_{32}^2$ through atmospheric and accelerator data; and $\theta_{13}$ through

**Table 12.2** Neutrino oscillation parameters measured at various experiments. $\nu_\ell(\bar{\nu}_\ell) \to \nu_{\ell'}(\bar{\nu}_{\ell'})$ denotes disappearance when $\ell' = \ell$ and appearance when $\ell' \neq \ell$

| Seminal discoveries and insights | Oscillation parameter | Data | | | |
|---|---|---|---|---|---|
| | | Solar $\nu_e \to \nu_e$ | Reactor $\bar{\nu}_e \to \bar{\nu}_e$ | Atmospheric $\nu_\mu \to \nu_\mu$ | Accelerator $\nu_\mu(\bar{\nu}_\mu) \to \nu_\ell(\bar{\nu}_\ell)$ |
| Ray Davis *et al.* report solar $\nu_e$ deficit.[26] | $\theta_{12}$ | Super-K[27] SNO[28, 29] | | | |
| | $\Delta m_{21}^2$ | | (LBL) KamLAND[30] | | |
| Super-Kamiokande finds atm $\nu_\mu$ deficit.[16] | $\theta_{23}$ | | | Super-K[31] | |
| | $\Delta m_{31}^2$ | | | | K2K[17] MINOS[32] |
| Comparison of solar and KamLAND data suggest $\theta_{13} > 0$.[33, 34] | $\theta_{13}$ | | (SBL) Double Chooz[35] Daya Bay[36] RENO[37] | | (LBL) T2K[38] MINOS[39] |

short-baseline reactor and long-baseline accelerator data.

**Table 12.3**, which is based on the table presented in Ref. [40], summarizes the oscillation parameter values derived from a global fit of all applicable experimental data. Here one notes in particular that $\theta_{23}$ is consistent with maximal mixing; $\theta_{13} > 0$; and CP-violating phase $\delta \sim 3\pi/2$.

**Table 12.3** Neutrino oscillation parameter summary.[40] NO denotes normal ordering, IO inverted ordering.

| Parameter | Ordering | Best fit $\pm 1\sigma$ | $3\sigma$ |
|---|---|---|---|
| $\Delta m^2_{21}$ $[\times 10^{-5}\mathrm{eV}^2/c^4]$ | -------- | $7.60^{+0.19}_{-0.18}$ | $^{+.058}_{-.049}$ |
| $\lvert\Delta m^2_{31}\rvert$ $[\times 10^{-3}\mathrm{eV}^2/c^4]$ | NO | $2.48^{+0.05}_{-0.07}$ | $^{+0.17}_{-0.18}$ |
| | IO | $2.38^{+0.05}_{-0.06}$ | $^{+0.16}_{-0.18}$ |
| $\sin^2\theta_{12}$ | ------- | $0.323^{+0.016}_{-0.016}$ | $^{+0.052}_{-0.045}$ |
| $\sin^2\theta_{23}$ | NO | $0.567(0.467^a)^{+0.032}_{-0.028}$ | $^{+0.076}_{-0.075}$ |
| | IO | $0.573^{+0.025}_{-0.043}$ | $^{+0.067}_{-0.170}$ |
| $\sin^2\theta_{13}$ | NO | $0.0234^{+0.0020}_{-0.0020}$ | $^{+0.0060}_{-0.0057}$ |
| | IO | $0.0240^{+0.0019}_{-0.0019}$ | $^{+0.0057}_{-0.0057}$ |
| $\delta$ | NO | $1.34\pi^{+0.64\pi}_{-0.38\pi}$ | $^{+0.66\pi}_{-1.34\pi}$ |
| | IO | $1.48\pi^{+0.34\pi}_{-0.32\pi}$ | $^{+0.52\pi}_{-1.48\pi}$ |

[a] Local minimum.

## 12.11 Failure of the $3\times 3$ model

The simple $2\times 2$ configuration discussed in **Sec. 12.9** suffices to illustrate the principle of lepton-number-conserving oscillation with massless neutrinos. To describe the real world of three lepton families—a 6-D world in which lepton number is conserved—it would seem natural to assume, by extension of (12.57), a $3\times 3$ configuration consisting of $3^2$ flavor eigenstates $\lvert\nu_{\ell\to\ell'}\rangle$ ($\ell,\ell' = e,\mu,\tau$), each such state represented by three lepton-number bearing mass-squared eigenstates $\lvert\ell\rangle\lvert\ell j\rangle$ ($\ell = e,\mu,\tau;\ j = 1,2,3$). The model is formally expressed by (12.9) with $N = 3$, the flavor and mass-squared eigenstates connected by three matrix equations of the form [extension of (12.46) and (12.47) to three families]:

$$\begin{pmatrix} |\nu_{\ell\to e}\rangle \\ |\nu_{\ell\to\mu}\rangle \\ |\nu_{\ell\to\tau}\rangle \end{pmatrix} = U \begin{pmatrix} |\ell\rangle|\ell 1\rangle \\ |\ell\rangle|\ell 2\rangle \\ |\ell\rangle|\ell 3\rangle \end{pmatrix}, \quad (\ell = e, \mu, \tau), \tag{12.74}$$

where the $3\times3$ unitary mixing matrix $U$ is given by (12.68). Unfortunately, although the $3\times3$ model places no restrictions on the solar ($\theta_{12}$) mixing angle, it is strongly at variance with the SBL reactor results ($\theta_{13}$). Consequently it cannot be considered a final theory of mixing. Nevertheless we shall develop the model here as a precursor and guide to the more complete theory to follow.

The $3\times3$ oscillation problem has seven unknowns: three squared masses $m_j^2$ ($j = 1,2,3$) three mixing angles $\theta_{12}$, $\theta_{13}$ and $\theta_{23}$ and the CP-violating phase $\delta$. There are also five equations of constraint: two mass-squared differences

$$m_2^2 - m_1^2 = \Delta m_{21}^2 \tag{12.75}$$

and, assuming normal ordering,

$$m_3^2 - m_1^2 = \Delta m_{31}^2, \tag{12.76}$$

and three homogeneous equations of the form (12.21) expressing the masslessness of the three flavor neutrinos:

$$m_{\nu_\ell}^2 = \sum_{j=1}^{3} |U_{\ell j}|^2 m_j^2 = 0, \quad (\ell = e, \mu, \tau). \tag{12.77}$$

To find the $m_j^2$, we first write down from (12.77) all three homogeneous equations corresponding to $\ell = e, \mu$ and $\tau$, and then add them together (this is equivalent to summing the $|U_{\ell j}|^2$ over $\ell$). This gives an equation in the mass-squared unknowns independent of all rotation angles and independent of $\delta$:

$$\sum_{\ell=e,\mu,\tau} m_{\nu_\ell}^2 = m_1^2 + m_2^2 + m_3^2 = 0\,. \tag{12.78}$$

Equations (12.75), (12.76) and (12.78) can now be solved for the three $m_j^2$:

$$\left.\begin{aligned} m_1^2 &= \frac{1}{3}\left(-\Delta m_{21}^2 - \Delta m_{31}^2\right) \\ m_2^2 &= \frac{1}{3}\left(2\Delta m_{21}^2 - \Delta m_{31}^2\right) \\ m_3^2 &= \frac{1}{3}\left(-\Delta m_{21}^2 + 2\Delta m_{31}^2\right) \end{aligned}\right\}. \tag{12.79}$$

To connect $\theta_{12}$ and $\theta_{13}$ we substitute mass-squared expressions (12.79) into (12.77) with $\ell = e$. This gives

$$\left(1-3s_{12}^2c_{13}^2\right)\Delta m_{21}^2 + \left(1-3s_{13}^2\right)\Delta m_{31}^2 = 0\,. \tag{12.80}$$

Given the observationally-known ratio $\Delta m_{21}^2/\Delta m_{31}^2 = 0.031$, we see at once that whatever the value of $s_{12}^2$ may happen to be, the $3\times 3$ model predicts $s_{13}^2 \sim 1/3$. Since in reality $s_{13}^2 \cong 0.023$ the $3\times 3$ model does indeed fail to describe accurately the real world of neutrino oscillation.

Nevertheless, it is of interest to see what the model says about mixing angle $\theta_{23}$. Take from (12.77) the two homogeneous equations corresponding to $\ell = \mu,\tau$ and subtract one from the other. This yields

$$a\cos 2\theta_{23} + b\sin 2\theta_{23} = 0 \tag{12.81}$$

where

$$\left.\begin{aligned} a &= c_{13}^2\Delta m_{31}^2 - \left(c_{12}^2 - s_{12}^2 s_{13}^3\right)\Delta m_{21}^2 \\ b &= s_{13}\cos\delta\sin 2\theta_{12}\Delta m_{21}^2 \end{aligned}\right\} \tag{12.82}$$

Solving (12.81) for $s_{23}^2$ we get the more interpretable result

$$\sin^2\theta_{23} \cong \frac{1}{2}\left(1-\frac{b^2}{a^2}\right) = 0.5 - O\left(10^{-17}\right), \tag{12.83}$$

which describes maximal mixing. Although $3\times 3$ model fails as a whole, it at least reveals the origin of maximal mixing in the atmospheric anomaly: it is attributable to the masslessness of flavor neutrinos $\nu_\mu$ and $\nu_\tau$.

## 12.12 Introducing a fourth flavor, *s*: the $3\times(3+1)$ oscillation model

Now there is nothing wrong in principle with the $3\times 3$ model. It's just that Nature chooses to ignore it, having in mind a value for $\theta_{13}$ other than the one asked for by the model. Does this mean that the model must be abandoned? No, it means only that the model requires expanding for more flexibility, and this is readily accomplished. One need only add a fourth flavor, *s*, to the model, leading to an additional term in (12.80) and permitting the l. h. s. to vanish while accommodating Nature's value for $\theta_{13}$. The model is again formally expressed by (12.9), this time with $N=4$. Instead of (12.74) we now have the following connection between flavor and mass-squared eigenstates:

$$\begin{pmatrix} |\nu_{\ell\to e}\rangle \\ |\nu_{\ell\to\mu}\rangle \\ |\nu_{\ell\to\tau}\rangle \\ |\nu_{\ell\to s}\rangle \end{pmatrix} = U \begin{pmatrix} |\ell\rangle|\ell 1\rangle \\ |\ell\rangle|\ell 2\rangle \\ |\ell\rangle|\ell 3\rangle \\ |\ell\rangle|\ell 4\rangle \end{pmatrix}, \quad (\ell = e,\mu,\tau). \qquad (12.84)$$

where $U$ is a $4\times 4$ unitary mixing matrix. Here we see that there are in fact *three* additional flavor states *s*, namely the $|\nu_{\ell\to s}\rangle$, each carrying its own lepton number $L_\ell$, $\ell = e$, $\mu$ or $\tau$. Equation (12.84) thus represents three sets of four equations, each set carrying its own lepton number $L_\ell$. Such a model may reasonably called the *$3\times(3+1)$ oscillation model*—reasonable because in an $M^4\times\bar{M}^2$ world with 3 lepton families and 3+1 neutrino flavors ($e,\mu,\tau$ and *s*), we have a total of $3\times(3+1)$ flavor states to work with, as each neutrino flavor may carry lepton number $L_e$, $L_\mu$ or $L_\tau$.

The three additional flavor states are labeled *s* because they are *sterile*, meaning that—because there are only three lepton families—they are incapable of entering into weak interactions. Now sterile neutrinos occur as well the standard theory of oscillation reviewed in

**Sec. 12.10**. In that standard setting they are routinely invoked to account for the observed LSND,[41] MiniBooNE[42] and reactor[43] anomalies—oscillation effects unexplainable within the bare $3\nu$ paradigm. Appropriately, the standard theory augmented by $n$ sterile neutrinos is generally termed a $3+n$ theory. Such neutrinos—so-called light sterile neutrinos, $\nu_s$—do not appear in the Standard Model, but "are present in most extensions of the standard model, and in principle can have any mass." [44] It should be understood that the sterile neutrinos $\nu_s$ adopted by standard oscillation theory cannot be identified with the $|\nu_{\ell\to s}\rangle$ flavor states of our $3\times(3+1)$ theory. The reason is this: Both standard oscillation theory and the SM extensions giving rise to light sterile neutrinos are theories constructed in 4-D $M^4$. Consequently the $\nu_s$ occurring in these theories know nothing of pseudomomentum $p_5$ and the origin of mass in dimension $x^6$. But as we can see from (12.84), the flavor state $|\nu_{\ell\to s}\rangle$ is proportional to eigenstate $|\ell\rangle$ of operator $\hat{p}_5$, which means that it carries pseudomomentum $p_5$, thus ensuring conservation of $L_\ell$ while in oscillation with the other three states having that same lepton number. Moreover, the masses $m_j^2$ $(j=1,2,3,4)$ of which the $|\nu_{\ell\to s}\rangle$ are composed cannot have just "any mass." As we shall see, those masses—including $m_4^2$—get their values from established spectral patterns in dimension $x^6$. The three sterile flavor states $|\nu_{\ell\to s}\rangle$ are thus unique to the $3\times(3+1)$ model. They, like the other six states of the form $|\nu_{\ell\to\ell'}\rangle$ $(\ell\neq\ell')$ are products of oscillation, not of the Standard Model or any of its extensions.

## 12.13 The four-flavor oscillation problem

*12.13.1 Linearized mixing matrix.* The $4\times4$ mixing matrix $U$ is, on the other hand, the same for both the standard and present $3\times(3+1)$ theory of oscillation. Here we adopt what has become in effect the standard parameterization of matrix $U$:[45]

$$U=\begin{pmatrix}1&0&0&0\\0&1&0&0\\0&0&c_{34}&s_{34}\\0&0&-s_{34}&c_{34}\end{pmatrix}\begin{pmatrix}1&0&0&0\\0&c_{24}&0&s_{24}\\0&0&1&0\\0&-s_{24}&0&c_{24}\end{pmatrix}\begin{pmatrix}1&0&0&0\\0&c_{23}&s_{23}e^{-i\delta_3}&0\\0&-s_{23}e^{i\delta_3}&c_{23}&0\\0&0&0&1\end{pmatrix}$$
$$\times\begin{pmatrix}c_{14}&0&0&s_{14}\\0&1&0&0\\0&0&1&0\\-s_{14}&0&0&c_{14}\end{pmatrix}\begin{pmatrix}c_{13}&0&s_{13}e^{-i\delta_2}&0\\0&1&0&0\\-s_{13}e^{i\delta_2}&0&c_{13}&0\\0&0&0&1\end{pmatrix}\begin{pmatrix}c_{12}&s_{12}e^{-i\delta_1}&0&0\\-s_{12}e^{i\delta_1}&c_{12}&0&0\\0&0&1&0\\0&0&0&1\end{pmatrix}\tag{12.85}$$

where, as in (12.68), $s_{ij}=\sin\theta_{ij}$, $c_{ij}=\cos\theta_{ij}$, the $\theta_{ij}$ are the rotation angles in the *i-j* planes and the $\delta_k$ denote possible CP-violating phases. The oscillation problem now has thirteen unknowns: four mass-squared values ($m_j^2$, $j=1,2,3,4$); six mixing angles ($\theta_{12},\theta_{13},\theta_{14},\theta_{23},\theta_{24},\theta_{34}$); and three CP-violating phases ($\delta_1,\delta_2,\delta_3$). In addition, there are six equations of constraint: three independent mass-squared differences

$$m_2^2-m_1^2=\Delta m_{21}^2\tag{12.86}$$
$$m_3^2-m_1^2=\Delta m_{31}^2\tag{12.87}$$
$$m_4^2-m_1^2=\Delta m_{41}^2\tag{12.88}$$

and, from (12.21) with $N=4$, three zero mass conditions

$$m_{\nu_\ell}^{\ 2}=\sum_{j=1}^{4}\left|U_{\ell j}\right|^2 m_j^{\ 2}=0,\quad (\ell=e,\mu,\tau).\tag{12.89}$$

Note that there is no fourth zero-mass condition; that is because (12.9) and (12.89) refer to the mass of the flavor neutrino $\nu_\ell$ belonging to lepton family $\ell$, and there are only three such families.

Now we intend to solve Eqs. (12.86-89) for the four unknown $m_j^2$. This will at the same time yield values for the unknown angles $\theta_{24},\theta_{34}$ as well as a modification of the maximal mixing result (12.83). As we expect the values of $s_{14}^2$, $s_{24}^2$ and $s_{34}^2$ to be small in relation to unity, the analysis can be much simplified by neglecting all terms in the elements of matrix $U$ quadratic and higher in $s_{14}$, $s_{24}$ and $s_{34}$.[46] For comparison with the failed $3\times3$ model discussed above, we do *not* assume $s_{13}$ to be small. We are then left with the linearized form

$$U = \begin{pmatrix} U_{e1} & U_{e2} & U_{e3} & U_{e4} \\ U_{\mu 1} & U_{\mu 2} & U_{\mu 3} & U_{\mu 4} \\ U_{\tau 1} & U_{\tau 2} & U_{\tau 3} & U_{\tau 4} \\ U_{s1} & U_{s2} & U_{s3} & U_{s4} \end{pmatrix}$$

$$\cong \begin{pmatrix} c_{12}c_{13} & c_{13}s_{12}e^{-i\delta_1} & s_{13}e^{-i\delta_2} & s_{14} \\ -c_{23}s_{12}e^{i\delta_1} - c_{12}s_{13}s_{23}e^{i(\delta_2-\delta_3)} & c_{12}c_{23} - s_{12}s_{13}s_{23}e^{i(-\delta_1+\delta_2-\delta_3)} & c_{13}s_{23}e^{-i\delta_3} & s_{24} \\ s_{12}s_{23}e^{i(\delta_1+\delta_3)} - c_{12}s_{13}c_{23}e^{i\delta_2} & -c_{12}s_{23}e^{i\delta_3} - s_{12}s_{13}c_{23}e^{i(-\delta_1+\delta_2)} & c_{13}c_{23} & s_{34} \\ U_{s1} & U_{s2} & U_{s3} & 1 \end{pmatrix} \tag{12.90}$$

where

$$\begin{aligned} U_{s1} &= s_{12}e^{i\delta_1}\left(c_{23}s_{24} - s_{23}s_{34}e^{i\delta_3}\right) - c_{12}\left[c_{13}s_{14} - s_{13}e^{i\delta_2}\left(s_{23}s_{24}e^{-i\delta_3} + c_{23}s_{34}\right)\right] \\ U_{s2} &= -c_{12}\left(c_{23}s_{24} - s_{23}s_{34}e^{i\delta_3}\right) - s_{12}e^{-i\delta_1}\left[c_{13}s_{14} - s_{13}e^{i\delta_2}\left(s_{23}s_{24}e^{-i\delta_3} + c_{23}s_{34}\right)\right]. \\ U_{s3} &= -s_{13}s_{14}e^{-i\delta_2} - c_{13}\left(s_{23}s_{24}e^{-i\delta_3} - c_{23}s_{34}\right) \end{aligned}$$

*12.13.2 Solving for the* $m_j^2$. Let us first write down from (12.89) the three zero-mass equations corresponding to $\ell = e,\ \mu,\ \tau$:

$$c_{12}^2c_{13}^2m_1^2 + s_{12}^2c_{13}^2m_2^2 + s_{14}^2m_4^2 = 0 \tag{12.91}$$

$$\begin{aligned} &\left(s_{12}^2c_{23}^2 + s_{13}s_{23}c_{23}\sin 2\theta_{12}\cos\delta + c_{12}^2s_{13}^2s_{23}^2\right)m_1^2 \\ &\quad + \left(c_{12}^2c_{23}^2 - s_{13}s_{23}c_{23}\sin 2\theta_{12}\cos\delta + s_{12}^2s_{13}^2s_{23}^2\right)m_2^2 \\ &\quad + c_{13}^2s_{23}^2m_3^2 + s_{24}^2m_4^2 = 0 \end{aligned} \tag{12.92}$$

$$\begin{aligned} &\left(s_{12}^2s_{23}^2 - s_{13}s_{23}c_{23}\sin 2\theta_{12}\cos\delta + c_{12}^2s_{13}^2s_{23}^2\right)m_1^2 \\ &\quad + \left(c_{12}^2s_{23}^2 + s_{13}s_{23}c_{23}\sin 2\theta_{12}\cos\delta + s_{12}^2s_{13}^2c_{23}^2\right)m_2^2 \\ &\quad + c_{13}^2c_{23}^2m_3^2 + s_{34}^2m_4^2 = 0 \end{aligned} \tag{12.93}$$

where we have defined

$$\delta = -\delta_1 + \delta_2 + \delta_3 . \tag{12.94}$$

Adding together the three zero-mass equations we obtain a single equation independent of $\delta$:

$$m_1^2 + m_2^2 + m_3^2 + \sigma m_4^2 = 0 . \tag{12.95}$$

where we have defined

$$\sigma = s_{14}^2 + s_{24}^2 + s_{34}^2 . \tag{12.96}$$

Equation (12.95) may be solved simultaneously with (12.86-88), yielding the four $m_j^2$:

$$\left.\begin{aligned} m_1^2 &= -\frac{1}{3+\sigma}\left(\Delta m_{21}^2 + \Delta m_{31}^2 + \sigma \Delta m_{41}^2\right) \\ m_2^2 &= \Delta m_{21}^2 + m_1^2 \\ m_3^2 &= \Delta m_{31}^2 + m_1^2 \\ m_4^2 &= \Delta m_{41}^2 + m_1^2 \end{aligned}\right\} . \tag{12.97}$$

For illustration let us assume normal ordering, i.e., $\Delta m_{21}^2$, $\Delta m_{21}^2$, $\Delta m_{21}^2 > 0$ . Then from the first of Eqs. (12.97), $m_1^2 < 0$. In addition suppose it assumed that $m_2^2 < m_3^2 < 0$. Then to satisfy (12.95) we must have $m_4^2 > 0$. The $m^2$ spectral structure corresponding to these assumptions is depicted in **Fig. 12.2**. It is easy to show that this picture is an accurate one. For the condition that $m_3^2 < 0$ is equally well expressed by the inequality $m_1^2 < -\Delta m_{31}^2$, which, utilizing (12.97), may be written

$$\sigma \Delta m_{41}^2 > (2+\sigma)\Delta m_{31}^2 - \Delta m_{21}^2 .$$

As we shall see below, the l. h. s. of this expression is of the order $0.1\ \mathrm{eV}^2/c^4$. The inequality is thus well satisfied. Furthermore, the condition that $m_4^2 > 0$ is equally well expressed by the inequality $m_1^2 > -\Delta m_{41}^2$, which from (12.97) can be written

$$3\Delta m_{41}^2 > \Delta m_{21}^2 + \Delta m_{31}^2 .$$

As the l. h. s. of this expression is of the order $1\ \mathrm{eV}^2/c^4$, it too is well satisfied. **Fig. 12.2** thus can be considered a valid depiction of the real-world $m^2$ spectrum (though not to scale).

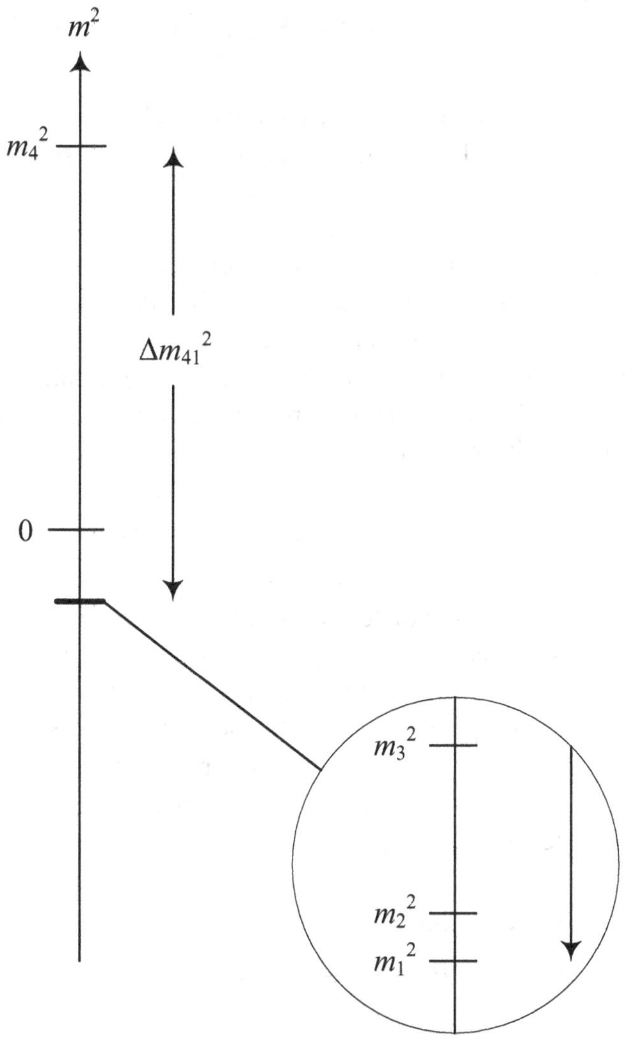


**Fig. 12.2.** Depiction of the mass-squared eigenvalues $m_j^2, j = 1,2,3,4$, assuming normal ordering and $m_3^2 < 0$. Not to scale.

*12.13.3 Zero-mass equation for the unknowns* $s_{14}$, $s_{24}$, $s_{34}$. Let us now replace the $m_j^2$ in the zero-mass equation for $\nu_e$, Eq. (12.91), with their values given by (12.97), assuming $\sigma \ll 1$. This yields an expression in the unknown variables $s_{14}$, $s_{24}$, $s_{34}$:

$$\begin{aligned}&\left(1-3s_{12}^2c_{13}^2\right)\Delta m_{21}^2+\left(1-3s_{13}^2\right)\Delta m_{31}^2\\&\qquad+\left[\left(s_{24}^2+s_{34}^2\right)-2s_{14}^2\right]\Delta m_{41}^2=0\end{aligned}. \tag{12.98}$$

This is just (12.80) with an extra term contributed by flavor $s$, permitting the mass of $\nu_e$ to vanish, provided the extra term is of suitable sign and magnitude.

*12.13.4 Calculation of mass-squared difference* $m_4^2 - m_1^2 = \Delta m_{41}^2$. Eq. (12.98) connects the three parameters $s_{j4}^2$, but does so in terms of the unknown mass-squared difference $\Delta m_{41}^2$. However, unlike $\Delta m_{21}^2$ and $\Delta m_{31}^2$—differences that have to be determined empirically—$\Delta m_{41}^2$ is inferable from the $3\times(3+1)$ formalism. To calculate its value we first note that, by definition, from (12.12b),

$$\Delta m_{41}^2c^2 = 2G_e\Delta g_{e41} = 2G_\mu\Delta g_{\mu41} = 2G_\tau\Delta g_{\tau41} \tag{12.99}$$

where the three frequency differences $\Delta g_{\ell 41}$ are those required to yield for each $\ell$ the same constant $\Delta m_{41}^2$. But as noted in our discussion of (12.12b), the system of three lepton families is self-consistent. That is to say, each family appears to be informed about the structure of the other two. Specifically, the system of three families already has in inventory the six known frequency differences $\Delta g_{\ell ij}$ defined by the following two sets of equations:

$$\left.\begin{aligned}\Delta m_{21}^2c^2 &= 2G_e\Delta g_{e21} = 2G_\mu\Delta g_{\mu21} = 2G_\tau\Delta g_{\tau21}\\ \Delta m_{31}^2c^2 &= 2G_e\Delta g_{e31} = 2G_\mu\Delta g_{\mu31} = 2G_\tau\Delta g_{\tau31}\end{aligned}\right\}. \tag{12.100}$$

Now suppose it is assumed that these six existing frequency differences are the *only* ones available for creating mass-squared differences $\Delta m_{ij}^2$ within the three-family system. Then, in principle, any one of the three $\Delta g_{\ell 41}$ could be equivalent to any one of the six $\Delta g_{\ell 21}$, $\Delta g_{\ell 31}$, leading to 18 possible values of $\Delta m_{41}^2$. These 18 are displayed in **Table 12.4**. Actually, 6 of these values are degenerate, and so the actual number of options is 12, not 18. But even this number is further reducible. For we recall that the sole function of flavor $s$ is to yield vanishing mass for the electron neutrino. This means of the 6 existing frequency differences, only 2 can be considered effective for defining $\Delta m_{41}^2$, namely $\Delta g_{e21}$ and $\Delta g_{e31}$, as these are the only differences attached to the electron family's 5-sphere of radius $G_e$. But of these two remaining differences, $\Delta g_{e21}$ has by far the greater defining power: for instance if $\sin^2\theta_{13}=0$ (which is not far from the case), then $\Delta g_{\ell 31}$ appears not at all in the oscillation probabilities (12.72). Consequently it is safe to assume that it is $\Delta g_{e21}$ that defines $\Delta m_{41}^2$. The optional $\Delta m_{41}^2$ values stemming from it are those displayed in the highlighted row of **Table 12.4**. The first option shown

**Table 12.4** Calculated optional values of $\Delta m_{41}^2\ \left[\mathrm{eV}^2/c^4\right]$[a]

| Substitute for $\Delta g_{\ell 41}$ | Representations of $\Delta m_{41}^2$ from Eq. (12.99) | | |
|---|---|---|---|
| | $2G_e\Delta g_{e41}/c^2$ | $2G_\mu\Delta g_{\mu 41}/c^2$ | $2G_\tau\Delta g_{\tau 41}/c^2$ |
| $\Delta g_{e21}=\dfrac{\Delta m_{21}^2c^2}{2G_e}$ | $\Delta m_{21}^2$ | $\dfrac{G_\mu}{G_e}\Delta m_{21}^2=1.636$ | $\dfrac{G_\tau}{G_e}\Delta m_{21}^2=457.7$ |
| $\Delta g_{\mu 21}=\dfrac{\Delta m_{21}^2c^2}{2G_\mu}$ | $\dfrac{G_e}{G_\mu}\Delta m_{21}^2=3.53\times10^{-9}$ | $\Delta m_{21}^2$ | $\dfrac{G_\tau}{G_\mu}\Delta m_{21}^2=0.0214$ |
| $\Delta g_{\tau 21}=\dfrac{\Delta m_{21}^2c^2}{2G_\tau}$ | $\dfrac{G_e}{G_\tau}\Delta m_{21}^2=1.25\times10^{-11}$ | $\dfrac{G_\mu}{G_\tau}\Delta m_{21}^2=2.7\times10^{-7}$ | $\Delta m_{21}^2$ |
| $\Delta g_{e31}=\dfrac{\Delta m_{31}^2c^2}{2G_e}$ | $\Delta m_{31}^2$ | $\dfrac{G_\mu}{G_e}\Delta m_{31}^2=53.4$ | $\dfrac{G_\tau}{G_e}\Delta m_{31}^2=15{,}045$ |
| $\Delta g_{\mu 31}=\dfrac{\Delta m_{31}^2c^2}{2G_\mu}$ | $\dfrac{G_e}{G_\mu}\Delta m_{31}^2=1.1\times10^{-7}$ | $\Delta m_{31}^2$ | $\dfrac{G_\tau}{G_\mu}\Delta m_{31}^2=0.70$ |
| $\Delta g_{\tau 31}=\dfrac{\Delta m_{31}^2c^2}{2G_\tau}$ | $\dfrac{G_e}{G_\tau}\Delta m_{31}^2=0.41\times10^{-9}$ | $\dfrac{G_\mu}{G_\tau}\Delta m_{31}^2=0.88\times10^{-5}$ | $\Delta m_{31}^2$ |

[a]Values of $\Delta m_{21}^2$ and $\Delta m_{31}^2$ from **Table 12.3**; values of the $G_\ell$ from **Table 4.1**.

is the uninteresting one, $\Delta m_{21}^2$. The third option, *viz.*, 457.7 is technically viable. However, as indicated in **Table 12.5**, global analysis of a variety of experiments puts $\Delta m_{41}^2 \sim 1$, which obviously eliminates 457.7.

| **Table 12.5** Global analysis of sterile neutrino hypothesis | | | |
|---|---|---|---|
| Source | $\Delta m_{41}^2 \left[\mathrm{eV}^2/c^4\right]$ | $\lvert U_{e4}\rvert^2 \cong s_{14}^2$ | $\lvert U_{\mu 4}\rvert^2 \cong s_{24}^2$ |
| J. Kopp *et al.*[47] | 0.93 | 0.0225 | 0.0289 |
| C. Guinti *et al.*[48] | 1.6 | 0.03 | 0.013 |

The only option remaining is the value 1.636, which happens to agree perfectly with the global $\Delta m_{41}^2$ value reported by the authors of Ref, [48], **Table 12.5**. On that basis we shall assume in the calculations to follow that $\Delta g_{\mu 41} = \Delta g_{e21}$ and hence that

$$\Delta m_{41}^2 = 1.636\ \mathrm{eV}^2/\mathrm{c}^4. \tag{12.101}$$

It is interesting to see that, although we have arrived at this value on general grounds, each of the other values displayed in **Table 12.4**—with the possible exception of 0.70—is either too small or too large to represent parameter $\Delta m_{41}^2$.

From (12.99) we see once again that the system of neutrino flavors $\ell$ forms a self-communicating whole: it provides the frequency difference $\Delta g_{\ell 41}$ required to generate for each $G_\ell$ the same constant $\Delta m_{41}^2$. Algebraically, as shown in **Fig. 12.3**, the three datum pairs $(G_\ell, \Delta g_{\ell 41})$, $\ell = e, \mu, \tau$, represent points on a hyperbola of semitransverse axis $(\Delta m_{41}^2)^{1/2} c$.

*12.13.5 Numerical values of* $s_{24}$ *and* $s_{34}$. Let us now return to Eqs. (12.92) and (12.93) and subtract one equation from the other. This gives

$$a\cos 2\theta_{23} + b\sin 2\theta_{23} + c = 0, \tag{12.102}$$

in which

$$\left.\begin{aligned} a &= c_{13}^2 \Delta m_{31}^2 - \left(c_{12}^2 - s_{12}^2 s_{13}^2\right)\Delta m_{21}^2 \\ b &= s_{13}\sin 2\theta_{12}\cos\delta\Delta m_{21}^2 \\ c &= \left(s_{34}^2 - s_{24}^2\right)[\Delta m_{41}^2 - (\Delta m_{21}^2 + \Delta m_{31}^2)/3] \end{aligned}\right\}. \tag{12.103}$$

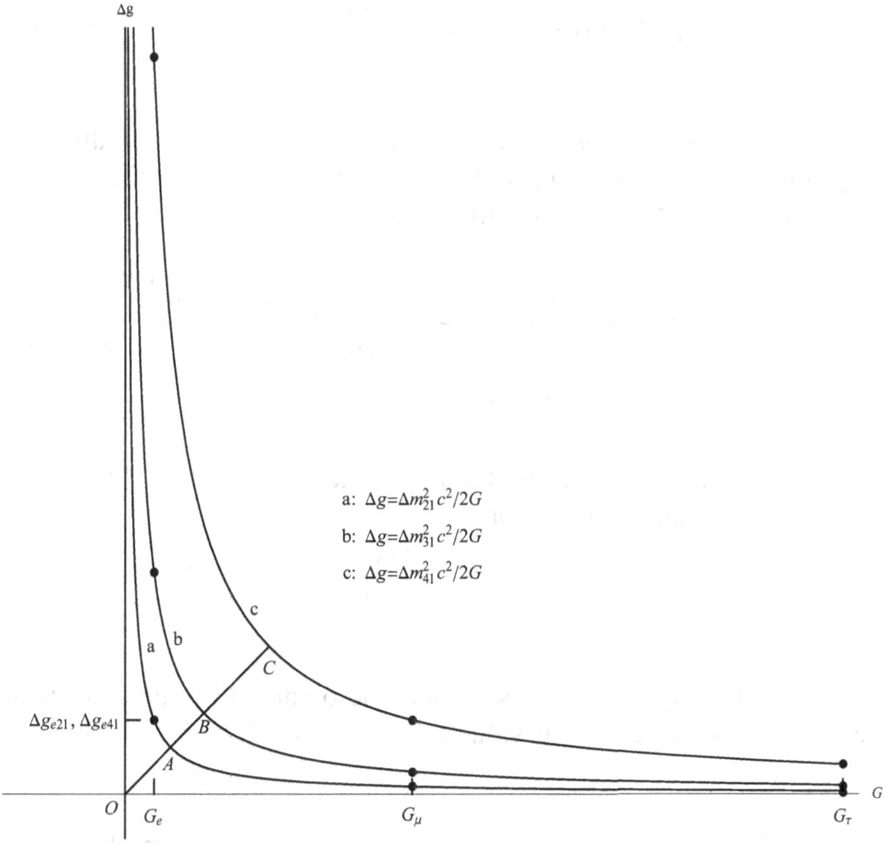


**Fig. 12.3.** This figure is a continuation and enlargement of **Fig. 12.1**. Momentum pairs $(G_e, \Delta g_{e41})$, $(G_\mu, \Delta g_{\mu 41})$ and $(G_\tau, \Delta g_{\tau 41})$ are here depicted as points on hyperbola "c" of semitransverse axis $OC = (\Delta m_{41}^2)^{1/2} c$. Hyperbola "c" is defined on the assumption that, as shown, $\Delta g_{\mu 41} = \Delta g_{e21}$.

We now solve (12.102) for $\sin^2\theta_{23}$, a standard parameter defining oscillation between $\nu_{\mu\to\mu}$ and $\nu_{\mu\to\tau}$:

$$\sin^2\theta_{23} = \frac{\left(a^2+b^2+ac\right)-\sqrt{\left(a^2+b^2+ac\right)^2-\left(a^2+b^2\right)\left(a+c\right)^2}}{2\left(a^2+b^2\right)} \qquad (12.104)$$

Since $\Delta m_{21}^2 \ll \Delta m_{21}^2$, $b^2 \ll a^2$ and $m_4^2 \cong \Delta m_{41}^2$, Eq. (12.104) simplifies to

$$\sin^2\theta_{23} \cong 0.5 + \frac{c}{2a} \cong 0.5 + \frac{\Delta m_{41}^2}{2\Delta m_{31}^2}\left(s_{34}^2 - s_{24}^2\right). \qquad (12.105)$$

Now the coefficient $\Delta m_{41}^2 / 2\Delta m_{31}^2 = 347$, so $s_{24}^2$ and $s_{34}^2$ can differ in magnitude by no more than about $(0.57-0.50)/347 = 0.0002$. Then to a good approximation we have from (12.98)

$$\begin{aligned} s_{24}^2 \cong s_{34}^2 \cong s_{14}^2 &- \frac{\left(1-3s_{12}^2c_{13}^2\right)\Delta m_{21}^2 + \left(1-3s_{13}^2\right)\Delta m_{21}^2}{2\Delta m_{41}^2} \\ &= s_{14}^2 - 0.0007 \end{aligned} \qquad (12.106)$$

For illustration, suppose from **Table 12.5** we take the Ref. [48] value $s_{14}^2 = 0.030$. Then from (12.106)

$$\left.\begin{aligned} s_{24}^2 &\cong 0.0293 \\ s_{34}^2 &\cong 0.0293 \end{aligned}\right\}. \qquad (12.107)$$

This value for $s_{24}^2$ is very close to the one obtained by the authors of Ref. [47], **Table 12.5.** Our value for $s_{34}^2$ is a prediction of the present theory.

*12.13.6 Maximal mixing spoiled.* Eq. (12.105) is our formula for the mixing parameter $s_{23}^2$. It shows that maximal mixing can occur, provided $s_{24}^2 = s_{34}^2$. On the other hand, given the magnitude of the factor $\Delta m_{41}^2 / 2\Delta m_{31}^2$ , the smallest difference between $s_{24}^2$ and $s_{34}^2$ drives the mixing away from maximal. And it appears from the experimental results presented in **Table 12.3** that that is exactly what has occurred.

It should be emphasized that maximal or almost-maximal mixing follows directly from the zero-mass conditions (12.89), as was shown already at the end of **Sec. 12.11**. The very presence of maximal mixing, or an approximation to it, represents *prima facie* evidence for the masslessness of flavor neutrinos. And the departure from maximal mixing can be considered evidence for the existence of sterile neutrino flavor *s*, as can the existence of the LSND, MiniBooNE and reactor anomalies. These, however, are all side effects. The central significance of the sterile state is that it enables the vanishing of the mass of the electron neutrino.

## 12.14 Summary and a comment on IceCube

The Principle of True Representation yields a new picture of neutrino mixing and oscillation, a picture substantially at variance with standard neutrino phenomenology. By way of summary, **Table 12.6** reviews the defining steps of our $3\times(3+1)$ theory of oscillation and compares them with the corresponding steps in the standard 3+1 theory.

**Table 12.6** Standard 3+1 and present $3\times(3+1)$ theories of oscillation compared

| Feature | 3+1 theory | $3\times(3+1)$ theory |
|---|---|---|
| Spacetime platform | 4D $M^4$ | 6D $M^4\times\bar{M}^2$ |
| Origin of mass-squared eigenstates | Standard Model Lagrangian | Complex waveform in $\bar{M}^2$ |
| Nature of neutrino | Unresolved. May be Majorana or Dirac. | All $\nu_\ell$ carry a $U(1)$ charge $G_\ell$, and thus must be Dirac particles. |
| Mass of flavor neutrino $\nu_\ell$ | Oscillation implies non-vanishing mass. Defined quantum mechanically as probable value $\sum\lvert U_{\ell j}\rvert^2 m_j$. | Mass-squared $m_\ell^2$ is proportional to neutrino electric charge. Hence all $\nu_\ell$ have vanishing mass |
| Why do neutrinos oscillate? | No reason. Optional. They just do. | Neutrinos must oscillate to exist. Required to generate fifth and sixth probability current densities $s_5$, $s_6$. |
| Sign of mass-squared eigenvalues, $m_j^2$ | All positive. | Both positive and negative values required to yield zero mass of flavor neutrino. |
| Why are there no right-handed neutrinos? | If neutrinos are massive, right-handed ones should exist. Yet none are observed. | Right-handed tachyonic states have no positive norm. But tachyonic states are required for zero mass. Hence right-handed neutrinos do not exist. |

| Feature | 3+1 theory | $3\times(3+1)$ theory |
|---|---|---|
| Conservation of lepton number, $L_\ell$ | Neutrino oscillation violates conservation of $L_\ell$, and is the only known leptonic process to do so | $L_\ell$ is conserved in all processes, including oscillation. To ensure this, all mass-squared eigenstates carry $L_\ell$. So if there are 3 lepton families and 3+1 flavors, then the total number of flavor states is $3\times(3+1)$. The extra states do not violate the known width of $Z^0$. |
| Charged lepton partners of the extra flavor states. | There are no extra flavor states. | Such partners must exist and in principle are detectable. |
| Measured mixing parameters | Values of parameters measured assuming standard 3+1 theory apply equally well in $3\times(3+1)$ theory. | |
| Function of sterile flavor, s | Invoked to explain LSND, MiniBooNE and reactor anomaly. | Required to permit mass of the electron neutrino to vanish. LSND, MiniBooNE and reactor anomaly arise as side effects. |
| Mass-squared difference $\Delta m_{41}^2$ | Not given theoretically. Must be measured. | Given theoretically. Value 1.6 confirmed by measurement. |
| Value of parameter $s_{14}^2$ | Must be measured. | Must be measured. |
| Value of parameter $s_{24}^2$ | Must be measured. | Calculable from $3\times(3+1)$ theory. Value 0.029 confirmed by measurement. |
| Value of parameter $s_{34}^2$ | Must be measured | $s_{34}^2 \cong s_{24}^2$ predicted. |
| Mixing of $\nu_\mu$ and $\nu_\upsilon$ | Measured to be almost maximal. Unpredicted. | Predicted to be almost maximal. Departure from maximal due to slight difference between $s_{24}^2$ and $s_{34}^2$. |

As shown above, a sterile neutrino $\nu_s$ is required to give vanishing $\nu_e$ mass. Very recently, however, the IceCube collaboration reported finding no evidence for a light sterile neutrino, specifically one having our calculated mass-squared value (12.101).[49] Can the theory developed here be reconciled with the IceCube result? Possibly so, and here is how. As stated at the end of **Sec. 12.13**, the three sterile states $\left|\nu_{\ell\to s}\right\rangle$, like the other six states $\left|\nu_{\ell\to\ell'}\right\rangle$ ($\ell \neq \ell'$), are products, not of the Standard Model, but of the process of neutrino wave propagation. Thus the mixing angles $\theta_{ij}$ in (12.90), too, are products of, or arise in association with, wave propagation. This opens up the possibility that the $\theta_{ij}$ are *energy dependent*, their relative values always adjusted to maintain unitarity of the mixing matrix—a feature unavailable to the standard theory of neutrino mixing in $M^4$. Now the IceCube result is based on the failure to find a dip at 3 TeV in the frequency-of-event vs. energy curve for atmospheric muon neutrinos. But this energy is 5 orders of magnitude larger than that of the muons involved in the LSND experiment, and 3 orders larger than those involved in MiniBooNE. So if the $\theta_{ij}$ are energy dependent, and if the $\theta_{i4}$ tend to 0 with increasing energy, then the IceCube result and the $\nu_e$ appearance results of LSND and MiniBooNE may be reconciled. Of course, according to (12.80) and (12.98), to maintain vanishing $\nu_e$ mass at those energies where $\theta_{i4} \to 0$, one requires $\sin^2\theta_{13} \approx 1/3$! Notice that this possible explanation of the null result of IceCube makes sense only if flavor neutrinos are massless. If they were not, then the $\theta_{ij}$ would be frame dependent in addition to being energy dependent.

## Notes and references


[1] B. Kayser, *The Physics of Massive Neutrinos* (World Scientific, Singapore, 1989), p. 11.

[2] B. Kaiser and R. N. Mohapatra, "The Nature of Massive Neutrinos," in *Current Aspects of Neutrino Physics*, edited by D. O. Caldwell (Springer, Berlin, 2001), pp. 17-38.
F. Boehm and P. Vogel, *Physics of Massive Neutrinos* (Cambridge U. Press, Cambridge, 1992).

[3] P. Vogel, "Double Beta Decay: Theory, Experiment and Implications," in *Current Aspects of Neutrino Physics*, edited by D. O. Caldwell (Springer, Berlin, 2001),p. 178.

[4] D. E. Groom, *et al.*, "Neutrino Mass," The European Physical Journal **C15**, 1 (2000).

[5] C. Guinti, "Theory of Neutrino Oscillations," arXiv:hep/ph/0409230 (20 Sep 2004).
[6] W. C. Haxton, "Neutrino Oscillations and the Solar Neutrino Problem," in *Current Aspects of Neutrino Physics*, edited by D. O. Caldwell (Springer, Berlin, 2001), pp. 65-88.
[7] Y. Wang and Z-z. Xing, "Neutrino Masses and Flavor Oscillations," arXiv:1504.06155 [hep-ph] 23 Apr 2015.
[8] Super-Kamiokande Collaboration, K. Abe, *et al.*, "Solar neutrino results in Super-Kamiokande-III," Phys. Rev. **D83**, 052010 (2011), arXiv:1010.0118.
[9] SNO Collaboration, A. W. P. Poon, "Neutrino Observations from the Sudbury Neutrino Observatory," arXiv:nucl-ex/0110005 (7 Oct 2001).
[10] J. G. Learned, "The Atmospheric Neutrino Anomoly: Muon Neutrino Disappearance," in *Current Aspects of Neutrino Physics*, edited by D. O. Caldwell (Springer, Berlin, 2001), pp. 89-130.
[11] MINOS Collaboration, P. Adamson *et al.*, "Measurement of the neutrino mass splitting and flavor mixing by MINOS," Phys. Rev. Lett. **106**, 181801 (2001), arXiv:1103.0340 (2 Mar 2011).
[12] Super-Kamiokande Collaboration, S. Fukuda, *et al.*, "Tau Neutrinos Favored over Sterile Neutrinos in Atmospheric Muon Neutrino Oscillations," Phys. Rev. Lett. **85**, 3999 (2000), arXiv:hep-ex/0009001 (1 Sep 2000).
[13] See Refs. 38-41 of **Chapter 1**.
[14] K. A. Olive *et al.*, "The number of light neutrino types from collider experiments," Chin. Phys. **C38**, 090001 (2014) (URL: http//pdg.lbl.gov)
[15] LSND Collaboration, A. Aguilar, *et al.*, "Evidence for Neutrino Ocillations from the Observation of $\overline{\nu}_e$ Appearance in a $\overline{\nu}_\mu$ Beam," Phys. Rev. D **64**, 112007 (2001). arXiv:hep-ex/0104049 (27 Apr 2001).
[16] Super-Kamiokande Collaboration, Y. Fukuda *et al.*, "Evidence for oscillation of atmospheric neutrinos," Phys. Rev. Lett. **81**, 1562 (1998); arXiv:hep-ex/9807003.
[17] K2K Collaboration, M. H. Ahn *et al.*, "Measurement of Neutrino Oscillation by the K2K Experiment,"Phys. Rev. **D 74**, 072003 (2006); arXiv:hep-ex/0606032.

[18] MINOS Collaboration, P. Adamson *et al.*, "Measurement of the neutrino mass splitting and flavor mixing by MINOS," Phys. Rev. Lett. **106**, 181801 (2011); arXiv:1103.0340 [hep-ex].
[19] The OPERA Collaboration, N. Agafonova *et al.*, "Discovery of tau neutrino appearance in the CNGS neutrino beam with the OPERA experiment," arXiv:1507.01417 [hep-ex] 6 Jul 2015.
[20] B. Pontecorvo, "Mesonium and anti-mesonium," J. Exptl. Theoret. Phys. **33**, 549 (1957) and Sov. Phys. JETP **6**, 429 (1957).
[21] B. Pontcorvo, "Inverse beta processes and nonconservation of lepton charge," J. Exptl. Theoret. Phys. **34**, 247 (1958) and Sov. Phys. JETP **7**, 172 (1958).
[22] Z. Maki, M. Nakagawa and S. Sakata, "Remarks on the Unified Model of Elementary Particles," Prog. Theor. Phys. **28**, 870 (1962).
[23] For a thorough review of experimental methods, findings and future prospects, see Y. Wang and Z-z. Xing, "Neutrino Masses and Flavor Oscillations," arXiv:1504.06155 [hep-ph] 23 Apr 2015.
[24] C. Guinti and C. W. Kim, *Fundamentals of Neutrino Physics and Astrophysics* (Oxford University Press, New York, 2007).
[25] M. Huang *et al.*, "Dependence of Mixing Angles and CP-Violating Phase on Mixing Matrix Parametrizations," arXiv:1108.3906 [hep-ph] 19 Aug 2011.
[26] R. Davis, Jr., D. S. Harmer and K. C. Hoffman, "Search for Neutrinos from the Sun," Phys. Rev. Lett. **20**, 1205 (1968).
[27] Super-Kamiokande Collaboration, Y. Fukuda *et al.*, "Measurements of the solar neutrino flux from Super-Kamiokande's first 300 days," Phys. Rev. Lett **81**, 1158 (1998). arXiv:hep-ex/9805021.
[28] SNO Collaboration, Q. R. Ahmad *et al.*, "Direct Evidence for Neutrino Flavor Transformation from Neutral-Current Interactions in the Sudbury Neutrino Observatory," Phys. Rev. Lett. **89**, 011301 (2002). arXiv:nucl-ex/0204008.
[29] SNO Collaboration, B. Aharmim *et al.*, "Electron Energy Spectra, Fluxes, and Day-Night Asymmetries of $^{8}B$ Solar

Neutrinos from the 391-day Salt Phase SNO Data Set," Phys. Rev. **C72**, 0555502 (2005). arXiv:nucl-ex/0505021.
[30] KamLAND Collaboration, S. Abe *et al.*, "Precision Measurement of Neutrino Oscillation Parameters with KamLAND," Phys. Rev. Lett. **100**, 221803 (2008). arXiv:0801.4589.
[31] Super-Kamiokande Collaboration, Y. Ashie *et al*, "Measurement of atmospheric neutrino oscillation parameters by Super-Kamiokande I," Phys. Rev. D **71**, 112005 (2005). arXiv:hep-ex/0501064.
[32] MINOS Collaboration, P. Adamson *et al.*, "An improved measurement of muon antineutrino disappearance in MINOS," Phys. Rev. Lett. **108**, 191801 (2012). arXiv:1202.2772.
[33] A. B. Balantekin and D. Yilmaz, "Contrasting solar and reactor neutrinos with a non-zero value of $\theta_{13}$," J. Phys. G **35**, 075007 (2008). arXiv:0804.3345 [hep-ph].
[34] G. L. Fogli, E. Lisi, A. Marrone, A. Palazzo and A. M. Rotuno, "Hints of $\theta_{13} > 0$ from global neutrino data analysis," Phys. Rev. Lett. **101**, 141801 (2008). arXiv:0806.2649 [hep-ph].
[35] Double Chooz Collaboration, Y. Abe *et al.*, "Indication for the disappearance of reactor antineutrinos in the Double Chooz experiment," Phys. Rev. Lett. **108**, 131801 (2012). arXiv:1112.6353.
[36] Daya Bay Collaboration, F. P. An *et al.*, "Observation of electron-antineutrino disappearance at Daya Bay," Phys. Rev. Lett. **108**,171803 (2012). arXiv:1203.1669.
[37] RENO Collaboration, Soo-Bong Kim *et al.*, "Observation of Reactor Electron Antineutrino Disappearance in the RENO experiment," Phys. Rev. Lett. **108**, 191802 (2012). arXiv:1204.0626.
[38] T2K Collaboration, K. Abe *et al.*, "Observation of Electron Neutrino Appearance in a Muon Neutrino Beam," Phys. Rev. Lett. **112**, 161802 (2014). arXiv:1311.4750.
[39] MINOS Collaboration, P. Adamson *et al.*, "Improved search for muon-neutrino to electron-neutrino oscillation in MINOS," Phys. Rev. Lett. **107**, 181802 (2011). arXiv:1108.0015.
[40] D. V. Forero, M. Tórtola, and J. W. F. Valle, "Neutrino oscillations refitted," arXiv:1405.7540 [hep-ph] 29 May 2014.

[41] LSND Collaboration, A. Aguilar *et al.*, "Evidence for Neutrino Oscillations from the Observation of $\bar{\nu}_e$ Appearance in a $\bar{\nu}_\mu$ Beam," Phys. Rev. D **64**, 112007 (2001). arXiv:hep-ex/0104049, 27 Apr 2001.
[42] MiniBooNE Collaboration, A. A. Aguilar-Arevalo, *et al.*, "Improved Search for $\bar{\nu}_\mu \rightarrow \bar{\nu}_e$ Oscillations in the MiniBooNE Experiment," Phys. Rev. Lett. **110**,161801 (2013). arXiv:1303.2588 [hep-ex] 12 Mar 2013.
[43] G. Mention, *et al.*, "The Reactor Antineutrino Anomaly," arXiv:1101.2755 [hep-ex] 23 Mar 2011.
[44] K. N. Abazajian, *et al.*, "Light Sterile Neutrinos: A White Paper," arXiv:1204.5379 [hep-ph] 18 Apr 2012.
[45] O. Yasuda, "Sensitivity to sterile neutrino mixings and the discovery channel at a neutrino factory," arXiv:1004.2388 [hep-ph] 14 Apr 2010.
[46] The U matrix is displayed unapproximated in A. Donini, *et al.*, arXiv:0812.3703 [hep-ph] 24 Jul 2009.
[47] J. Kopp, *et al.*, "Sterile Neutrino Oscillations," JHEP **1305**, 050 (2013). arXiv:1303.3011 [hep-ph] 27 Mar 2014.
[48] C. Giunti, *et al.*, "Pragmatic View of Short-Baseline Neutrino Oscillations," Phys. Rev. **D88**, 073008 (2013). arXiv:1308.5288 [hep-ph] 6 Nov 2013.
[49] M. G. Aartsen *et al.*, "Searches for Sterile Neutrinos with the IceCube Detector," Phys. Rev. Lett. **117**, 071801, 8 August 2016.

# Chapter 13

# Charged lepton mass formula

## 13.1 Introduction

The present chapter has two principal aims: to calculate the masses of the charged leptons and to explain why there are three generations of these fundamental particles.

In the Standard Model of electroweak interactions[1-4] the masses of the quarks and charged leptons are supposed to be generated by Yukawa coupling to the vacuum Higgs field $v$,.[5,6] However, because the coupling constants $g_f$ are arbitrary, the masses coupled to them are not predicted. To calculate the masses it seems that one must first find the coupling constants. So far, the many efforts to accomplish this have proved unsuccessful.[7] However, as pointed out in **Sec. 4.1**, the production of mass-like terms in the SM Lagrangian by spontaneous symmetry breaking does not prove that mass is generated by coupling to the Higgs field. Indeed, it is possible that the masses of the fundamental fermions are generated by a mechanism entirely unrelated to gauge symmetry. We suggest, in other words, that the fermions may come to the Lagrangian already equipped with mass. These preëxisting masses would then enter the Lagrangian via the coupling constants, $g_f = \sqrt{2}m_f / v$. Our objective here is to describe this previously undiscovered mechanism and show how it generates with reasonable accuracy the observed mass spectrum of the fundamental fermions. The significance of this result is underscored by some uncertainty as to whether the Higgs particle has actually been observed.[8] If the Higgs mechanism is not operating, then an alternative mechanism of some kind obviously must exist to give the fermions (and gauge bosons) mass.

Our calculation of the charged-fermion mass spectrum assumes that the fundamental fermions can be described as objects propagating in six-dimensional, external spacetime $M^4 \times \overline{M}^2$. As we showed in **Secs. 4.3** and **4.4** the equation of motion of the free charged fermion may be written

$$\left[\gamma^{\mu}\left(\hat{p}^{6}-\gamma^{5}\hat{p}_{5}\right)\hat{p}_{\mu}+\widehat{m^{2}c^{2}}\right]\chi=0 \qquad (13.1)$$

where $\hat{p}_a = i\hbar\partial_a$, $\widehat{m^2} = \left(\hat{p}_5^{\,2} - \hat{p}_6^{\,2}\right)/c^2$ is the mass-squared operator and the four-component spinor [see (4.39)]

$$\chi(x, x^5, x^6) = \psi(x)\exp\left[-i\left(p_5 x^5 - G x^6\right)/\hbar\right], \tag{13.2}$$

and where $G$ is the radius of the 5-momentum shell in 6-space. The demand that (13.1) remain invariant under 5-dimensional Lorentz transformations forbids infinitesimal rotations in the $\mu$-5 planes, implying invariance of the fifth, time-like component of momentum, $p_5$. Applying the mass-squared operator to spinor (13.2) we find for the rest mass $m$ of any charged lepton or free quark

$$m^2c^2 = G^2 - p_5^{\,2} \tag{13.3}$$

Let us consider the case of the charged leptons, for which [see (4.88)]

$$p_5 = G - m_e c \tag{13.4}$$

where $m_e$ is the mass of the electron, and for the electron itself, $p_5 = 0$ [see (4.74)]. In analogy with (13.4) we define for the general rest mass $m$

$$d = G - mc\,, \tag{13.5}$$

where $d$ is the invariant spacetime interval shown in **Fig. 13.1**. Eliminating $G$ between (13.4) and (13.5) we obtain

$$m = m_e + \Delta U/c^2\,, \tag{13.6}$$

Where $\Delta U = (p_5 - d)c$. Equation (13.6) dictates the direction of our method of calculation. It states that the extra mass of any charged lepton above that of the electron derives from the difference $\Delta U$ between the invariants $p_5$ and $d$. We shall show that this difference is in fact electromagnetic in origin. As the electromagnetic form of $\Delta U$ turns out to be a function of mass $m$, relation (13.6) ultimately yields an algebraic expression for the lepton mass spectrum.

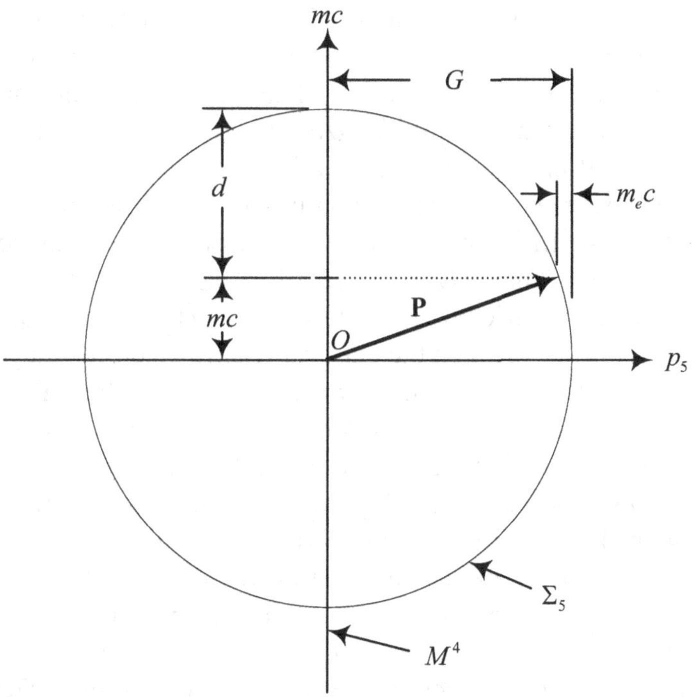


**Fig. 13.1.** Five-momentum vector $\mathbf{P} = \mathbf{e}_a p^a = \mathbf{e}_\mu p^\mu + \mathbf{e}_5 p^5$ describing motion of free charged lepton in $M^4 \times \bar{M}^2$. Unit vectors $\mathbf{e}_a$ satisfy $\mathbf{e}_a \cdot \mathbf{e}_b = g_{ab}$, ($a = 0,1,2,3,5$). As shown, resultant vector $\mathbf{P}$ and its transverse component $\mathbf{e}_\mu p^\mu$ lie in the plane of the paper. Vector $\mathbf{e}_\mu p^\mu$ is of length *mc*, and the vertical axis is so labeled. Vector $\mathbf{P}$ is of length $p^6 = G$, its tip lying in the 5-D momentum shell $\Sigma_5$. $M^4$ = 4-D hyperplane. Invariant interval $d = G - mc$.

## 13.2 Charged lepton mass and self-energy

To carry out this program we are going to identify $\Delta U$ with the difference between the *classical* electromagnetic self-energies of the isolated lepton computed in two different coordinate frames in 6-dimensional $M^4 \times \bar{M}^2$. In light of the successes of mass renormalization $m = m_0 + \delta m$ in quantum electrodynamics,[9] our reliance on classical self-energy to solve the mass hierarchy problem may seem at best an anachronism, and at worst, at odds with the field-theoretic picture. But that is not the case, for the following reasons. (1) The self-energy $\delta m$ is a radiative correction, in magnitude very much smaller than the uncorrected, bare mass $m_0$. But $\Delta U / c^2$, viewed as a correction term, is very much *larger* than the uncorrected mass $m_e$ appearing in (13.6). Clearly, then, $\Delta U$ lies far beyond the reach of mass renormalization, and probably beyond field-theoretic methods generally. (2) $\delta m$ is unobservable. Thus, even if it could somehow be made much larger in magnitude, $\delta m$ could not be identified with the observable self-energy difference $\Delta U$. (3) $\delta m$ is defined in four dimensions, whereas $\Delta U$ is defined in six. The difference is decisive; for it is from the fifth dimension $x^5$ that most of the energy comprising $\Delta U$ comes. (4) That $\Delta U$ is essentially classical in nature is already shown by its geometric representation, $(p_5 - d)c$. In sum, in our view, there is no conflict between the classically defined electromagnetic self-energy difference $\Delta U$ and the $\delta m$ from quantum field theory. Like the classical and quantum versions of mechanics, they are simply different entities, with different areas of relevance and applicability.

We define the classical self-energy as follows. Starting with the Lagrangian density given by (10.55), we can define for the charged lepton a mechanical energy-momentum tensor $\Theta^{\mu\nu}$. For a particle at rest, the space integral over $\Theta^{00}$ yields the mechanical rest energy, $mc^2$. Similarly, one may also define a stress-tensor $T^{\mu\nu}$ for the electromagnetic field generated by the charged lepton. The integral over $T^{00}$ yields the energy $U$ of the surrounding electromagnetic field. The total energy, or classical *self-energy*, *W*, of the isolated lepton is given by the sum of the mechanical and electromagnetic energies:[1]

$$W = mc^2 + U \tag{13.7}$$

Now the total self-energy is not an observable, because *U* is not. No experiment can distinguish between the mechanical and electromagnetic terms on the r. h. s. of this expression.[11] It is nonetheless formally well-defined, and if we could find a second, independent expression for *W*, one also involving mass *m*, we could then equate the two expressions, providing a single equation in the unknown *m*. A second expression for *W* is in fact available, and is determined as follows.

In **Sec. 4.3** we argued that the 6-dimensional Dirac equation (4.7) (with $\bar{\gamma}^5 = \gamma^5$ and $\gamma^6 = I$)

$$\left(\gamma^{\mu}\hat{p}_{\mu}+\gamma^{5}\hat{p}_{5}+I\hat{p}_{6}\right)\chi=\left(\gamma^{\mu}\hat{p}_{\mu}+\gamma^{5}p_{5}-G\right)\chi=0 \tag{13.8}$$

could not be used to describe the motion of charged leptons because its associated Hamiltonian is, except when $p_5=0$, non-Hermitian. We were then led to adopt the modified equation of motion (13.1), whose associated Hamiltonian is properly Hermitian. We observe, however, that (13.8) remains form-invariant under all 5-dimensional Lorentz transformations, including those forbidden by (13.1). In consequence, it is possible, for plane-wave solutions of (13.8), to choose a new coordinate frame $K'(x_0',\mathbf{x}',x_5')$, generated by a *finite* (i.e., non-infinitesimal) rotation of the original coordinate frame $K(x_0,\mathbf{x},x_5)$, such that in $K'$ the fifth component of momentum vanishes: $p_5\rightarrow p_5'=0$. Assuming $K'$ and the charged lepton of mass *m* both to be at rest with respect to *K*, , i.e., $\mathbf{p}'=\mathbf{p}=0$, the coordinate transformation required is the real, circular rotation in the $x^0$- $x^5$ plane shown in **Fig. 13.2**:

$$\left.\begin{aligned} x_0' &= x_0\cos\vartheta + x_5\sin\vartheta \\ \mathbf{x}' &= \mathbf{x} \\ x_5' &= -x_0\sin\vartheta + x_5\cos\vartheta \end{aligned}\right\}, \tag{13.9}$$

where $\sin\vartheta=p_5/G$. This yields in place of (13.8) the transformed equation of motion

$$\left(\gamma^{\mu}\hat{p}_{\mu}'-G\right)\chi'=0, \tag{13.10}$$

whose associated Hamiltonion in $K'$ is now Hermitian. Our objective is to define in $K'$ a self-energy $W'$ corresponding to the self-energy (13.7) defined in *K*. This requires some care because, as emphasized in *Sec. 3.9.6*, coordinate $x^5$ represents not a second time dimension, but an extra spatial dimension whose metrical signature happens to be time-like. Thus $x_0'$ as given in (13.9) does not represent a pure time coordinate, but rather a mixture of time ($x_0$) and space ($x_5$). And correspondingly, the momentum-space coordinate $p_0'$ does not represent pure energy, nor does the constant *G* in (13.10) represent pure rest mass. For a particle at rest in $K'$ ($\mathbf{p}'=0$), the coordinate $p_0'$ is given by

$$p_0'=G=m_e c+p_5, \tag{13.11}$$

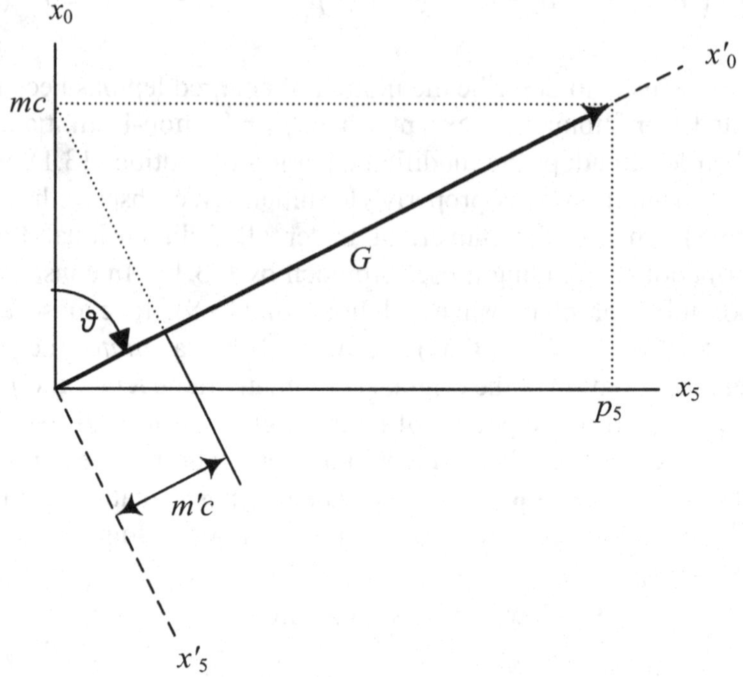


**Fig. 13.2.** Reference frame $K'(x_0', \mathbf{x}', x_5')$ is rotated by real angle $\vartheta$ from reference frame $K(x_0, \mathbf{x}, x_5)$. Vector of length $G$ represents the 5-momentum vector of a charged lepton of mass $m$ at rest in both $K$ and $K'$ ,i.e., $\mathbf{p}' = \mathbf{p} = 0$. Interval $m' = m\cos\vartheta$ denotes the projection of interval $m$ from the $x_0$ axis of $K$ onto the $x_0'$ axis of $K'$.

where we have used (13.4). This form suggests that, for a particle at rest in $M^4$, the momentum part of $p_0'$ is $p_5$ and the energy part is $m_e c$. In other words, the rest energy in $K'$ is $m_e c^2$. This can be seen in another way. Take a particle at rest in $K$ and $K'$, and consider the normal projection $m'$ of rest mass $m$ from the $x_0$ axis of $K$ onto the $x_0'$ axis of $K'$ (see **Fig. 13.2**). Using (13.3) and (13.4), the projection $m' = m\cos\vartheta$ $= m^2 / G$ can be manipulated into the form

$$m' = \frac{m_e}{1 - \dfrac{m_e}{m}\dfrac{v_5}{c}}, \tag{13.12}$$

where $v_5 = p_5/m$, the fifth component of velocity [see Eq. (10.20)]. Thus, contrary to what one might expect, the projection $m'$ is a relativistic mass, not a rest mass; the rest mass in this expression is $m_e$. Hence corresponding to (13.7), we have for the self-energy in $K'$:

$$W' = m_e c^2 + U', \tag{13.13}$$

where $U'$ is the energy of the classical electromagnetic field generated by the charged lepton in frame $K'$.

Our final step consists in equating the expressions for $W$ and $W'$ given by (13.7) and (13.13). This yields (13.6) with $\Delta U$ now given by the electromagnetic self-energies in $K'$ and $K$:

$$\Delta U = U' - U. \tag{13.14}$$

It should be emphasized that the reference frames $K$ and $K'$ represent physically distinct realms of spacetime, permanently separated by a real, finite rotation angle, $\vartheta$. Each realm even demands its own wave equation, specifically (13.1) and (13.10), equations that cannot be transformed into one another by infinitesimal rotation. In such a setting one can only *assume* the equality of $W$ and $W'$ as we have done to obtain (13.14). But as the two total self-energies simply refer to the same particle observed at rest in two different frames of reference, the assumption would seem to be a good one.

## 13.3 Self-energy in the *K*-frame

We are well-positioned to evaluate the self-energies $U$ and $U'$, having studied already in **Chapter 6** the extension of Maxwell electrodynamics to 6-dimensional spacetime. The energy density $T^{00} = \varpi_0 c$ of the electromagnetic field generated by a charged lepton in frame $K$ is given by Eq. (6.30). The fields $\mathbf{E}$, $\mathbf{B}$, $E_5$ and $\mathbf{B}_5$ on which $T^{00}$ depends are obtained by differentiating the 5-potential $A^a$ in accordance to (6.23). The $A^a$ are the retarded potentials defined by (6.39) and (6.40). To evaluate these we first need to insert the 5-current components $J^a$, and therein lies a problem.

The current components $J^a$ are given by Eqs. (10.44), (10.52) and (10.53), and the internal distribution $D_r$ by (10.35). Unfortunately $D_r$

is analytically too complicated to permit an exact evaluation of the retarded potentials. Now the reason $D_r$ is complicated is that it reflects the classical helical structure of the heavier leptons $\mu$ and $\tau$ as depicted in **Fig. 10.3**. If we could ignore the helical structure, replacing the helix by its centerline, then the internal structures of $\mu$ and $\tau$ would presumably resemble that of the electron, apart from radius $a_\ell$. One could then employ instead of (10.35) the very much simpler form of $D_r$, Eq. (10.37a):

$$D_r(\xi) = -\frac{1}{4\pi\rho}\delta'(\rho - a_\ell), \tag{13.15}$$

with $a_\ell$ given by (10.16). The question is, is this permissible? According to the data presented in **Table 10.1**, the helices in question appear as thin, thread-like structures, closely approximating (except for orbital radius $a_\ell$) the purely circular orbit of the electron's Zitterbewegung. Thus one can assume that the error in adopting (13.15) for all charged leptons is small. Our final results will bear out this assumption.

Because the point charge comprising a charged lepton undergoes Zitterbewegung, the instantaneous currents appearing in (6.39) and (6.40) are in principle unspecifiable. The best one can do is to define the internal currents in a time-averaged sense, as we did to obtain the current components (10.44), (10.52) and (10.53) [see comment following Eq. (10.44)]. The time-averaged currents, when used in (6.39) and (6.40) in place of the instantaneous currents, yield time-averaged potentials. Carrying out the indicated integrations using (13.15), we obtain for charged lepton at rest in $K$ the potentials

$$A_0 = -\frac{\mu_0}{4\pi}\frac{ec}{r} \tag{13.16.a}$$

$$\mathbf{A} = -\frac{\mu_0}{4\pi}\boldsymbol{\mu}\times\nabla\left(\frac{1}{r}\right) \tag{13.16.b}$$

$$A_5 = -\frac{\mu_5}{4\pi}\frac{p_5}{m}\frac{e}{r} \tag{13.16.c}$$

where $r = |\mathbf{x}| \le a$, $\boldsymbol{\mu} = (-e\hbar/2m)\hat{\mathbf{s}}$ is the spin magnetic moment and $a = \hbar/2mc$. All potential components vanish interior to the volume of the Zitterbewegung, $r<a$, and thus remain finite as long as $a$ is finite. (Note that we omit unnecessary subscripts on both $m$ and $a$.)

The static fields $\mathbf{E}$, $\mathbf{B}$, $E_5$ and $\mathbf{B}_5$, obtained by differentiating potentials (13.16) via (6.23), are given by

$$\left.\begin{aligned} \mathbf{E} &= -\frac{e}{4\pi\varepsilon_0}\frac{\mathbf{x}}{r^3} \\ \mathbf{B} &= \frac{\mu_0}{4\pi}(\boldsymbol{\mu}\cdot\nabla)\frac{\mathbf{x}}{r^3} \\ E_5 &= 0 \\ \mathbf{B}_5 &= -\frac{\mu_5}{4\pi}\frac{ep_5}{m}\frac{\mathbf{x}}{r^3} \end{aligned}\right\},\ r\geq a \tag{13.17}$$

where the dielectric constant $\varepsilon_0 = 1/\mu_0 c^2$. Note that the derivatives of the potentials are all taken outside the volume of the Zitterbewegung, $r\geq a$. Strictly speaking, the fields $\mathbf{E}$, $\mathbf{B}$ and $\mathbf{B}_5$ interior to that volume are undefined.

From (13.17) we note also that field $\mathbf{B}_5$ is proportional to field $\mathbf{E}$:

$$\mathbf{B}_5 = \frac{\mu_5 p_5}{m}\cdot\varepsilon_0\mathbf{E}. \tag{13.18}$$

Consequently, the final term of the energy density $T^{00} = \varpi_0 c$, Eq. (6.30), can be written in terms of the fifth component of the electromagnetic momentum density, $\varpi_5$, Eq. (6.35):

$$\begin{aligned} \varpi_5 &= \varepsilon_0\mathbf{E}\cdot\mathbf{B}_5 \\ &= \frac{\mu_5 p_5}{m}(\varepsilon_0\mathbf{E})^2 \end{aligned} \tag{13.19}$$

Inserting the fields (13.17) into (6.30) and integrating over the infinite space $r\geq a$ we obtain the finite electromagnetic self-energy

$$U = \alpha mc^2 + \frac{2}{3}\alpha mc^2 + \frac{p_5}{2m}\int\varpi_5 dV$$

$$=\frac{5}{3}\alpha mc^2+\frac{p_5}{2m}p_5^{\mathrm{em}}, \tag{13.20}$$

where $\alpha = e^2/4\pi\varepsilon_0\hbar c$ is the fine-structure constant, $p_5^{\mathrm{em}} = \int \varpi_5 dV$ is the fifth component of the *electromagnetic* momentum, and $p_5$ is the fifth component of the *mechanical* momentum, Eq. (13.4). The first term on the r. h. s. of the first line of (13.20) is the static electrical self-energy and the second term is the magnetic self-energy due to the magnetic dipole induced by spin. The third term is the self-energy coming from the field generated by potential $A_5$; it is proportional to the product of the fifth components of the mechanical and electromagnetic momenta. This third term is positive, provided $\mu_5 > 0$. However, the value of parameter $\mu_5$ is not known, and so $p_5^{\mathrm{em}}$ is not known either. We shall undertake a wave-mechanical determination of $p_5^{\mathrm{em}}$ shortly.

## 13.4 Self-energy in the $K'$-frame

Having found the form of self-energy $U$ in the $K$-frame, it is easy to deduce the corresponding self-energy $U'$ in the $K'$-frame. The equation of motion (13.10) of any charged lepton in $K'$ is structurally similar to the equation of motion (13.1) of an electron in $K$. The self-energy of the electron in $K$ is, from (13.20) (with $m = m_e$ and $p_5 = 0$), $U = (5/3)\alpha m_e c^2$. Thus, putting $m_e c \to G$ we obtain for the electromagnetic self-energy $U'$ of the generic charged lepton in the $K'$-frame

$$\begin{aligned} U' &= \frac{5}{3}\alpha Gc \\ &= \frac{5}{3}\alpha\frac{m^2+m_e^2}{2m_e}c^2 \end{aligned}, \tag{13.21}$$

where we have expressed $G$ in terms of $m$ utilizing (13.3) and (13.4) [see also (4.91)]. There is no contribution to $U'$ corresponding to the second term in the second line of (13.20), because, in $K'$, we have $p_5' = 0$. Note that the radius $a'$ of the Zitterbewegung in $K'$ is $\hbar/2G$, a dimension which, for all charged leptons heavier than the electron, is orders of magnitude smaller than the corresponding radius in $K$.

## 13.5 Energy spectrum of an electron trapped in inner spacetime $N^2$

We now have in hand the general forms of the self-energies *U* and $U'$. It remains, however, to put the factor $p_5^{\text{em}}$ appearing in (13.20) in terms of known quantities. Because we are seeking a *spectrum* of mass values *m*, it is clear that the momentum $p_5^{\text{em}}$ must somehow be quantized. This suggests that we look for an operator having $p_5^{\text{em}}$ as its eigenvalue. To see how such an operator might arise, let us examine more closely the structure of (13.20). The strength of the combined electric and magnetic parts of the self-interaction energy *U* is, relative to energy $mc^2$, expressed by the electromagnetic coupling constant $5\alpha/3$. Similarly, the strength of the part of *U* contributed by potential $A_5$ is, relative to energy $p_5^{\text{em}}c$, expressed by the coupling constant $p_5/2mc$. The energies $mc^2$ and $p_5^{\text{em}}c$ thus play comparable roles in the expression for *U*, although they entail different coupling constants. Now, for the particle at rest in *K*, the energy $mc^2$ is an eigenvalue of the Dirac Hamiltonian. This suggests that the operator we are looking for is a second Hamiltonian, one whose eigenvalue is the energy $p_5^{\text{em}}c$. But in *K* there exists only one Hamiltonian, namely the Dirac Hamiltonian, generalized to six dimensions. Thus, if a second Hamiltonian exists at all, it must come from still higher dimensions.

But we know already of a manifold of still-higher dimensions, namely the internal Minkowskian spacetime $N^2$ invoked in **Sec. 10.7** to account for the negative probability density of the lepton's Zitterbewegung. This 2-dimensional inner spacetime is parameterized by one temporal dimension $u_0 = c\tau$ and one bounded spatial dimension *u*, $-W/2 \le u \le W/2$. Its product with 6-D external spacetime generates the 8-D slab geometry $M^4 \times \bar{M}^2 \times N^2$ shown in **Fig. 10.12**. Our discussion of the Zitterbewegung in **Sec. 10.7** already has the charged leptons performing classical oscillatory motion within $N^2$, as depicted in **Figs. 10.13, 14** and **16**. We propose now to describe this motion wave-mechanically.

Our main task is to find an equation motion. The first step in this direction is to identify the particle whose motion the equation represents. There is only one plausible candidate, namely, the electron, the lightest of the charged leptons. We imagine it trapped in a one-dimensional well of width *W* in $N^2$. The internal states of the heavier leptons can then be identified with excited states of the trapped electron, and their masses $m$ related to the eigenvalues $E = p_5^{\text{em}}c$ of the internal Hamiltonian.

Second, we need to specify the potential well and the boundary conditions at the walls. Such specifications are of course conjectural, as we have no direct access to the internal spacetime. Nevertheless, we can find physically meaningful solutions by assuming a square confining well of width $W = h/2m_ec$ and depth $V_0$, as shown in **Fig. 13.3**, and by adopting *classical* boundary conditions:

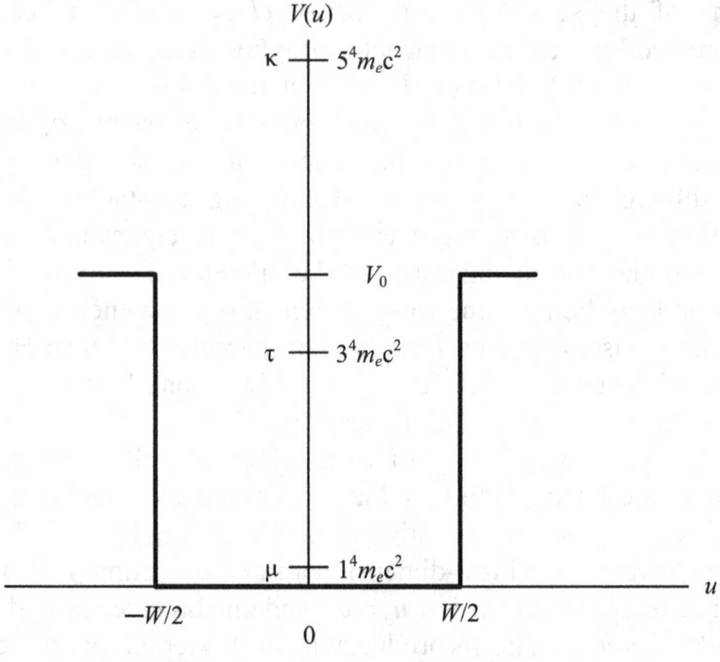


**Fig. 13.3** One-dimensional square well of width $W$ and depth $V_0$ in the internal subspace $N^2$. Energy levels $\mu$ and $\tau$ correspond to leptonic bound states of an electron trapped in the well. Energy level $\kappa$ lies above $V_0$ and is unsupported by the well.

(i) For $0 \le E \le V_0$, the electron in the well is absolutely confined (perfectly rigid walls), i.e., the state function vanishes at the walls $|u| = W/2$.

(ii) For $E > V_0$ the electron is free and does not feel the presence of the well.

Third, we need to identify the source of the internal equation of motion. This is not hard to do. For although we are asking for two equations of motion, one external and one internal, there is only one available wave function, namely the function $\chi$ defined by (13.2). Thus the internal equation of motion must somehow emerge from the external equation of motion, (13.1). Obviously for this to occur, $\chi$ must, in addition to the six external coordinates $(x^{\mu},\ x^5,\ x^6)$, also be a function of the internal coordinates $(u_0,\ u)$. Assuming this to be the case, $\chi$ then spans 8-D $M^4 \times \bar{M}^2 \times N^2$. The only things needed now are internal operators to act on the internal coordinates of $\chi$. To this end, we define two mutually commuting Hermitian operators $\hat{\omega}$ and $\hat{\Omega}$ acting on $\chi$ through the coordinates $u_0$ and $u$, respectively. Then in wave equation (13.1) we make the operator replacement

$$\hat{p}_6 \to \hat{p}_6 + \hat{\omega} - \hat{\Omega}, \tag{13.22}$$

so that our single wave equation (13.1) now embodies operators acting on all eight coordinates of $M^4 \times \bar{M}^2 \times N^2$. Now the charged leptons are known to be states of definite mass; their wave functions are eigenstates of the mass-squared operator with corresponding eigenvalues $m^2 = (G^2 - p_5^2)/c^2$. Hence for the charged leptons we have the condition

$$\left(\hat{\omega} - \hat{\Omega}\right)\chi = 0. \tag{13.23}$$

We recall in passing that the flavor neutrino, although not a particle of definite mass, nevertheless suffers Zitterbewegung, and like the charged lepton, performs an oscillatory motion within $N^2$. Thus condition (13.23) applies to the neutrino as well; otherwise the neutrino mass would not sum to 0.

Now we have to define the operators $\hat{\omega}$ and $\hat{\Omega}$. They should provide from (13.23) a conserved probability current density in $N^2$. We recall that, as noted in *Sec. 10.7.1*, there exists no upper bound on velocity in $N^2$. This eliminates operators giving an equation of the Klein-Gordon form, which is a relativistic equation, and which anyway is not guaranteed to produce positive-definite probability density. We conclude that in (13.23) we must have

$$\left.\begin{aligned} \hat{\omega} &= i\hbar\frac{\partial}{\partial u_0} \\ \hat{\Omega} &= -\frac{\hbar^2}{m_e c}\frac{\partial^2}{\partial u^2} \end{aligned}\right\}, \tag{13.24}$$

which together yield an equation of the Schrödinger form. We then have a conserved probability density in $N^2$ with components

$$\left.\begin{aligned} P &= |\chi|^2 \\ S &= -\frac{i\hbar}{m_e}\left(\chi^*\frac{\partial\chi}{\partial u} - \frac{\partial\chi^*}{\partial u}\chi\right) \end{aligned}\right\}. \tag{13.25}$$

One might think that in (13.23) we now have our equation of motion, but that is incorrect. For if we put $\hat{\omega}\chi = 0$, corresponding to the ground state of the trapped electron, and also to the state of the flavor neutrino, then to satisfy boundary condition (i) above, we must have $\chi = 0$. But this means, absurdly, that the particle does not exist. Equation (13.23) suffices to calculate probability density components $P$ and $S$, but does not constitute an equation of motion as such.

To generate a proper equation of motion we define $\chi$ in terms of a potential function, $\phi$:

$$\chi = \left(\hat{\omega} + \hat{\Omega}\right)\phi. \tag{13.26}$$

Putting this into (13.23) and dividing by $m_e$ we obtain a fourth-order equation in $\phi$ of the form

$$\hat{H}\phi = \hat{E}\phi, \tag{13.27}$$

where $\hat{H}$ is the Hamiltonian operator

$$\hat{H} = \frac{\hat{\Omega}^2}{m_e} = \frac{\hbar^4}{m_e^3 c^2}\frac{\partial^4}{\partial u^4} \tag{13.28}$$

and $\hat{E}$ is the energy operator

$$\hat{E} = \frac{\hat{\omega}^2}{m_e} = -\frac{\hbar^2}{m_e}\frac{\partial^2}{\partial u_0^2}. \tag{13.29}$$

Although quantum equation (13.27) appears unorthodox, the theory does provide a conventional probabilistic interpretation in terms of the wave function $\chi$ and current components (13.25). However, it is the function $\phi$, not $\chi$, that determines the energy eigenvalues *E*. In particular, setting $\hat{E}\phi = 0$ we obtain from (13.27) the ground state wave function

$$\phi(u) = A\left(\frac{W^2}{4} - u^2\right), \tag{13.30}$$

where *A* is a normalization constant. This inverted parabola is the state function of the unexcited electron, with $W = h/2m_e c$. It is also the ground state of flavor neutrino $\nu_\ell$, with $W = h/2G_\ell$ [see (10.9) and (10.139)]. The corresponding probability amplitude is, from (13.26), $\chi(u) = \text{const.}$, $|u| \le W/2$, confirming that the electron is in the well, even in the unexcited ground state. This physically plausible result suggests that we have arrived at a good equation of motion. We proceed now to solve it and get the energy eigenvalues *E*.

To do this we note first that because $\phi$ never undergoes the squaring operation, it must stand on its own as a physical descriptor. This has two significant consequences. First, $\phi$ must be real-valued. Second, it must be symmetrical in $u_0$, because there is no preferred direction in time. Equation (13.27) thus admits solutions of the form

$$\phi(u_0, u) = \Phi(u)\cos\beta u_0, \tag{13.31}$$

where $\Phi$ is a real-valued function of *u* and $\beta = (m_e E)^{1/2}/\hbar$. This gives the one-dimensional eigenvalue equation

$$\frac{\hbar^4}{m_e^3 c^2}\frac{d^4\Phi}{du^4} = E\Phi. \tag{13.32}$$

For the symmetrical well assumed above, (13.32) yields both even and odd parity solutions. However, just as there is no preferred direction in time, there is no preferred direction in space, and *all physically realizable (observable) charged-lepton states must have even parity.* This condition screens out as unobservable one-half the mass spectrum generated by (13.32). However, we do *not* discard the solutions having odd parity, as one of them, despite being unobservable, will figure crucially in our calculation of the $c$ and $t$ quark masses in the next chapter.

The even parity solutions of (13.32) are

$$\Phi(u) = A\cos\frac{n\pi u}{W}, \qquad n = 1,\ 3,\ 5\ \cdots \tag{13.33}$$

and the odd parity solutions are

$$\Phi(u) = B\sin\frac{n\pi u}{W}, \qquad n = 2,\ 4,\ 6\ \cdots \tag{13.34}$$

In either case the energy $E$ is given by

$$E = \left(\frac{\lambda_C}{2W}\right)^4 n^4 m_e c^2 . \tag{13.35}$$

Now in Sec. *10.7.1*, Eq. (10.140), we concluded on the basis of Zitterbewegung dynamics that the well-width $W = \lambda_C / 2$, which fixes the energy eigenvalues (13.35). Before going any further it is of interest to show how this same value arises from the quantum dynamics developed in the present chapter. To determine it we first calculate the average momentum of the bound electron. From (13.26), (13.31) and (13.33) the normalized probability amplitude is

$$\chi(u_0, u) = \sqrt{\frac{2}{W}}\cos\frac{n\pi u}{W}\cdot e^{-i\beta u_0} . \tag{13.36}$$

This amplitude $\chi$ is in fact an eigenstate of the momentum operator $\hat{\Omega}$, with eigenvalue $p$:

$$p = \langle \hat{\Omega} \rangle = \left( \frac{\lambda_C}{2W} \right)^2 n^2 m_e c, \quad n = 1,\ 3,\ 5 \cdots \tag{13.37}$$

Note that, because $\hat{\Omega}$ involves the second-order derivative, the average momentum does not vanish as it does for a particle in a symmetrical well in ordinary 3-space. This is consistent with the momentum distribution displayed in **Fig. 10.14**. The momentum $p$ is evidently that of an electron executing cyclic, back and forth motion within the well, the linear concomitant of the electron's orbital Zitterbewegung in 3-space. For the first excited state $(n = 1)$ it is natural to assume that the path length of one complete cycle of back-and forth movement is equivalent to the Compton wavelength of the electron; i.e., $2W = \lambda_C$. Then $p = m_e c$ and the phase integral $\oint p dq = m_e c \cdot 2W = h$. So the peculiar quantum formalism developed in this chapter does indeed reproduce the well width found in **Chapter 10**.

The bound-state energies (13.35) can now confidently be assumed to be

$$E = n^4 m_e c^2 . \tag{13.38}$$

## 13.6 Lepton mass formula

With the identification $E = p_5^{\text{em}} c$, (13.20) and (13.38) give for the self-energy in the $K$-frame:

$$\begin{aligned} U &= \frac{5}{3} \alpha m c^2 + \frac{p_5 m_e c}{2m} n^4 \\ &= \frac{5}{3} \alpha m c^2 + \frac{m^2 - m_e^2}{4m} c^2 n^4 \end{aligned}, \tag{13.39}$$

where we have used the relation $p_5 = (m^2 - m_e^2) c / 2m_e$, obtained by eliminating $G$ between (13.3) and (13.4). Inserting (13.39) and (13.21) into (13.14), we obtain from (13.6) a cubic equation in the lepton mass $m$:

$$(m - m_e) \left( \frac{10\alpha}{3} m^2 - \gamma m_e m - n^4 m_e^2 \right) = 0 \tag{13.40}$$

where

$$\gamma = n^4 + 4 + \frac{10\alpha}{3}. \qquad (13.41)$$

One solution of (13.40) is $m = m_e$; in this theory the electron mass is an input parameter whose numerical value is not predicted. The quadratic piece of (13.40) has one positive root:

$$m = \frac{3}{20\alpha}\left(\gamma + \sqrt{\gamma^2 + \frac{40\alpha}{3}n^4}\right)m_e, \qquad n = 1,\ 3,\ 5\cdots \qquad (13.42)$$

This, finally, is our formula for the charged lepton mass spectrum, in units of $m_e$. It is a function solely of the fine structure constant $\alpha$ and well energy quantum number *n*.

## 13.7 Conclusion: the charged lepton mass spectrum

Formula (13.42) yields the calculated masses $m_{\ell\text{calc}}$ given in **Table 13.1**.

**Table 13.1** Calculated vs. average experimental values of the charged-lepton masses, $m_{\ell \neq e}$. Calculated masses are obtained from Eq. (13.42) with $m_e = 0.511\ \text{MeV}/c^2$ and $\alpha = 7.297 \times 10^{-3}$. Average experimental masses are those cited in Ref. 12 (Particle Data Group). Odd-parity particles (in parentheses) do not exist and are listed for reference only. Even-parity particles below the heavy line, e.g. $\kappa$, also do not exist, as they are unsupported by the finite well depth. See **Fig. 10.3**.

| Internal energy index, *n* | Parity of internal wave function $\Phi(u)$ | Flavor of charged lepton $\ell$ | Calculated mass, $m_{\ell\text{calc}}$ $(\text{MeV}/c^2)$ | Average experimental mass, $m_{\ell\exp}$ $(\text{MeV}/c^2)$ | $\%\Delta = \frac{m_{\ell\text{calc}} - m_{\ell\exp}}{m_{\ell\exp}} \times 100$ |
|---|---|---|---|---|---|
| 1 | Even | $\mu$ | 105.651 | 105.658 | −0.007 |
| 2 | Odd | $(\sigma)$ | (421.071) | Non-existent | ----- |
| 3 | Even | $\tau$ | 1786.6 | 1776.82 | 0.550 |
| 4 | Odd | $(\iota)$ | (5463.0) | Non-existent | ----- |
| 5 | Even | $\kappa$ | 13 214.8 | Non-existent | ----- |
| ⋮ | ⋮ | ⋮ | ⋮ | ⋮ | |

We conclude as follows:

(1) The calculated muon and tau masses, $m_{\mu\text{calc}} = 105.651\ \text{MeV}/c^2$ and $m_{\tau\text{calc}} = 1786.6\ \text{MeV}/c^2$, are in reasonable agreement with the currently-accepted average experimental values.[12] The small differences $\Delta = m_{\ell\,\text{exp}} - m_{\ell\text{calc}}$ are in all likelihood due to the approximate form (13.15) adopted for the spatial density of the Zitterbewegung—although one cannot be certain of this explanation. It should be emphasized that the mass formula (13.42) contains no free parameters, nor does it rely on field-theoretic ideas. So there is little if any room for *ad hoc* adjustment of our results.

(2) After $\tau$ the next even-parity lepton in the series, $\kappa$, with a predicted mass of 13.2 $\text{GeV}/c^2$, is not observed, nor are any other leptons observed more massive than $\tau$. This apparent termination of the series at $\tau$ is explained if the well depth $V_0$ lies in the range [use (13.38) with $n = 3,\ 5$]

$$3^4 m_e c^2 \le V_0 \le 5^4 m_e c^2 . \tag{13.43}$$

As we can see from **Fig. 10.3** a well of this limited depth supports no bound states more massive than $\tau$. Why $V_0$ should have the particular value it does, terminating the series at $\tau$, is not known. The existence of three generations of leptons, as dictated by the ratio $V_0 / m_e c^2$, appears to be an accident of spacetime geometry.

(3) The lepton labeled $\sigma$ has odd parity and is thus unobservable. In particular it does not couple to the photon or get produced in $e^- + e^+$ experiments. Nevertheless, as we shall see in the next chapter, this unobservable state is, as mentioned above, essential to our calculation of the quark masses $m_c$ and $m_t$. For that reason one might more accurately refer to $\sigma$ as the "hidden" (or "dark") lepton. It exists only in the sense that it connects to the quark masses. The unobservable odd-parity lepton, $\iota$, has no apparent function and is listed for reference only.

(4) Expanding (13.42) in powers of $\alpha$, we obtain to order $\alpha^{-1}$ the approximate form

$$m \cong \frac{3m_e}{10\alpha}\left(n^4+4\right), \quad n=1,\ 3,\ \cdots \tag{13.44}$$

For $n=1$ we obtain $m_\mu \cong 3m_e/2\alpha$, confirming an approximate mass formula noted long ago on empirical grounds.[13,14] An iterative formula yielding masses similar to those given by (13.42) was obtained by Barut[15] by assuming that the quantized energy of the magnetic self-interaction of any charged lepton, when added to its rest energy, gives the mass of the next lepton in the series. This procedure does not predict the hidden odd-parity lepton $\sigma$, and consequently sheds no light on the origin of the quark masses, nor does it suggest any mechanism for terminating the series at $\tau$.

(5) According to the Standard Model of particle physics, the electron acquires its inertial mass $m_e$ by interaction with the scalar Higgs field; see comments in *Sec. 4.4.3*. However, Eq. (10.140)—and the argument following Eq. (13.37)—shows that the mass of the electron is determined not by quantum field theory but by spacetime geometry. For we can write that equation in the form

$$m_e = \frac{h}{2Wc}. \tag{13.45}$$

Thus, on the assumption that geometry is fundamental, and that the constants $h$ and $c$ are externally determined, we see that the mass of the electron is fixed by a geometric feature of inner spacetime $N^2$, namely, its width $W$. To account for the observed mass of the electron, one has only to assume that $W=1.213\times10^{-12}$ m. In addition, the connection (13.45) between $m_e$ and $W$ explains why all electrons have the same mass. Worth mentioning also is the fact that a non-observable, extra-dimensional width $\Delta u = W$ gives rise to an observable physical mass, $m_e$. This is perhaps the simplest possible example of the dictum laid down in the first paragraph of **Chapter 1**: that an unobservable external (noumenal) world may leave traces of its existence in the part of the world that one *can* observe.

## References

[1] S. L. Glashow, Nucl. Phys. **22**, 579 (1961).
[2] A. Salam and J. C. Ward, Phys. Lett. **13**, 168 (1964).
[3] S. Weinberg, Phys. Rev. Lett. **19**, 1264 (1967).
[4] A. Salam, in *Elementary Particle Physics: Relativistic Groups and Analyticity (Nobel Symposium No. 8)*, edited by N. Svartholm (Almqvist and Wiksell, Stockholm, 1968).
[5] P. W. Higgs, Phys. Rev. Lett. **12**, 132 (1964); Phys. Rev. **145**, 1156 (1966).
[6] F. Englert and R. Brout, Phys. Rev. Lett. **13**, 321 (1964); G. S. Guralnik, C. R. Hagan and T. W. Kibble, Phys. Rev. Lett. **13**, 585 (1964).
[7] Z. Berezhiani, "Fermion Masses and Mixing GUT," arXiv:hep-ph/9602325.
[8] A. Belyaev *et al.*, "Technicolor Higgs boson in the light of LHC data," Phys. Rev. D **90**, 035012 (2014).
[9] F. Mandl and G. Shaw, *Quantum Field Theory* (John Wiley & Sons, New York, 1984), p. 225.
[10] V. Weisskopf, Phys. Rev. **56**, 72 (1939). Reprinted in *Selected Papers on Quantum Electrodynamics*, edited by J. Schwinger, (Dover, New York, 1958).
[11] W. Heitler, *The Quantum Theory of Radiation* (Oxford, London, 1954), pp. 27-34.
[12] J. Beringer *et al.*, (Particle Data Group), Phys. Rev. D **86**, 010001 (2012). URL: http//pdg.lbl.gov.
[13] Y. Nambu, Progr. Theoret. Phys. **7**, 595 (1952).
[14] H. Primakoff, in *Nuclear and Particle Physics at Intermediate Energies*, edited by J. B. Warren (Plenum, New York, 1976).
[15] A. O. Barut, Phys. Rev. Lett. **42**, 1251 (1979).

# Chapter 14

# Bare quark mass formulas

## 14.1 Defining 5-momentum shell radius $G_q$ for quarks

Our aim now is to calculate the bare quark masses. We shall employ the theory of free-fermion propagation in $M^4 \times \bar{M}^2$ as developed in **Secs. 4.3** and **4.4** and as employed in the previous chapter in our calculation of the charged-lepton masses. [See Eqs. (13.1) and (13.2)]. Of course, real quarks are not free, but are confined inside hadrons. The theory of free quarks in six dimensions is therefore a false theory: it does not describe quarks as they are found in Nature. Fortuitously, out of the free-particle formalism there emerges a phenomenological constant $R$ whose value van be linked to the energy of confinement. The constant $R$ appears to provide a sufficient representation of quarks in confinement to allow us to compute the bare quark masses using free-particle theory.

All bare quark masses $m_q$ are computed from the mass formula (4.36):

$$m_q^2 c^2 = G_q^2 - p_5^2 \tag{14.1}$$

where $G_q$ is the radius of the 5-momentum shell corresponding to quark flavor $q$ and, reproducing Eq. (4.101),

$$p_5 = BR + (B+Q)m_e c\,. \tag{14.2}$$

Here $B$ is the baryon number ($=1/3$ for quarks), $R$ is the above phenomenological constant, and the electric charges $Q_{u,c,t} = +2/3$ and $Q_{d,s,b} = -1/3$. For reference we quote again the corresponding expression for leptons, Eq. (13.4):

$$p_5 = L_\ell G_\ell + Q m_e c\,, \tag{14.3}$$

where $L_\ell$ is the lepton family number, $G_\ell$ is the radius of the five-momentum shell defining family $\ell$, and the electric charges $Q_{\nu_\ell} = 0$ and $Q_{\ell^-} = -1$. To calculate masses from (14.1) we need to know the value of constant $R$ and, for each quark flavor $q$, the form of the radius $G_q$.

As for constant $R$, we recall that in **Sec. 4.12** we proposed two optional values, one based on the mass of the $\pi$ meson, the other on the mass of the muon. Here we are going to adopt the second of the two options, writing for $R$ in (14.2)

$$R = m_\mu c = 105.7 \text{ MeV}/c. \tag{14.4}$$

This is taken to be the energy of confinement of quark $q$ trapped in a hadron. As we shall see, it yields quark masses in good agreement with those cited by the Particle Data Group.

Next we turn to the problem of defining $G_q$. Since $p_5$ is known for each quark flavor from (14.2), the problem is solved if we can define $G_q$ in terms of $p_5$. To do this we shall assume that

$$G_u = G_d \equiv G_{u,d}. \tag{14.5}$$

This means, in accordance with the physical definition of *family* given in **Sec. 4.4**, that quark flavors *u* and *d* belong to the same family of fermions. The relationship between quarks *u* and *d* in momentum space is thus the same as between leptons $\nu_\ell$ and $\ell^-$: the members of each pair belong to a common 5-momentum shell of radius *G*, and for each pair the difference in their $p_5$ values equals $m_e c$. Explicitly, from (14.2), $p_5(u) - p_5(d) = m_e c$, and from (14.3), $p_5(\nu_\ell) - p_5(\ell^-) = m_e c$. This at once suggests that we define for quarks *u* and *d*, in analogy with the lepton formula (14.3),

$$p_5 = G_{u,d} + Q m_e c. \tag{14.6}$$

However, this cannot be right, because for $Q = +2/3$, Eq. (14.6) when substituted into (14.1) yields an imaginary mass. The lepton-like formula (14.6) therefore must be modified to render it applicable to quarks. But the only parameter in this expression open to modification is the charge *Q*. As we are dealing with quarks, we assume that *Q* can be modified in steps of one-third. Thus we write in place of (14.6)

$$p_5 = G_{u,d} + \left[Q - \left(N - \frac{M}{3}\right)\right] m_e c, \quad M = 0,\ 1,\ 2;\ N = 1,2,3\cdots \quad (14.7)$$

in which expression the effect of the term $(N - M/3)$ is to draw down the effective electric charge. Equating the right-hand sides of (14.7) and (14.2) we obtain

$$G_{u,d} = \frac{R}{3} + \left(N + \frac{1-M}{3}\right) m_e c \quad (14.8)$$

where we have put $B = 1/3$. This is our modified expression for the radius of the 5-momentum shell defining the first family of quarks. Integers *M* and *N* are not given *a priori* and have to be selected to give appropriate mass values.

## 14.2 Masses of the light quarks *u*, *d* and *s*

Let us try the minimal values $M = 0$ and $N = 1$. Then (14.8) becomes

$$G_{u,d} = \frac{R}{3} + \frac{4}{3} m_e c\,. \quad (14.9)$$

Using this, (14.4) and (14.2) in (14.1), we obtain

$$m_u = 3.49\ \mathrm{MeV}/c^2 \quad (14.10)$$

and

$$m_d = 6.96\ \mathrm{MeV}/c^2\,. \quad (14.11)$$

These values are listed below in **Table 14.1** together with the corresponding Particle Data Group (PDG) current quark masses, the latter normalized at the renormalization scale $\mu = 1\ \mathrm{GeV}/c^2$.[1] (See also **Table 14.1A** of the **Appendix**.) They are also depicted graphically in **Fig. 14.1** in relation to the PDG ranges of values. The calculated values are well within the ranges for these masses recommended by the PDG. Our calculated masses give the average value

$$\frac{m_u + m_d}{2} = 5.23 \text{ MeV}/c^2 \tag{14.12}$$

and ratio

$$\frac{m_u}{m_d} = 0.50, \tag{14.13}$$

both of which are also in good agreement with the PDG estimates. They are listed in **Table 4.2** and depicted graphically in comparison to PDG values in **Fig. 14.2**.

Now what about *s*, the third light quark? Its $p_5$ value is the same as that of *d*, so the only way of obtaining a mass different from $m_d$ is by replacing the shell radius $G_{u,d}$ with a new radius $G_s$. What then is $G_s$? To answer this, we note that, remarkably, the radius $G_{u,d}$ of (14.9) depends on the electron mass, $m_e$, although weakly in relation to constant *R*. This seems to say that the *u* and *d* quarks owe their existence to the electron, as their masses are coupled through $G_{u,d}$ to the electron mass. This in turn suggests the possibility that each quark mass derives, in part, from coupling to one of the lepton masses. To formalize this idea we need only replace $m_e$ in (14.8) with the generic lepton mass $m_\ell$, resulting in the general *G*-form

$$G_q = \frac{R}{3} + \left(N + \frac{1-M}{3}\right) m_\ell c, \quad M = 0,\ 1,\ 2; \quad N = 1, 2, 3 \cdots \tag{14.14}$$

We then have a one-to-one correspondence between quark mass $m_q$ and lepton mass $m_\ell$. Of course, for each such correspondence we still have to pick *M* and *N*.

To test this idea, let us assume that $m_s$ is coupled to $m_\mu$. Then for the *s* quark

$$G_s = \frac{R}{3} + \left(N + \frac{1-M}{3}\right) m_\mu c, \quad M = 0,\ 1,\ 2; \quad N = 1, 2, 3 \cdots \tag{14.15}$$

Let again try the minimal values $M = 0$ and $N = 1$, as these values were good for *u* and *d*. Inserting (14.15) and (14.2) (with $Q = -1$) into (14.1) we obtain $m_s = 172.6 \text{ MeV}/c^2$, which, according to **Table 14.1** and **Fig. 14.1**, falls well above the currently accepted range of masses cited by the PDG. With the maximum *M*-value and the same $N = 1$ we obtain

$m_s = 99.4 \text{ MeV}/c^2$, which is well *below* the acceptable range. Thus we are obliged to take $M = 1$ and $N = 1$. This gives

$$G_s = \frac{R}{3} + m_\mu c \tag{14.16}$$

and then

$$m_s = 136.4 \text{ MeV}/c^2, \tag{14.17}$$

a value only slightly above the accepted PDG range. The mass ratio

$$\frac{m_s}{m_d} = 19.6 \tag{14.18}$$

sits almost exactly in the middle of the accepted range, while the ratio

$$\frac{m_s}{\left(\frac{m_u + m_d}{2}\right)} = 26.0 \tag{14.19}$$

falls a little below the accepted range. See **Table 14.2** and **Fig. 14.2**.

## 14.3 Masses of the heavy quarks *c* and *b*

A pattern has emerged: the mass of each down-type quark appears to be coupled to the mass of one of the charged leptons; specifically, $m_d$ is coupled to $m_e$, and $m_s$ to $m_\mu$. (See **Table 14.1A** of the **Appemdix.**) By inference, then, $m_b$ is coupled to $m_\tau$. Eq. (14.14) becomes for the *b* quark

$$G_b = \frac{R}{3} + \left(N + \frac{1-M}{3}\right) m_\tau c, \tag{14.20}$$

where for $m_\tau$ we assume the accepted experimental value $1776.82 \text{ MeV}/c^2$ (see **Table 13.1**.). With $M = 0$ and $N = 2$, we have

$$G_b = \frac{R}{3} + \frac{7}{3} m_\tau c, \tag{14.21}$$

and then from (14.1)

$$m_b = 4.18\ \text{GeV}/c^2, \tag{14.22}$$

a value centered precisely on the PDG range for $m_b$.

We have now run out of observed leptons, with the up-type quark masses $m_c$ and $m_t$ still unaccounted for. It is at this point that the "hidden" (unobservable) charged lepton $\sigma$ discussed in **Sec. 13.7** comes into play. Let us assume that $m_c$ couples to $m_\sigma$. Then (14.14) becomes for the $c$ quark

$$G_c = \frac{R}{3} + \left( N + \frac{1-M}{3} \right) m_\sigma c \tag{14.23}$$

Here we pick $M = 1$ and $N = 3$, giving

$$G_c = \frac{R}{3} + 3m_\sigma c\,. \tag{14.24}$$

Then with $m_\sigma = 421.071$ MeV/$c^2$ (again see **Table 13.1**), we have from (14.1)

$$m_c = 1.30\ \text{GeV}/c^2\,. \tag{4.25}$$

This value sits on the upper boundary of the PDG range for $m_c$.

## 14.4 Mass of the ultra-heavy top quark, *t*

The top quark is a special case. We assume that its mass is defined by coupling to the mass of one of the charged leptons, as this appears to be the case for the other five quarks. However, there are no leptons remaining to provide a linear coupling relation of the form (14.14). The only way to accommodate the top quark in our calculations is to assume it to belong to a fermion family that includes one of the known leptons—the unobservable $\sigma$ lepton to be precise. In other words we assume that $t$ belongs to the same 5-momentum shell as lepton $\sigma$, so that

$$G_t = G_\sigma = \frac{m_\sigma^2 + m_e^2}{2m_e} c\,, \tag{14.26}$$

where the second equality is obtained by eliminating $p_5$ between (13.3) and (13.4) [see also (4.91)]. In the case of the top quark, the dependence of $G_q$ on lepton mass is quadratic rather than linear. With $m_\sigma = 421.071\ \text{MeV}/c^2$ we obtain from (14.1)

$$m_t = 173.8\ \text{GeV}/c^2, \tag{14.27}$$

a value well within the PDG range for direct measurement of the top mass.

**Table 14.1** Calculated bare quark mass $m_q$ vs. Particle Data Group (PDG) quark mass listings.[1] In these listings, the *u*, *d* and *s* masses are "current quark masses" normalized here at renormalization scale $\mu = 1\ \text{GeV}/c^2$, obtained by multiplying the PDG listed values (normalized at $\mu = 2\ \text{GeV}/c^2$) by 1.35. The *c* and *b* masses are "running mass" values. The *t* masses are those observed in $p\bar{p}$ collisions. Indices *N* and *M* determine the numerical coefficient of lepton mass $m_\ell$ in the 5-momentum shell radii $G_q$, given by Eq. (14.14). Calculated quark masses are found by inserting these radii, together with the $p_5$ values of Eq.(14.2), into Eq. (14.1). In these equations, $R = m_\mu c = 105.7\ \text{MeV}/c$ and $m_e = 0.511\ \text{MeV}/c^2$.

| Flavor $q$ | Charge $Q$ | Coupled Lepton $\ell$ | Index $N$ | Index $M$ | Calculated Mass $m_q c^2$ | PDG Listing |
|---|---|---|---|---|---|---|
| $u$ | $2/3$ | $e$ | 1 | 0 | 3.49 MeV | 2.43-4.05 |
| $d$ | $-1/3$ | $e$ | 1 | 0 | 6.96 MeV | 6.08-7.16 |
| $c$ | $2/3$ | $\sigma$ | 3 | 1 | 1.30 GeV | 1.25-1.3 |
| $s$ | $-1/3$ | $\mu$ | 1 | 1 | 136.4 MeV | 121.5-135. |
| $t$ | $2/3$ | $\sigma$ | N.A. | N. A. | 173.8 GeV | 172.-174.4 |
| $b$ | $-1/3$ | $\tau$ | 2 | 0 | 4.18 GeV | 4.15-4.21 |

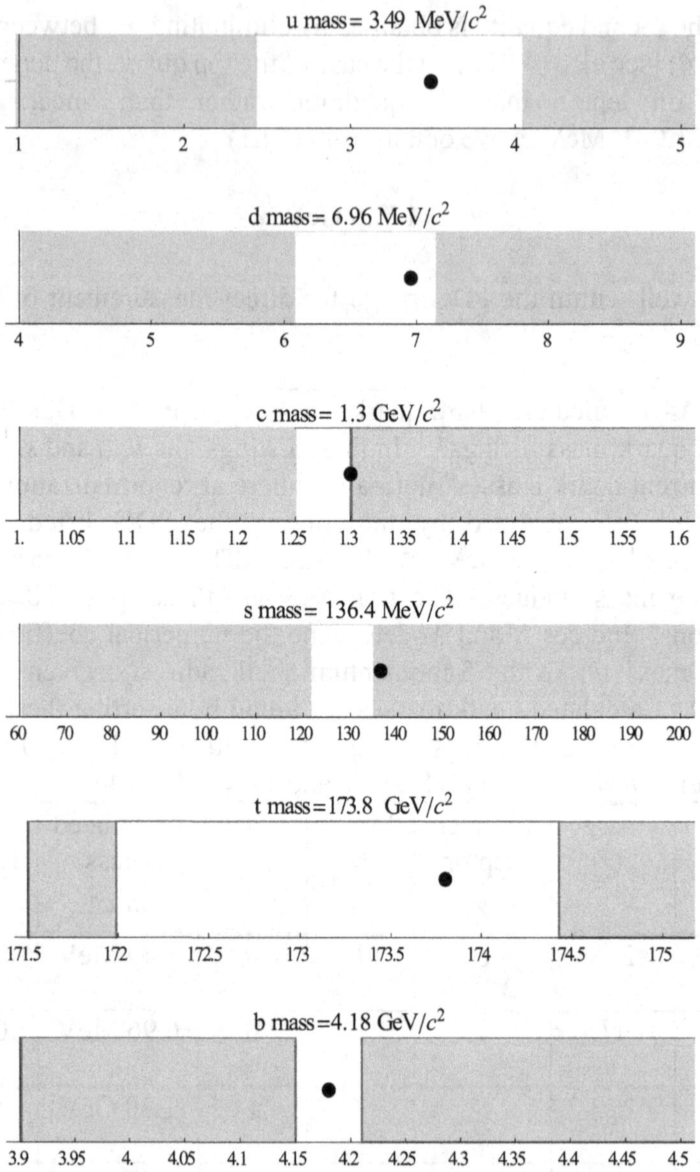


**Fig. 14.1** Graphical presentation of the six calculated quark masses and corresponding PDG ranges listed in **Table 14.1**.

**Table 14.2** Average mass $(m_u + m_d)/2$ and quark mass ratios using calculated bare quark masses listed in **Table 14.1**, compared with Particle Data Group (PDG) quark listings.[1] The *u*, *d* and *s* masses are "current quark masses" normalized at renormalization scale $\mu = 1\ \mathrm{GeV}/c^2$, obtained by multiplying the PDG listed values (normalized at $\mu = 2\ \mathrm{GeV}/c^2$) by 1.35.

| Quantity | Calculated Value | PDG Listed Value |
|---|---|---|
| $(m_u + m_d)/2$ | $5.23\ \mathrm{MeV}/c^2$ | $4.46\text{-}5.67\ \mathrm{MeV}/c^2$ |
| $m_u / m_d$ | 0.50 | 0.38-0.58 |
| $m_s / m_d$ | 19.6 | 17-22 |
| $m_s / [(m_u + m_d)/2]$ | 26.0 | 26.5-28.5 |

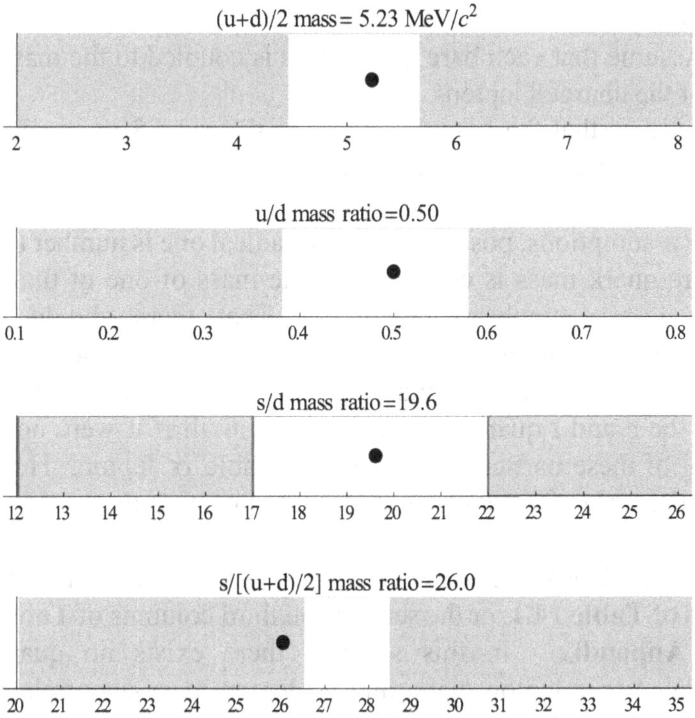


**Fig. 14.2** Graphical presentation of the calculated average mass, mass ratios and corresponding PDG ranges listed in **Table 14.2**.

## 14.5 Conclusions and comments

Our theory of fermion propagation in $M^4 \times \bar{M}^2$ has given a reasonably good account of the quark mass spectrum. Although the calculated $s$ mass falls marginally outside the acceptable range of values listed by the PDG, as does the ratio $m_s / \left[ (m_u + m_d)/2 \right]$, all other masses and ratios can be said to agree well with the listed PDG values.

To achieve this within the framework of a theory of free particles, a number of *ad hoc* steps had to be taken. These were, in summary:

(1) Estimate constant $R$ by relating it to an energy of confinement.
(2) Assume that quarks $u$ and $d$ belong to the same 5-momentum shell in $M^4 \times \bar{M}^2$.
(3) Adopt *Ansatz* (14.7), an expression of quark-lepton universality.
(4) Introduce integers $M$ and $N$, the only free parameters in the theory.
(5) Assume that each bare quark mass is coupled to the mass of one of the charged leptons.
(6) Assume that the $t$ quark belongs to the same 5-momentum shell as the unobservable $\sigma$ lepton.

Of these assumptions, possibly the most radical one is number (5)—that each bare quark mass is correlated to the mass of one of the charged leptons $e$, $\mu$, $\sigma$ (unobservable) or $\tau$. That there should exist a connection between quark and lepton masses is by no means obvious, but it appears that the quark mass spectrum is unpredictable without it. Indeed, the $c$ and $t$ quarks would not exist at all if it were not for the coupling of these particles to the unobservable $\sigma$ lepton. Here is yet another example of an unobserved world impacting the world that one *can* observe.

The quark/lepton correlations are indicated in the first and third columns of **Table 14.1**, or the second and third columns of **Table 14.1A** of the **Appendix**. In this scheme, there exists no quark mass uncorrelated to a lepton mass, and no lepton mass uncorrelated to at least one quark mass. The system is closed. The underlying correlation mechanism is, however, not understood. Clearly the need exists for a deeper, more complete theory than the one presented here, one that explains from first principles the apparent connection between lepton and quark masses, and one in which the quark mass spectrum is predicted in a more automatic fashion.

In **Fig. 14.3**, the six quark masses are depicted graphically—but not to scale—as projections of the 5-momentum vectors of length $G_q$ onto the 4-dimensional hyperplane, $M^4$.

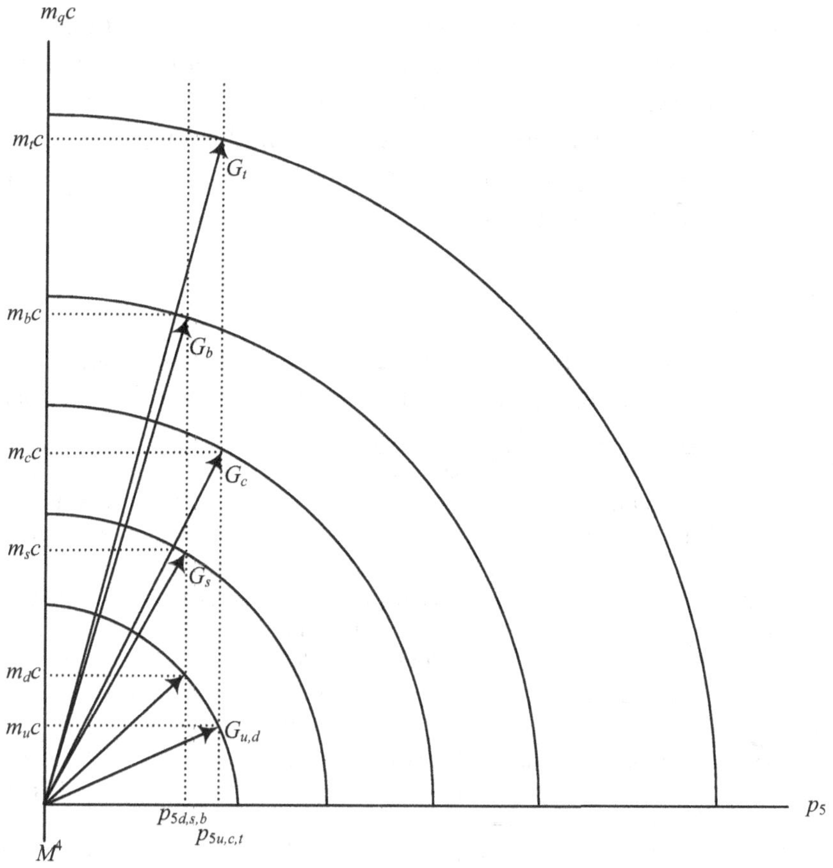


**Fig. 14.3** Free quark masses $m_q$ depicted as normal projections of 5-momentum vectors of length $G_q = (p_5^2 + m_q^2 c^2)^{1/2}$ onto the 4-D hyperplane $M^4$. Quarks $d$, $s$, $b$ have pseudomomentum $p_{5d,s,b} = R/3$ while $\underline{u}$, $c$, $t$ have $p_{5u,c,t} = R/3 + m_e c$. Of the six quarks, only $u$ and $d$ belong to the same 5-momentum shell in $M^4 \times \bar{M}^2$. Relative vector lengths and masses are not to scale.

This figure illustrates an unexpected feature of quark family structure. In our physical theory of flavor as presented in **Sec. 4.4**, fermion family $f$ was defined as the set of all particles belonging to the same 5-momentum shell of radius $G_f$. By that definition, *u* and *d* belong to the same family of quarks. However, by that same definition, from the figure we see that *c* and *s* do not belong to the same family, nor do *t* and *b*, since all four flavors belong to different 5-momentum shells. This arrangement differs markedly from the Standard Model picture, which has (*u*, *d*), (*c*, *s*) and (*t*, *b*) forming three identical families (except for mass) based on their transformation properties under electroweak *SU*(2). Thus one should not expect the SM to predict quark masses; the group-theoretic pairing of quarks in the theory of weak interactions has little to do with the family structure inherent in quark mass generation.

## Appendix

Here we present an alternate formulation of **Table 14.1**, with quarks separated into types up and down. Up types *c* and *t* would not exist at all if it were not for the unobserved $\sigma$ lepton.

**Table 14.1** Calculated bare quark mass $m_q$ vs. Particle Data Group (PDG) quark mass listings.[1] In these listings, the *u*, *d* and *s* masses are "current quark masses" normalized here at renormalization scale $\mu = 1\ \text{GeV}/c^2$, obtained by multiplying the PDG listed values (normalized at $\mu = 2\ \text{GeV}/c^2$) by 1.35. The *c* and *b* masses are "running mass" values. The *t* masses are those observed in $p\bar{p}$ collisions. Indices *N* and *M* determine the numerical coefficient of lepton mass $m_\ell$ in the 5-momentum shell radii $G_q$, given by Eq. (14.14). Calculated quark masses are found by inserting these radii, together with the $p_5$ values of Eq.(14.2), into Eq. (14.1). In these equations, $R = m_\mu c = 105.7\ \text{MeV}/c$ and $m_e = 0.511\ \text{MeV}/c^2$.

| Type | Flavor $q$ | Coupled Lepton $\ell$ | Index $N$ | Index $M$ | Calculated Mass $m_q c^2$ | PDG Listing |
|---|---|---|---|---|---|---|
| Down $Q = -1/3$ | $d$ | $e$ | 1 | 0 | 6.96 MeV | 6.08-7.16 |
| | $s$ | $\mu$ | 1 | 1 | 136.4 MeV | 121.5-135. |
| | $b$ | $\tau$ | 2 | 0 | 4.18 GeV | 4.15-4.21 |
| Up $Q = 2/3$ | $u$ | $e$ | 1 | 0 | 3.49 MeV | 2.43-4.05 |
| | $c$ | $\sigma$ | 3 | 1 | 1.30 GeV | 1.25-1.3 |
| | $t$ | $\sigma$ | N.A. | N. A. | 173.8 GeV | 172.-174.4 |

## Reference

[1] K. A. Olive *et al.*, Particle Data Group, Chin. Phys. **C38**, 090001 (2014). (URL: http://pdg.lbl.gov.

# PART VI

# THE STRUCTURE OF SPACETIME

# Chapter 15

# Why are there three spatial dimensions?

## 15.1 Introduction

We have seen in some detail how the structure of spacetime actively contributes to the behavior and structure of matter. For example:

- Owing to the fifth dimension $x^5$, only those reactions can occur that conserve the fifth momentum, $p_5$ (*Sec. 3.9.7*).
- From the conservation of $p_5$ and charge $Q$ it follows that, in all processes, lepton number and baryon number are separately conserved (**Sec. 4.13**).
- The existence of timelike dimension $x^5$ permits massless neutrinos to oscillate (**Chapter 12**).
- The spatial width $W$ of inner spacetime $N^2$ determines the electron mass, $m_e$ (**Sec. 13.7**).

In this chapter we see how the electron returns the favor: we shall show that the electron requires for its existence precisely three spatial dimensions, thus fixing the dimensionality of Minkowski $M^{N+1}$. And since, as we saw in **Chapters 13** and **14**, all matter ultimately connects to the electron, it follows that the existence of matter of any sort will be possible if and only if $N = 3$.

In their monumental exposition of the anthropic principle, Barrow and Tipler recall in detail the history of efforts to explain or underwrite the fact that the world has, or appears to have, three spatial dimensions.[1] These efforts fall into two main categories. The first we might call the *formal* approach. It argues from the form of the laws of physics, showing, for example, that only in three dimensions can there be stable planetary orbits, stable atoms and molecules or even proper dimensional analysis. The second approach is *anthropic*. It argues that only in a world of three spatial dimensions can there exist conditions hospitable to the growth of intelligent life; and since such life exists, the world necessarily has three spatial dimensions. In contrast, the approach taken here is *existential*. It argues that nothing material can exist at all unless $N=3$.

## 15.2 The electron in $N$ spatial dimensions

Our method is particularly simple. We are going to calculate the spatial distribution $D$ of the electron's Zitterbewegung—the microscopic orbital motion of a massless point $P$ moving at speed $c$—in Euclidian $E^N$, and then show that only when $N=3$ is the orbit stable and physically viable. The starting point for these calculations is the propagation law Eq. (10.22), rewritten here to exhibit both specialization to the electron ($k_5=0$) and generalization to $N$ spatial dimensions:

$$e(x)=f(x)*e(x) \tag{15.1}$$

where $e(x)$ represents either the probability or charge amplitude of the electron, $f(x)$ is the propagator of (3.79) and (3.81) generalized to $N+1$ spacetime dimensions

$$f(x)=\frac{1}{(2\pi)^{N+1}}\int F(k)e^{-ik\cdot x}d^{N+1}k\,, \tag{15.2}$$

in which

$$F(k)=\begin{cases}\exp\left(-i\xi^5\sqrt{k_6{}^2-k^2}\right), & k^2\le k_6{}^2\\ \exp\left(-\xi^5\sqrt{k^2-k_6{}^2}\right), & k^2>k_6{}^2\end{cases} \tag{15.3}$$

and $*$ denotes the $(N+1)$-dimensional convolution operation. Here $k$ and $x$ denote, respectively, the $(N+1)$-vectors $(k^0,\mathbf{k})$ and $(x^0,\mathbf{x})$, where boldface vectors $\mathbf{k}$ and $\mathbf{x}$ each have $N$ components in Euclidian $E^N$. For the electron, $k^6=-k_6=m_ec/\hbar$. Now, as in (10.24), we take an electron at rest, with

$$e(x)=Ae^{-ik^6x^0}\,, \tag{15.4}$$

where $A$ is a constant spinor in $N+1$ dimensions. Then from (15.1)

$$\begin{aligned} e(x) &= \int f(\xi) e(x-\xi) d^{N+1}\xi \\ &= \int f(\xi) e^{ik^6\xi^0} d\xi^0 d^N \boldsymbol{\xi} \cdot e(x) \\ &= \int D(\boldsymbol{\xi}) d^N \boldsymbol{\xi} \cdot e(x) \end{aligned} \tag{15.5}$$

where the spatial distribution $D(\boldsymbol{\xi})$ is given by the temporal inverse Fourier transform of propagator $f$:

$$D(\boldsymbol{\xi}) = \int f(\xi^0, \boldsymbol{\xi}) e^{ik^6\xi^0} d\xi^0 . \tag{15.6}$$

We now replace $f$ in the integrand of this expression by its Fourier integral representation (15.2). After integrating over $\xi^0$ and $k^0$ we obtain

$$D(\boldsymbol{\xi}) = \frac{1}{(2\pi)^N} \int e^{-i\xi^5\sqrt{\mathbf{k}^2}} \cos(\mathbf{k}\cdot\boldsymbol{\xi}) d^N\mathbf{k} . \tag{15.7}$$

The real part of this expression is the internal probability distribution $D_r$ of the Zitterbewegung in $E^N$. Note that, by (15.5), the integral over the imaginary part vanishes.

**$N = 1$** For this case the real part of (15.7) is

$$\begin{aligned} D_r(\xi) &= \frac{1}{2\pi} \int_{-\infty}^{\infty} \cos(|k|\xi^5) \cos(k\xi) dk \\ &= \frac{1}{2}\delta(\xi - \xi^5) + \frac{1}{2}\delta(\xi + \xi^5) \end{aligned} \tag{15.8}$$

This distribution is plotted in **Fig. 15.1,** along with for comparison the plots for higher dimensions. Clearly the massless point $P$ performs no motion at all, being confined probabilistically to the two $\delta$-functions. And as there is no internal motion, the Zitterbewegung produces zero rest energy, contradicting wave function (15.4), which prescribes rest energy $m_e c^2$. Moreover, a particle of zero mass cannot be at rest. The Zitterbewegung in this case is non-functional and we conclude that the electron cannot exist in $E^1$.

**$N = 2$** The two-dimensional case carries with it a long history of scientific speculation, in part inspired by the fictional writings of Edward Abbot Abbot (1884)[2] and A. K. Dewdney (1984)[3]. But could such a universe actually exist? The radial distribution of the electron's Zitterbewegung in 2-D is, from (15.7),

$$\begin{aligned} D_r(\rho) &= \frac{1}{(2\pi)^2}\int_0^\infty kdk\cos(k\xi^5)\int_0^{2\pi}\cos(k\rho\cos\phi)d\phi \\ &= \frac{1}{2\pi}\frac{\partial}{\partial\xi^5}\int_0^\infty \sin(k\xi^5)J_0(k\rho)dk \\ &= \frac{1}{2\pi}\frac{\partial}{\partial\xi^5}\left[\frac{U(\xi^5-\rho)}{\left(\xi_5^2-\rho^2\right)^{1/2}}\right] \qquad\qquad , \qquad (15.9) \\ &= \frac{1}{2\pi}\left[-\frac{\xi^5 U(\xi^5-\rho)}{\left(\xi_5^2-\rho^2\right)^{3/2}}+\frac{\delta(\rho-\xi^5)}{\left(\xi_5^2-\rho^2\right)^{1/2}}\right] \end{aligned}$$

where $U(x)$ is the unit step function defined by Eq. (10.33). This distribution is plotted in **Fig. 15.1**. In radial structure it resembles to a degree the corresponding distribution of the electron in three dimensions, **Fig. 10.4** (and also **Fig. 15.1**). In particular, one branch of the distribution goes algebraically negative, indicating that point $P$ oscillates within internal spacetime $N^2$ as described in **Sec. 10.7**. Note, however, that this negative branch is of finite width, and within this branch at radii $\rho < \xi^5$ the orbital paths are non-reëntrant. This means that, as in the cases of leptons $\mu$ and $\tau$, whose Zitterbewegung distributions in 3-D also feature zones of finite width (see **Figs. 10.5** and **10.6**), the electron in 2-D is unstable and subject to decay. But there is no material particle lighter than the electron it can decay to, and the reaction $e^- \rightarrow \nu_e\bar{\nu}_e$, which does conserve $p_5$, does not conserve charge. So we must conclude that neither the electron nor anything else can exist in $E^2$. The Flatland universe remains purely fictional.

**$N = 3$** For ease of comparison, we reproduce here and in **Fig. 15.1** the electron's Zitterbewegung distribution in $E^3$:

$$
\begin{aligned}
D_r(\rho) &= \frac{1}{(2\pi)^3}\int_0^\infty k^2 dk \cos(k\xi^5)\int_0^{2\pi} d\phi \int_0^\pi \cos(k\rho\cos\theta)\sin\theta d\theta \\
&= -\frac{1}{2\pi^2\rho}\frac{\partial}{\partial\rho}\int_0^\infty \cos(k\xi^5)\cos(k\rho)dk \qquad (15.10)\\
&= -\frac{\delta'(\rho-\xi^5)}{4\pi\rho}
\end{aligned}
$$

The distribution is of zero width and consequently all orbits are reëntrant, ensuring the existence of a stable, ground state lepton, *e*.

$\boldsymbol{N=4}$ Just as the possibility of a world of two spatial dimensions has attracted scientific interest, so has a world of four dimensions.[4, 5] The question for us here is, can matter actually exist in $E^4$? From (15.7)

$$
\begin{aligned}
D_r(\rho) &= \frac{1}{(2\pi)^4}\int_0^\infty k^3 dk \cos(k\xi^5) \\
&\quad \times \int_0^{2\pi} d\phi \int_0^\pi \sin\theta d\theta \int_0^\pi \cos(k\rho\cos\gamma)\sin^2\gamma d\gamma \\
&= -\frac{1}{4\pi^2\rho}\frac{\partial}{\partial\rho}\int_0^\infty \cos(k\xi^5) J_0(k\rho) k dk \qquad (15.11)\\
&= -\frac{1}{4\pi^2\rho}\frac{\partial}{\partial\rho}\left[-\frac{\xi^5 H(\xi^5-\rho)}{\left(\xi_5^2-\rho^2\right)^{3/2}} + \frac{\delta(\rho-\xi^5)}{\left(\xi_5^2-\rho^2\right)^{1/2}}\right] \\
&= \frac{1}{4\pi^2\rho}\left[\frac{3\rho\xi^5 H(\xi^5-\rho)}{\left(\xi_5^2-\rho^2\right)^{5/2}} - \frac{2\xi^5\delta(\rho-\xi^5)}{\left(\xi_5^2-\rho^2\right)^{3/2}} - \frac{\delta'(\rho-\xi^5)}{\left(\xi_5^2-\rho^2\right)^{1/2}}\right]
\end{aligned}
$$

This distribution, pictured in **Fig. 15.1**, contains a branch of finite width, corresponding to which the orbital paths are non-reëntrant. The electron in 4-D is thus unstable and, as in the case of the electron in 2-D, is subject to fatal decay. We conclude that matter cannot exist in 4-dimensional $E^4$.

**$N = 5$** From (15.7) we have

$$\begin{aligned} D_r(\rho) &= \frac{1}{(2\pi)^5}\int_0^\infty k^4 dk \cos(k\xi^5) \\ &\times \int_0^{2\pi} d\phi \int_0^\pi \sin\theta d\theta \int_0^\pi \sin^2\gamma d\gamma \int_0^\pi \cos(k\rho\cos\delta)\sin^3\delta d\delta \\ &= -\frac{2}{8\pi^3\rho^2}\left(\frac{1}{\rho}\frac{\partial}{\partial\rho} - \frac{\partial^2}{\partial\rho^2}\right)\int_0^\infty \cos(k\xi^5)\cos(k\rho)dk \\ &= \frac{1}{8\pi^2\rho^2}\left[-\frac{\delta'(\rho-\xi^5)}{\rho} + \delta''(\rho-\xi^5)\right] \end{aligned} \quad . \quad (15.12)$$

This distribution, shown in **Fig. 15.1**, is perhaps less easily evaluated than the ones already considered. All five peaks comprising the distribution are of zero width and so there is no question of instability attributable to non-reëntrant orbits. (The peaks are shown here broadened for visibility.) In that respect the present case is similar to that of $N = 3$. The decisive element here is the particle's intrinsic spin. This is readily computed by generalizing the spin integral (10.72) to five spatial dimensions. We find that the $\delta'$ and $\delta''$ terms of the distribution have, respectively, associated spins of 1/3 and 2/3, ostensibly yielding a total spin of 1. Now we know from the $N = 3$ case that the spin of a $\delta'$ distribution is 1/2, not 1/3, and so there is an immediate contradiction in the interpretation of the Zitterbewegung in $N = 4$. But even more importantly, we remind ourselves that $D_r(\rho)$ is a *probability* distribution, meaning that the massless point $P$ comprising the Zitterbewegung must sometimes be in an orbit defined by the $\delta'$ term (spin 1/3) and sometimes in an orbit defined by the $\delta''$ term (spin 2/3). Such a picture obviously entails changing from one orbital type to the other during the lifetime of the particle. But switching from one orbital type to the other is forbidden, as to do so would violate conservation of spin momentum. Clearly then, because the Zitterbewegung cannot be negotiated without violating conservation of spin, matter in 5-dimensional $E^5$ cannot exist.

Alternatively, we can compute the particle's electric charge. This is just the integral over $D_r(\rho)$. We find that the charges corresponding to

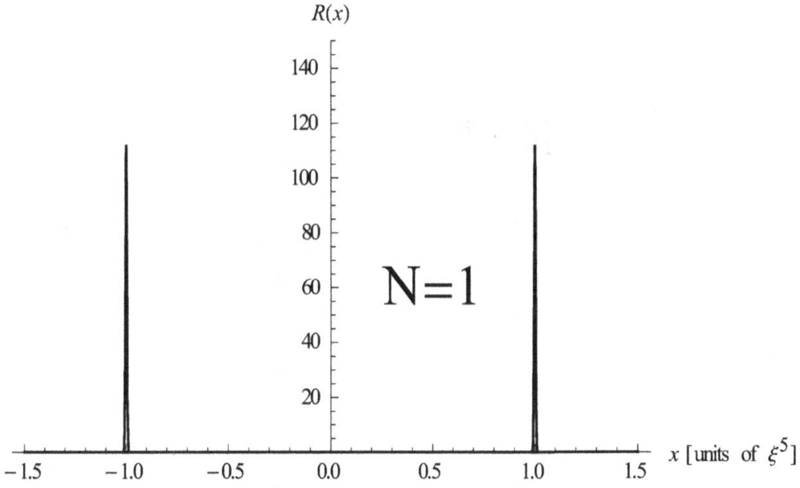
R(x)
140
120
100
80
60
40
20
N=1
−1.5
−1.0
−0.5
0.0
0.5
1.0
1.5
x [units of ξ^5]

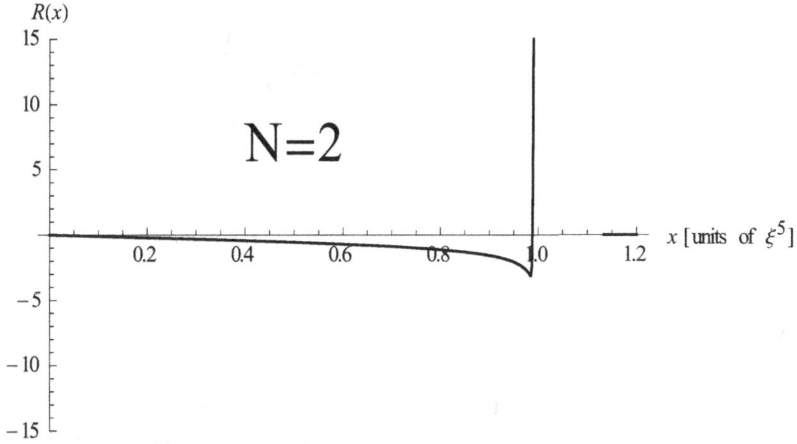
R(x)
15
10
5
−5
−10
−15
N=2
0.2
0.4
0.6
0.8
1.0
1.2
x [units of ξ^5]

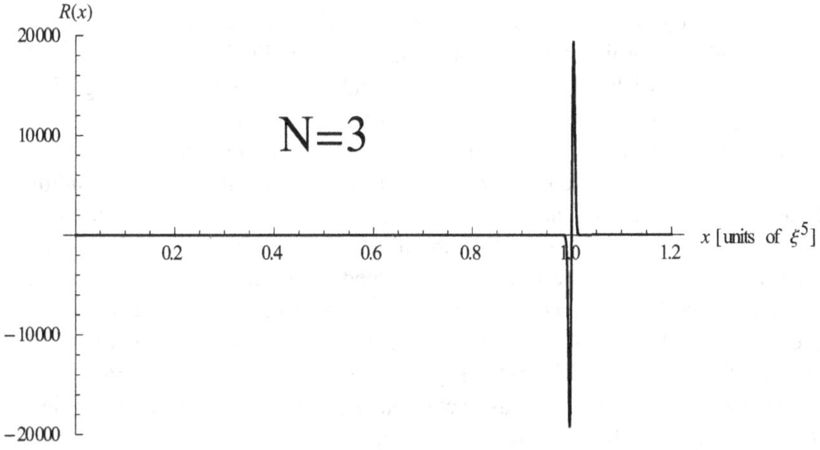
R(x)
20000
10000
−10000
−20000
N=3
0.2
0.4
0.6
0.8
1.0
1.2
x [units of ξ^5]

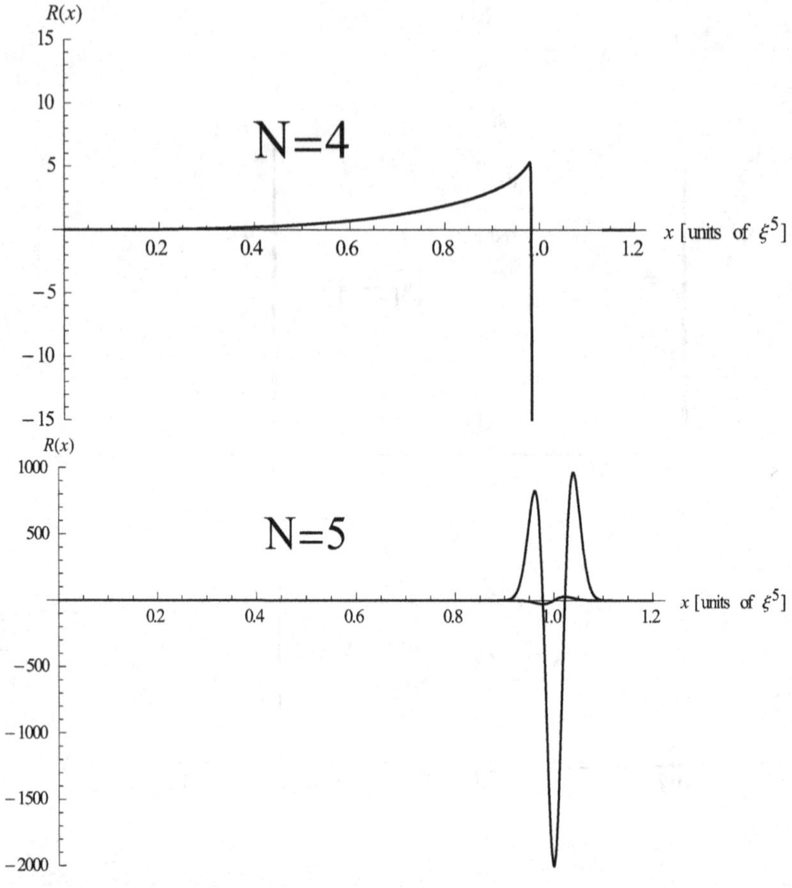


**Fig. 15.1** Plots of the radial probability distribution, $R(x) = \Omega\rho^{N-1}D_r(\rho)d\rho/dx$, generated by massless point *P*, whose motion constitutes the Zitterbewegung of the electron in *N* spatial dimensions. Here $\rho$ and $\Omega$ are the radial coordinate and global solid angle in Euclidian $E^N$, $x$ is the dimensionless variable $\rho/\xi^5$ and $\xi^5$ is the radius of the Zitterbewegung. The isolated peaks of ***N*=1** signify an absence of internal motion, thus precluding the development of rest mass and the existence of the electron in $E^1$. The smeared-out distributions of ***N*=2** and **4** portend non-reëntrant orbits, leading to decay of the particle and precluding the existence of the electron in either $E^2$ or $E^4$. The five peaks of ***N*=5** are of zero width, which precludes decay; however, the switching between orbits defined by the two components of the distribution violates both conservation of spin momentum and electric charge, precluding existence of the electron in $E^5$. The peaks of ***N*=3**, however, are of zero width, ensuring absolute stability of the particle, and the execution of the Zitterbewegung violates no conservation laws. Among Euclidian geometries, $E^3$ alone supports the existence of the electron.

the first and second terms of the distribution are 1/3 and 2/3, respectively, yielding a total charge of 1. Thus the changing of one orbital type to the other violates conservation of charge and is therefore forbidden. Again we are forced to conclude that matter in $E^5$ cannot exist.

## 15.3 Conclusion

Although we have studied in detail the electron's Zitterbewegung in five cases only, some generalizations are possible. Thus for $N$ even, the distribution $D_r$ will always include zones of finite width, leading to non-reëntrant orbits and eventual decay of the electron. We conclude matter cannot exist in an even-dimensional world.

In the case of $N$ odd, the distribution will always consist of the $\delta$-function and its derivatives, so that non-reëntrance of orbit is never an issue. However, for odd $N > 3$, the Zitterbewegung cannot proceed without violating conservation of both spin and charge, and in such universes matter cannot exist.

We conclude that only in three spatial dimensions can matter exist.

## References

[1] J. D. Barrow and F. J. Tipler, *The Anthropic Cosmological Principle* (Oxford University Press, New York, 1986), p. 258 ff.
[2] E. A. Abbott, *Flatland: A Romance in Many Dimensions* (Dover Thrift Edition, New York, 1992).
[3] A. K. Dewdney, *The Planiverse: Computer Contact with a Two-Dimensional World* (Corpericus, 2000).
[4] Rudy Rucker, *The Fourth Dimension: Toward a Geometry of Higher Reality* (Houghton Mifflin, Boston, 1984).
[5] Michio Kaku, *Hyperspace* (Anchor Books, New York, 1995).

# Chapter 16

# Self-imaging and the spacetime origins of fractional charge and the fine structure constant

## 16.1 Where does charge come from?

In **Chapter 4** (see especially **Secs. 4.4, 4.9** and **4.11**) it was argued that, in the expansion of a self imaging field $\chi(x^{\mu}, x^5, x^6)$ in eigenstates of the fifth momentum operator $i\hbar\partial_5$, the expansion index $-n$ represents the values of electric charge $Q$ in the case of leptons, and the values of $Q+B$ in the case of quarks, with $B$ denoting baryon number. As $-n$ is an integer, these expansions explain the quantization of both $Q$ and $Q+B$. What the expansions do *not* explain, however, is where the unit charge $-e$ comes from, and why the quarks come in one-third integral multiples of that charge. Now it is true that Grand Unification via the group $SU(5)$ predicts $Q_d = Q_{e^-}/3$ and $Q_u = -2Q_d$, ostensibly explaining why it is that $Q_p = -Q_{e^-}$.[1, 2] But the authority of this result is undermined by the fact that the $SU(5)$ GUT also predicts decay of the proton, a process that not only has not been observed, but is disallowed according to our Principle of True Representation (see **Sec. 4.13**). And so if the GUT is not an active law of Nature, then there remain still open the questions of the origin of charge and how it happens to come in both integral and fractional forms.

The first hint of an alternative to the GUT explanation of charge comes with our discovery in **Sec. 6.7** that the vacuum may be filled with a constant scalar Maxwell field $\phi$ having the same units as the electric and magnetic fields $\mathbf{E}/c$ and $\mathbf{B}$. If the field $\phi$ actually exists, then the vacuum is not innocent of charge and must in some way embody it. A second and more convincing clue is the fact that charge is an invariant. This immediately disqualifies Minkowski $M^4$ as the place of origin of charge. For intervals $\Delta x^0$ and $\Delta x$ in $M^4$ vary according to the state of motion of the observer and consequently can have no functional relation to the invariant $Q$. Evidently to connect charge to spacetime we shall have assume the presence of a special manifold, an observer-independent one within which charge can be defined and supported independently of $M^4$. But we know already of such a manifold, namely

$\bar{M}^2 \times N^2$, an invariant subspace of our total 8-D world geometry $M^4 \times \bar{M}^2 \times N^2$. We know that all fundamental particles possess pseudomomentum $p_5$ and thus occupy $\bar{M}^2$. We also know from our studies of lepton and quark mass (**Chapters 13** and **14**) that all fundamental particles occupy $N^2$ as well. Moreover, as $\bar{M}^2 \times N^2$ attaches to $M^4$ by direct product, it at once offers a place for the creation of invariant charge and renders that charge accessible to relativistic interactions in $M^4$.

But how exactly is charge to be manifested in $\bar{M}^2 \times N^2$? A fundamental claim of this book has been that all fundamental particles—fermions and bosons—can be described as self-imaging objects in 8-D spacetime. Among these objects, the heavy leptons and quarks are electrically charged. Thus it would not be too surprising to find that charge, too, is self-imaging. To get self-imaging, whether of electric charge or anything else, one needs a wave equation, in particular one whose wave function is expandable in eigenstates of the momentum operator corresponding to the principal direction of propagation. And so, to pursue the self-imaging idea, we are led to seek an appropriate wave equation in $\bar{M}^2 \times N^2$.

## 16.2 Wave equation for charge

Assuming such an equation is possible, on what variables should it depend? The dimensions available are $x^5, x^6 \in \bar{M}^2$ and $u^0, u \in N^2$. We know from **Chapter 3** [see in particular Eq. (3.72)] that dimension $x^6$ is dedicated to the production of rest mass and energy. Since charge is separate from mass and energy, the wave equation describing its self-imaging properties should not depend on $x^6$. In addition, because $u$ is bounded ($|u| \le W/2$), wave propagation in that direction cannot occur and dimension $u$ also can be omitted from the equation of motion. We conclude that our wave equation should be of the Helmholz form

$$\left( \frac{\partial^2}{\partial {u_0}^2} + \frac{\partial^2}{\partial {x_5}^2} + \kappa^2 \right) \phi(u^0, x^5) = 0 . \qquad (16.1)$$

Here constant $\kappa \equiv 2\pi/\lambda$, where $\lambda$ is the wave length of the monochromatic wave field $\phi$. Note that, although no time dimension drives this equation, it must still show a characteristic length, namely

$\lambda$, as otherwise it would not be a wave equation.

There are two additional points of interest here. First, although $u^0$ and $x^5$ are both metrically time-like, they behave in (16.1) as if they were spatial dimensions: wave propagation takes place in the $u^0$ - $x^5$ plane and in planes $u = \text{const.}$ parallel to it.

Second, although $u^0$ is a legitimate time axis, albeit internal and hidden, there is no null cone in this 3-dimensional subspace of $\bar{M}^2 \times N^2$. This is a simple consequence of the fact $u^0$ and $x^5$ have the same metrical signature. As depicted in **Fig. 16.1 (a)**, a world line

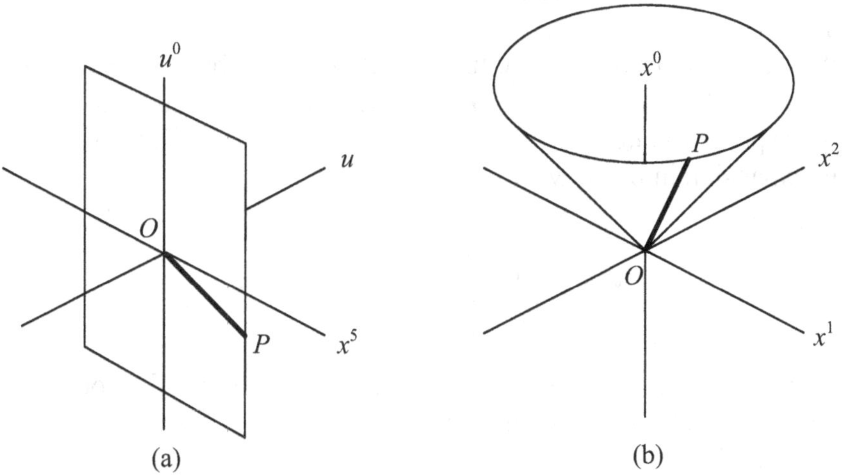


**Fig. 16.1.** Comparison of two 3-D spacetime diagrams. (a) Dimensions $u^0$, $u$ and $x^5$ depict spacetime manifold $\bar{M}^2 \times N^2$, omitting pseudotemporal dimension $x^6$, to which the axes shown are perpendicular. Although metrically time-like, $u^0$ and $x^5$ play the role of space. And since both have the same signature, there is no "null cone" in $\bar{M}^2 \times N^2$. All world lines, such as *OP*, drawn from the origin and lying in the $u^0$-$x^5$ plane, are lines of simultaneity. (b) Dimensions $x^0$, $x^1$ and $x^2$ depict Minkowski $M^4$, omitting dimension $x^3$, to which the axes shown are perpendicular. Shown also is the null cone. In $M^4$, all luxonic world lines, such as *OP*, lie in the null cone.

drawn from the origin may take any direction in the $u^0$ - $x^5$ plane. Such lines can be considered *lines of simultaneity*, a reasonable characterization, since no time dimension is present to order succession of events. The state of affairs in $\bar{M}^2 \times N^2$ is to be contrasted with that in $M^4$: in $M^4$ there exists a null cone and the world line of a luxon necessarily lies in it. See **Fig. 16.1 (b)**.

Given this characterization of world lines in $\bar{M}^2 \times N^2$, the process we have been calling "wave propagation" in fact occurs all at once, filling the entire space from $x^5 = -\infty$ to $+\infty$ simultaneously with full charge amplitude. Consequently the term "propagation" is, while descriptively useful, a misnomer. There is no actual propagation in $\bar{M}^2 \times N^2$. The waves defined by (16.1) are static and eternal. Properly speaking, Eq. (16.1) describes not a process of propagation but of occupation. Nevertheless, because of its descriptive utility, we may occasionally employ the term "propagation" if it is unlikely to cause confusion.

We can easily find the wave function $\phi$ produced by a line source of length $W$ on the $u$ axis; see **Fig. 16.2**. The source is assumed to emit at a single spatial frequency $1/\lambda$. Function $\phi$ can then be written in terms of an expansion in plane waves:

$$\phi(u^0, x^5) = \int_{-\infty}^{\infty} e^{-2\pi i(X_5 x^5 + U_0 u^0)} dU^0 . \qquad (16.2)$$

Inserting this into (16.1) we obtain the on-shell condition $X_5^{\ 2} + U_0^{\ 2} - 1/\lambda^2 = 0$, which may be solved for $X^5$:

$$X_5 = \begin{cases} \sqrt{1/\lambda^2 - U_0^{\ 2}}, & |U_0| \le 1/\lambda \\ -i\sqrt{U_0^{\ 2} - 1/\lambda^2}, & |U_0| > 1/\lambda \end{cases} . \qquad (16.3)$$

Putting this into (16.2) and integrating, we find[3]

$$\phi(u^0, x^5) = \frac{\pi x^5}{i\lambda} \frac{H_1^{(2)}\left(\frac{2\pi}{\lambda}\sqrt{u_0^{\ 2} + x_5^{\ 2}}\right)}{\sqrt{u_0^{\ 2} + x_5^{\ 2}}}, \qquad (16.4)$$

where $H_1^{(2)}(z) = J_1(z) - iY_1(z)$ is a Hankel function of order 1. As shown in **Fig. 16.2**, the wave front is a circular cylinder.

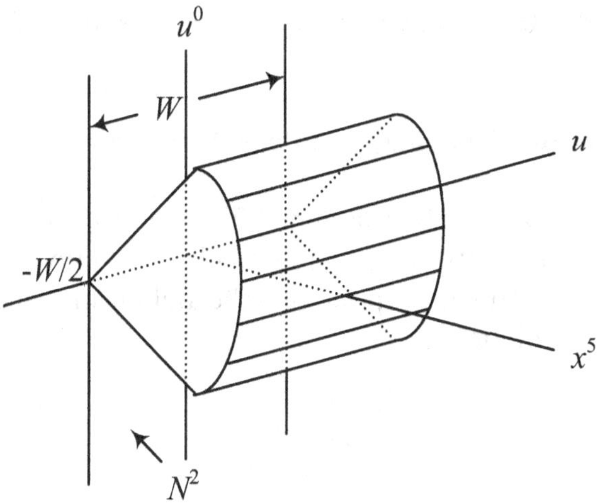


**Fig. 16.2** Line source of length $W$ on the $u$ axis of inner spacetime $N^2$ produces an expanding wave field whose wave front is a circular cylinder.

For many problems in diffraction and interference—including the present one of self-imaging charge—an approximate form of integral (16.2) suffices. If the angular spread $\Delta(\lambda U^0)$ in $\phi$ is small, so that $\lambda^2 U_0^{\ 2} \ll 1$, then the on-shell condition (16.3) approximates to the quadratic form $X^5 \cong 1/\lambda - \lambda U_0^{\ 2}/2$. Putting this into (16.2) we obtain

$$\begin{aligned} \phi(u^0, x^5) &= e^{-2\pi i x^5/\lambda} \int_{-\infty}^{\infty} e^{\pi i \lambda x^5 U_0^{\ 2} - 2\pi i U_0 u^0} dU^0 \\ &= \frac{e^{i\pi/4 - 2\pi i x^5/\lambda}}{\sqrt{\lambda x^5}} e^{-\frac{i\pi}{\lambda x^5} u_0^{\ 2}} \end{aligned} \tag{16.5}$$

This expresses $\phi$ in the *parabolic approximation*, so-called because the circular cylinder wave front of (14.4) is approximated by the parabolic one of (16.5). Its Fourier transform $\Phi$, read off directly from the integrand of the first line of (16.5), is

$$\begin{aligned} \Phi(U_0, x^5) &= \int_{-\infty}^{+\infty} \phi(u^0, x^5) e^{2\pi i U_0 u^0} du^0 \\ &= e^{-2\pi i x^5/\lambda + \pi i \lambda x^5 U_0^{\ 2}} . \end{aligned} \tag{16.6}$$

Note that, at $x^5 = 0$, the wave function $\phi$ reduces to the delta function $\delta(u^0)$.

## 16.3 In-focus (Fourier) images: integral charge

The subspace $\bar{M}^2 \times N^2$ is now assumed filled with a charge wave field, $q(u^0, x^5)$. Let $q(u^0, 0)$ denote the field distribution in the plane $x^5 = 0$. Then by Huygens' principle, the field $q$ in planes $x^5 > 0$ is given by the convolution

$$q(u^0, x^5) = \int_{-\infty}^{+\infty} q(t, 0)\phi(u^0 - t, x^5)dt \tag{16.7}$$

where, in the parabolic approximation, $\phi$ is the wave field (16.5). By the convolution theorem (16.7) may also be written in terms of a Fourier transform

$$q(u^0, x^5) = \int_{-\infty}^{+\infty} Q(U_0, 0)\Phi(U_0, x^5)e^{-2\pi i U_0 u^0} dU_0 \tag{16.8}$$

where

$$Q(U_0, 0) = \int_{-\infty}^{+\infty} q(u^0, 0)e^{2\pi i U_0 u^0} du^0 \tag{16.9}$$

and $\Phi$ is given by (16.6).

We now define the source field $q(u^0, 0)$ to be a phase-coherent, periodic function of $u^0$, extending from $u^0 = +\infty$ to $u^0 = -\infty$. Such a system is known to be self-imaging. Interference between the multitude of secondary wavelets $\phi$ emitted from the source plane produces, in a plane parallel to and at a distance $\Delta x^5$ downstream from it, an amplitude distribution proportional to that of the source. In fact, downstream from the source, there are formed in parallel planes an infinite number of such images, spaced axially apart by $\Delta x^5$. In the literature, images formed in this way are known as Fourier images.[4] To derive the interval $\Delta x^5$, we write the source distribution $q(u^0, 0)$ as the Fourier series

$$q(u^0, 0) = \frac{1}{a}\sum_{h=-\infty}^{+\infty} C_h e^{-2\pi i h u^0 / a}, \tag{16.10}$$

where $a$ is the period of the source field and the coefficients

$$C_h = \int_{-a/2}^{a/2} q(u^0,0)e^{2\pi ihu^0/a}du^0 . \tag{16.11}$$

The Fourier transform of the source field is, from (16.9), the $\delta$-function array

$$Q(U_0,0) = \frac{1}{a}\sum_{h=-\infty}^{\infty} C_h\delta(U_0 - h/a). \tag{16.12}$$

Substituting for $Q$ and $\Phi$ in the integrand of (16.8) using (16.12) and (16.6) yields the series

$$q(u^0,x^5) = \frac{e^{-2\pi ix^5/\lambda}}{a}\sum_{h=-\infty}^{+\infty} C_h e^{\pi ih^2\lambda x^5/a^2 - 2\pi ihu^0/a} . \tag{16.13}$$

So, for positive integers $p$, in planes $x^5 = 2pa^2/\lambda$ perpendicular to the $x^5$ axis we obtain interference images—Fourier images—proportional to the original distribution (16.10):

$$q(u^0,x^5) = e^{-4\pi ipa^2/\lambda^2} q(u^0,0) \tag{16.14}$$

As illustrated in **Fig. 16.3**, the spacing between successive images is

$$\Delta x^5 \equiv D = \frac{2a^2}{\lambda}, \tag{16.15}$$

a result first deduced by Rayleigh.[5]

The result (16.14) can be generalized, giving the relation between the field amplitudes in any two planes separated by distance $D$. In (16.13) we replace $x^5$ by $x^5 + D$. We then find that

$$q(u^0,x^5 + D) = e^{-4\pi ia^2/\lambda^2} q(u^0,x^5) \tag{16.16}$$

So, apart from a phase factor, the charge amplitude is periodic in $x^5$, with period $D$. The amplitude in any plane $x^5 = \text{const.}$ is the Fourier image of the amplitude in the plane distance $D$ upstream from it.

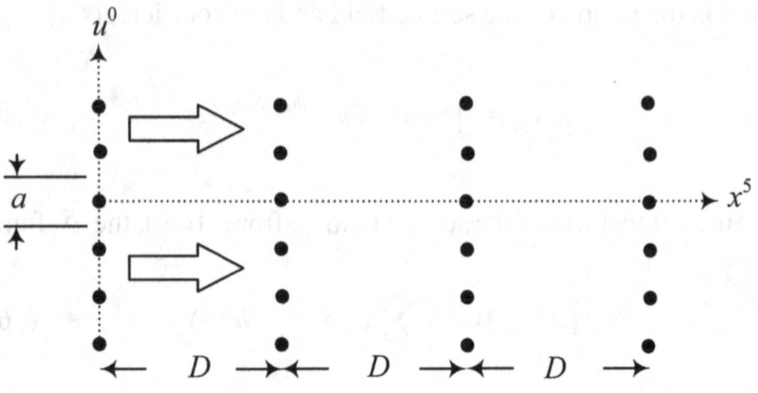


**Fig. 16.3** Interference (Fourier) images of an infinite, wave-emitting periodic source occupying the plane $x^5 = 0$. Spaced at regular intervals $\Delta x^5 \equiv D = 2a^2 / \lambda$, the images are exact images of the source. Here $a$ is the repeat interval of the source and $\lambda$ the emission wavelength. The periodic structure depicted consists of an infinite lattice of $\delta$-functions with complex amplitude factors $A$. With $\mathrm{Re}(A) = 1$, the lattice points represent the integer charge of the positron.

Now if we were doing ordinary wave optics in $M^4$, the phase factors appearing in (16.14) and (16.16) would be of marginal interest. One would square the amplitude to obtain the intensity, giving in the case of (16.14) an exact intensity image of the source. But we are not doing ordinary wave optics. Because there is no time dimension driving wave equation (16.1), there is no energy either. Consequently there is no cause to square the amplitude. Indeed, since charge carries both magnitude and sign, *physical charge is to be identified with the real part of the complex charge amplitude, q.* Thus we require exact self-imaging, not of intensity, but of amplitude, and to achieve this the exponential factors appearing in (16.14) and (16.16) must equal 1. We thus have the following restriction on the ratio of repeat interval $a$ and wavelength $\lambda$:

$$2a^2 / \lambda^2 = N, \qquad (16.17)$$

where $N$ is a positive integer.

There is a yet second demand to be made on the exponential factor in (16.16). With the assumption that charge originates in spacetime, one still expects spacetime to exhibit zero net charge. To meet this expectation, the argument of the exponential must be such as to provide zero net (integrated) charge in each period of length $D$ of the charge amplitude, $q$. Accordingly, in (16.13) we replace $x^5$ by $x^5 + D/2$. We then find that

$$q\left(u^0, x^5 + \frac{D}{2}\right) = e^{-N\pi i} q\left(u^0 - \frac{a}{2}, x^5\right). \tag{16.18}$$

So if integer $N$ in (16.16) is *odd*, then the amplitude in the plane located at $x^5 + D/2$ is just the negative of the amplitude in the plane located at $x^5$, the latter amplitude shifted laterally by one-half the source period, $a$.[6] With $N$ an odd integer, spacetime is assured to be charge neutral.

The above results are general. They hold good for the self-imaging of any infinite, periodic, phase-coherent source field, $q(u^0, 0)$. The specific source distribution required to generate the real word of particle-bearing charge is shown in **Fig. 16.4**. It comprises an infinite

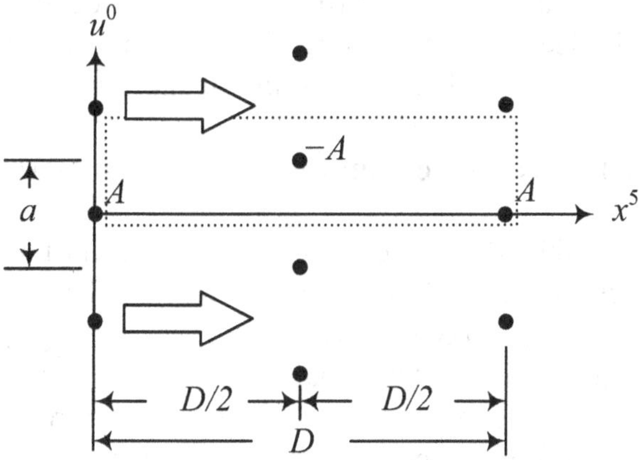


**Fig. 16.4** By interference, a lattice of $\delta$-functions of amplitude $A$ in plane $x^5 = 0$ images itself with amplitude $A$ in the plane $x^5 = D = 2a^2/\lambda$, where $a$ is the period of the lattice and $\lambda$ the wavelength of the propagating field. In accordance with Eq. (16.18) an intermediate image of amplitude $-A$ is formed as well in the plane $x^5 = D/2$, displaced laterally relative to the original lattice by $a/2$. The dotted-line box encloses one $a \times D$ period of the doubly-periodic field amplitude. The box contains one charge of amplitude $A$ and one of amplitude $-A$. With $\mathrm{Re}\,A = 1$, the two reconstructed lattice points represent the integer charges of the positron and electron, respectively. The net charge enclosed by the box is zero.

lattice of points of amplitude $A\delta(u^0 - ha)$, $h = \cdots, -1, 0, +1, \cdots$. Here $A$ is a complex constant $|A|\exp(i\varphi)$ whose amplitude and phase are to be determined, subject to the condition

$$\mathrm{Re}\, A = |A| \cos\varphi = 1 . \tag{16.19}$$

Since according to (16.16) the charge amplitude is periodic in $x^5$, it suffices to describe the image structure within one period $D$ downstream from the source. As shown in the figure, by interference alone, an exact image of amplitude $A$ is produced in the plane $x^5 = D$. This is the Fourier image of the source. In accordance with (16.18) an intermediate image of amplitude $-A$ is formed as well, midway between the source and Fourier image; it is displaced laterally by one-half the source repeat interval, $a$. Thus within one $a \times D$ period of the doubly-periodic field amplitude there are formed two charges of opposite sign. In the light of(16,19), these are to be identified with the integer charges of the positron and electron. Moreover, the two charges sum to zero, respecting the assumed charge-neutrality of spacetime.

## 16.4 Out-of-focus (Fresnel) images: one-third integral charge

We turn now to the interference formation of one-third integral charge. Like the integer charge of the electron, the fractional charges are formed in planes intermediate between successive Fourier-image planes. In the optical literature these intermediate—or out-of-focus—patterns are termed *Fresnel images*,[7] As we shall see, the real parts of the amplitudes of these intermediate patterns represent the charges of both quarks and antiquarks. Moreover, these charges are grouped just as they are found in the proton, neutron and their respective antiparticles. In turn, the proton, neutron, electron and antiparticle charges are grouped as they are found in the deuterium atom and its respective antiparticle.

To show all this we require a description of the Fresnel image amplitudes. Again, because the charge amplitude is periodic in $x^5$, we limit our attention to patterns formed within one period $D$ downstream from the source. The distance $x^5$ from the source plane to any Fresnel-image plane between the source and first Fourier image may be expressed in the form

$$x^5 = \beta D = 2\beta a^2 / \lambda , \tag{16.20}$$

where $\beta$ is a pure number—a fractional distance—defined for the range

$0 < \beta < 1$.

As in the case of the Fourier images, the Fresnel-image amplitudes are given by the convolution integral (16.7). However, for computing the Fresnel images, the Fourier series representation of the source (16.10) yields an unnecessarily complicated result. A better approach is to write the source field as a convolution of the *unit-cell function* $c(u^0)$ and a *lattice-generating function* $l(u^0)$:

$$q(u^0,0) = c(u^0) * l(u^0) \tag{16.21}$$

where, as depicted in **Fig. 16.4**,

$$c(u^0) = A\delta(u^0) \tag{16.22}$$

and

$$l(u^0) = \sum_{h=-\infty}^{\infty} \delta(u^0 - h/a). \tag{16.23}$$

In terms of these functions (16.7) becomes

$$\begin{aligned} q(u^0,x^5) &= c(u^0) * l(u^0) * \phi(u^0,x^5) \\ &= \frac{Ae^{i\pi/4}}{\sqrt{\lambda x^5}} e^{-2\pi i x^5/\lambda} \sum_{h=-\infty}^{\infty} \int \delta(t-ha) e^{-\pi i(t-u^0)^2/\lambda x^5} dt \\ &= \frac{|A| e^{i(\varphi+\pi/4)}}{\sqrt{\lambda x^5}} e^{-2\pi i x^5/\lambda - \pi i u_0^2/\lambda x^5} \sum_{h=-\infty}^{\infty} e^{-\pi i a^2 h^2/\lambda x^5 + 2\pi i h a u_0/\lambda x^5} \end{aligned} \tag{16.24}$$

We proceed to calculate the Fresnel image amplitudes for the following values of parameter $\beta$:

$$\beta = \frac{1}{6}, \frac{2}{6}, \frac{3}{6}, \frac{4}{6}, \frac{5}{6} \tag{16.25}$$

As will be shown, the first two $\beta$-values yield the charges of the quarks, and the last two the charges of the antiquarks. The middle value, reproduces the charge of the electron already calculated above, and serves a check on this alternative method of calculation. For the first three values of (16.25) we write $\beta = 1/n$, eventually letting $n=6$, 3 and 2.

For these three cases (16.25) becomes, with $x^5 = D/n = 2a^2/n\lambda = N\lambda/n$,

$$q(u^0, D/n) = \sqrt{\frac{n}{2}}|A|\frac{e^{i(\varphi+\pi/4)}}{a}e^{-2\pi iN/n - n\pi iu_0^2/2a^2}\sum_{h=-\infty}^{\infty}e^{-n\pi ih^2/2 + n\pi ihu_0/a}. \quad (16.26)$$

We now write $h = 2H + U$, where $H$ and $U$ are integers, and replace the single sum over $h$ with the double sum over $H$ and $U$:

$$\sum_{h=-\infty}^{\infty} \to \sum_{H=-\infty}^{\infty}\sum_{U=0}^{1}.$$

Eq. (16.26) can then be written as a sum of $\delta$-functions:

$$q(u^0, D/n) = \sum_{H=-\infty}^{\infty} Q_H \delta\left(u^0 - H\frac{a}{n}\right) \quad (16.27)$$

where

$$Q_H = |A|\frac{e^{i(\varphi+\pi/4)-2\pi iN/n}}{\sqrt{2n}}e^{-\pi iH^2/2n}\left(1+e^{-\pi in/2+\pi iH}\right)$$

$$= |Q_H| \arg Q_H \quad , \quad (16.28)$$

$$|Q_H| = \sqrt{\frac{2}{n}}|A|\left|\cos\left[\frac{\pi}{2}\left(H-\frac{n}{2}\right)\right]\right|, \quad (16.29)$$

$$\arg Q_H = \varphi + \frac{\pi}{4} - \frac{2\pi N}{n} - \frac{H^2\pi}{2n} + \arctan\left\{\frac{\sin\left[\pi\left(H-\frac{n}{2}\right)\right]}{1+\cos\left[\pi\left(H-\frac{n}{2}\right)\right]}\right\}, \quad (16.30)$$

and $N$ is an odd integer. The $\delta$-functions in (16.27) represent repeated images of the source, spaced apart from each other by $\Delta u^0 = a/n$. Let us test this formula with the one case for which we already know the answer, namely, $\beta = 3/6 = 1/2$. Putting $n=2$ into (16.28) we obtain the values listed in **Table 16.1**.

**Table 16.1** Coefficients $Q_H$ for $\boldsymbol{\beta} = \mathbf{1/2}$

| $H$ | $\lvert Q_H \rvert$ | $\arg Q_H$ |
|---|---|---|
| 0 | 0 | ------- |
| 1 | $\lvert A \rvert$ | $\varphi - \pi$ |

These values reproduce precisely the intermediate image (16.18) illustrated in **Fig. 16.4.** Our formula (16.28) appears to be in working order. Interestingly it shows that the lateral shift of the image of the source by $a/2$ is an illusion created by the vanishing of coefficients $Q_H$ for even $H$.

We next consider the cases $n=6,\ 3$. We see that there is a problem evaluating the $Q_H$, because integer $N$ is unknown, apart from its being odd. To fix the value of $N$—or more relevantly the value of the exponential in which it appears—we shall demand net charge neutrality, not only for the full period $0 < x^5 \le D$, but for the half-period $0 < x^5 \le D/2$ as well. In anticipation of this we define $N = 2M - 1$, with $M$ an integer to be determined by the charge-neutrality condition. We then obtain for $n=6$ the coefficients listed in **Table 16.2.**

**Table 16.2** Coefficients $Q_H$ for $\boldsymbol{\beta} = \mathbf{1/6}$

| $H$ | $\lvert Q_H \rvert$ | $\arg Q_H + 4\pi M/6$ |
|---|---|---|
| 0 | 0 | ------- |
| 1 | $\lvert A \rvert/\sqrt{3}$ | $\varphi + \pi/6$ |
| 2 | 0 | ------- |
| 3 | $\lvert A \rvert/\sqrt{3}$ | $\varphi - \pi/2$ |
| 4 | 0 | ------- |
| 5 | $\lvert A \rvert/\sqrt{3}$ | $\varphi + \pi/6$ |

Similarly, for $n=3$ we obtain the coefficients listed in **Table 16.3.**

**Table 16.3** Coefficients $Q_H$ for $\boldsymbol{\beta}=\mathbf{1/3}$

| $H$ | $\lvert Q_H \rvert$ | $\arg Q_H + 4\pi M/3$ |
|---|---|---|
| 0 | $\lvert A \rvert/\sqrt{3}$ | $\varphi+\pi/6$ |
| 1 | $\lvert A \rvert/\sqrt{3}$ | $\varphi-\pi/2$ |
| 2 | $\lvert A \rvert/\sqrt{3}$ | $\varphi-\pi/2$ |

Now what about the last two values of (16.25), $\beta=2/3$ and $5/6$? By (16.18) the amplitudes in these planes are just the negatives of the amplitudes in planes $\beta=1/6$ and $1/3$, respectively, but shifted laterally by $\Delta u^0=a/2$. And so the amplitudes in all five intermediate planes defined by (16.25) are accounted for.

### 16.5 Final calculation of charge values

Now let us assume as a trial case that *M is an integer multiple of* 3. Then the exponential factor containing *M* becomes unity for all *n*. From the data of **Tables 16.1-3**, by taking the real parts of the amplitudes, we find for the total charge in the half-period $0<x^5\le D/2$

$$Q_{\text{TOT}}=\lvert A\rvert\left\{-\cos\varphi+3\cdot\frac{1}{\sqrt{3}}\left[\cos\left(\varphi+\frac{\pi}{6}\right)+\cos\left(\varphi-\frac{\pi}{2}\right)\right]\right\}. \tag{16.29}$$

As stated above this total charge must vanish, providing charge neutrality of spacetime. We see from the plot of **Fig. 16.5** that $Q_{\text{TOT}}$ vanishes for both $\varphi=-\pi/6$ and $5\pi/6$. To satisfy (16.19) one is obliged to take $\varphi=-\pi/6$. We then have

$$\lvert A\rvert=1/\cos(-\pi/6)=2/\sqrt{3},$$

leading to the charge values listed in **Table 16.4.**

In the Fresnel image plane $\beta=1/6$ we find in one period $0\le u^0<a$ the 1/3-integral order charges of the three quarks forming the proton. In

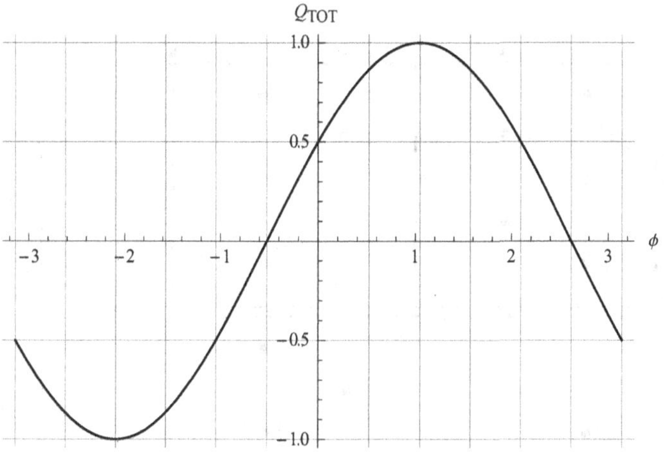


**Fig. 16.5**. Plot of the total charge $Q_{TOT}$ in the half-period $0 < x^5 \leq D/2$ as a function of the phase $\varphi$ of the source amplitude *A*. Vertical grid lines are spaced at intervals $\pi/6$. $Q_{TOT}$ vanishes at $\varphi = -\pi/6$ and $5\pi/6$.

**Table 16.4** Charge $Q$ at Fresnel image points in the full-period $0 \leq u^0 < a \times$ half-period $0 < x^5 \leq D/2$

| H | $Q = \lvert Q_H \rvert \cos(\arg Q_H)$ | | |
|---|---|---|---|
| | $\beta$=1/6 | $\beta$=1/3 | $\beta$=1/2 |
| 0 | 0 | 2/3 | 0 |
| 1 | 2/3 | -1/3 | -1 |
| 2 | 0 | -1/3 | |
| 3 | -1/3 | | |
| 4 | 0 | | |
| 5 | 2/3 | | |

plane $\beta = 1/3$ we have the charges of the three quarks comprising the neutron. And in plane $\beta = 1/2$ we have the charge of the electron, already shown in **Fig. 16.4.** The charges in planes $\beta = 2/3,\ 5/6,\ 1$ are the negatives of those in planes $\beta = 1/6,\ 1/3,\ 1/2$, but shifted laterally by $a/2$. All image points, together with their corresponding charge values, occurring within one $a \times D$ period of the doubly periodic amplitude pattern, are shown in **Fig. 16.6**. The charges of all physical particles appear to be accounted for, confirming that in taking *M* to be an integer multiple of 3, we have guessed correctly.

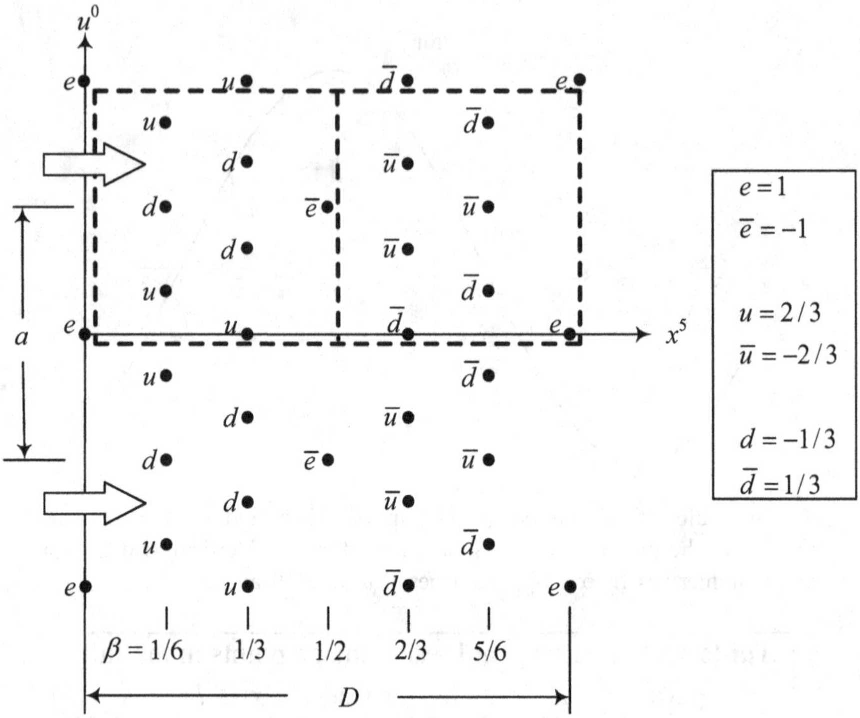


**Fig. 16.6** A wave-emitting periodic lattice of unit charge $+1$ in plane $x^5 = 0$ creates two types of interference pattern: (1) an exact self-image (Fourier image) in plane $x^5 = D$; and (2) intermediate (Fresnel) images in planes $x^5 = \beta D$, where the fractional distances $\beta = 1/6$, $1/3$, $1/2$, $2/3$ and $5/6$. Each image point is labeled by its charge, the code for which is given in the box at the right of the figure. The overall interference pattern is doubly-periodic, with period $D$ in the $x^5$ direction and period $a$ in direction $u^0$. The dashed $D \times a$ rectangle encloses one period of the doubly-periodic pattern. In the left half of this dashed rectangle we find in plane $\beta = 1/6$ images of charge $2/3$, $-1/3$ and $2/3$, i.e., the charges of the quarks making up the proton. In plane $\beta = 1/3$ we have images of charge $2/3$, $-1/3$ and $-1/3$, i.e., the charges of the quarks comprising the neutron. In plane $\beta = 1/2$ is a single image of charge $-1$, i.e., charge of the negatively charged heavy leptons and of the $W^-$ boson. Altogether, in this half-period one has all the charge elements of an atom of deuterium. The charge images in the right half of the dashed $D \times a$ rectangle are the negatives of those in the left half, and shifted in the $u^0$ direction by one half-period $a/2$. Thus, in the right half one has all the elements comprising the antiproton, antineutron, antileptons, the $W^-$ boson and antideuterium atom. The net charge in either half of the dashed $D \times a$ rectangle is zero.

## 16.6 Lattice structure of $M^4 \times \bar{M}^2 \times N^2$ spacetime

We have shown that a process of self-imaging in the subspace $\bar{M}^2 \times N^2$ can reproduce the charge values observed (or in the case of quarks, assumed to exist) in Minkowski $M^4$. In particular, in intermediate planes defined by (16.25), the process gives rise to the fractional charges of the quarks. And yet herein lies a problem. For Fresnel images are formed as well in planes other than those defined by (16.25), and the charges corresponding to them are not found in the known physical world. For example, for $n=4$, Eqs. (16.28-30) yield two image points within the dashed box at $\beta = 1/4$; their charges are $(1 \mp \sqrt{3})/2\sqrt{3}$. Since these charges—and those of other Fresnel images—are not seen, they must somehow be filtered out, rendered unobservable. The question is, how exactly does the filtering come about?

One way—and possibly the only way—is to assume that the product space $M^4 \times (\bar{M}^2 \times N^2)$ is not continuous, but instead forms an infinite lattice of $M^4$ hyperplanes perpendicular to (hyper)direction $\bar{M}^2 \times N^2$. As shown in **Fig. 16.7**, the lattice spacing is $D/6$, each hyperplane of

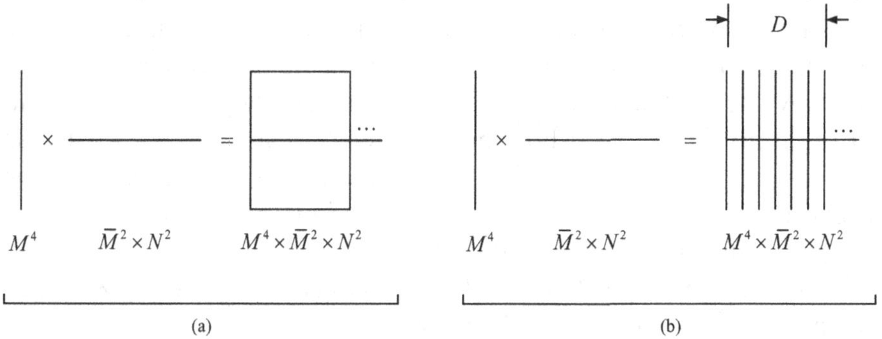


**Fig. 16.7** Forming the product space $M^4 \times (\bar{M}^2 \times N^2)$. (a) Conventional product, giving continuous 8-D manifold. (b) Product yields a discrete lattice of planes spaced $D/6$ apart, each plane coinciding with one of the charge-bearing planes of $\bar{M}^2 \times N^2$ defined by (16.25) and depicted in **Fig. 16.6.**

the lattice coinciding with one of the charge-bearing planes of $\bar{M}^2 \times N^2$ defined by (16.25). Note that in this lattice geometry wave propagation in the $x^5$ direction proceeds no differently than in the continuous geometry, thereby preserving momentum $p_5$, its invariance and conservation in all physical processes. The only effect of the lattice structure is to render unobservable in $M^4$ the images of charge formed between the lattice planes by the self-imaging process. Note also that the $M^4$ hyperplane shown in **Fig. 3.5** should now be replaced by a lattice of planes equivalent to that depicted in **Fig. 16.7 (b)**

## 16.7 Spacetime origin of the fine structure constant

An oddity of the self-imaging process is that the charge values generated by Eqs. (16.28-30) are independent of the repeat interval $a$ and propagation wavelength $\lambda$. Instead they depend only on the *ratio* of these parameters as defined via Eq. (16.17)—which leads us to inquire about the value of the integer $N$ defined by that equation. We know that, by constraining the ratio $a/\lambda$, it ensures exact self-imaging of charge amplitude. We know too that it must be odd to ensure the charge neutrality of spacetime; thus we write $N = 2M - 1$, where $M$ is a positive integer. Finally, we know that to generate physically-relevant charge values we must have $M = 3P$, where $P$ is a positive integer. Hence $N = 3(2P) - 1$. Now this a highly specialized form, and the question is whether it has any significance apart from its role in the determination of charge. For one thing we note that for certain values of $P$, $N$ "is an Eisenstein prime with no imaginary part and a real part of the form $3n-1$."[8] For another, we see that for $P = 23$ (also an Eisenstein prime), $N = 137$, very nearly the reciprocal of the Sommerfeld fine structure constant $\alpha = 1/137.036$. Could it be that $\alpha$ originates in extra-dimensional spacetime?

Let us combine (16.15) and (16.17) to obtain the ratio $D/\lambda$. Then on the assumption that $N = 137$ we have

$$\frac{D}{\lambda} = N = 137\,. \qquad (16.32)$$

If we can show that this is true, that $D/\lambda$ has the value 137, then we will have taken a significant step to showing that $\alpha$ is a bi-product of the structure of spacetime.

To do this we shall need the values of both $D$ and $\lambda$. As for $D$, we recall from the discussion of Eq. (10.35) that the radius of the electron's

Zitterbewegung, namely $a_e = \hbar / 2m_e c$ [see (10.15)], is in fact a wave propagation distance $\xi^5$ directed along coordinate $x^5$. During propagation of its guiding wave through this distance, point $P$ completes one circular orbit of circumference $\lambda_C / 2$ about the centroid of its internal motion, where $\lambda_C$ is the Compton wavelength of the electron. Reëntrance of the internal motion thus takes place over a propagation distance $2a_e$, and we may reasonably identify this distance with the Fourier-image repeat interval $D$ shown in **Fig. 16.6**:

$$D = \hbar / m_e c. \tag{16.33}$$

That is, the Fourier image distance $D$ is to be identified with the reduced Compton wavelength of the electron.

To get at wavelength $\lambda$ we propose first to define a fundamental length $\lambda_0$ by means of the following equation:[9]

$$\frac{e^2}{4\pi\varepsilon_0\lambda_0} = m_e c^2 \tag{16.34}$$

The l.h.s. of this relation is the work done in moving a first point electron in from infinity to a distance $\lambda_0$ from a second point electron. This energy is then equated to the rest energy of the electron. Note that this defining equation for $\lambda_0$ is a classical one, involving entities already connected to the structure of spacetime, namely the electron charge $-e$, mass $m_e$ [see (13.45)] and the permittivity of free space, $\varepsilon_0$ [see (6.76)]. Note also that $\lambda_0$ is a distance in ordinary Euclidian $E^3$, not in the extra manifold $\bar{M}^2 \times N^2$.

Now let us write down from (16.33) and (16.34) the ratio $D / \lambda_0$. This yields precisely the reciprocal fine-structure constant:

$$\frac{D}{\lambda_0} = \left( \frac{e^2}{4\pi\varepsilon_0 \hbar c} \right)^{-1} = \alpha^{-1}. \tag{16.35}$$

Combining (16.32) and (16.35) we can write

$$\frac{D}{\lambda} = \frac{\lambda_0}{\lambda}\frac{D}{\lambda_0} \equiv \nu\alpha^{-1} = 137 \tag{16.36}$$

where

$$\nu = \frac{\lambda_0}{\lambda} = \frac{\alpha}{137^{-1}} = 0.9997\,. \tag{16.37}$$

Here the constant $\nu$ has the formal appearance of an index of refraction, a ratio of the wavelengths of a propagating wave field in two different transparent media. In such an interpretation, the numbers $\alpha$ and $137^{-1}$ represent dimensionless propagation velocities in the two media. The two media in question are, of course, $M^4$ and $\bar{M}^2 \times N^2$. We conclude that there are in reality *two* fine- structure constants: one of them, $137^{-1}$, defined in $\bar{M}^2 \times N^2$; and the other, $\alpha$, defined in $M^4$. The two are linked numerically by the refractive index $\nu$ defined by (16.37). If one regards $137^{-1}$ as fundamental, then $\alpha$ can be said to arise from that reciprocal prime number, and the perennial mystery of the origin of $\alpha$ is resolved.

**Notes and references**


[1] F. Halzen and A. D. Martin, *Quarks and Leptons: An Introductory Course in Modern Particle Physics* (John Wiley & Sons, New York, 1984), p. 352.

[2] G. Kane, *Modern Elementary Particle Physics* (Perseus, Cambridge, MA, 1993), p. 278.

[3] W. Magnus and F. Oberhettinger, *Formulas and Theorems for the Functions of Mathematical Physics* (Chelsea, New York, 1949), pp. 117, 118.

[4] J. M. Cowley and A. F. Moodie, “Fourier Images: I – The Point Source,” Proc. Phys. Soc. B **70**, 486 (1957).

[5] Lord Rayleigh, Phil. Mag. **11**, 196 (1881).

[6] The lateral shift between interference patterns separated axially by *D*/2 was reported by B. Cook, “Interference Patterns of Ultrasonic Optical Gratings,” J. opt. Soc. Am. **53**, 429 (1963).

[7] J. T. Winthrop and C. R. Worthington, “Theory of Fresnel Images I. Plane Periodic Objects in Monochromatic Light,” J. Opt. Soc. Am. **55**, 373 (1965).

[8] See Wikipedia articles “Eisenstein prime” and “137 (number).”

[9] See Wikipedia article “Fine-structure constant.”

# Chapter 17

# Dual spacetime, dual time series and reduction of the state vector

But at my back I always hear
Time's wingèd chariot hurrying near;
And yonder all before us lie
Deserts of vast eternity.

ANDREW MARVEL
TO HIS COY MISTRESS

Dualisms of one kind or another lie at the heart of this book. We began in **Part I** with the metaphysical dualism of appearance and reality—of phenomenon and noumenon. Our goal of unifying that fundamental dualism led to the Principle of True Representation and formulation of the Law of Laws, Eq. (2.6). Next we encountered the dualism of the quantum microworld, that of wave and particle. Wave-particle duality, as noted in *Sec. 3.9.3*, is implicit already in the Law of Laws, and has formed the basis for much of our work up to this point, in particular the descriptions in **Part IV** of the internal structure of the field quanta. In the present chapter we introduce a new dualism, that of *dual spacetime*. Its existence and structure, hidden until now, follow directly from the six-dimensional Dirac equation introduced in **Chapter 4.** As we shall see, dual spacetime appears to resolve an ancient question on the nature of time, while throwing new light on yet another dualism, namely, the quantum state vector's expression of the "possible" before measurement, and of the "actual" after measurement.[1] It proves central, too, in our later discussion in **Part VII** of the most mysterious and contentious dualism of all, that of mind and brain.

## 17.1 Algebraic transformation of the Dirac equation in six dimensions

The uncovering of dual spacetime requires two steps: First, we transform the six-dimensional Dirac equation so that it *looks* four-dimensional. Then we show that the transformed equation admits of two, and only two, physically valid finite rotations in the $x^0$-$x^5$ plane, namely, rotations of $\pm\pi/2$. The original and one or the other of the two finitely-rotated frames represent separate realities and, superimposed, form dual spacetime.

In six dimensions the Dirac equation reads

$$\left[\left(\hat{p}^6+\gamma^5\hat{p}_5\right)\gamma^\mu\hat{p}_\mu-\widehat{m^2c^2}\right]\chi\left(x^\mu,x^5,x^6\right)=0, \tag{17.1}$$

where $\widehat{m^2}$ is the mass-squared operator (4.36). As the momenta $p^5=p_5$ and $-p_6=p^6\equiv G$ are invariants, the 6-D wave function $\chi$ is expressible in terms of a 4-D wave function $\psi\left(x^\mu\right)$:

$$\chi\left(x^\mu,x^5,x^6\right)=\psi\left(x^\mu\right)\exp\left[-i\left(p_5x^5+p_6x^6\right)/\hbar\right]. \tag{17.2}$$

Putting this into (17.1) we obtain

$$\left[\left(G+\gamma^5 p_5\right)\gamma^\mu\hat{p}_\mu-m^2c^2\right]\psi(x^\mu)=0. \tag{17.3}$$

This equation, while dependent on a reduced number of dimensions, is still embedded in 6-D $M^4\times\bar{M}^2$, and the presence in it of the metric operator $G+\gamma^5 p_5$ is evidence of this. In fact, owing to $\gamma^5$, neither (17.1) nor (17.3) are covariant under space inversion $\mathbf{x}\rightarrow-\mathbf{x}$, except in the case of the electron, for which $p_5=0$. This is a reflection of the fact that, in six dimensions, all leptons (except the electron) carry an intrinsic handedness, one that cannot be transformed away by Lorentz boost. The source of this handedness is the *Zitterbewegung*, which generates for the neutrino, and for the heavy leptons mu and tau, the helical internal geometries depicted classically in **Figs. 10.1** and **10.3**. The electron carries no intrinsic handedness and accordingly the Dirac equation describing its motion conserves parity. One can say that (17.3) describes accurately the wave motion of all leptons while respecting the internal geometries of these fundamental particles.

At the same time we know that the behavior in external fields of the charged leptons mu and tau is well-described by the standard 4-D Dirac equation. Therefore (17.3), while good as it stands, should be transformable into an equation of motion that at least *looks* four-dimensional. This is readily accomplished. A transformation with Hermitian matrix $S$ is made such that

$$\psi \rightarrow S\psi = \Psi \tag{17.4}$$

where $S^{-1}S = I$. Applying this to (17.3) we see that we get the standard form of the Dirac equation

$$\begin{aligned}&\left(\gamma^{\mu}\hat{p}_{\mu} - mc\right)\Psi(x^{\mu})\\&= \left(\gamma^{0}\hat{p}_{0} - \boldsymbol{\gamma}\cdot\hat{\mathbf{p}} - mc\right)\Psi(x^{\mu}) = 0\end{aligned} \tag{17.5}$$

provided that

$$S\gamma^{\mu}\left(G - \gamma^{5}p_{5}\right)S^{-1} = mc\gamma^{\mu}, \tag{17.6}$$

and where $m \neq 0$. A transformation matrix meeting condition (17.6) is

$$\left.\begin{aligned}S &= a - b\gamma^{5}\\S^{-1} &= a + b\gamma^{5}\end{aligned}\right\}, \tag{17.7}$$

where

$$\left.\begin{aligned}a &= \sqrt{(G + mc)/2mc}\\b &= \sqrt{(G - mc)/2mc}\end{aligned}\right\}. \tag{17.8}$$

Note that the transformed equation (17.5) now conserves parity, thereby masking the internal helical geometries of neutrinos and of mu and tau. Geometrically, transformation matrix $S$ projects the 5-D equation (17.3) onto the 4-D hyperplane. This is illustrated in **Fig. 17.1** for a charged lepton, $\ell^{-}$. The extra-dimensional information is now contained in the new wave function $\Psi$. Proof of this is that all six components of the conserved probability current density, Eqs. (4.55-57), remain intact. In terms of $\Psi$ these become

$$\left.\begin{aligned}s^{\mu} &= mc^{2}\tilde{\Psi}\gamma^{\mu}\Psi\\s^{5} &= p^{5}c\tilde{\Psi}\Psi\\s^{6} &= p^{6}c\tilde{\Psi}\Psi\end{aligned}\right\}. \tag{17.9}$$

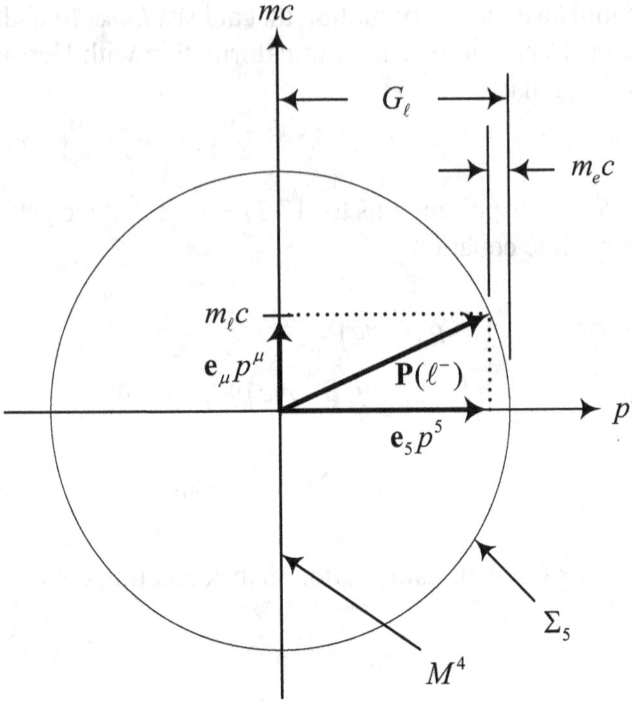


**Fig. 17.1.** Five-momentum vector $\mathbf{P} = \mathbf{e}_a p^a = \mathbf{e}_\mu p^\mu + \mathbf{e}_5 p^5$ defines motion of free charged lepton $\ell^-$ in $M^4 \times \bar{M}^2$. Unit vectors $\mathbf{e}_a$ satisfy $\mathbf{e}_a \cdot \mathbf{e}_b = g_{ab}$, ($a = 0,1,2,3,5$). As shown, resultant vector $\mathbf{P}$ and its transverse component $\mathbf{e}_\mu p^\mu$ lie in the plane of the paper. Vector $\mathbf{e}_\mu p^\mu$ is of length $m_\ell c$ and accordingly the vertical axis indicates mass. Vector $\mathbf{P}$ is of length $p^6 = G_\ell$, its tip lying in the 5-D momentum shell $\Sigma_5$. $M^4$= 4-D hyperplane. Five-dimensional wave equation (17.3) describes the motion of a particle characterized by the 5-momentum vector $\mathbf{P}$. Four-dimensional wave equation (17.5)—the result of Hermitian operator $S$ acting on spinor $\psi$—describes the same particle characterized by the $p^\mu$-component of $\mathbf{P}$, namely $p^\mu \mathbf{e}_\mu$. The apparent reduction in dimensionality of the wave equation is made possible by the fact that $p_5$ is invariant. Despite the reduction in dimensionality, the identities of neither $p^5$ nor $p^6$ are lost. The particle retains all components of the conserved probability 6-current, Eqs. (17.9).

The point is that, although the 6-D wave equation has been transformed into one that *looks* 4-D, the neutrinos and heavy charged leptons mu and tau still require six dimensions for their full description.

## 17.2 Finite rotation transformation

We propose now to examine the properties of the transformed wave equation (17.5) under rotation in the $x^0$-$x^5$ plane; see **Fig. 17.2**. This is a circular rotation, not a relativity rotation, because both coordinates are metrically time-like. Nevertheless, there is only one time, namely $x^0$. So in carrying out the rotation we are mixing time with a quasi-spatial coordinate, $x^5$, whose signature happens to be time-like.

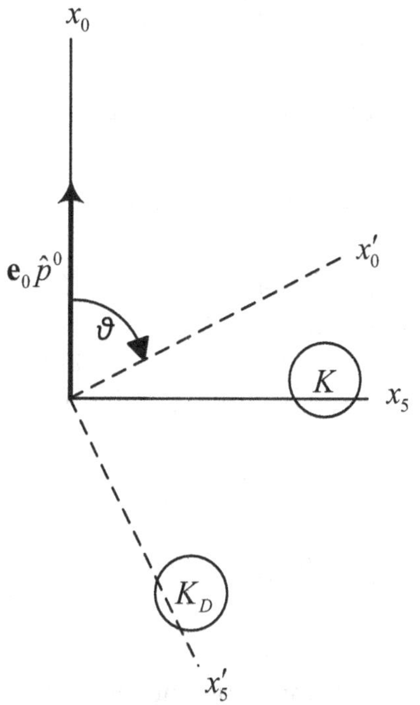


**Fig. 17.2** Coordinate frame $K_D(x_0', \mathbf{x}, x_5')$ is obtained by rotating frame $K(x_0, \mathbf{x}, x_5)$ about the **x** axis (into the paper) through an angle $\vartheta$. Of the four components of momentum operator $\hat{\mathbf{P}} = \mathbf{e}_\mu \hat{p}^\mu$ (not shown) only $\mathbf{e}_0\hat{p}^0$ acquires a new representation in $K_D$. Subscript $D$ stands for "dual frame."

Let us call the original reference frame $K(x_0, \mathbf{x}, x_5)$ and the frame obtained by rotation about the **x** axis $K_D(x_0', \mathbf{x}, x_5')$. (Since $p^6 = G_\ell$ is an invariant, one buried in mass $m_\ell$, we omit for brevity reference to $x^6$ in our original and rotated frames, $K$ and $K_D$.) Here the subscript $D$ stands for "dual frame," the significance of which term will emerge below. The Dirac equation (17.5) will be form invariant under rotation angle $\vartheta$

if we define

$$\Psi(x_0,\mathbf{x}) \to \Psi'(x_0',\mathbf{x}',x_5') = \mathbb{R}\Psi(x_0,\mathbf{x}), \tag{17.10}$$

where

$$\left.\begin{aligned} x_0 &= x_0' \cos\vartheta - x_5' \sin\vartheta \\ \mathbf{x} &= \mathbf{x}' \\ x_5 &= x_0' \sin\vartheta + x_5' \sin\vartheta \end{aligned}\right\} \tag{17.11}$$

and

$$\mathbb{R} = \exp\left(\frac{\vartheta}{2}\gamma^0\gamma^5\right). \tag{17.12}$$

Applying these definitions to (17.5) we obtain

$$\left[(\gamma^0 \cos\vartheta - \gamma^5 \sin\vartheta)\hat{p}_0 - \boldsymbol{\gamma}\cdot\hat{\mathbf{p}} - mc\right]\Psi' = 0, \tag{17.13}$$

where we have used the relation $(\mathbb{R}^{-1})^2 = \cos\vartheta - \gamma^0\gamma^5 \sin\vartheta$. Now the piece of the original momentum operator $\hat{\mathbf{P}} = \mathbf{e}_a\hat{p}^a$ lying in the $x^0$-$x^5$ plane of frame $K$ has components $(\hat{p}_0, 0)$, as shown in **Fig. 17.2**. These same components expressed in terms of frame $K_D$ are, by the chain rule,

$$\hat{p}_0 = \hat{p}_0' \cos\vartheta - \hat{p}_5' \sin\vartheta \tag{17.14}$$

and

$$\hat{p}_5 = 0 = \hat{p}_0' \sin\vartheta + \hat{p}_5' \cos\vartheta, \tag{17.15}$$

where $\hat{p}_0 = i\hbar\partial/\partial x^0$, $\hat{p}_0' = i\hbar\partial/\partial x'^0$, etc. Replacing $\hat{p}_0$ in (17.13) by means of (17.14), then eliminating the $\vartheta$ dependence by means of (17.15), we obtain

$$\left(\gamma^0\hat{p}_0' - \boldsymbol{\gamma}\cdot\hat{\mathbf{p}} + \gamma^5\hat{p}_5' - mc\right)\Psi' = 0 \tag{17.16}$$

This is the representation of Dirac equation (17.5) in frame $K_D$ for arbitrary $\vartheta$. It has the appearance of a 5-D equation of motion, as one would expect, but there are several things wrong with it. (1) As neither $\hat{p}_0'$ nor $\hat{p}_5'$ is invariant, each operator being dependent on $\hat{p}_0$ through (17.14) and (17.15), spacetime no longer *looks* four-dimensional. (2) More seriously, as in the case of Eq. (4.7) (see **Sec. 4.3**), the Hamiltonian associated with (17.16) is not Hermitian. Consequently that equation, for arbitrary $\vartheta$, cannot be considered a valid equation of motion. (3) Nor can it be transformed into one by multiplying by a metric operator of the form $(mc+\gamma^5\hat{p}'_5)$ or $(mc+\gamma^0\hat{p}'_0)$, again for the reason that neither operator $\hat{p}'_5$ nor $\hat{p}'_0$ is invariant. (4) Most obviously, (17.16) fails to demonstrate form invariance of Dirac equation (17.5) under the operations indicated in (17.10).

## 17.3 Dual spacetime

However, these many shortcomings disappear in the special case that $\vartheta=+\pi/2$ or $-\pi/2$. For we then have from (17.15), $\hat{p}'_0=0$ and (17.16) becomes

$$\left(\gamma^5\hat{p}'_5-\boldsymbol{\gamma}\cdot\hat{\mathbf{p}}-mc\right)\Psi'(x'^5,\mathbf{x})=0\,. \tag{17.17}$$

And so in these frames, and only in these frames, we obtain an equation of motion that is formally the same as (17.5). Our 4-D Dirac equation (17.5) does indeed exhibit form invariance, but only between the two frames $K$ and $K_D$ differing by rotation angle $\pm\pi/2$. The frames $K$ and $K_D$ therefore represent separate realms of spacetime. They are superimposed one upon the other, two frames mutually inaccessible by infinitesimal rotation, comprising a composite or *dual spacetime*, $K+K_D$. We call frame $K_D$ the *dual* of frame $K$.

Now while Eqs. (17.5) and (17.17) are formally covariant, they differ markedly in their physical interpretations. As for (17.5), we note that wave function $\Psi(x^\mu)$ is independent of the fifth dimension $x^5$. Thus, as already depicted in **Fig. 3.5**, and again in **Fig. 17.3** below, $\Psi(x^\mu)$ represents the probability amplitude of an infinite line current running parallel to the $x^5$ axis of frame $K$, where the current density components $(s^0,s^k,s^5)$ are given by (17.9). The equation of motion in any $M^4$ cross section of this infinite current is dependent on the time, *ct*, and energy $p_0$. The particle momentum perpendicular to $M^4$ is the invariant pseudomomentum, $p_5$.

In contrast, the wavefunction $\Psi'(x'^5,\mathbf{x})$ of (17.17) is independent of the time, *ct*. Thus, as depicted in **Fig. 17.3**, $\Psi'(x'^5,\mathbf{x})$ represents the probability amplitude of an infinite line current running parallel to the $x'^0$ axis of frame $K_D$. Applying transformation (17.10) to the currents of (17.9), we find for the current density components in $K_D$

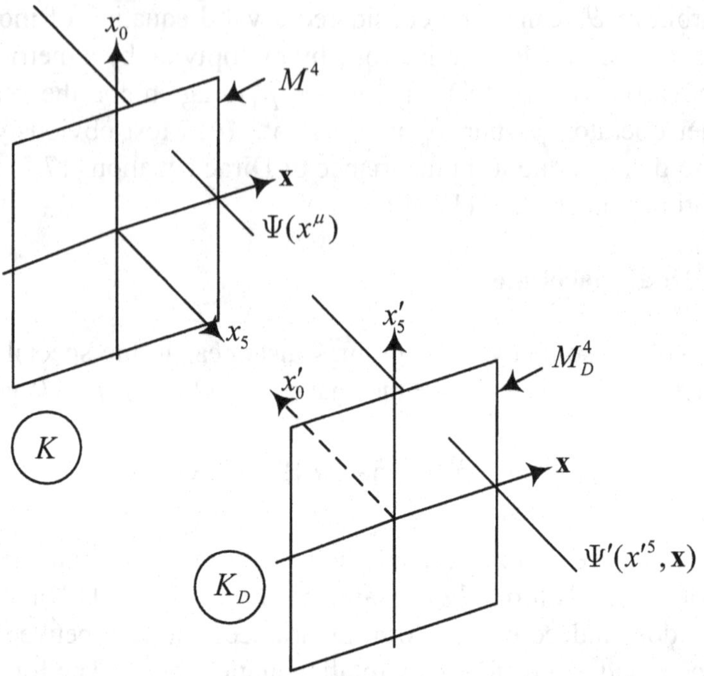


**Fig. 17.3** Dual spacetime consisting of superimposed frames $K$ and $K_D$, shown here separated for clarity. $K_D$ is obtained from $K$ by a $-\pi/2$ rotation in the $x^0$-$x^5$ plane. Wave function $\Psi(x^\mu)$ constitutes an infinite line current running parallel to the $x^5$ axis in $K$. Its intersection with hyperplane $M^4$ represents the probability density of an object in ordinary 4-space. Similarly, wave function $\Psi'(x'^5, \mathbf{x})$ constitutes an infinite line current running parallel to the $x'^0$ axis in $K_D$. Its intersection with dual hyperplane $M_D^4$ represents the probability density of the same object in dual 4-space.

$$\left.\begin{aligned} s^0 &\Rightarrow s'^5 = c\Psi'^\dagger\Psi' \\ s^k &\Rightarrow s'^k = c\Psi'^\dagger\gamma^5\gamma^k\Psi' \\ s^5 &\Rightarrow s'^0 = cp^5\Psi'^\dagger\gamma^5\Psi' \\ s^6 &\Rightarrow s'^6 = cp^6\Psi'^\dagger\gamma^5\Psi' \end{aligned}\right\}, \qquad (17.18)$$

where we have assumed a negative rotation angle $\vartheta = -\pi/2$. To arrive at (17.17) the energy operator $\hat{p}_0 = i\hbar\partial/\partial x^0$ of (17.5) has been transformed away and replaced by the (metrically time-like) momentum operator $\hat{p}'_5 = i\hbar\partial/\partial x'^5$. Thus, the "equation of motion" in any *dual* $M_D^4$ cross section of this current gives the probability amplitude as a function of pseudospatial dimension, $x'^5$, and pseudomomentum, $p'_5$. In its four-dimensional form, (17.17) describes *a world* $M_D^4$ *without time and without energy*, these variables having been replaced by $x'^5$ and $p'_5$. Interestingly, since $x'^0$ is perpendicular to $M_D^4$, the particle "momentum" perpendicular $M_D^4$ to is the invariant energy $p'_0$.

The existence of dual spacetime means that every physical event receives a dual description, one in frame *K* and its *dual* in frame $K_D$. Thus the track of an object in the physical world is at once described by a world line in *K* and a *dual* world line in $K_D$, the former parameterized by time $ct$, the latter by pseudospatial $x'^5$. This is illustrated in **Fig. 17.4**, where the two frames are shown separated for clarity. For defi-

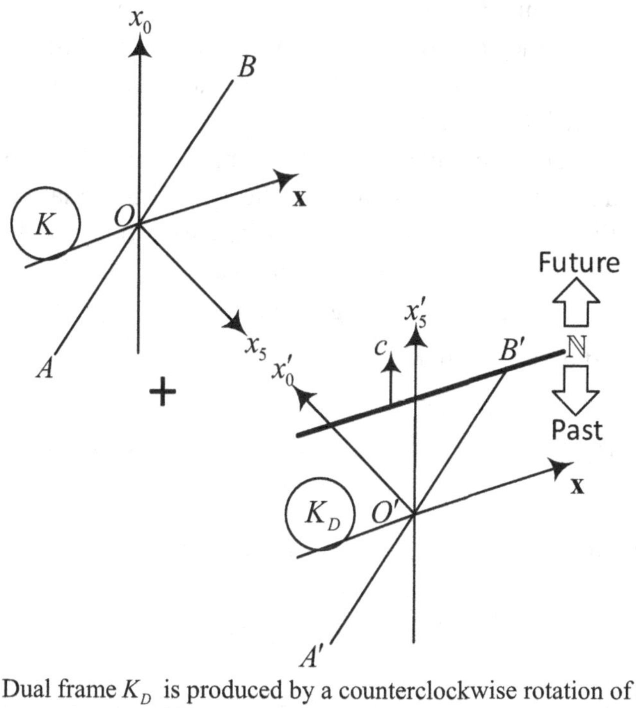


**Fig. 17.4** Dual frame $K_D$ is produced by a counterclockwise rotation of frame *K* through an angle of $-\pi/2$ about the **x** axis. As a result of the rotation, time $x^0$ is transformed into the quasi-spatial (metrically time-like) dimension $x'^5$. Superimposed, the two frames comprise *dual spacetime* $K+K_D$. Line *A'O'B'* in $K_D$ is the *dual* of world line *AOB* in *K*. The line labeled ℕ in frame $K_D$ represents the *moving present moment* or *now*. This line is in reality an infinite three-dimensional hyperplane oriented perpendicularly to the $x'^5$ axis and moving in a direction parallel to the $x'^5$ axis at the speed of light, *c*. The sum total of all such dual world lines up to the present moment is sometimes referred to as a *growing block universe*; see *Sec. 17.4.2.*

niteness we have chosen $\vartheta = -\pi/2$, which makes $x'^5$ run parallel to $x^0$, and $x'^0$ antiparallel to $x^5$. Note that $K$ and $K_D$ coincide geometrically in Euclidian $E^3$; the two frames are distinguished by the differing parametrizations in the time-like domain.

We should note that there is nothing special about frame *K*. It is one out of a three-fold infinite set of inertial frames, each member of which moves at some constant velocity. (We ignore the effects of acceleration, gravitational or otherwise.) Thus, referring to **Fig. 17.4,** the inclination of world line *AB* from the time axis of frame *K* will vary depending on frame *K*'s velocity relative to the object whose world line is *AB*. And correspondingly, the inclination of world line $A'B'$ from the $x'^5$ axis of frame $K_D$ will vary *its* inclination so as to remain the dual of world line *AB*. Since frame $K_D$ derives from *K* by finite rotation, it remains permanently connected to *K*, whichever member of the set of inertial frames *K* may happen to be.

For simplicity we have inferred the existence of dual spacetime by means of a rotation transformation of the 4-D Dirac equation (17.5). We note that the same transformation is applicable to the 6-D form (17.1), although in this case one has in place of (17.15) the relation $\hat{p}_5 = \hat{p}'_0 \sin\vartheta + \hat{p}'_5 \cos\vartheta \neq 0$. Again taking $\vartheta = -\pi/2$ we have for the transformed equation in $K_D$

$$\left[\left(\hat{p}^6 + \gamma^0 \hat{p}'_0\right)\left(\gamma^5 \hat{p}'_5 - \boldsymbol{\gamma}\cdot\hat{\mathbf{p}}\right) - \widehat{m'^2}c^2\right]\chi'\left(x'^0, \mathbf{x}, x'^5, x^6\right) = 0\,, \quad (17.19)$$

where the mass squared operator is now given by

$$\widehat{m'^2} = (\hat{p}_6{}^2 - \hat{p}'^2_0)^{1/2}/c^2\,. \quad (17.20)$$

So by rotation the roles of $x^0$ and $x^5$ are reversed. In particular in $K_D$ it is energy $p'_0$ that is invariant.

Also, it should be mentioned that one can include external electromagnetic fields in these equations by means of the minimal couplings $(\hat{p}_0, \hat{\mathbf{p}}) \to (\hat{p}_0 - qA_0, \hat{\mathbf{p}} - q\mathbf{A})$ in *K*; and $(\hat{p}'_5, \hat{\mathbf{p}}) \to (\hat{p}'_5 - qA'_5, \hat{\mathbf{p}} - q\mathbf{A})$ in $K_D$. Note the field transformation relation $A'_5(x'^5, \mathbf{x}) = A_0(x^0, \mathbf{x})$, which follows by analogy with (17.14), with $\vartheta = -\pi/2$.

## 17.4 The dual nature of time

With dual spacetime are associated two distinct conceptions of *time*. They are not new, having been long compared in the philosophic literature, but take on new significance in the context of dual spacetime.

*17.4.1 The B-Series.* In relativity theory,[2]

> *Time* ($ct$) is the dimension of the physical universe which, at a given place, orders the sequence of events.

Time in this sense is variously called *objective*,[3] *conceptual*[4] or *physical*[5] time. Physical time is the time of third-person physics in frame $K$ and the time that a clock measures. The defining feature of this objective depiction of the universe is that nothing moves. In the objective world there are only physical occurrences, laid out in a static system of spatial and temporal coordinates, with intervals between them. Such a universe is called a *block universe*,[6, 7] because the events comprising it are displayed monolithically, in one great motionless landscape. In the block universe, temporal succession is expressed by the so-called "B-series" *earlier-simultaneous-later*,[8] a series applicable to each pair of events in 4-D frame $K$. Thus, in **Fig. 17.4**, event $A$ is earlier than origin $O$, event $B$ is later than $O$, and all points on the $\mathbf{x}$-axis are simultaneous with $O$ and each other. There is nothing special about the individual events making up the series. There are no privileged moments in time, just as there are no privileged locations in space. Events in frame $K$ are distinguished temporally solely by their B-series relationships with other events.

Now as was emphsized above, every physical event in dual spacetime is described twice, once in $K$ and once in $K_D$. Thus there exists in $K_D$ a *dual* block universe, a universe identical to the one in $K$, but parametrized not by time $ct$, but by pseudospatial coordinate $x'^5$. Then by analogy with the preceding definition of time,

> *Pseudospatial dimension* $x_5'$ is the dimension of the physical universe which, at a given place, orders the sequence of dual events.

This dual block universe is no less objective than the block universe in frame *K*. Moreover, the dual events of which it is composed enjoy the same B-series relationships as the events of frame K.

*17.4.2 The A-Series.* Viewed separately, dimensions *ct* and $x'^5$ represent coequal categories of existence, on the same plane as space and causality. Both are ordering parameters, and both are *relational*, i.e., time would not exist without events and $x'_5$ would not exist without dual events. But let us now consider the connection between them. The connection is given automatically by the first of transformation equations (17.11), with $\vartheta = -\pi / 2$:

$$x'^5 = ct \,. \tag{17.21}$$

This is an equation of motion. In it, time *t* is the same classical Newtonian universal time appearing in the Robertson-Walker metric, (9.43). As depicted in **Fig. 17.4,** it describes the motion of a 3-D hyperplane $\mathbb{N}$ of infinite extent oriented perpendicularly to the $x'^5$ axis and moving in a direction parallel to the $x'^5$ axis at the speed of light, *c*. This moving infinite hyperplane is the realization of what is commonly referred to as the *moving present moment*, or *now*.[5] The existence of a moving present moment signifies that the world is in a constant state of becoming—a *growing block universe*[9]—and $\mathbb{N}$ is its leading edge. One thus has, in addition to the B-Series, the so-called "A-Series" of temporal succession: *past-present-future*.[8] This series is exhibited explicitly in **Fig. 17.4**. One may also speak, in hybrid fashion, of past block time and future block time, both of which blocks appear explicitly in frame $K_D$, separated by the moving present moment.

The idea of a moving present moment is usually dismissed as nonsensical, first because in ordinary spacetime it implies a form of motion measured in seconds per second; and secondly because its direction is unspecifiable.[6, 7] In the literature its presumed unreality is signaled through the use of such terms as "the specious present"[10] and the "the myth of passage."[11] But in the context of dual spacetime the present moment emerges as a legitimate element of reality—a direct consequence of the finite transformation of one coordinate frame into another, measured in terms of ordinary units of velocity and directed along the $x'^5$ coordinate axis. It serves as well to define the famous *arrow of time*,[12, 13] represented by the "Future" arrow in **Fig. 17.4**.

It should be stressed that, historically, doubts about the legitimacy of the "now" arose, not in physics, but in the context of psychology and conscious awareness. For that reason the A-Series is sometimes referred to as *subjective* or *perceptual* time.[3] But dual spacetime as described here exists whether or not there exist conscious beings. Thus the moving present moment defined by (17.21) is to be considered an *objective* temporal element of reality. It is not, in other words, a product of awareness. It is, however, as we shall see in **Part VII** of this book, a precondition for it.

Which brings us to the question of the future. Is it already determined or not? If there were no frame $K_D$, then one could argue that, since space and time form a 4-D continuum, and since events can be spread out endlessly in all directions in space, then why not also in time? This is the block universe idea. It suggests the possibility that, "Objectively, past, present and future must be equally real."[7] In such an objective view, the future already exists, and nothing can be done to change it. But if spacetime has in fact the dual structure disclosed here, then in frame $K_D$ (but not in $K$) there exists a boundary between past and future block times—the moving present moment. Consequently, since frame $K_D$ is inaccessible to observing apparaus in $K$, we cannot yet objectively say whether the future in $K_D$ is open or closed. At this stage in our study of time, the future in $K_D$ must simply be considered unknown.

To repeat, the moving present moment, as indicated in **Fig. 17.4,** is found in frame $K_D$ only. And that is precisely why the "now" appears nowhere in the relativistic formulation of physics in frame $K$. But that does not mean that it is without consequence for physics. As we shall now show, it figures indispensably in the resolution of one of the standing mysteries of quantum mechanics.

## 17.5 Reduction of the state vector

In the opening paragraph of this chapter reference was made to the dual behavior of the quantum-mechanical state vector. In this closing section we intend show how that behavior couples to, and is enabled by, the dual spacetime discussed above. There are two distinct ways of exemplifying the state vector's dual behavior: one by way of a *single* particle, which may occupy at once two or more separate paths in spacetime, as in the famous two-slit experiment; another by way of *two* particles which interact and whose individual quantum states, as a result of the interaction, become *entangled.*

*17.3.1 The single particle.* The state vector of a quantum system $S$ is conventionally denoted $|\psi\rangle$. It is expressible in terms of the orthonormal eigenvectors $|\phi_n\rangle$ of an observable $\hat{O}$:

$$|\psi\rangle = \sum_n \psi_n |\phi_n\rangle. \tag{17.22}$$

Here the $|\phi_n\rangle$ represent "possible" (or potential) states of the system with associated eigenvalues $\lambda_n$, and the coefficients $\psi_n = \langle\phi_n|\psi\rangle$ are their probability amplitudes. The vector $|\psi\rangle$ is assumed normalized so that the sum of the probabilities of all possible states is unity:

$$\langle\psi|\psi\rangle = \sum_n |\psi_n|^2 = 1. \tag{17.23}$$

In speaking of the dual behavior of the state vector, we are referring to its characterization of $S$ before and after a measurement on $S$ is performed.[14] (1) Before measurement, $|\psi\rangle$ evolves causally in time in accordance to the Schrödinger equation:

$$i\hbar\frac{\partial}{\partial t}|\psi\rangle = H|\psi\rangle, \tag{17.24}$$

where $H$ is the Hamiltonian of the system. (2) When a measurement on $S$ with respect to observable $\hat{O}$ is made, vector $|\psi\rangle$ reduces via projection into a particular direction in Hilbert space, say that defined by basis vector $|\phi_N\rangle$; in other words,

$$|\psi\rangle \rightarrow \psi_N |\phi_N\rangle, \tag{17.25}$$

where, because probability is conserved, one now has in place of (17.23),

$$|\psi_N|^2 = 1. \tag{17.26}$$

Thus by reduction of the state vector, the merely "possible" has become "actual."

But therein lies a paradox. For suppose that, prior to collapse, the possible states comprising superposition (17.22) are separated spatially, with the distances between them arbitrarily large. Upon collapse, all but one of the $\langle \phi_n | \psi \rangle$ must instantly vanish. How is such correlated behavior across the superposition possible? It appears on the face of it to violate the principles of special relativity. Hence the *paradox of collapse*.

To see how this paradox is resolved in dual spacetime, we shall work through in detail an elementary yet illuminating example. Referring to **Fig. 17.5**, a beam of light containing a single photon impinges on a half-silvered mirror N, whereupon it is split into two component beams of equal intensity, one reflected into the $x$ direction, the other transmitted into the $y$ direction. If our single photon is one of a pair produced simultaneously by parametric downconversion, then its sister can be used to set the time of the single photon's splitting into two beams. Photodetectors A and B are placed in the paths of the beams at $x = a$ and $y = b$, respectively. We assume $a < b$, as shown. According to quantum mechanics our single photon occupies both beams simultaneously.[15] Thus the observable in this arrangement is the identity of the beam in which the photon is found upon measurement. Let $|\psi\rangle_x$ and $|\psi\rangle_y$ denote the eigenvectors corresponding to this observable. We then have the representation

$$|\psi\rangle = c_x |\psi\rangle_x + c_y |\psi\rangle_y, \tag{17.27}$$

where the coefficients

$$c_r = {}_r\langle \psi | \psi \rangle, \quad r = x, y. \tag{17.28}$$

Immediately after reflection and transmission by the beam splitter, but before a measurement is made, the coefficients $c_x = c_y = 1/\sqrt{2}$. At the moment of measurement, one of the $c_r$ must vanish, while the other becomes unity. The vectors $|\psi\rangle_x$ and $|\psi\rangle_y$ can in turn be expanded in the eigenvectors $|x\rangle$ and $|y\rangle$ of the position operator $\hat{r}$ with eigenvalues $x$ and $y$, respectively. This gives from (17.27)

$$|\psi\rangle = c_x \int \psi_x |x\rangle \, dx + c_y \int \psi_y |y\rangle \, dy, \tag{17.29}$$

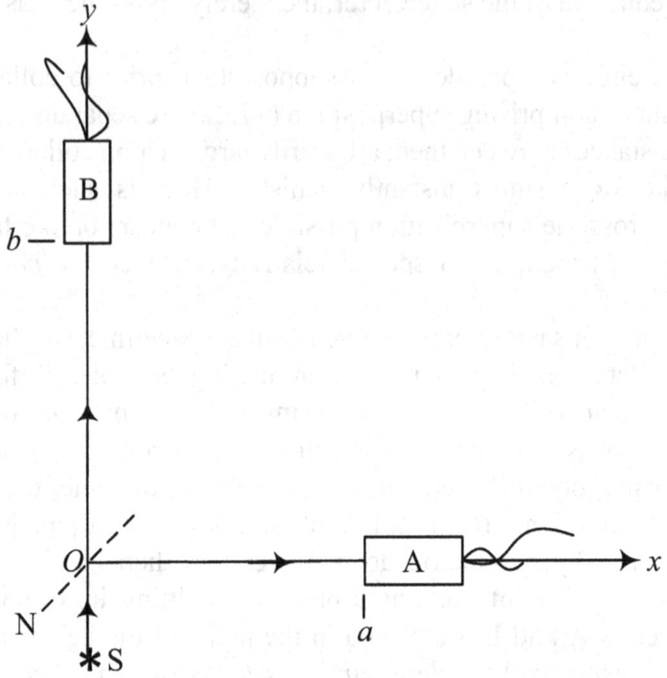


**Fig. 17.5** Set-up illustrating reduction of the state vector. Source S emits a photon which is split into two beams by beam splitter N, the reflected beam being sent off in direction $x$ and the transmitted one in direction $y$. If this single photon is one of a pair generated by parametric downconversion, its sister can be used to mark the moment that the single photon hits the beam splitter. Photodetectors A and B are placed in the paths of the two beams at $x = a$ and $y = b$, respectively. It is assumed that $a < b$, so that reduction always occurs as a result of measurement at detector A, whether or not the photon is detected there.

where in (17.29) the discrete summation of (17.22) has been replaced by an integral over the continuous variables $x$ and $y$, and orthonormality of the basis states now means $\langle x' | x \rangle = \delta(x - x')$. The "coefficients" $\psi_x$ and $\psi_y$ in this expansion are just the quantum mechanical wave functions

$$\psi_r = \langle r | \psi \rangle_r = \psi(ct - r), \qquad r = x,\, y\,. \qquad (17.30)$$

The wave function $\psi(ct - r)$ describes a normalized minimum wave packet[16]:

$$\begin{aligned}\psi(ct-r) &= \frac{1}{\left[2\pi(\Delta x)^2\right]^{1/4}}\int_{-\infty}^{\infty}\left[2\sqrt{\pi}(\Delta x)e^{-(\Delta x)^2(k-k_0)^2}\right]\frac{e^{-ik(ct-r)}}{\sqrt{2\pi}}dk \\ &= \frac{1}{\left[2\pi(\Delta x)^2\right]^{1/4}}\exp\left[-\frac{(ct-r)^2}{4(\Delta x)^2}-ik_0(ct-r)\right]\end{aligned}\quad , \quad (17.31)$$

in which $(\Delta x)^2 = \langle r^2\rangle - \langle r\rangle^2$ and $k_0 = 2\pi/\lambda_0$, $\lambda_0$ being the central radiation wavelength. This wave packet does not spread. The structure of the corpuscular photon, whose position is defined probabilistically by (17.30), is described by Eqs. (11.45-48). The radius of the physical photon's Zitterbewegung is, according to (11.48), $\lambda_0/2\pi$. Hence a reasonable measure of the width of the wave packet is $\Delta x \sim \lambda_0/2\pi$. Then to form the miminum packet one requires a spread in wave number $\Delta k = 1/2\Delta x \sim \pi/\lambda_0$.

Before measurement the vector (17.29) satisfies the Schrödinger equation for a single photon[17]

$$i\hbar\frac{\partial}{\partial t}|\psi\rangle = H|\psi\rangle = \hbar c\sqrt{-\nabla^2}\,|\psi\rangle . \quad (17.32)$$

And so the first part of the state vector's dual nature is properly exhibited.

As for the second part, we mention again that, because prior to measurement the photon is in both beams at once, the probability of its being detected at detector A is ½; and likewise the probability of its being detected at detector B is ½. However, because $a < b$, it is invariably the measurement made at A that causes collapse of the state vector. Suppose that the photon is actually detected at A. The state vector reduces to

$$|\psi\rangle \to \int \psi(\xi)|a+\xi\rangle d\xi , \quad (17.33)$$

the $c_r$ having changed immediately from their original values to $c_x = 1$ and $c_y = 0$. Note that, in contradistinction to (17.25), the reduction

(17.33) involves integration over a small bundle of the continuously-variable eigenstates of operator $\hat{r}$. In this way the full width of the wave packet is subjected to measurement.

Now what about the case where, upon measurement, the photon is *not* detected at A? We know beforehand the time $t_A$ of arrival of the wave packet at A: it is $a/c$. At time $t_A$ the integration appearing on the r.h.s. of (17.32) is carried out, effecting a measurement at A. When no photon is detected, the $c_r$ change immediately from their original values to $c_x = 0$ and $c_y = 1$, and the state vector reduces to

$$|\psi\rangle \to \int \psi(ct - y)|y\rangle dy. \tag{17.34}$$

And so the second part of the photon's dual nature is exhibited as well.

There are several points of interest here.

(1) Collapse occurs as a result of a measurement made at A, *whether or not the photon is detected there.*
(2) This must mean that the wave packet is physically real, and no mere computational device, as otherwise its arrival at A could not effect collapse of the state vector. [See *Sec. 3.9.1.*]
(3) The state vector is an organic whole, as otherwise it could not act in the coordinated fashion that it does, as in collapse. The physical photon, being in both beams at once before measurement, is the glue that holds it together.
(4) The coefficients $c_r$ change their values suddenly in accordance to the result of measurement at A. They do so even though the wave packets to which they are attached may be far from each other, with no known means of communication between them.

So we have in hand a straightforward, single-particle illustration of the paradox of collapse. Clearly the paradox is unresolvable in ordinary $M^4$, where the speed of communication presumably cannot exceed the speed of light. But in dual spacetime, there is no such limitation.

To show this, let us refer to **Fig. 17.6,** which depicts once again dual spacetime $K + K_D$. We omit dimensions $x^5$ and $x'^0$, which play no direct role in state vector collapse. The lines OA and OP in frame $K$ depict the world lines of our two component beams up to the time of measurement

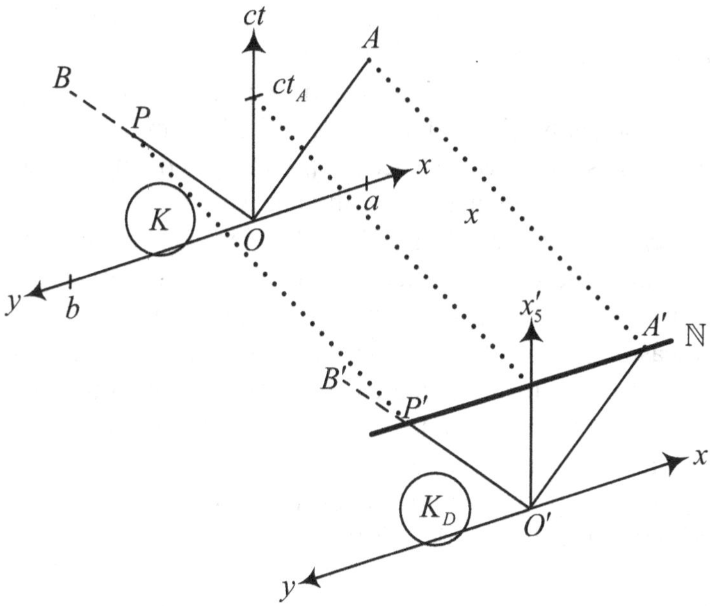


**Fig. 17.6** Illustrating resolution of the paradox of collapse. Depicted is dual spacetime $K + K_D$, omitting dimensions $x^5$ and $x'^0$. Lines OA and OP in frame *K* denote the world lines of a single photon present at once in two beams originating by means of a beam splitter at *O*, one beam traveling in the *x* direction, the other in the *y* direction. The two directions, which in the experimental set up are perpendicular to each other, are for ease of depiction shown here as opposite directions on the same axis. A measurement of the arrival or non-arrival of the photon is performed at A, leading to the collapse of the state vector. If the photon is detected at A, then world line OP stops at P. If it is not detected at A, then world line OP continues, terminating at detector B. Because the interval AP is spacelike, this coordinated behavior between opposing world lines is unexplainable in terms of frame K alone. However, the moving present moment—line ℕ (actually a 3-D hyperplane)—in frame $K_D$ connects the duals of world lines OA and OP, namely, lines O′A′ and O′P′, and the interval A′P′ is a line of simultaneity. Thus at the time of measurement, points A′ and P′ are in causal contact, irrespective of the spatial distance between them, resolving the paradox of collapse.

at detector A. World line OP stops at P if the photon is detected at A. It continues on to detector B if the photon is not detected at A. (To avoid an unnecessary complication of the figure, directions *x* and *y* are represented as opposite directions on the same axis.) Because the interval AP in frame *K* is spacelike, no signal announcing the detection or non-detection of the photon at A can pass from A to P. And so instantaneous collapse is unexplainable in terms of frame $K$ alone. In frame $K_D$, however, the moving present moment—3-D hyperplane $\mathbb{N}$—connects the duals of world lines OA and OP, namely lines O′A′ and O′P′. Since these dual world lines are parametrized not by time $ct$, but by the quasi-spatial dimension $x'^5$, the interval A′P′ in $K_D$ is a line of simultaneity. The points A′ and P′ are thus in causal contact, no matter the spatial distance between them. And so the paradox of collapse is resolved by virtue of dual spacetime. It is a case of division of labor: Before measurement, state vector $|\psi\rangle$ evolves causally, parametrized in frame $K$ by time $ct$ in accordance to the Schrödinger equation (17.27). When a measurement is made, $|\psi\rangle$ reduces to one of its constituent eigenvectors, an instantaneously-occurring process parametrized in frame $K_D$ by the pseudo-spatial coordinate $x'^5$. To the above four points of interest, we may add a fifth:

(5) If state vector reduction is intrinsic to quantum mechanics, and the reduction is enabled by dual space time, then dual space time is intrinsic to quantum mechanics. QM, at least in its present form, could not exist without it.

*17.3.2 Entangled particles.* Quantum entanglement presents an equally dramatic manifestation of the paradox of collapse.[18] Here one may have two particles arising from a common origin whose measurable properties, owing to a conservation law, are correlated, even after separation. The experimental arrangement used by Aspect *et al.* to show that the correlations cannot be explained deterministically in terms of hidden variables provides a case in point.[19, 20] In this arrangement, a radiative cascade of Calcium-40 emits two photons, $\nu_x$ and $\nu_y$, heading off in opposite directions, namely in directions $x$ and $y = -x$, as shown in **Fig. 17.7**. As the final state of the atomic cascade has spin 0, the two

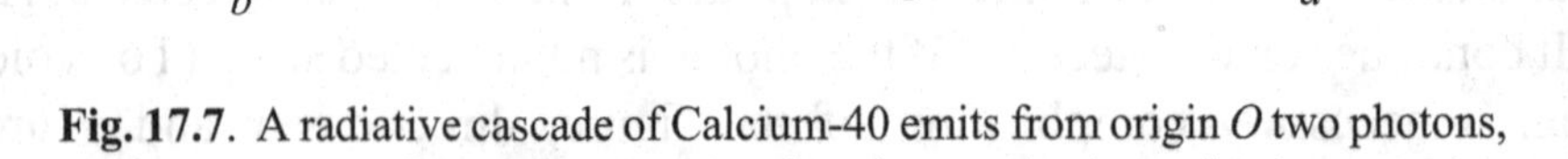

**Fig. 17.7.** A radiative cascade of Calcium-40 emits from origin $O$ two photons, $\nu_x$ and $\nu_y$, the former heading in direction $x$, the latter in the opposing direction y.

photons likewise have net spin 0, meaning that they are at once right-handed (R) with probability ½ and left-handed (L) with probability ½ . Thus the observable in this experiment is the handedness of the circular polarization. Accordingly, polarization measuring devices A and B are placed in the paths of the two beams at $x = a$ and $y = b$, respectively, where it is assumed that $a < b$, as shown. Let $|R\rangle$ and $|L\rangle$ denote the eigenvectors corresponding to the two possible states of circular polarization. In terms of these the state vector $|\psi\rangle$ of the total system expands to

$$|\psi\rangle = c_R|R\rangle + c_L|L\rangle, \tag{17.35}$$

where the coefficients

$$c_P = \langle P|\psi\rangle, \quad P = R,\ L. \tag{17.36}$$

Immediately after photon production, and before a measurement is made, the coefficients $c_R = c_L = 1/\sqrt{2}$. At the moment of measurement, one of the $c_P$ must vanish, while the other becomes unity. Eqs. (17.35) and (17.36) are the analogs of (17.27) and (17.28).

Now the eigenvectors $|R\rangle$ and $|L\rangle$ each are subject to representation in terms of the circular-polarization basis states of the individual photons. Let the basis states belonging to $\nu_x$ be denoted $|R\rangle_x$ and $|L\rangle_x$ and those of $\nu_y$, $|R\rangle_y$ and $|L\rangle_y$. In terms of these the state vector (17.35) of the emitted photon pair before measurement becomes

$$|\psi\rangle = c_R|R\rangle_x \otimes |R\rangle_y + c_L|L\rangle_x \otimes |L\rangle_y, \tag{17.37}$$

where $\otimes$ denotes the tensor product. This is an *entangled state*, so-called because it cannot be written in the product-state form $|P\rangle_x \otimes |Q\rangle_y$. It expresses the fact that, owing to conservation of angular momentum, the two photons are locked together, forming a composite whole; whatever happens to one of the photons immediately affects the other. The four individual states $|P\rangle_r$ ($P = R,\ L$; $r = x,\ y$) each are expandable in the coordinate representation, much as in (17.29). However, in the present case each polarization has its own basis:

$$\left.\begin{aligned} |R\rangle_r &= \int \rho_r |r\rangle_R\, dr \\ |L\rangle_r &= \int \lambda_r |r\rangle_L\, dr \end{aligned}\right\}, \quad r = x,\ y \tag{17.38}$$

with ${}_{P'}\langle r'|r\rangle_P = \delta_{PP'}\delta(r-r')$, ${}_{P'}\langle x|y\rangle_P = 0$ ($P'$, $P = R$, $L$) and

$$\left.\begin{aligned} \rho_r &= {}_R\langle r|R\rangle_r \\ \lambda_r &= {}_L\langle r|L\rangle_r \end{aligned}\right\} = \psi(ct-r), \qquad r = x, y. \tag{17.39}$$

In the latter expression, $\psi(ct-r)$ is the wave packet defined by (17.32). In terms of the expansions (17.38), Eq. (17.37) becomes

$$\begin{aligned} |\psi\rangle &= c_R \iint \rho_x |x\rangle_R \, dx \otimes \rho_y |y\rangle_R \, dy + c_L \iint \lambda_x |x\rangle_L \, dx \otimes \lambda_y |y\rangle_L \, dy \\ &= \iint \psi(ct-x)\psi(ct-y)\left(c_R |x\rangle_R \otimes |y\rangle_R + c_L |x\rangle_L \otimes |y\rangle_L\right) dxdy \end{aligned} \tag{17.40}$$

Before measurement the vector $|\psi\rangle$ in the form (17.40) satisfies the Schrödinger equation for two photons:

$$i\hbar \frac{\partial}{\partial t}|\psi\rangle = (H_x + H_y)|\psi\rangle = \hbar c\left(\sqrt{-\partial_x{}^2} + \sqrt{-\partial_y{}^2}\right)|\psi\rangle. \tag{17.41}$$

Now what happens when the polarization is measured? Since $a < b$, the measurement inducing collapse of the state vector necessarily occurs at device A. If the polarization measured there is found to be right-handed, say, the coefficients (17.37) immediately become $c_R = 1$ and $c_L = 0$, and the state vector (17.40) reduces to the *separable* form

$$|\psi\rangle = \int \psi(\xi)|a+\xi\rangle_R \, d\xi \otimes \int \psi(ct-y)|y\rangle_R \, dy. \tag{17.42}$$

This indicates that photon $\nu_y$, which is still in flight, is now definitely right-handed as a result of the measurement of $\nu_x$'s state of polarization at A. And so we have once again the paradox of collapse: how does $\nu_y$, which may be far distant from A, "know" that its polarization, which previously could be either R or L with equal probability, is now to be right-handed?

The resolution is the same as in the single-particle case discussed previously. In fact, exactly the same diagram, **Fig. 17.6**, applies. In this figure, which depicts dual spacetime $K + K_D$, lines OA and OP in frame $K$ depict, respectively, the world lines of photons $\nu_x$ and $\nu_y$ up to the time of polarization measurement at A. Upon measurement of $\nu_x$ 's state of polarization at A, $\nu_y$ 's state of polarization immediately reduces at P to that found for $\nu_x$ . World line OP continues on to detector B, where $\nu_y$ 's state of polarization is found to be the same as that of $\nu_x$ . Because the interval AP in frame $K$ is spacelike, no signal informing P of $\nu_x$ 's state of polarization at A can pass from A to P. And so instantaneous collapse is unexplainable in terms of frame $K$ alone. In frame $K_D$, however, the points $\mathrm{A}'$ and $\mathrm{P}'$ are connected by the moving present moment $\mathbb{N}$ and are automatically in causal contact, no matter the spatial distance between them. And so the paradox of collapse is again resolved by way of dual spacetime. To the above five points of interest, let us add a sixth:

(6) The concept of entanglement arises only when we regard the system as consisting of two separate photons. Indeed one needs to to invoke the entangled state to show that before measurement $|\psi\rangle$ obeys the Schrödinger equation (17.41). The phenomenon of collapse, however, even for a system two particles, has nothing to do with entanglement. This is shown by Eq. (17.35), where the values of the coefficients $c_R$ and $c_L$ after collapse are determined only by the state of polarization measured at A. The state of the collapsed system as a whole is determined not by entanglement but by conservation of angular momentum. Entanglement is an expression of the conservation law.

In summary, we find that the paradoxical aspect of state-vector collapse is resolved in dual spacetime, where the moving present moment in frame $K_D$ places all points of the state vector in causal contact. We have considered two examples of collapse—the single photon occupying two paths at once, and two correlated photons having two polarizations at once. In the first example the realized outcome of measurement is dependent on conservation of probability of location; in the second case the outcome of measurement is dependent on conservation of angular momentum. In that light, one sees that the issue of collapse is not one of how many particles there are, but of the essential unity of the state vector over its spatial extent.

## Notes and references


[1] W. Heisenberg, *Physics and Philosophy* (Harper & Row, New York, 1958), pp. 54-55.

[2] "Time," *McGraw-Hill Dictionary of Physics and Mathematics*, D. N. Lapades, Ed. (McGraw-Hill, New York, 1978), p. 987.

[3] "Time," *Dictionary of Philosophy*, D. D. Runes, Ed. (Littlefield, Adams & Co., Ames, IA, 1959), p. 318.

[4] H. Margenau, *The Nature of Physical Reality* (McGraw-Hill, New York, 1950), pp. 137, 139.

[5] G. J. Whitrow, *The Nature of Time* (Penguin, Harmondsworth, Middlesex, England, 1975), p. 144.

[6] H. Price, *Time's Arrow and Archimedes' Point*: *New Directions for the Physics of Time* (Oxford University Press, Oxford, 1996), pp. 12-16.

[7] P. Davies, "That Mysterious Flow," *Scientific American*, Sept. 2002, pp. 40-47.

[8] R. Scruton, *Modern Philosophy* (Penguin, New York, 1994), pp. 366-68.

[9] C. D. Broad, *Scientific Thought* (Routledge & Kegan Paul, London, 1923), Ch. II, esp. pp. 59, 66. For additional references, see the Wikipedia article, "Growing Block Universe."

[10] Anonymous (E. Robert Kelly), *The Alternative: A Study in Psychology* (Macmillan, London, 1882); William James, *The Principles of Psychology, Vol. 1* (H. Holt, New York, 1890).

[11] D. C. Williams, "The Myth of Passage," *The Journal of Philosophy* **48**, 457-72 (1951).

[12] H. Price, *ibid.*, pp. 16-17.

[13] P. Davies, *The Physics of Time Asymmetry* (U. of California Press, Berkeley, 1974), pp. 3, 22; *About Time: Einstein's Unfinished Revolution* (Touchstone, New York, 1995).

[14] J. von Neumann, *Mathematical Foundations of Quantum Mechanics* (Princeton U. Press, Princeton, NJ, 1955), Chapters V and VI.

[15] P. A. M. Dirac, *The Principles of Quantum Mechanics*, 4th Ed. (Oxford U. Press, London, 1958), pp. 8,9.

[16] L. I. Schiff, *Quantum Mechanics* (McGraw-Hill, New York, 1955), pp. 54-59.

[17] S. S. Schweber, *An Introduction to Relativistic Quantum Field Theory* (Row, Peterson, Evanston, IL, 1961), p. 116.
[18] A. Einstein, B. Podolsky and N. Rosen, "Can Quantum-Mechanical Description of Physical Reality Be Considered Complete?" *Phys. Rev.* **47**, 777 (1935).
[19] A. Aspect, P. Grangier and G. Roger, "Experimental Tests of Realistic Local Theories via Bell's Theorem," *Phys. Rev. Lett.* **47**, 460 (1981).
[20] A. Aspect, J. Dalibard and G. Roger, "Experimental Test of Bell's Inequalities Using Time-Varying Analyzers," *Phys. Rev. Lett.* **49**, 1804 (1982).

# PART VII

# THE HARD PROBLEM OF CONSCIOUSNESS

# Chapter 18

## Consciousness and the world

Finally we come to the problem of consciousness. Although for the moment we lack even a good definition of it, we nevertheless know that consciousness is a part of physics. We know this because, as was noted in *Sec. 1.2.2*, the inner life of the conscious being occupies a special frame of reference—the privileged first-person perspective—and frames of reference constitute the basic furniture of physical theory. What we do *not* know, and may never fully grasp, is how the special frame of reference comes to be occupied by a conscious Ego. It may well constitute a noumenal fact of creation, an impenetrable mystery to be accepted without explanation.[1]

And yet, at one time, gravitation too was a mystery. Consider Isaac Newton's quandary. He had discovered the inverse square law, and in the *Principia* (1687)[2] presented it as straightforward empirical fact, an algebraic fit to the results of observation of the planetary orbits.[3] He had no idea why the law had that particular form. Nor in particular could he imagine how the gravitational force between massive bodies might be transmitted without an intervening mechanical medium. He was disturbed by and could not accept the idea of action at a distance,[4] and so for Newton, gravitation was a mystery.

And mysterious it remained until, two centuries later, along came Albert Einstein, whose general theory of relativity (1916)[5] showed that the gravitational force could be attributed to the curvature of spacetime and moreover, for weak fields, was well-described by the form predicted by Newton. Impenetrable mystery solved.

Note that Einstein's solution to the mystery of gravitation came by way of a radically altered conception of space and time. It was a two-step process. First came his special theory of relativity, in which a 4-D spacetime manifold replaced the absolute space and time of Newton.[6] Then came the general theory,[5] in which curved space replaced the flat space of the special theory, while retaining the latter's relativistic features. With the new formalism in hand, one could now, with tenacity, read out exactly what Nature had in mind when it came to gravity.

## 18.1 The easy vs. hard problem of consciousness

The point being that mysteries are made to be solved. The mystery or problem of consciousness is sometimes described as having an easy part and a hard part.[7] Roughly speaking, the easy part deals with the functioning of a physical system, *viz*., the brain. The hard part is to explain how objective brain functions get to be accompanied by subjective experience. The problem of gravitation, too, had "easy" and "hard" parts, the former (discovering the inverse square law) solved by Newton, the latter (explaining the spacetime origin of the law) by Einstein. Neuroscience today yields an understanding of consciousness comparable to the inverse square law of Isaac Newton. It is an understanding gained, not from studies of conscious experience as such, but from the *neural correlates* of conscious experience, i.e., patterns of neuron firings.[8] And so one has an accumulation of neurological data, a data-set analogous to the planetary-orbital data available to Newton, but without any understanding of the inner subjective mechanism producing it. Getting to that understanding is the hard problem of consciousness and is our driving concern here.

But we do have a head start. For we have noted already the connection between conscious experience and a physical, yet special, frame of reference. This suggests that the hard problem of consciousness, like that of gravitation, will require for its solution a radically revised conception of spacetime.

This is not a new idea. For some time now the philosopher Colin McGinn has advocated an expansion of our idea of spatiality as the key to resolving the problem of consciousness.[9, 10] McGinn's view is not far removed from ideas expressed much earlier by the philosophers C. D. Broad and H. H. Price.[11] McGinn's idea runs something like this. Brain and mind go together. Because the brain occupies space, it appears that mind too must in some sense occupy space if it is to connect up with brain. But space, as currently understood, has no place for mind. If it did, the neurosurgeons would have found it. Therefore, assuming that the mind exists and has to be somewhere, we can conclude that the standard conception of space is radically incomplete. McGinn's proposal, as well as those of Broad and Price, thus foreshadow the thrust of the present investigation.

In the writer's view, the easy/hard distinction, while usefully dramatizing the mysterious nature of subjectivity, is ultimately misleading. It corresponds roughly to an unrealistic classical picture of the world, where observer and observed are neatly distinguished. In classical physics, the interaction between measurement apparatus and the thing observed either can be ignored or compensated for, allowing one to measure, within the limits of experimental accuracy, properties inherent to the object alone, e.g., its position, velocity or angular momentum. In the real physical world, however, especially at the quantum level, the act of measurement necessarily disturbs the thing observed, and the distinction between measurement apparatus and the thing observed becomes problematic. One then speaks of an irreducible total system consisting of object, apparatus and their interaction. In much the same way we may, to a first (classical) approximation, speak separately of mind and brain, and the corresponding problems—hard and easy—that they present. In the end, however, we are going to have to account for mind/brain interaction—if indeed such interaction takes place. If it does, then the easy problem will no longer be easy. It will have evolved to an even harder one of accounting for the total irreducible system consisting of mind, brain and their interaction.

## 18.2 Geometry of the first-person perspective

The first step in our study of consciousness is to define its connection to the geometry of spacetime. A picture hinting at the relationship of consciousness to reality a whole was given in **Fig. 2.1.** This figure depicts the empirical world B of phenomena flanked on one side by the external world A of noumena, and on the other by the internal world C of percepts. It is this latter internal world that is ultimately to be associated with consciousness. Let us briefly review how the structure of spacetime, including its dimensionality, arises from this overall picture of reality. We shall then find it straightforward to identify the first-person frame of reference. Having done that, we will see how consciousness is to be defined and how it operates within the spacetime structure of the world.

(1) As was shown in **Sec. 2.2**, single-particle physical laws (such as the Maxwell and Dirac equations) are the expressions of a state of consistency between the facts of the external world A and the

representations given to them in empirical world B. This is the projection theory, so-named because in the formalism a linear operator $\hat{R}$ projects the facts of external reality A into the empirical world B of representations.

(2) As shown in **Sec. 3.5**, the projection process is a physical process, implying the existence of two extra dimensions, $x^5$ and $x^6$. Coordinate $x^5$ is infinite in extent and metrically time-like, although it acts as if it were an extra spatial dimension. Coordinate $x^6$ is likewise infinite, but is metrically space-like and plays the role of an extra temporal dimension. Together, $x^5$ and $x^6$ form a 2-D Minkowski space: $x^5$, $x^6 \in \bar{M}^2$, where the bar reminds us that the metrical signatures are opposite to the ones normally associated with space and time. Total spacetime is then the 6-D product space $M^{4+2} = M^4 \times \bar{M}^2$. We may note two additional things in passing: (a) The number of spatial dimensions in Minkowski $M^4$ is 3, because only in three dimensions can matter exist (**Chapter 15**). (b) There exists also an inner Minkowski spacetime, $N^2$ (see *Sec. 10.7.1*), but this internal manifold has no apparent bearing on consciousness and will not concern us here.

(3) A study of the symmetry properties of the Dirac equation reveals that to each inertial reference frame $K(x^0, \mathbf{x}, x^5)$ in $M^{4+2}$ there exists a dual frame, $K_D(x'^0, \mathbf{x}, x'^5)$, related to $K$ by a $90°$ rotation in the $x^0$-$x^5$ plane. (See **Chapter 17**, especially **Fig. 17.3**.) To each external *event* $E$ in $K$ there corresponds a dual event or *internal representation* $E'$ in $K_D$. The physical world is thus described twice: once in $K$ as a function of time $x^0 = ct$, and once in $K_D$ as a function of the rotated time-like fifth coordinate $x'^5 = x'_5$.

(4) The rotation transformation taking $K$ into $K_D$ defines also the moving present moment. This objective expression of temporal succession is found in $K_D$ only. (See *Sec. 17.4.2.*)

Now a key attribute of reference frame $K_D$ is its uniqueness. No frame other than $K_D$ derived from $K$ by rotation in the $x^0$-$x^5$ plane has associated with it a valid Dirac wave equation. Because the rotation is finite, $K_D$ is inaccessible to third-person observers in $K$. But inaccessibility to third-person observers is the hallmark of the first-person frame of reference. Frame $K_D$ thus provides precisely what we are looking for: a geometric basis for the subjective, first-person perspective of conscious experience. Moreover, it even includes a mechanism for the subjective experience of time passing: the moving present moment. In short, we conclude that, in the absence of any plausible alternatives, *$K_D$ is the geometric home of conscious experience.*

It bears repeating that our deduction in **Chapter 17** of the existence of frame $K_D$ required no assumptions about the nature of consciousness or conscious experience. Rather, its existence was discovered through a purely formal consideration: the inherent symmetry of the Dirac equation in $M^{4+2}$. Thus, *frame $K_D$ precedes conscious experience.* Its prior existence is what renders subjective conscious experience possible in the first place. Fortuitously, it also makes the problem of consciousness accessible to the formal manipulations of theoretical physics.

Note also that we are speaking here of a purely geometric dualism, neither a substance dualism[12] nor a property dualism[13]. The material substance of the world is given but once, but receives a different description (i.e., timelike parametrization) depending on whether it is formulated in $K$ or $K_D$. In the case of fermions the two descriptions are embodied formally in Eqs. (17.5) and (17.17).

## 18.3 The relativistic origin of self

In 4-D relativity, a frame of reference $K$ is a coordinate system for specifying the spatial and temporal coordinates of events. In the formulation of this theory the presence of an *observer* moving with $K$ is implicitly assumed. This observer may or may not be conscious; a dedicated machine will do. In any case, it is the observer who, using rods and clocks, establishes the coordinates of events as seen from the point of view of $K$. It is true that one may sometimes refer to a frame $K$ without mentioning the observer, as for example in such phrases as "the rest frame of the particle" or the "laboratory frame." But always implicit is the possible presence of an observer should one be required for taking measurements. From an empirical point of view, the frame $K$ has meaning only if it is, or can be, occupied by an observer.

Consider now frame $K$'s dual frame, $K_D$. On grounds of consistency, just as we associate a potential observer with each inertial frame $K$, we may do so as well for frame $K_D$. However, as we have seen, $K_D$ is inaccessible to third-person observers. We may thus reasonably conclude that *the "observer" to be associated with $K_D$ (assuming one is present) is none other than the "Self," the first-person subject of conscious experience.* This does not mean that the mystery of conscious experience is now suddenly solved. It means only that the concept of Self is already countenanced by relativity in extra dimensions.[14]

## 18.4 Internal representations

Obviously the observer in $K_D$, should one be present, must have something to observe. What exactly is this something? We claim that, at a fundamental level, *what the self in $K_D$ observes are internal*[15] *representations*[16] *of external events.* It could hardly be otherwise, as such representations are the only real things present to observe. To illustrate, suppose that an observing self occupies frame $K_D$ of **Fig. 18.1**. A material point in frame $K$ moves at constant velocity, its motion

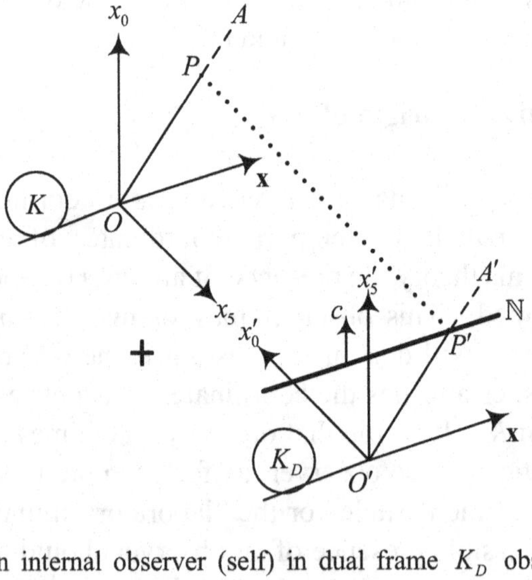


**Fig. 18.1** An internal observer (self) in dual frame $K_D$ observes internal representation $O'A'$, the dual of world line $OA$. The moving present moment, $\mathbb{N}$, divides $O'A'$ into three temporal components—past, present and future. Since $K_D$ is parametrized by pseudo-spatial $x_5'$, not $ct$, the observer has access to all three components of the temporal order.

described by world line $OA$ in $K$. What the self in $K_D$ observes is not $OA$ directly, but its internal representation $O'A'$, an objective element of reality we have previously referred to as the *dual* of world line $OA$. Note, however, that representation $O'A'$ is intersected by the moving present moment, $\mathbb{N}$, which exists in frame $K_D$ only. Thus, representation $O'A'$ divides into three constituent parts—past, present and future. Since $K_D$ is parametrized by pseudo-spatial dimension $x_5'$ rather than time $ct$, we see that the first-person observing self has unobstructed access to all three parts of the temporal order. Such is clearly *not* the case for the third-person observer in frame $K$. For that observer sees only the present point $P$, in accordance with the transformation equation (17.20). On the basis of this analysis of internal representations, one guesses that a proper description of consciousness will entail a tripartite structure linked to the temporal order—past, present and future.

## 18.5 Consciousness defined

Having with these few definitions brought the notion of conscious experience within the scope of physical theory, one can now venture to say what is meant by consciousness. Surely it does no good to define it circularly as the state of being conscious, or just as unhelpfully, as a state of awareness or sentience. For states are things that can be switched on and off, like a beam of light. Instead, we seek to understand consciousness for itself alone, as a category of existence distinct from the cortical switching mechanism to which it may be bound. For our purposes, we shall define consciousness operationally as *the capacity of the individual or self (Ego) to read, understand and act upon internal representations of external events.* The part of the individual that does the reading, understanding and acting is called the *mind*, and the feeling that accompanies these internal activities (or the awareness of or attendance to them) is called *conscious experience*. Traditionally, mind stands in sharp distinction to the body. For whereas the body and its brain are accessible to objective study, the mind and conscious experience are not: they are hidden within the individual and unavailable for public, third-person viewing. Thus consciousness, as the basis of mind, seems doubly obscure and resistant to systematic study. However, as we shall see, the operational definition given here, while limited in scope—leaving out, for example, thought and emotion—entails everything one needs to know to undertake the study of consciousness from a physical point of view. Note that the tripartite structure (reading, understanding, acting) mirrors, as anticipated above, the temporal order (past, present, future). In the following paragraphs we shall attempt to throw a little more light on the significance and terms of our definition of consciousness.

*18.5.1 External events.* External events are physical occurrences in the mind-independent external world. Here are five examples of external events: (a) solar photons scattered by the features of a natural landscape; (b) pressure waves in air generated by a loudspeaker; (c) sugar molecules in milk chocolate; (d) $NH_3$ molecules in ammonia gas; (e) a nail through the sole of one's shoe. Each is an external event available for detection by one or another of the sense organs serving the five sensory modalities: (a) visual, (b) auditory, (c) gustatory, (d) olfactory and (e) somatosensory. In each of these examples, the detection process converts the incoming signal—a stream of photons incident on the retina, for instance—into electrical pulses that are then sent along various nervous and neural pathways to dedicated sensory transmitting areas in the cerebral cortex. There, neurons are caused to fire. The firing neurons are themselves to be considered external events, although secondary ones. They serve to encode the original events—those events external to the sensory apparatus.

*18.5.2. Internal pictures and qualia.* Somehow—we have no idea how this works—neuron firings in the cerebral cortex cause the internal pictures experienced in perception. Thus, for example, corresponding to the external events listed above, the following internal pictures may be experienced: (a) the *sight* of the blue sky and green grass of the visible landscape; (b) the *sound* of a Bach fugue; (c) the sweet *taste* of milk chocolate; (d) the choking *smell* of ammonia gas; and (e) the *pain* induced by a nail penetrating the skin of one's foot. Each of these internal pictures (or conscious states) has a subjective, qualitative feel to it. They define "what it is like"[17] to be the individual experiencing them. In the philosophy of mind, such qualitative states are called *qualia*.[18] Qualia comprise (along with the will, as discussed below) the entire content of conscious experience in the primitive sense in which we are defining it. What qualia are ontologically—i.e., their mode of existence—and how they are formed, remain for now a complete mystery. Without qualia, though, there would be no content and no consciousness to talk about. As the philosopher John R. Searle sees it, "...the problem of qualia is not just an aspect of the problem of consciousness; it *is* the problem of consciousness."[19]

For the first 11 billion years of the life of the Universe, there existed no light or color, no sounds, tastes, odors or pains. For these phenomena are qualia—the feel of internal representations—and exist only in the minds of conscious beings. Sometime after the emergence of life on Earth about 3.8 billion years ago there occurred an act of creation quite as profound, if not as energetic, as the primordial fireball. This was the creation of qualia. In conscious experience we have a second creation, one Colin McGinn has aptly called the Soft Shudder.[20]

*18.5.3 Conscious will.* If reading internal pictures represents the *grasping* side of conscious experience, then acting on them by way of motor response—bodily movement or speech, for example—represents its *shaping* side. To the extent that it is conscious, this shaping activity may be called an act of *conscious will.* The act of will forms a part of our definition of consciousness, not only for reasons of symmetry,[21] but more importantly, because we *feel* that we are free to act. This feeling of freedom helps to define, along with our perceptions of the external world, "what it is like" to be human. For that reason the writer considers it indispensable to the definition of consciousness. Conscious will forms a key part of the hard problem of consciousness.

Now we may *think* we have free will, but the question is whether or not we really do. Many workers in psychology and cognitive science argue that we do not—that voluntary action is illusory.[22] But until the problem of consciousness is actually solved, that skeptical conclusion seems at best premature, and very likely wrong. For if the shaping side of conscious experience is illusory, then, by symmetry alone, one should expect the grasping side to be illusory too. But few would argue that the visual percept, for example, is an illusion.[23] So, if free will appears from the evidence to be an illusion, then we would do well to question our current assumptions about the nature of consciousness.

*18.5.4 The role of understanding.* At first glance, it seems that the element of "understanding" might, without undue harm, be omitted from our definition of consciousness. It seems, in a way, an inessential refinement. For many actions—such as withdrawing one's hand from a hot stove—are taken reflexively and unconsciously. In such cases, the action is unaccompanied by conscious understanding. On the other hand, understanding often forms a vital bridge between the experiences of grasping and shaping. Ideally, if time is available, one does not act until one has formed an understanding of the current state of affairs. Moreover, there is something it feels like to understand, as the 'aha!' experience vividly illustrates. The inclusion of understanding confers, in the writer's opinion, a conceptual completeness to our definition of consciousness. As we shall see in the next chapter, it is the role of the cortex to provide understanding.

*18.5.6 The Ego.* The individual mind or self, the one who experiences, may be thought of as an internal observer and agent. What she observes and acts upon are qualia. In her mental life she is, in a sense, the sum of her qualia. Now talk of an internal observer naturally raises the nightmare of the homunculus—a little man in the head, or ghost in the machine, who perceives and guides mental activity. This fellow was allegedly disposed of long ago by the philosopher Gilbert Ryle, who observed that a second homunculus would be required to receive and interpret the mental contents of the first homunculus, and a third to monitor the second one and so on, leading to an infinite regress of homunculi.[24] And yet, common sense suggests that when experience is present, someone—call her an internal observer—is at home, having the experience. Understanding the origin of self is part of the hard problem of consciousness. Thus, because the hard problem has not been solved, it seems premature to reject categorically the existence of at least some kind of internal observer.

## 18.6 Is consciousness reducible?

We have undertaken to solve the problem of consciousness within the framework of physical theory. We have made a start, but one may reasonably ask, how far can we go with this? Typically, to explain something in physical terms is to make it understandable by identifying the causal chain or principle giving rise to it. For instance, we can explain air pressure as the result of air molecules colliding with the walls of a container; we can explain the discrete spectrum of atomic hydrogen as a result of hydrogenic electrons jumping from stationary orbits of higher energy to ones of lower energy; and we can explain the shape of a soap film as the result of its seeking the shape of lowest potential energy.

The explanatory process just described is often called *reductive explanation,* or simply *reduction.*[25] Whenever a given state of affairs A admits of a complete reductive explanation in terms of an underlying state of affairs B, one says that A reduces to B. Now what about consciousness? Can it be reduced to, or explained in terms of, some underlying principle or mechanism, in the way that air pressure reduces to the force of molecules striking the walls of a container?

The answer is no, it cannot, and there are two reasons for this. First, in accordance to our definition, consciousness deals not with events in the external world, but with the reading, understanding and acting upon internal representations of those external events. Because the four known forces of nature and the laws governing them refer solely to events in the external world, and are functions of time $ct$, it follows that consciousness and its relationship to internal pictures cannot be explained in terms of known physical laws. This result is sometimes expressed by asserting that even the most complete account of the structure and function of the living brain cannot explain why these purely physical attributes may be accompanied by conscious experience. The apparent impossibility of deriving inner experience from external structure and function has become known as the *explanatory gap.*[26] It is what makes the experiential side of the problem of consciousness "hard."

The second reason that consciousness cannot be explained causally—i.e., as the effect of one thing pushing on another—is that consciousness as we have defined it is not a thing or process at all. Rather, it an abstract *capacity*—the capacity of the self to have *experience* of a certain kind, namely, *conscious* experience, the experience of reading, understanding and acting upon internal pictures. Since consciousness is tied to no particular physical structure, and because conscious beings are themselves part of the physical universe, we may generalize our original definition to read that consciousness is the capacity of the *universe itself* to manifest conscious experience.

This reformulation emphasizes our goal of drawing consciousness into the realm of physical theory.

Viewed in this universal light, consciousness begins to look like an independent category of existence, closely similar, in fact, to the dimension time. Metrically, time orders the sequence of events at each point of space. Yet, in a deeper sense, time may also be understood as the *capacity* of the world for succession, a capacity that exists at each point of space whether or not there happen to be present matter and energy in motion. If this were not the case, time would have to be reinvented whenever there arose at a given point a new sequence of events. And so it is with consciousness. The capacity for conscious experience exists at each point of space, whether or not there happen to exist living, conscious beings. If this were not the case, conscious experience would have to be reinvented wherever and whenever there arose living beings. But, on a naturalistic (materialistic) world-view, such reinvention could only come about from the action of matter itself, a possibility already ruled out by the logic of the explanatory gap. Hence consciousness, like time, is to be considered a fundamental category of existence and, as such, closed to explanation in terms of deeper laws or processes.

## 18.7 Mind-brain dualism

According to our definition of consciousness, the capacity for conscious experience hinges on a fundamental image-object dualism: the dualism of internal representations and external events. Now internal representations exist in the mind (certainly they are found nowhere else), whereas external events are encoded in physical brain states (neuron firings). Moreover, although internal representations and external events are obviously correlated, neither side of the relationship reduces in an explanatory sense to other. For example, the experience of seeing a blue sky no more reduces to a brain state than does the meaning of the word "cat" reduce to the arrangement of alphabetic characters on a printed page. This is the message of our previous discussion of the explanatory gap. Thus the image-object dualism inherent in our definition of consciousness leads at once to a dualism of mind and brain. Whatever ontological distinction—i.e., difference in mode of being—exists between internal representations and external events, the same distinction applies to mind and brain. This is hardly the place to review the anguished history of attempts to relate mind and brain—the so-called mind-body problem. For now, it is enough to note that, in contemporary philosophy and neuroscience, almost no one supports dualism. The prevailing argument against it is that it "does not work."[27] That is, there seems to be no natural mechanism by means of which mind and brain, conceived as ontologically distinct entities, could interact and influence one another. In particular, the production by a non-physical mind of the action potentials necessary to initiate motor activity would violate the principle of conservation of energy.[28, 29] And yet, if consciousness is a real category of the world and properly characterized by the definition set forth above, then a dualism of mind and brain *of some kind* must obtain. That does not mean that biological naturalism—the view that brains cause conscious experience—is false. It means only that we do not yet understand the natural order that allows this to happen. That is the real mystery of consciousness, and the main piece of it.

## 18.8 Summary and comment

By the "problem of consciousness" is meant the hard problem, not the (so-called) easy one. The following set of related questions may serve to summarize our formulation of the problem so far:

- How can *subjective* centers of experience arise in a centerless material world governed by *objective* physical laws?
- How does conscious *mind* emerge from material *brain*?
- How do *internal pictures* (qualia) arise from *external brain events*?
- How does *conscious will*, if it is real, bring about *motor response*?
- In what sense is there an *observing self* in the *machine*?

The striking thing about this summary is that each question expressing some facet of the hard problem is phrased in terms of a pair of ontologically distinct elements, i.e., the irreducible halves of a dualism: subjective-objective; mind-brain; internal-external; will-response; observer-machine. Each dualism grows naturally out of our starting definition of consciousness. So if our definition is good, then dualism follows. Dualism may or not be true (which in the latter case would imply that our definition is false), but at least it logically follows. It also comports with the commonsense view—the view that we *seem* to be more than flesh and bone. (See the Wallace Stevens epigraph at the beginning of this book.) In any case, based on our discussion so far, dualism is not obviously false. The hard problem of consciousness is the problem of discovering the origin and physical basis of these fundamental dualisms.

Already we know where *not* to look for a solution. From our discussion above on the explanatory gap we have concluded that the subjective side of conscious experience lies beyond the scope of the known physical laws in 4-D spacetime. This result wields considerable critical power. It means that *none* of the many proposed explanations of conscious experience based solely on four dimensions can be true. All are invalid in principle. This includes all theories of consciousness in four dimensions based on:

- *Physicalism (identity theory)*—the theory that mental states and physical brain states are identical.
- *Functionalism*—the idea that mental states are physical and arise owing to a propitious functional organization of the brain.
- *Computation*—the notion that computer software of the right design confers consciousness on the hardware.
- *Complexity*—the idea that consciousness might emerge from any physical system, given sufficient complexity.
- *Information*—a measure of that which alters previous knowledge.
- *Quantum mechanics*—the physics of microscopic nature.
- *Quantum gravity*—a quantum field theory of gravity analogous to quantum electrodynamics.
- *Neurophysiology*—the study of the structure and function of physical brain.

All explanations of consciousness based on these ideas and disciplines must fail because they cannot manifest a first-person perspective, the foundational characteristic of conscious experience. As argued in *Sec. 1.2.2*, the principle of relativity forbids it. John Searle in the *Mystery of Consciousness*[30] disarms all such explanations employing a mixture of general logic and common sense. The relativity principle drives a final nail in the materialist coffin.

Although it is implicit in the preceding remarks, it is worth stating explicitly that *no man-made machine operating exclusively on the basis of known physical laws can truthfully claim an inner conscious life.* Machine consciousness is impossible in principle. There can be no felt experiences of blue sky or Bach fugues in the minds of man-made machines. For no machine, operating under known physical laws can manifest the first-person perspective required to experience internal representations. Or more accurately, no machine operating under known laws can manifest an *occupant* of the first-person frame of reference. The principle of relativity forbids it. That does not mean that a machine could not someday be built that fools its human interrogators into thinking that it is conscious. It is just that the machine could not be aware of what it was doing.

Living systems, because they are made of matter and perform genetically programmed functions, are machines. Some of them may in fact be conscious—cats, dogs and ourselves, for example. However, for the reason set forth above, conscious living systems cannot be synthesized in the laboratory. This is a radical claim. It means that even if one could build an artificial brain that replicated the natural brain in every detail, neuron for neuron, it would still not be conscious. The best that one could do is to create an imitation of the living machine, molecule for molecule, absent its mind, a *zombie*.[31] To understand why this is so one has to get past the fallacious naturalist idea that frame $K$, together with what goes on in it, is fundamental. It is not. The fundamental reality is dual spacetime $K+K_D$, and what goes on in dual frame $K_D$ (conscious experience) can neither be created nor managed by third-person workers in frame $K$.

## Notes and references


[1] See discussion of the "new mysterianism," a term introduced by O. Flanagan, *The Science of Mind*, 2nd Ed. (MIT Press, Cambridge, MA, 1991), pp. 312-314.

[2] Isaac Newton, *Philosophiæ Naturalis Principia Mathematica* (London, 1686/1687). In Latin. Published in English 1728.

[3] S. Weinberg, "Newtonianism and today's physics,"in *300 Years of Gravitation*, S. Hawking and W. Israel, Eds. (Cambridge U. Press, Cambridge,1987), pp 5-16.

[4] See J. Berkovitz, "Action at a Distance in Quantum Mechanics," in *Stanford Encyclopedia of Philosophy*, E. N. Zalta, Ed. (Winter 2008 Edition). Online.

[5] A. Einstein, "The Foundation of the General Theory of Relativity," Annalen der Physik **49**, 1916; reprinted in *The Principle of Relativity* (Dover), pp. 109-164.

[6] A. Einstein, "On the Electrodynamics of Moving Bodies," Annalen der Physik **17**, 1905; reprinted in *The Principle of Relativity* (Dover), pp. 37-65.

[7] D. J. Chalmers, *The Conscious Mind: In Search of a Fundamental Theory* (Oxford U. Press, Oxford, UK, 1996), pp. v, vi.

[8] *Neural Correlates of Consciousness*, T. Metzinger, ed. (MIT Press, Cambridge, MA, 2000).

[9] C. McGinn, "Consciousness and Space," in *Conscious Experience*, T. Metzinger, ed. (Imprint Academic, Thorverton, UK, 1995), pp. 149-163.

[10] C. McGinn, *The Mysterious Flame* (Basic Books, New York, 1999), Chapter 4.

[11] For a discussion of these theories plus references, see J. R. Smythies, "Aspects of Consciousness," in *Beyond Reductionism*, A. Koestler and J. R. Smythies, eds. (Beacon Press, Bos-ton, MA, 1969, pp. 233-257.

[12] J. Kim, *Philosophy of Mind, 2nd Ed.* (Westview Press (Perseus), Cambridge, MA, 2006), Chapter 2. Substance dualism: the notion that mind and body are two ontologically distinct substances.

[13] J. Kim, *ibid.*, pp. 50-52. Property dualism holds that, although the world is made up of one kind of substance, that same substance can exhibit both physical (bodily) and mental properties.

[14] Note that even if $K_D$ is unoccupied by an observing self, it may still hold meaning for third-person observers in $K$. For $K_D$ contains the moving present moment, which, as we saw in **Sec. 17.5**, accounts for the collapse of the quantum-mechanical state vector. This in turn suggests—since third-person observers need

not be conscious—that consciousness has little to do with collapse of the state vector, apart from its role in specifying and setting up the method of measurement. All that is needed for collapse is the act of measurement, and this is carried out not by a mind but a machine. The present account of measurement thus contradicts "von Neumann's far-reaching contention that it is impossible to formulate a complete and consistent theory of quantum mechanical measurement without reference to human consciousness." See M. Jammer, *The Philosophy of Quantum Mechanics* (John Wiley, New York, 1974), pp. 474-482.

[15] The qualification "internal" is needed here to distinguish representation in $K_D$ from those produced by operator $\hat{R}$ at the boundary between worlds A and B of **Fig. 2.2**. The latter representation, which we have denoted $\chi'$, might be called "external" representations because they, like the facts they represent, depend on the time *ct*, not $x_5'$.

[16] J. Kim, *op. cit.*, pp 24-25. The term *representation* is standardly employed for the description of mental contents. We use it here as a synonym for *event dual.*

[17] The term originates in the famous essay of T. Nagel, "What is it like to be a bat?" in *Mortal Questions* (Cambridge U. Press, Cambridge, UK, 1979), pp 165-180.

[18] J. Kim, *op. cit.*, pp 224-229.

[19] J. R. Searle, *The Mystery of Consciousness* (A New York Review Book, New York, 1997), pp. 28-29.

[20] C. Mc Ginn, *op. cit.* (*The Mysterious Flame*), p. 15.

[21] S. Blackmore, *Consciousness: An Introduction* (Oxford U. Press, Oxford, UK, 2004), p. 13.

[22] D. M. Wegner, *The Illusion of Conscious Will* (MIT Press, Cambridge, MA, 2002).

[23] But see A. Noë, "Is the Visual World a Grand Illusion?," J. of Consciousness Studies **9**, 1 (2002).

[24] G. Ryle, *The Concept of Mind* (U. Chicago Press, Chicago, IL, 1949), pp 30-31.

[25] J. Kim, *op. cit.*, Chapter 10.

[26] J. Levine, "Materialism and Qualia: The Explanatory Gap," Pacific Philosophical Quarterly **64**, 354-361 (1983); *Purple*

*Haze: The Puzzle of Consciousness* (Oxford U. Press, Oxford, UK, 2001), 76-80.

[27] S. Blackmore, *op. cit*., p. 13.

[28] D. Dennett, *Consciousness Explained* (Back Bay Books, New York, 1991), pp. 34, 35.

[29] D. L. Wilson, "Mind-Brain Interaction and Violation of Physical Laws," in *The Volitional Brain: Towards a Neuroscience of Free Will*, B. Libet, A. Freeman & K. Sutherland, eds. (Imprint Academic, Thorverton, UK, 1999), pp. 185-200.

[30] J. R. Searle, *op. cit*., Chapters 2-7.

[31] D. J. Chalmers, *op. cit*., p. 94.

# Chapter 19

# Mind-brain interaction in dual spacetime

This next-to-final chapter deals with the interaction of mind and brain. It assumes, and is wholly dependent upon, the existence of dual spacetime $K + K_D$ as detailed in the previous two chapters. We shall take our resolution in **Chapter 17** of the mystery of state-vector collapse as direct evidence of the reality of dual spacetime. Additional, if circumstantial, evidence for dual spacetime is the existence mind itself, whose privileged status in the world implies the existence somewhere of a privileged frame of reference, one we assume to be dual frame, $K_D$. The essential idea is that objective brain occupies frame $K$, while brain dual and subjective mind occupy $K_D$, implying duality of mind and brain. Our aim is to show how these two ontologically disparate elements interact, out of which interaction arises a conscious being or self.

There are two forms of interaction to consider: one connecting external receptor-created pictures to corresponding internal percepts; the other connecting an internal visualized motor objective to a corresponding motor response. We are going to explore, in other words, the reciprocal operations of *perception* and *will*. These operations, we should note, correspond to two of the three functional components of our definition of consciousness, **Sec. 18.4**: the reading and acting upon internal representations of external events. These are the fundamental grasping and shaping functions of the conscious being. The third component of our definition—understanding—serves *inter alia* as guiding link between grasping and shaping. Understanding appears to be generated by the neocortex and prefrontal cortex areas of the brain working as a whole.[1-3]

**Fig. 19.1** presents an overall view of the conscious being (CB) situated within dual spacetime. To depict the flow of information within the CB, the two frames are shown separated from each other; actually they are superimposed, with a common origin. In such a picture the CB's total being, shown bounded by the dotted rectangle, appears bi-bisected by an impenetrable crosshatched barrier between $K$ and $K_D$.

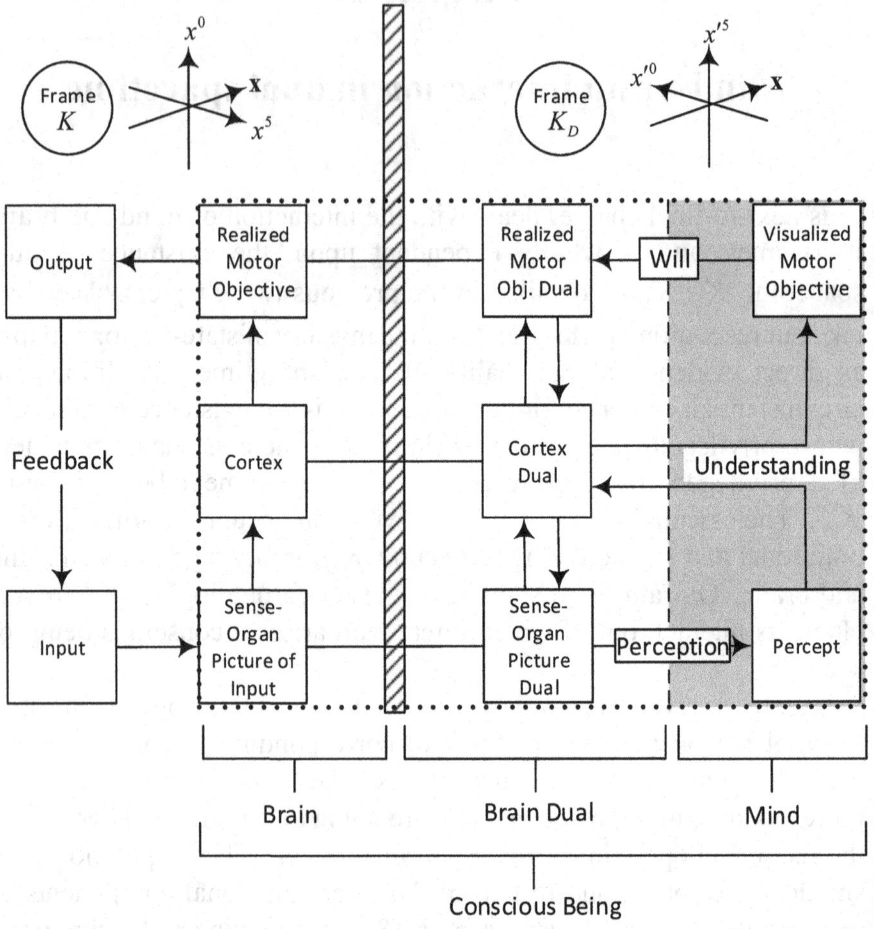


**Fig. 19.1** Information flow in the conscious being (CB). Her total being, shown bounded by the dotted rectangle, occupies at once two frames of reference, *viz.*, third-person frame $K$ and first-person frame $K_D$, the former containing her brain, the latter her brain dual and mind. The two frames are shown separated by an impenetrable barrier, denying third-person observers access to the mind of CB. This barrier is precisely division 2 of **Fig. 2.1**. Information flow in the *brain* is unidirectional, following the arrow of time. In the *brain dual,* however, it is effectively bidirectional, for in frame $K_D$ time $ct$ has been replaced by pseudospatial dimension $x'^5$. Mind-brain dualism entails two forms of interaction: *perception* and *will.* The former links brain to *percept*; the latter *visualized motor objective* to *brain. Understanding*, arising from cortical processing, runs parallel to perception and will.

While her brain lies in frame *K*, her brain dual and mind (shown shaded) lie in $K_D$. Notably, information flow in the *brain* is unidirectional, following the arrow of time. This is normal causality. In the *brain dual,* however, the flow is bidirectional, because in frame $K_D$ time *ct* is replaced by pseudo-spatial dimension $x'^5$. The CB's mind, being within $K_D$, is privileged and inaccessible to third-person observers in frame *K*.

As indicated in the figure, perception begins with a stimulus *input.* The energy of the input activates an appropriate receptor, or sense organ, such as the retina or cochlea, creating an objective, physical representation or *picture* of the input. This sense-picture is duly projected electrochemically to appropriate receiving areas of the *cortex,* e.g., visual or auditory, where it is expressed in dispersed, coded form, as a kind of hologram of the original picture. This encoded picture is duplicated in the brain dual, parametrized not by time *ct* but by dimension $x'^5$. *Perception* is the mind-brain interaction required to reconstruct from the cortical hologram the original picture in the form of an interpretable, understandable *percept.*

Also as indicated in the figure, voluntary motor activity begins with a *visualized motor objective. Will* is the mental operation required to project the visualized objective—conceived as a kind of mental picture (analogous to the percept)—into the motor cortex (analogous to the sensory cortex), where it is subsequently converted into *realized motor activity.* The consequent pushing and pulling on the external world by motor activity represents the *output* of CB.

Perception is the means by which the CB becomes conscious of stimulus *inputs* from the empirical world of frame *K*. Similarly, will is the means by which she consciously imposes an *output* of action on that same empirical world. As indicated in the figure, output feeds input, yielding a conscious machine under negative feedback control.

Such are the main features of mind-brain interaction and the conscious being. But the mere display of these features in a block diagram tells little about how it all actually works. It remains to examine in detail the following:

- How does the mind reconstruct internal pictures (percepts) from cortical (holographic) representations of the external world? I.e., How does perception work?
- How do percepts come to be associated with qualia? In fact, what *are* qualia? How physically do they arise?

- How does the mind instigate motor activity from a mere visualized motor objective? That is, how does the will work?
- Are there subjective states analogous to qualia associated with the act of will?
- What is the role of *energy* in these interactions?
- How does the self arise in the first place?

## 19.1 The visual percept

In **Sec. 18.3** it was claimed that what the self "sees" are the internal representations of external events. This claim is by itself nothing new—just the familiar story of the phenomenal content of conscious experience correlating in some fashion to neural and nervous activity. What *is* new, however, is the telling of this story from a relativistic standpoint. Special relativity in dual spacetime offers a glimpse inside the realm of conscious experience. In particular, it can tell us how the internal images comprising the phenomenal content of conscious experience are formed, and also *where* and *when* they are formed.

We shall focus here on the visual experience, vision being the most critical and well-studied of the senses. The other sensory modalities and corresponding percepts can be dealt with in a parallel manner.

*19.1.1 Naturalistic (third-person) account of the visual experience.* An overall view of the neurobiologic visual process as presently understood is given schematically in **Fig. 9.2.**[3-5] In brief, light collected by the cornea and lens falls on the rod-cone mosaic of the retina, where it is converted into electrical pulses which are sent along the optic nerve and sub-cortical visual pathways (optic chasm and lateral geniculate nucleus) to the visual cortex. Most such signals enter the visual cortex at the *primary visual cortex* (V1) and are then routed to specific feature processing areas in the *prestriate cortex* (V2-V8). The routing from V1 is so managed as to create two data streams, the *ventral stream* ("vision for perception") and the *dorsal stream* ("vision for action"). The two streams ultimately converge upon the *lateral prefrontal cortex* for final processing.

The neural activity just outlined is generally presumed to cause the conscious visual experience. And yet, how can this be? The brain has taken a perfectly formed retinal image and projected it into the volume

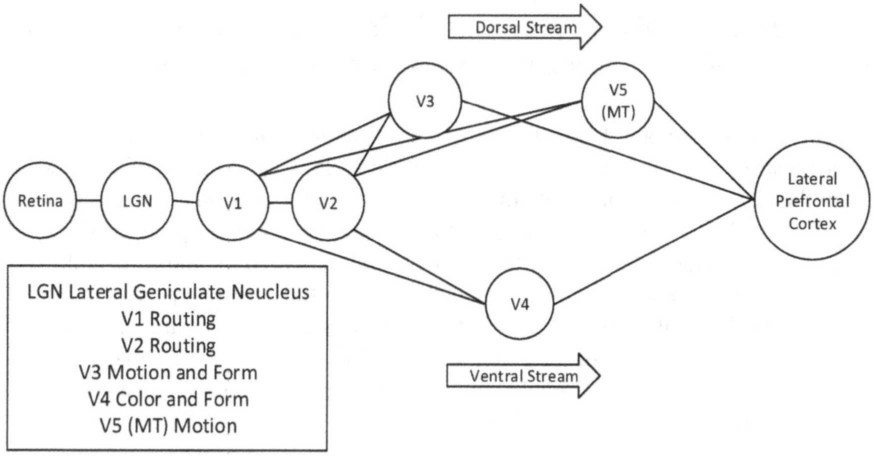


**Fig. 19.2** Projection of the retinal signal into feature-processing areas V1-V5 of the prestriate cortex. These areas support two data streams, the dorsal and the ventral, which eventually converge on the lateral prefrontal cortex for final processing.

of the cortex, where it is radically taken apart and no longer resembles the retinal image. In particular, the geometric or topologic content of the input is now badly distorted. If the brain's objective is to create a visual experience, why does the brain not manage it directly? Why does it first divide the retinal image into separate features? The answer is clear. At the retinal level, the optical image is uninterpreted. If the readers of this book were mobile cameras or non-conscious robots, a clear, uninterpreted retinal image of words on a page would suffice. But as we were apparently meant to be conscious, visual images have to mean something. We have to *understand* them. And the way that Nature has worked out to create understanding is to project the retinal image into the data processing streams depicted in **Fig. 19.2.**[4] Owing to these, the percipient can understand, for example, the visual image as it relates to motor activity.

And so, in principle, and all thanks to neurobiology, we have the possibility of understanding, but what about the visual image itself? How, where and when is it formed? The problem of recombining the projected features of the retinal image (form, light/color, motion) to create a unified, accurate representation of the empirical world external to us is called the *binding problem*.[4, 6] Here is John Eccles' pessimistic assessment of the problem, in conversation with Karl Popper[7]:

> You will remember, Karl, we were talking about that [problem of image reconstruction] when we were up at the castle looking up to the beautiful view of the head of Lake Como with the mountains, with the boats in the water, with all the villages around the lakeside, to the mountains rising up on all sides. Here is a wonderful picture of the most varied kind, all in the most incredibly fine detail, all in the clear air. Somehow from the fine punctate picture in our retina an integrated picture is eventually experienced as a result of all the processing in the brain of the coded transmission from the retina. It comes to us in this picture of vivid delight, and it seems to me that never can we get this completion on the neurophysiological level.

This expression of the binding problem was made in 1974, more than forty years ago at this writing. As a committed dualist, Eccles considered brain neither capable of image reconstruction nor the seat of conscious experience. However, despite Eccles' preeminence in the field, and for the reasons cited in *Sec. 18.4.3*, the neuroscience community has long rejected dualism of any sort, and from the 1970s onward a number of neurophysiological solutions to the binding problem have been advanced. Most of these exploit an assumed neural network of communication between the several feature processing areas into which the retinal image has been projected. Signal synchrony between those areas is supposed to restore the image to its original, unified form. As ingenious as such solutions are, it is easy to see, without going into network theory, that they cannot possibly do the two things they are supposed to do, namely, (a) reconstruct the retinal image and (b) create a unified visual experience for the brain's owner. There are at least three reasons why the neurophysiological explanations of binding must fail.

(1) Without form, the attributes color and movement have nothing to bind to. Thus form is fundamental, and the viability of a theory of binding reasonably can be judged by its ability to reconstruct form. The block diagram of **Fig. 19.3** depicts a visual process leading to the perception of form, formulated in naturalistic (non-dualistic, neurobiological) terms. In other words, it is assumed to take place in ordinary Minkowski $M^4$. An input image signal $I$ enters a first black box, where

it is acted upon by an operator $T$, projecting signal $I$ in coded form $TI$ into the various receiving areas of the cortex. The encoded signal $TI$

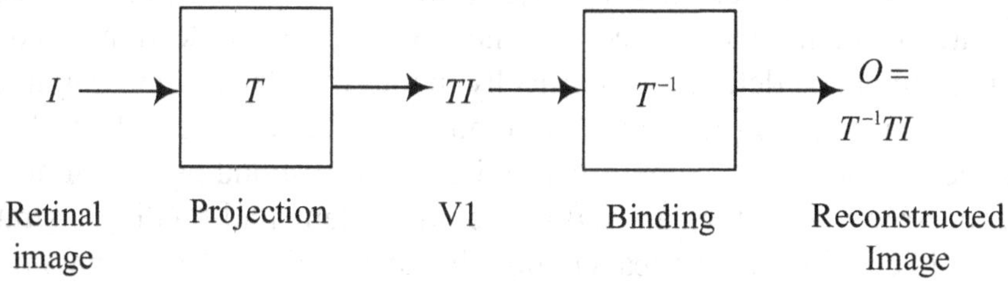


**Fig. 19.3** Information flow for the perception of form, according to standard neurophysiology. An input visual signal $I$ is acted upon by projection operator $T$, yielding a coded signal $TI$ distributed over the volume of the primary visual cortex, V1. To reconstruct the form of the original signal, coded signal $TI$ is acted upon by binding operator $T^{-1}$, yielding an output signal $O = T^{-1}TI$. If $T^{-1}T = 1$, then $O = I$.

then passes through a second black box, where it is operated on by a *binding* operator, $T^{-1}$, yielding an output signal, $O$:

$$O = T^{-1}TI. \tag{19.1}$$

Here $O$ is the putative perceived image, replicating the original retinal image $I$, provided that $T^{-1}T = 1$. But we can show that, in respect to form, there exists in the brain no such operator $T^{-1}$, thus precluding the reconstruction of form by purely neurophysiological means. Suppose that the input image $I$ consists of $N$ activated points of the retina. Let their locations in a Cartesian frame of reference be defined by coordinate vectors $\mathbf{x}_i$, $i = 1, \cdots, N$. The effect of $T$ operating on $I$ is to translate the activated points of the input to new positions in the visual cortex, namely to V1, where static form is recognized. From retinopic mapping studies[8, 9] it is known that the correspondence between the old and new positions is one-to-one.[4, 10] The effect of $T$ is thus to transform the original coordinate vectors to a new set of vectors $\mathbf{x}_i + \mathbf{a}_i$, $i = 1, \cdots, N$, where the $\mathbf{a}_i$ define the relative positions in 3-space of the cortical and original activation points. Following this, operator $T^{-1}$ acts on the distributed cortical image, yielding a final set of activation points defined by coordinate vectors $\mathbf{x}_i + \mathbf{a}_i + \mathbf{a}'_i$, $i = 1, \cdots, N$, where the $\mathbf{a}'_i$ define the relative locations of corresponding reconstructed and cortical activation points. So, for exact reconstruction of the retinal image, one requires $\mathbf{a}'_i = -\mathbf{a}_i + \mathbf{C}$, where $\mathbf{C}$ is an arbitrary constant displacement vector. However, as far as anyone knows, nowhere in the brain is there created or maintained a record of the $N$ vectors $\mathbf{a}_i$. Indeed, how could such vectors be determined cortically in the first place? One might call it the "calibration problem." Assuming it to be unresolvable by brain alone, it follows that there can exist no corresponding set of form-reconstructing vectors $\mathbf{a}'_i$, and hence no fundamental binding operator $T^{-1}$. In conclusion it appears impossible to reconstruct the retinal image in ordinary spacetime. Curiously, to suppose that a binding operator $T^{-1}$ does exist in $M^4$ is to suppose that it has a *purpose*, namely that of reconstructing the retinal image. One hardly expects a neurophysiological explanation of binding to be founded on teleology.

(2) There is an even stronger argument against purely neurophysiological binding. Assume that there really does exist a neurophysiological binding operator $T^{-1}$, and further that $T^{-1}T = 1$. Then (19.1) gives $O(\mathbf{x}) = I(\mathbf{x} + \mathbf{C})$, where $\mathbf{C}$ is the aforementioned constant vector. This says that the perceived output $O$ is nothing but a displaced version of the objective, third-person input. But this cannot be so, for nowhere in the brain is there found a third-person reconstructed image. Perceived images are first-person objects, unavailable for viewing by anyone but the percipient. Eq. (19.1) is thus ontologically inconsistent, as one side is third-person, the other first-person, refuting the assumed existence of operator $T^{-1}$.

(3) The two sides of (19.1) are inconsistent in another way as well. Clearly, without conversion of the input energy into light and color, there would be no visual experience. This means that the l. h. s. of (19.1)—the perceiving mental side—somehow entails light and color. But the r.h.s. of (19.1)—the brain side—knows nothing of light and color; it simply expresses the end result of a flow of electrochemical information from one area of the brain to another. The two sides of the equation are again seen to be incompatible—one side light, the other dark—reinforcing our claim that binding is not to be explained in neurophysiological terms alone.

*19.1.2 The visual experience in dual spacetime.* When it comes to accounting for the visual experience, dualism succeeds in precisely those areas where naturalistic neurobiology fails. Specifically, we can show (1) that perception in dual spacetime faithfully reconstructs form; (2) that the reconstruction has first-person ontology; (3) that it is experienced in light and color.

*A. Reconstruction of the visual image.* **Fig. 19.4** depicts the "Per-

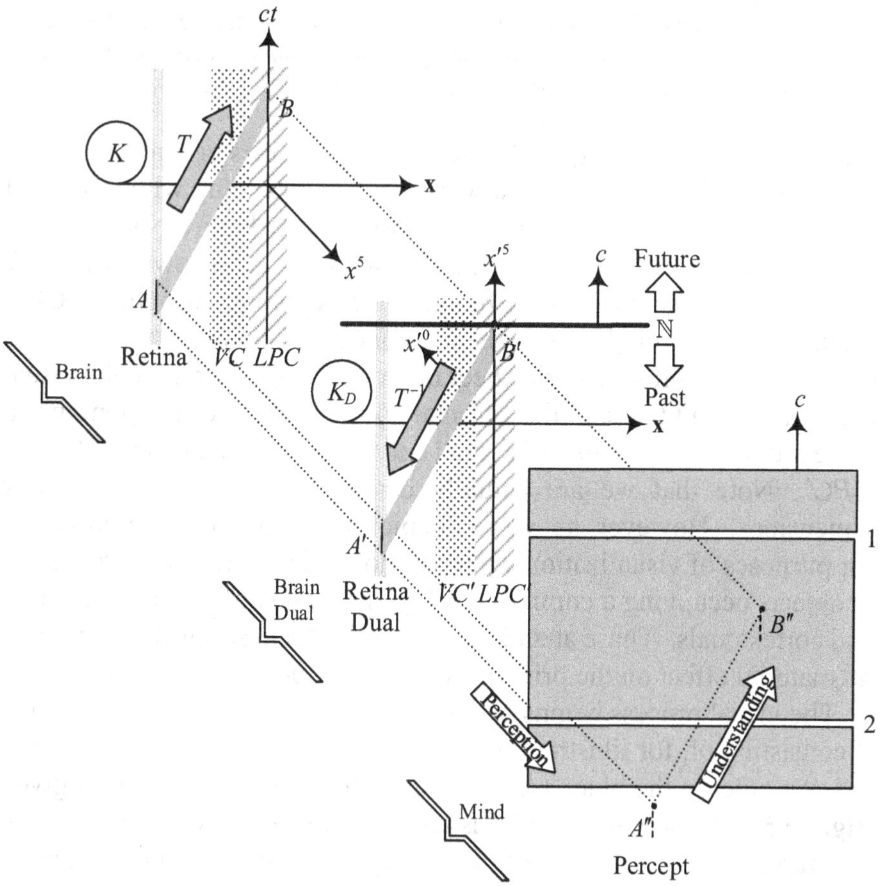


**Fig. 19.4** Creation of the visual percept in dual spacetime. Frame *K* contains the conscious being's (CB) brain, while dual frame $K_D$ contains both her brain dual and mind. The brain's world sheet in *K* consists of the retina, the visual cortex (VC) and the lateral prefrontal cortex (LPC). Corresponding to these areas in $K_D$ are the retina dual and the dual cortical areas $VC'$ and $LPC'$. Retinal image world line *A* projects to world line *B* in the cortex, creating world sheet *AB*. Shaded arrow labeled *T* symbolizes the projection operation. World lines *A*, *B* and world sheet *AB* are reproduced in dual frame $K_D$ as dual world lines $A'$, $B'$ and dual world sheet $A'B'$, respectively. The retina dual $A'$

reconstructs perfectly the form of retinal image $A$, and it is $A'$ that the CB sees in visual perception. Shaded arrow labeled $T^{-1}$, the inverse of operation $T$, symbolizes the reconstruction operation. An opaque screen, pierced by slits 1 and 2, blocks all but world lines $A'$ and $B'$ from projecting, point by point, into the mind of the CB. The screen is time-locked to the moving present moment, $\mathbb{N}$, and the resultant scanning of retinal dual world line $A'$ by slit 2 gives to percept $A''$ the visual attributes *motion* and *light/color*. The projection $B' \to B''$ from the cortex provides for understanding. World line $A''B''$ denotes a mental operation connecting past and present. Without it there could be no internal mental presence ("ghost in the machine").

ception" operation of **Fig. 19.1** in terms of world lines in dual spacetime, beginning with the retinal input world line $A$ and ending with the visual percept world line $A''$. Our task is to explain how the mind-brain interaction process $A \to A''$ actually works. As shown, frame $K$ contains the conscious being's brain, while $K_D$ contains both the CB's brain dual and mind. As we are concerned here with the visual experience, the brain's world sheet in $K$ consists of the retina, the visual cortex (VC) and the lateral prefrontal cortex (LPC). Corresponding to these areas in $K_D$ are the retina dual and the dual cortical areas $VC'$ and $LPC'$. Note that we are actually dealing with a problem in *five* dimensions. However, as we can depict only three of these on paper, for purposes of visualization we have shown the retina and cortex as 1-D objects occupying a common $1+1$-plane, and similarly for the retina and cortex duals. These anatomical distortions and simplifications have no material effect on the principles of perception now to be discussed.

The visual process begins in frame $K$ with a retinal image world line $A$ consisting of, for illustration, two moving points, one emitting in the red, the other in the blue. (A detailed picture of the two points is given **Fig. 19.5**; the scale of **Fig. 19.4** is too small to show them.) The retinal signal pulses emitted in world line $A$ project into the VC and from there to world line $B$ in the LPC, creating world sheet $AB$. The entire process of projection into the cortex is indicated by the shaded arrow labeled $T$. Here $T$ performs the same operation as the naturalistic projection operator $T$ of **Fig. 19.3**. World lines $A$, $B$ and world sheet $AB$ are reproduced in dual frame $K_D$ as world line duals $A'$, $B'$ and dual world sheet $A'B'$, respectively.

So much for projection of the retinal image into the cortex and cortex dual. It remains to reconstruct the original retinal image and present it to the CB in the form of an interpretable percept. This remaining operation requires three steps. The first is to recognize that, in the matter of form, the retinal image dual, $A'$, reproduces exactly the original retinal image, $A$. In addition, we recall that events in $K_D$ are parametrized not by time *ct*, but by the pseudo-spatial coordinate $x'^5$ (see *Sec. 17.4.1*). In other words, the dual world lines $A'$ and $B'$ are spatially, but not temporally, separated. Thus, world line $A'$ can in principle, without violating any causal laws, be reconstructed working backward from $B'$, as indicated in the figure by the shaded arrow labeled $T^{-1}$. On this basis, we claim that *what the CB sees in visual perception is retinal image dual* $A'$. Note that operator $T^{-1}$ performs the same operation as the reconstruction operator $T^{-1}$ of **Fig. 19.3**. However, because $T^{-1}$ now operates in $K_D$ rather than *K*, the reconstructed image $A'$ not only replicates perfectly the form of the original retinal image, but also has first-person ontology. Note that $A'$ is doubly hidden from third-person observers; for it exists not just in frame $K_D$, but in a place *earlier* than point $B'$ in the *LPC* dual, i.e., in past block time.

*B. Projection of retinal image into the mind.* The second step in the perception process is to make $A'$ and $B'$—$B'$ being the source of understanding—accessible to the conscious being. This means projecting both back into the CB's mind. But this happens automatically because, as discussed in **Sec. 17.3**, world lines $A'$ and $B'$ are cross sections in the $\mathbf{x}$-$x'^5$ plane of infinite line currents running parallel to the $x'^0$ axis. These currents are shown in **Fig. 19.4** as dotted lines passing through $A'$ and $B'$ and extending into the mind of the CB. But now a difficulty is encountered: it is not only $A'$ and $B'$ that get projected into the mind; the entire world sheet $A'B'$ connecting world lines $A'$ and $B'$ is projected as well. But as experienced conscious beings, we know perfectly well that one does not observe what takes place in the cortex. So some mechanism must exist to block all but world lines $A'$ and $B'$ from entering the mind of the CB. As third-person observers, we can never know what such a mechanism actually consists of or looks like, especially as it operates in past block time. Functionally, however, as depicted in the figure, it behaves as if it were an opaque screen, pierced by narrow slits 1 and 2, the former permitting passage of the current from $B'$, the latter the current from $A'$. (In reality the screen is $3+1$-dimensional, and the edge and slit 3-dimensional!)

*C. Introduction of the moving present moment.* But even after all this we still do not have perception. Thus far, we have been working in the static world of the dual block universe, and in that universe, although $A'$ accurately reproduces *form*, it cannot by itself exhibit the attributes *motion* and *light/color*. To generate these, a third and final step is required, namely, the introduction of *time* into frame $K_D$. The objective time *ct* of block time will not do, because objective time serves only to locate events within the static block, and anyway exists in frame *K*, not $K_D$. Instead, the temporal form required is the one we have previously identified as the moving present moment or now, $\mathbb{N}$. The present moment, which as noted in *Sec. 17.4.3* is found in frame $K_D$ only, is usefully conceived as the leading edge of a growing block universe, a universe whose present and (growing) past are fully determined, but whose (diminishing) future is unknown. The creation of the block universe grown to the point of containing both world lines $A'$ and $B'$ begins with $\mathbb{N}$ passing through $A'$ and ends with it passing through $B'$. This ending position is the one shown in the figure. Crucially, our cortex-blocking screen is time-locked to $\mathbb{N}$, so that both slits travel together parallel to the $x'^5$ axis at light speed *c*. Slit 1 coincides temporally with $\mathbb{N}$, and in the position shown gives entry into the mind of CB the current from $B'$, providing to CB the cortical image $B''$ necessary for understanding. At the same time, slit 2 gives entry into CB's mind the current from $A'$, setting the stage for the creation of percept $A''$, one exhibiting not only accurately-reconstructed form, but motion and light/color as well.

*D. Motion and color encoding.* Let us now look in detail at this final stage. The attributes *motion* and *light/color* are generated by the action of slit 2 cutting across the currents proceeding from dual world line $A'$ and into percept area $A''$ of the mind. The entire process as seen from the point of view of the CB is depicted in **Fig. 19.5**. This figure presents a detailed view of the dual retinal image world line $A'$ of **Fig. 19.4**, with the screen pierced by moving slit 2 projected upon it. The objects to be perceived visually are the retinal images of two moving points, one emitting in the red (r) and the other in the blue (b). The world lines of these retinal images in frame *K* (not shown) are reproduced in frame $K_D$ as dual world lines r and b, imaged upon CB's retina dual as shown. As explained previously, because dual world lines r and b are exact copies of the original world lines in frame *K*, the *form* of percept $A''$

accurately reproduces that of the original retinal image. To render percept $A''$ in light and color, the retina in $K$ is equipped with rods and cones, their duals repeated in frame $K_D$. For simplicity, only two kinds

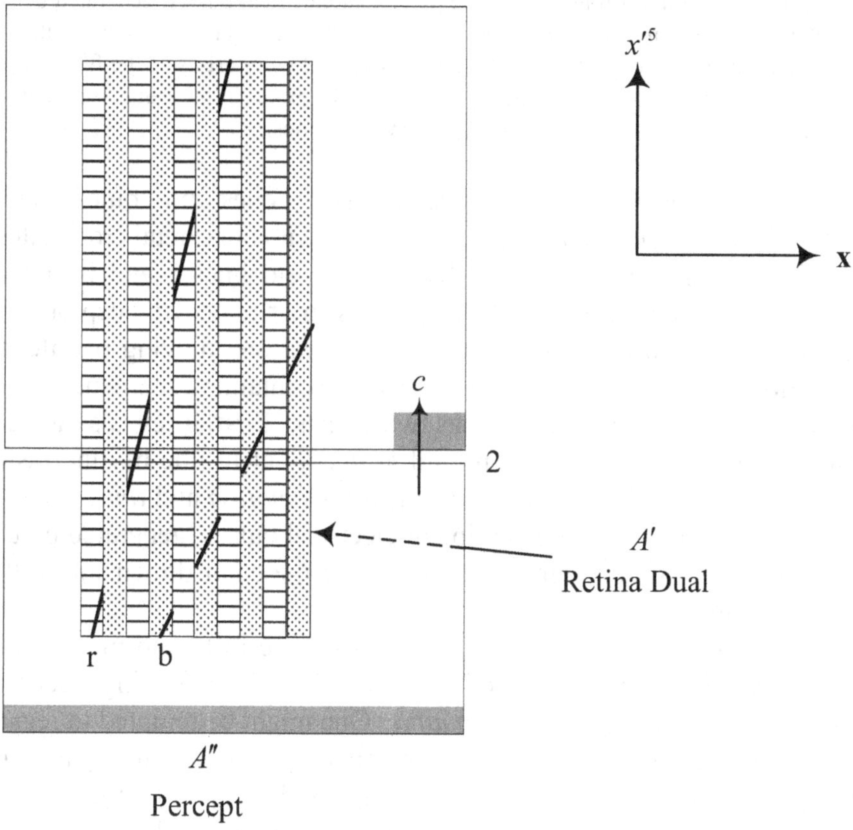


**Fig. 19.5** Detail of the process of visual perception depicted in **Fig. 19.4**, as seen from the point of view of the conscious being (CB). The objects to be perceived visually are the retinal images of two moving points, one emitting in the red (r) and the other in the blue (b). The world lines of these retinal images in frame $K$ (not shown) are reproduced in frame $K_D$ as dual world lines r and b, imaged upon CB's retina dual $A'$ as shown. Because dual world lines r and b are exact copies of the original world lines in frame $K$, the *form* of percept $A''$ accurately reproduces that of the original retinal image. To enable a rendering of percept $A''$ in light and color, the retina in $K$ is equipped with rods and cones, their duals repeated in frame $K_D$. For simplicity, only two kinds of visual receptor are depicted: "red" cones (horizontal cross hatch) and "blue" cones (dot pattern). The dual world lines r and b appear segmented because each is seen only by its color-matched receptor. Slit 2, time-locked to the moving present moment and moving parallel to direction $x'^5$ at speed $c$, scans

the retina dual containing segmented lines r and b. This scanning motion has two effects. First, it reveals the dual world lines to the percipient not all at once, but line-by-line, creating for her the twin illusions of *motion* and time passing. Second, it transmits to the percipient electrical pulses, generated by the two kinds of visual receptor, which encode the color dimension of the retinal image. It is postulated that the sensation of color arises as the felt side of a dualism; i.e., an associated *feel* of the pulse train, in the same way that particles arise in association with QM wave functions.

of visual receptor cell are considered: "red" cones and "blue" cones. The world sheets of the former are indicated with horizontal cross hatch, the latter with a dot pattern. The dual world lines r and b appear segmented because each is seen only by its color-matched receptor. Slit 2, time-locked to the moving present moment and moving parallel to direction $x'^5$ at speed $c$, scans the retina dual containing segmented lines r and b. This scanning motion has two effects. First, it reveals the dual world lines to the percipient not all at once, but in line-by-line raster fashion, creating for her the twin illusions of *motion* and *time passing*. Second, it transmits to the percipient electrical pulses, generated by the two kinds of visual receptor, which encode the color dimension of the retinal image.

*E. Origin of visual qualia.* And so we have form and motion, but how do we get the experience of color from the color-encoding electrical pulses? This is the question of *qualia*. One might be tempted to regard the pulse train as the *cause* of a corresponding quale, or felt quality of the pulse train. The causal relation implies, however, that qualia are somehow reducible to—and hence predictable from—the pulse train. But this does not work. For pulse trains, when expressed as functions of time $ct$, are public, whereas qualia are private. How could one construct a causal bridge between these ontologically disparate realms? Assuming it cannot be done, there seems only one way available to account for the quale of color, and that is to regard the pulse train and quale as comprising a physical dualism analogous to that of wave and particle. With electromagnetic radiation of frequency $\nu$ there is associated a photon of energy $E = h\nu$, where $h$ is Planck's constant. In the same way, with a pulse train of current distribution $i(x'^5 + C) = i(ct + C)$ (where C denotes a location in past block time) there is associated a subjective feel, the quale. Just as electromagnetic radiation exhibits a wave aspect or a particle aspect depending on experimental context (e.g., two-slit interference vs. the photoelectric effect), the phenomenon of conscious experience exhibits a pulse-train aspect or a feeling aspect depending on observational perspective (third-person description in frame $K$ vs. first-person experience in frame $K_D$). This argument from dualism, we hold, accounts for the experience of color.

Against this, one might argue that, if it is true that events in $K$ are replicated in $K_D$, and *vice versa*, should not the dualism current/quale be observable in $K$, just as the QM dualism wave/particle is observable in $K$? The answer is No, because the current/quale dualism occurs in *past* block time only, and the past is inaccessible in frame $K$. The perception of light and color is a first-person phenomenon having no third-person counterpart. The world outside the mind is dark. (Also there is no sound, pain, and so on.)

*F. In summary:* The conscious being (CB) is a perceiving machine residing in dual spacetime, $K + K_D$. Frame $K$ contains her brain, while frame $K_D$ contains both her brain dual and mind. Visual perception—and by extension, perception in all sensory modalities—is a dualistic process wherein the retinal image and related cortical processing data are projected into the mind of the CB. What the CB actually sees is the retinal image dual, which (1) replicates perfectly the form of the original retinal image, (2) already exists frozen in dual (past) block time, and furthermore (3) enjoys first-person status. As this image dual can be considered a cross section of an infinite line current, this same image projects automatically into the mind of the CB. An hypothesized opaque screen pierced by two narrow slits allows passage of (1) cortical data-processing currents required for understanding and (2) the image itself, while blocking those currents projecting from the cortex as a whole. The screen, time-locked to the moving present moment, scans both the cortical data-processing area and retinal image dual, creating the illusion of image motion. This scanning motion captures as well electrical currents from the rod-cone mosaic dual, encoding color and sending these into the mind of the CB. They are read out, however, not as electrical currents, but as color qualia, "particulate" companions of the electrical "wave function."

## 19.2 Free will

*Free will*, in the sense understood at least since Kant and Schopenhauer, is rational agency.[11] In contrast to determinism, it is the capacity of the conscious self, with the aid of reason, to make free choices, to decide without constraint or coercion to do one thing and not another. The sense of being free to choose may be the ultimate human experience, not passive but active, for some exhilarating, for others dreadful, and in any case marking a principal distinction between ourselves and those further down the evolutionary ladder. (Cat owners may dispute this.) Special relativity in dual spacetime has yielded for us a clear explanation of the grasping side of conscious experience—the special mind-brain interaction required for perception. We intend now to show that our dual geometry accommodates as well the shaping side of conscious experience—the capacity for free choice. We shall find that it tells us not only how the decision to act is formed, but also *where* and *when* it is formed.

*19.2.1 Naturalistic (third-person) assessment of free will.* Although free will is manifestly a first-person faculty, it is still open to investigation in third-person terms. Progress along these lines effectively began with Kornhuber and Deeke's discovery (1965) that a voluntary act, such as choosing to flex the fingers, is preceded by electrical brain activity measurable by electroencephalogram (EEG) scans at the top of the head.[12] This electrical activity, now called the *readiness potential* (RP) [from the German *Bereitschaftspotential* (BP)], was found to begin about 800 ms before the performance of the voluntary act. The finding that brain activity precedes movement is—apart from its longish duration—not surprising, as it only confirms the expected causal relation

$$\text{Excitation of motor cortex} \rightarrow \text{Movement of fingers} \qquad (19.2)$$

What the experiment does *not* reveal is what excites the motor cortex in the first place. In addition it does not tell at what point in relation (19.2) the decision to act occurs. To investigate this latter point, Benjamin Libet repeated the Kornhuber/Deeke experiment involving the flexion of the fingers or wrist, this time having the participants note, by observing the angular position of a rotating spot of light, the time at which they were aware of having made the decision to act.[13] The results were striking. They showed that, while the RP preceded finger motion by 550-1000 ms, depending upon how much preplanning went into the decision to act, the participants' conscious awareness of that decision preceded the act itself by a mere 200 ms, regardless of the amount of preplanning. Moreover, even this number, small as it is, reduces to 150 ms when an ubiquitous error in the subjects' reported estimate of the spot position is taken into account. In other words, the onset of the RP was found to precede the awareness to act by at least $550-150=400$ ms. The Libet experiment has been repeated many times, often taking advantage of advanced methods for observing brain activity—viz., by fMRI scanning[14, 15] and by recording the activity of single neurons[16]—but with generally similar results. Despite some pointed criticism of experimental methodology and interpretation of the data,[17] the neuroscience community as a whole appears to have concluded that, in the light of these experiments, one does not make free choices after all; that decisions are actually made unconsciously; that, in short, free will is an illusion.[18-20]

Yet Libet himself had doubts about such a conclusion. In 1999, he wrote:[21]

> However, we must recognize that the almost universal experience that we can act with a free, independent choice provides a kind of *prima facie* evidence that conscious mental processes can causatively control some brain processes. ... The intuitive feelings about the phenomenon of free will form a fundamental basis for views of our human nature, and great care should be taken not to believe allegedly scientific conclusions about them which actually depend upon hidden *ad hoc* assumptions.

In fact, it is not hard to spot the flaw in the RP argument against free will. **Fig. 19.6** depicts in block-diagram form the naturalistic, neurobiology-based process of decision making, where "naturalistic" means materialism in $M^4$. In overall structure the naturalistic description of volition is similar to the naturalistic process of visual perception depicted in **Fig. 19.3**. According to neurobiology, decision making begins with an unconscious motor objective input, $\overline{I}$. One imagines $\overline{I}$ to consist of a set of neurons in the lateral prefrontal cortex (LPC), primed to initiate unconsciously the performance of an intentional act, such as flexing the wrist. The input $\overline{I}$ enters a first black box, where it is operated on by an operator $U^{-1}$, projecting $\overline{I}$ in coded form $U^{-1}\overline{I}$ into the motor cortex. It is signal $U^{-1}\overline{I}$ that is detected in the Kornhuber-Deeke-Libet type experiments. Signal $U^{-1}\overline{I}$ then passes through a second black box, where it is operated on by a *reconstruction* operator, $U$ , yielding an output event, $\overline{O}$, e.g., flexion of the wrist:

$$\overline{O} = UU^{-1}\overline{I}\,. \tag{19.3}$$

Here $U$ plays the same reconstructing rôle that binding operator $T^{-1}$ is supposed to play in visual perception. However, unlike $T^{-1}$, operator $U$ actually exists and is responsible for sending signal $U^{-1}\overline{I}$ from the motor cortex down the spinal column and into the muscle receptors controlling motor activity.

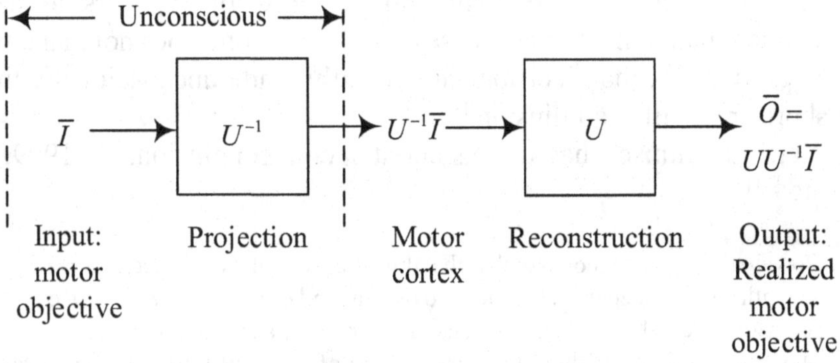


**Fig. 19.6** Information flow for the voluntary act, according to standard neurophysiology. An unconscious input motor objective $\overline{I}$ is acted upon by projection operator $U^{-1}$, yielding a coded signal $U^{-1}\overline{I}$ distributed over appropriate receptors in the motor cortex. To realize the motor objective, coded signal $U^{-1}\overline{I}$ is acted upon by reconstruction operator $U$ , sending $U^{-1}\overline{I}$ down the spinal column to muscle receptors, creating the realized motor objective, $\overline{O} = UU^{-1}\overline{I}$ .

Note that the conscious self makes no appearance in the diagram at all. In this deterministic interpretation of willed behavior, the self is at best an epiphenomenon, without causal power. However, it *thinks* it makes free decisions, and can estimate when, in relation to the completed act, it thinks these occur.

Now let us take a closer look at the motor objective input, $\overline{I}$ . It begins a causal chain leading ultimately to the realized motor objective, $\overline{O}$. But what, one may ask, are *its* causal antecedents? Since it is an intentional brain state, its cause must be another intentional brain state, and *that* caused by another such state, and so on, in infinite regress. But the RP, an observable expression of brain activity, is finite in duration, not infinite. This means that the backward chain of causation leading to $\overline{I}$—if there is one—eventually must stop. And if it stops, then the naturalistic explanation of decision making is refuted: for in a deterministic account of the world, there can be no uncaused causes.

One could try to get around this result by assuming that the decision to act and its corresponding motor objective, $\overline{I}$, have first-person ontology. While this assumption is entirely plausible, it also renders Eq. (19.3) ontologically inconsistent, the right side having now become first-person, the left side remaining third-person. Decision making in the conscious being appears unexplainable in deterministic terms alone.

*19.2.2 Free will in dual spacetime.* We have just shown that the deterministic account of volition does not work: for in that account, the unconsciously-wished-for behavior entails an uncaused cause. Moreover, it fails to admit the reality of the first-person perspective. In contrast, in the following formulation of volition in dual spacetime, we shall see that willed action (1) is initiated knowingly by the conscious self, (2) has first-person ontology, and (3) begins before generation of the RP. In short, we show that, in dual spacetime, free will is no illusion, but real.

As one can see from **Fig. 19.1**, if perception is the mind's input, then will is its output. Operationally, will is perception run in reverse. Suppose we adopt this as dictum, a guiding principle. The operation of will is then derivable from the perception operation depicted in **Fig. 19.4**. To carry this out, we proceed as follows: (a) let $x^0 \to -x^0$ and $x'^5 \to -x'^5$, reversing the temporal order; this means inverting the opaque screen as well, with slit 2 now becoming slit 3; (b) reverse the directions of all operations indicated by arrows ⇨ ; (c) let $\mathbf{x} \to -\mathbf{x}$, converting visual cortex into motor cortex; (d) replace the visual percept $A''$ with a motor objective *CD*. We then arrive at the process of voluntary action depicted in **Fig. 19.7**, where the operators $U$ and $U^{-1}$ are exactly those defined for the neurophysiologic process depicted in **Fig. 19.6**. Significantly, as a result of temporal inversion, the willed motor objective is *later* than point *B*—the source of understanding—whereas in perception, the percept is *earlier* than that point. We come to a remarkable conclusion: if perception finishes in past block time, then *the action of will begins in future block time.* In *Sec. 17.4.2* it was argued that, on the basis of what we knew then about

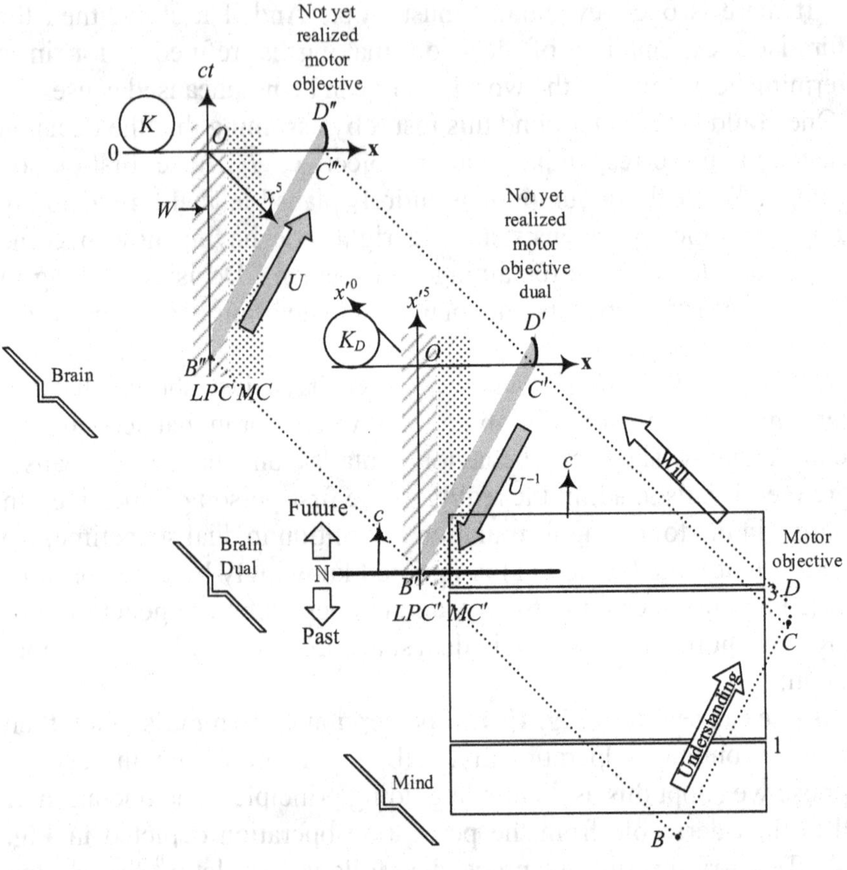


**Fig. 19.7** Action of the will in dual spacetime. Frame $K$ contains the conscious being's brain, while dual frame $K_D$ contains both brain dual and mind. The brain's world sheet in $K$ consists of the motor cortex $MC$ and lateral prefrontal cortex $LPC$. Corresponding to these areas in $K_D$ are the dual cortical areas $MC'$ and $LPC'$. Consciously willed behavior begins at present moment $B$, generating by way of understanding the *motor objective* world line $CD$. At the moment of decision $B$, world line $CD$ exists in future block time, unavailable for viewing by third-person observers. The opaque screen of **Fig. 19.4**, now shown pierced by a third slit 3, blocks all currents but those from $B'$ and world line $CD$ from entering or leaving the mind of the CB. Slit 3, time locked to the moving present moment, cuts across currents $CC'C''$ through $DD'D''$, generating motor data for encoding the motor objective in future block time by projection into the cortex via world sheet

$B'C'D'$, an operation symbolized by projection operator $U^{-1}$. Realization in frame $K$ of both the motor cortex and motor objective, an operation symbolized by operator $U$, is brought about by the moving present moment, $\mathbb{N}$, sweeping sequentially through world sheet $B'C'D'$ in $K_D$ at light speed $c$. Realization of the *MC* is detected by the RP; realization of the motor objective is detected by electromyogram and visually by third-person observers. In the Libet experiment, world line *CD* represents an intended flexion of the wrist. Time *W* is the subject's own estimate of when she chose to act, assuming that realization of the motor objective begins at time $ct = 0$.

time, the question whether the future was open or closed was unanswerable. We now see that, with the advent in nature of the conscious being, the future is in fact determined—in a small but significant way—by the conscious decision to act.

*A. Formation of the motor objective.* Referring to **Fig. 19.7**, projection $B' \to B$ from the cortex provides for *understanding*—the rational half of rational agency. The motivation for willed action arises in the present at cortical point *B*, generating by way of understanding a visualized *motor objective* world line *CD* —a wished-for or intended movement of a part of one's own body. The motor objective—which lies in the future of *B*—is a form of mental image and, like the visual percept, is unavailable for inspection by third-person observers. In fact, it is not explicitly "seen" in the mind's eye of the conscious self either. It nevertheless possesses causal power, inducing those cortical states ultimately to be revealed to third-person observers by the readiness potential. How can that be? How can an unobservable mental image result in detectable motor activity? To make this at least plausible, let us look at an optical analogy, illustrated in **Fig. 19.8**. The visual percept, which *is* seen by the conscious self, may be compared to the *real* image formed by a positive lens of an infinitely distant source. Analogously, the motor objective, which is *not* actually seen by the conscious self, may be compared to the *virtual* image formed by a negative lens of the same infinitely distant source. Just as the wave fronts emerging from the negative lens have causal power, so too do the currents $CC'C''$ and $DD'D''$ (and those currents in between), leading from the motor objective and into the brain dual and brain have causal power. With this optical analogy in mind, one can think of the motor objective as constituting a virtual rather than real internal image, but one nevertheless having the power to initiate motor activity.

Note that while *understanding* lies entirely within the mind, the motor objective world line *CD* —which lies in future block time, ahead of point B—projects out of the mind through slit 3 and thus properly

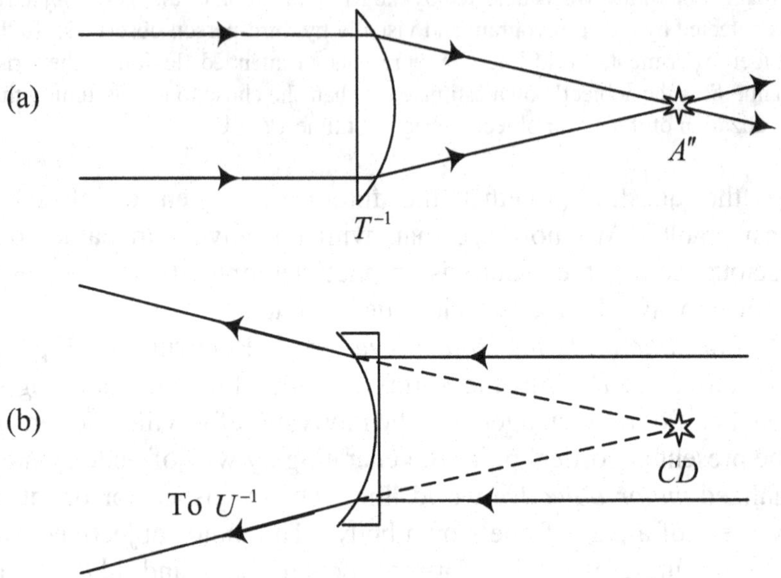


**Fig. 19.8** Optical analogs of two types of mental image. (a) Real visual percept $A''$ formed as in **Fig. 19.4** by reconstruction operator $T^{-1}$, here depicted as a positive lens. (b) Virtual motor objective $CD$ formed as in **Fig. 19.7** by wish to act, here depicted as a negative lens. $U^{-1}$ denotes projection into cortex induced by motor objective $CD$.

represents the beginning of the mind-brain interaction we call the *will*. In other words volition, as a physical interaction between mind and brain, begins at world line *CD*, not cortical point *B*. As we shall see in **Sec. 19.3**, the distinction is crucial when it comes to interpreting the Libet experimental findings in the context of dual spacetime.

Note also that world line *BC* denotes a mental operation connecting present and future. In its absence there can be no internal mental presence (i.e., the self).

*B. Projection of motor objective into brain dual.* In accordance with the mind-brain interaction labeled "Will" in both **Figs. 19.1** and **19.7**, the motor objective world line *CD* projects out of the mind and into the brain dual of the CB. Ultimately this projected world line *CD* is to become realized, an observable spacetime track $C''D''$ of motor activity in frame $K$. For this to occur, the motor objective first has to be encoded in the motor cortex. How, exactly, is this accomplished? Both form and motion data are required. To obtain these, we make use again of the hypothesized opaque screen depicted in **Fig. 19.4**. The screen is, however, now pierced by a third slit, labeled 3. Just as percept $A''$ of **Fig. 19.4** is generated point-by-point by the scanning action of slit 2, giving to $A''$ the attributes form and motion, motor objective world line $CD$ is projected not all at once, but point-by-point, owing to the scanning action of slit 3, sending form and motion data into the brain dual via currents $CC'C''$ through $DD'D''$. From this information, for each sequentially revealed point $P$ (not shown) of $CD$, there is created automatically a corresponding world line connecting point $P'$ (not shown) of realized motor objective dual $C'D'$ to the cortex dual. This projection into the cortex dual is symbolized by the projection operator $U^{-1}$, as shown. It occurs automatically because the state of the cortex dual has to match the realized motor objective dual meant to unfold from it. In exact coincidence with this activity there is created in frame $K$ a corresponding world line connecting point $P''$ (not shown) of realized motor objective $C''D''$ to the cortex. This is a reconstruction operation symbolized by operator $U$, as shown.

Two remarkable facts attend creation of the motor objective and its projection into the motor cortex dual: (1) Both operations exist in future block time, ahead of all eventual third-person (machine) observation of the RP. (2) Because in frame $K_D$ events are parametrized by pseudo-spatial $x'^5$, not temporal $ct$, *there is no expenditure of energy in carrying out these operations*, thereby eliminating the main argument against mind-brain dualism (see **Sec. 18.7**). Instead, although we do not prove this, the oppositely directed arrows associated with the operations "Understanding" and "$U^{-1}$" suggest that what is conserved in the action of the will is pseudomomentum $p'_5$ in frame $K_D$:

$$\Delta p'_5 = p'_5(\text{Understanding}) + p'_5(U^{-1}) = 0 \tag{19.4}$$

Such is the causal power of the virtual motor objective, setting up the state of the motor cortex required for realization of the motor objective. The mentally-created motor objective replaces the uncaused cause inherent to the deterministic account of volition. Moreover, it is created consciously and, as it is formed in future block time, enjoys first-person ontology.

*C. Realization of the motor objective.* **Fig. 19.7** depicts the moving present moment $\mathbb{N}$ passing through point $B'$ of the cortex dual. The projections of the motor objective $CD$ into the brain dual ($C'D'$) and brain ($C''D''$) are thus not yet realized and are so indicated. Realization is brought about by the movement of $\mathbb{N}$ through $K_D$ from point $B'$ to the endpoint $D'$ of dual world line $C'D'$. This movement of $\mathbb{N}$ has two observable effects. First, in accordance with reconstruction operator $U$, it converts the virtual motor objective, moment-by-moment, into a realized motor objective, i.e., real motor activity detectable by electromyogram and visually by third-person observers in frame $K$. Of course, the observed willed movement of the body is actually an illusion, the effect of $\mathbb{N}$ intersecting sequentially the points of fixed world line $C'D'$.

The second observable effect of the motion of $\mathbb{N}$ is that it converts the virtual motor cortex, moment-by moment, into a realized motor cortex, generating the RP measurable by third-person observers in frame $K$. As may be seen from **Fig. 19.7,** this readout of RP begins at cortical point $B''$, the origin of understanding. Not to belabor this, but the RP was present all along, starting at points $B'$ and $B''$, poised in future block time, ready to be realized and released for observation by the moving present moment. Note that this release proceeds exactly as described by naturalistic neuroscience, with no violation of energy conservation having occurred.

*D. Feeling of the wish to act.* The decision to act may be accompanied by a subjective feeling of freedom to perform the act. Could this feeling of freedom can be considered the quale of the will? The answer is: probably not—at least not in the sense of will and quale comprising a dualism analogous to that between wave and particle. Such an interpretation works plausibly well in the case of perception and quale, no doubt for the reason that the percept (visual or otherwise) is real. In the case of the will, however, the motor objective is virtual, and for such an image it is hard to imagine a particulate companion. More likely, the feeling of freedom is something one acquires cognitively, a deduction based on a history of success in one's ability to choose between one act and another.

*E. In summary:* The apparent symmetry between grasping and shaping, between the twin operations of perception and will, suggests that will can be understood as perception run in reverse. On this line of thought, if (passive) perception ends in past block time, then the (active) will begins in future block time. The essential, defining feature of the will is that the motor objective, together with the state of the cortex required for its realization, are generated consciously as one complete event in future block time. Moreover, absent the temporal dimension $x^0$, it all occurs without expenditure of energy, undermining the conventional argument against dualism. The state of the cortex required for realization of the motor objective is synthesized from form and motion data generated by moving slit 3 cutting across currents from the motor objective. Realization of the motor cortex, and of the motor objective itself, occurs as the moving present moment, $\mathbb{N}$, passes sequentially over these preëxisting elements of frame $K_D$, rendering them observable, moment by moment, in frame $K$.

## 19.3 Two experimental tests of mind-brain interaction in dual spacetime

> For once we have been told that the aim of science is to explain, and that the most satisfactory explanation will be one that is most severely testable and most severely tested, we know all we need to know as methodologists.
>
> Karl R. Popper[22]

The preceding accounts of perception and will in dual spacetime are telling indeed. For as we have seen, they have the ability to explain, among other things, perceptual binding, qualia and freedom of the will—phenomena apparently untouchable by standard neuroscience. But that does not mean that the accounts are automatically true. They of course need to be tested; and, as Karl Popper advises, tested severely. Test them we can and will do. But first we need predictions. It is hard to imagine testing perception and will separately. Even if we could, it would make no practical sense, because in a whole conscious being these operations exist not separately, but together, poised for mutual interaction. And so to formulate tests we must look at the whole being, with attention to the interaction between its grasping and shaping functions. Two distinct configurations of the whole being present themselves for testing. Call them I and II. On the side of volition, the two perform identically, each making a conscious decision to act—a flexion of the wrist, say—at a moment of the subjects' choosing. Where they differ is on the perceptual side. In I, the conscious being, focusing on an external clock, reads off the time at which she becomes aware of the decision to act. This is the Libet experiment. In II, she focusses visually, not on a clock, but on a *self-possessed* object, namely, the flexion of her own wrist. All this is easy to say, but not so easily pictured. Moreover, in neither case is it obvious what is to be tested. To identify the variables and make testable predictions we shall have to study in detail the logic flow in the two configurations.

*19.3.1 Configuration I: The Libet experiment in dual spacetime.* As the Libet experiment is already known to us, it makes sense to examine it first, to see how it is to be interpreted in a dualist perspective. **Fig. 19.9** depicts the logical flow in dual spacetime. The diagram looks complicated, as it involves the joining of the perceptual and volitional operations depicted separately in **Figs. 19.4** and **19.7**. It is nevertheless straightforwardly decoded by following the single-shafted arrows from points *A* to *F* on the side of Will; and then the double-shafted arrows from points *G* to *L* on the side of Perception. The triple-shafted arrow from the clock denotes an effective line of simultaneity. To avoid unnecessary clutter the slit-pierced screens shown in **Figs. 19.4** and **19.7** are omitted in **Fig. 19.9**.

*A.* In the figure, the motivation for willed action arises at cortical point *A*, generating by way of understanding a visualized *motor objective* (flexion of the wrist) world line at point *B* in future block time. Then, without expenditure of energy, the motor objective proceeds by force of will out of the mind to point *C* of the brain dual, whereupon it is projected by operator $U^{-1}$ to point *D* of the cortex dual. The dual cortical state so produced is duplicated precisely at point *E* of the cortex in frame *K*, whereupon, owing to the scanning motion of present moment $\mathbb{N}$, the realized motor objective is formed at point *F*, i.e., in accordance to Libet's convention, at the origin of time, $ct = 0$. The realization or reconstruction of the motor objective is symbolized by operator *U* acting along world line *EF*. It is along *EF* that the readiness potential is generated. That completes the shaping side of the Libet experiment.

*B*. In the grasping side of the Libet experiment, the subject is asked to note, by observing the angular position of a rotating spot of light (called here the "clock"), the time at which she first becomes aware of having made the decision to act.[13] Now as emphasized previously

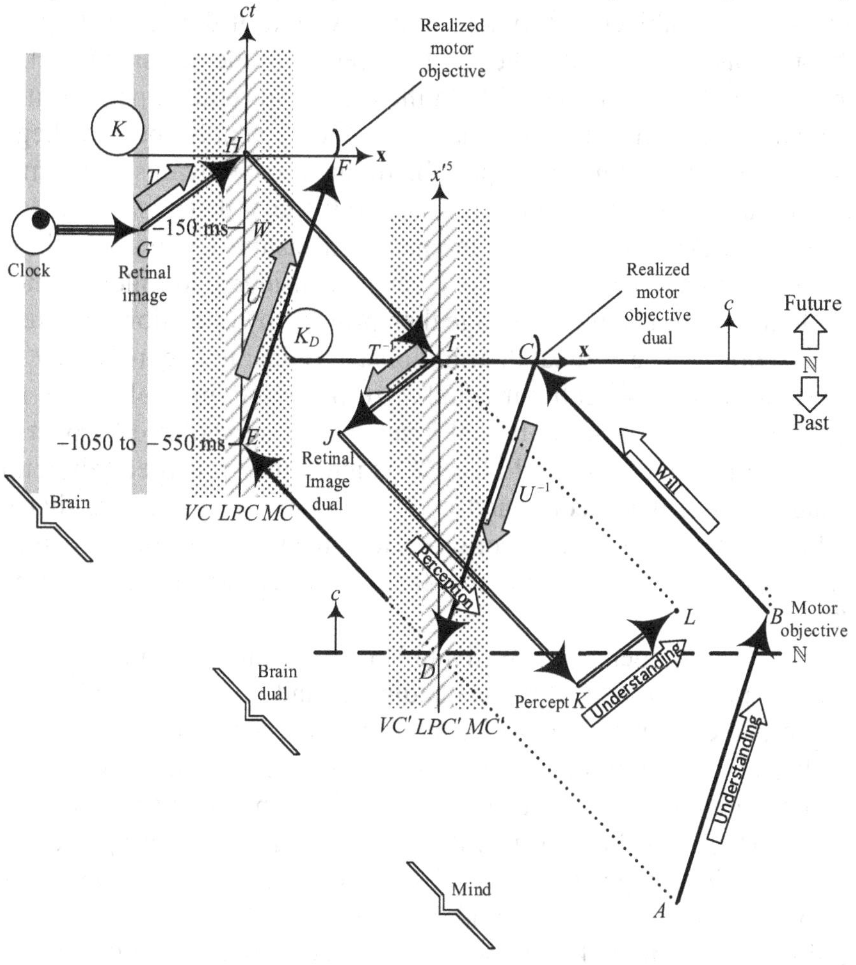


**Fig. 19.9** Configuration I: The Libet experimenet in dual spacetime. Single-shafted arrows trace action of the Will from points *A* to *F*. Double-shafted arrows trace the Perception operation from points *G* to *L*. Triple-shafted arrow, an effective line of simultaneity, represents the world line of light from clock face to retina. Temporally, the process begins with present moment ℕ passing through cortical dual point *D* (dashed line) and ends with its passing through cortical dual point *I* (solid line). *W* denotes the subject's estimate of the time she chose to act.

(seeSec.19.2.2 A.), the subject considers her decision to be made not at cortical point $A$, but at future block time point $B$, where the motor objective (a virtual flexion of wrist) is brought into being. When realized, the motor objective appears at point $F$ in frame $K$ at time, $ct = 0$. The subject endeavors, by observation of the clock, to report this time as the moment of her first awareness of the decision to act. In the figure, the world lines of both the clock and retinal image of it are shown as vertical stripes in frame $K$. The perception process begins with the clock face imaged upon the retina at $G$. Obviously the image is continually changing with time, but only one of these images is to be selected marking the subject's moment of decision. (As the signal from clock to retina moves at speed $c$, it can be considered to follow a line of simultaneity.) The retinal image then projects by operation $T$ to area $H$ in the cortex. The cortical distribution and retinal image in frame $K$ are reproduced automatically at $I$ and $J$, respectively, in dual frame $K_D$. The reconstruction of the retinal image at $J$ is symbolized by operator $T^{-1}$, acting along dual world line $IJ$, as indicated. It is precisely retinal image dual $J$, when continued back to point $K$ in the mind, that the subject sees in visual perception. Her understanding of it occurs at cortical point $L$, an extension into the mind of cortical point $H$ in frame $K$.

*C.* The two operations, Will and Perception, are linked cortically by their respective points of understanding, $A$ and $L$. What $A$ understands is the subject's moment of decision, which occurs at $B$; and what $L$ understands is the reading on the clock face perceived at $K$. Now points $B$ and $L$ are the mental images, respectively, of points $F$ and $H$ in frame $K$ and, moreover, line $HF$ is a line of simultaneity. Consequently, the clock reading imaged at $G$ and subsequently projected to $H$ and thence to percept point $K$ of the mind, represents the value of the subject's estimate of the time, $W$, at which she first becomes aware of the decision to act. What is $W$, relative to time $ct = 0$ ? Clearly, it is just the negative of the interval $c\Delta t$ between the moment of image formation at retina $G$ and that of image recognition at cortical point $H$. Or, in mental terms, it is the negative of the interval $\Delta x'^5 = c\Delta t$ between the visual percept at $K$ and cortical point of understanding $L$. In either case we have,

$$W = -c\Delta t . \tag{19.5}$$

Now the image processing time $c\Delta t$ in the human visual system is known to be about 150 ms.[23] (Note that we report $W$ in units of time rather than space, as if $c = 1$.) *Hence we predict that the subject's estimate of first awareness of the decision to act is* $W = -150$ *ms.* This, of course, is precisely the (corrected) $W$-value Libet found experimentally. A secondary, dependent prediction, following trivially from the fact that volition begins at $B$, not $A$, is that $W$ should have one and the same value, namely that given by (19.5), regardless of whether the decision to act is planned or spontaneous. That prediction, too, is confirmed by the results of the Libet experiment.

*19.3.2 Configuration II: Observation of one's own willed action.* In configuration I—the Libet experiment—an external device is required to observe and record the moment at which willed muscle activation occurs. The subject's reported $W$-time is defined with respect to that moment. In configuration II, depicted in **Fig. 19.10**, no external device is required. Instead, the subject observes visually her own willed action—a flexion of the wrist, for example. She then can compare mentally, without assistance, the moment of perceived action relative to the moment of her awareness of the wish to act. We inquire here as to what that relation must be in the context of dual spacetime. It might be called the *self-W-time.*

*A.* Comparison of **Figs. 19.9** and **19.10** shows that the action of Will is the same in both configurations I and II. It begins at understanding point $A$ of the mind and ends at realized motor objective point $F$ in frame $K$.

*B.* The Perception process begins with the subject's realized motor objective imaged onto the retina at $G$. (Again, as the signal from wrist to retina moves at speed $c$, it can be considered to follow an effective line of simultaneity.) From $G$, the perceptual process follows a path identical to the one taken in configuration I. The subject perceives visually her own wrist flexion at point $K$ in the mind. Her understanding of it occurs at cortical point $L$, an extension into the mind of cortical point $H$ of frame $K$.

*C.* As in configuration I, Will and Perception are linked cortically by their respective points of understanding, $A$ and $L$. Again, what $A$ understands is the subject's moment of decision, which occurs at $B$; however, in the present case, what $L$ understands is the flexion of her own wrist perceived at $K$. Now points $B$ and $K$ are mental images,

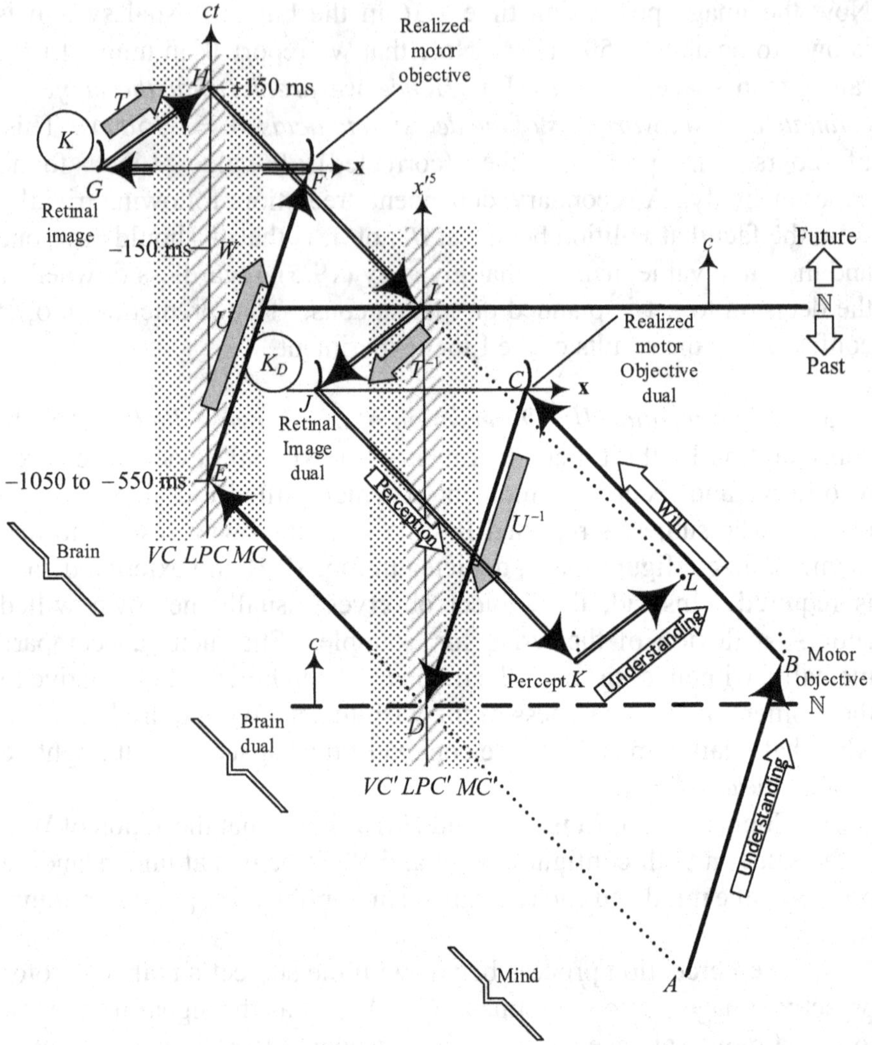


**Fig. 19.10** Configuration II: Observation of one's own willed action. Single-shafted arrows trace action of the Will from points *A* to *F*. Double-shafted arrows trace the Perception operation from points *G* to *L*. Triple-shafted arrow, an effective line of simultaneity, denotes the world line of light from realized motor objective (e.g. flexion of wrist) to retina. Temporally, the process begins with present moment ℕ passing through cortical dual point *D* (dashed line) and ends with its passing through cortical dual point *I* (solid line). For reference, *W* denotes the subject's estimate of the time she chooses to act when viewing external clock, as in **Fig. 19.9**.

respectively, of points *F* and *G* of frame *K* and, moreover, line *FG* is a line of simultaneity. Consequently, *we predict that the moment of visual perception at point K coincides precisely with, and is indistinguishable from, the moment B at which she first becomes aware of the decision to act.* There is no perceptible interval between them.

This prediction is easily tested: Gaze at your wrist and plan to flex it at the count of three. On carrying out this plan, you find that the visual image of your willed motor activity occurs simultaneously with the moment you choose to act. The two events occur in perfect coincidence: the self-*W*-time in dual spacetime is 0 ms. (This demonstration works equally well in the auditory channel: close your eyes, and substitute a snap of the fingers for the flexion of the wrist.)

## 19.4 Assessment

*19.4.1 What do these tests mean?* They mean, first of all, that dualism is testable. And not only testable, but severely so. Consider configuration I. According to standard, naturalistic neurobiology, "one would expect conscious will to appear before, or at least at the onset, of the RP, and thus command the brain to perform the intended act."[21] The fact that it does not do this, that the subjects' *W*-value occurs against expectations a mere 150 ms before muscle activation, occasions real surprise in the neuroscience community. In that light, and given the scientific (as opposed to folk) bias against dualism, our prediction of the *W*-value found experimentally by Libet and coworkers is surely unexpected. One might compare it to predicting correctly the rest mass of a new and unexpected fundamental particle. In the writer's view, it well corroborates the picture of mind-brain interaction in dual spacetime. It underscores, too, the danger of drawing untestable conclusions about the first-person experience of free will from third-person measurements of readiness potential. From the subject's report of *W*-time it does *not* follow that the decision to act is unconscious and that free will is an illusion. In fact her decision *must* be conscious if (1) the uncaused cause is to be avoided and (2) it is to have first-person ontology—a state of being that in principle cannot arise from material brain. In retrospect, and especially since everyone "knows" he has free will, the neuroscience community might have at least raised the possibility that, in light of the Libet result, naturalism may be false.

Configuration II yields an additional corroboration of dualism, predicting, as it does, an easily tested—and confirmed—coincidence between the appearance of conscious will and visual image of the intended act. But it does something more. It puts standard naturalistic neurobiology to the test as well. Suppose naturalism were asked to predict the self-$W$-value. What would it be? Since naturalist theory knows nothing of dual spacetime and assumes $3+1$ spacetime dimensions, the value in question can be found by inspection of the world lines shown in brain frame $K$ of **Fig. 19.10**. In the naturalistic picture, volition, when preplanned, begins unconsciously at point $E$, some 1050 ms before muscle activation at point $F$, where $ct=0$. Point $E$ marks as well the onset of the readiness potential, and $F$ its endpoint. The subject's first awareness of the wish to act occurs at Libet's $W$-time, i.e., at $W=-150\text{ ms}$, as shown. Her cognizance of the visual image of that act occurs at time $c\Delta t=+150\text{ ms}$. Thus, naturalistic theory predicts for the self-$W$-value, $-W+c\Delta t=300\text{ ms}$. The interval is small but discernable if it should exist. (It corresponds to the period of a metronome setting of 200 beats/min.) However, as we have seen, it does not exist. The actual interval experienced is 0 ms. *The monistic, naturalistic account of the interaction between volition and visual perception is thereby falsified. And by inference, biological naturalism itself is falsified.* This result comes as no surprise. For we have suspected since **Chapter 1** that Minkowski $M^4$ cannot on its own support the first-person perspective. We now have empirical confirmation of that conjecture.

To repeat, the error in the conventional interpretation of the Libet experiment is the assumption that naturalism must be true, leading to the untestable conclusion that free will is an illusion. It is worth noting that a logically similar error of interpretation occurs in neutrino physics as well. In his analysis of the end-point energy spectrum of beta-decay electrons, Enrico Fermi concluded that "the rest mass of the neutrino is either zero, or, in any case, very small in comparison to the mass of the electron."[24] Moreover, at present, all direct (kinematic) measurements of neutrino mass are consistent with zero, and the minimal Standard Model of particle physics in fact assumes the neutrino to be massless. In view of that background, the experimental finding that neutrinos oscillate between one flavor and another could be taken to mean that the standard theory of neutrino mixing—which states that only massive neutrinos can oscillate—is false. That is a logical option. But the particle physics community does not interpret it that way. Instead, it takes it to mean that neutrinos must have mass (even though the oscillation experiments measure mass-squared differences, not mass). Why does it take this latter view? Surely it does so out of a commitment to the standard scientific picture of the world—a material world of particles and fields interacting with each other in $3+1$ spacetime dimensions. The point is that rigid adherence to this standard picture is a block to progress, whether one is doing neuroscience or particle physics.

*19.4.2 Helical world-line structure of the self.* **Figure 19.10** reveals the world-line structure of the conscious being in the act of observing her own willed action. It depicts a typical activity, the first step, say, in the subject's reaching for an external object, such as a glass of water. Somewhere in that picture is the Ego, or self, the one who observes and takes action. The New Oxford American Dictionary defines the self as "a person's essential being that distinguishes them [sic] from others, especially as considered as the object of introspection or reflexive action." In the light of that serviceable definition, it makes little sense to separate the self from the whole being. The self is all that lies within the boundary of her being, distinguishing her from others, encompassing both mind and brain, neither of which, presumably, can exist usefully without the other.

The self, as suggested by the figure depicting it, is complicated, and rightly so. Consider what it entails:

(1) There is first of all the geometric setting, dual spacetime. The two disjoint frames, $K$ and $K_D$, provide for separation of brain and mind, accounting at once for the third-person status of brain and the privileged, first-person status of mind.

(2) Next come the complementary operations of Perception and Will. A real self has thoughts and emotions too, of course; but Perception and Will are the indispensable, foundational functions of the conscious self. Bound together by Understanding, they form the *minimum conscious unit*.

(3) Finally there is the present moment, $\mathbb{N}$, whose motion creates both a growing past and a diminishing future in block time. It brings to life the representations we call Percepts; and to realization the motor

objectives of the Will. Separating past from future, ℕ enables the projection of motor objectives into future block time, and the reading of percepts in past block time. Crucially, the projection of motor objectives conserves energy, as in future block time in $K_D$, energy does not even appear, it being replaced by pseudomomentum $p_5$.

Let us now follow once again the path defined by the arrows of **Fig. 19.10**. The sequence begins at Understanding point *A* and ends at Understanding point *L*. Note that *L* lies directly above *A*. The path as a whole thus can be considered one turn of a complicated helix of world lines of pitch *AL*. This helix is the signature of the conscious self, depending for its construction on the three elements listed above. Note that its projection onto the $M^4$ hyperplane (frame *K*) yields a planar figure consisting of three world lines, *EF* (nervous realization of the motor objective), *FG* (propagation of light) and *GH* (projection of retinal image into cortex). In such a figure the helical signature of self is lost, just as the identity of a cube, when projected to a square, is lost. Here we see in graphical terms the error of attempting to define the attributes of self (e.g., nature of the will) from third-person measurements in frame *K* (e.g., readiness potential).

Note too that, in principle, one could build a machine operating in accord with the world lines of frame *K*. Such a machine would duplicate exactly the neural and motor structure of a conscious being. And yet it would not be conscious. To be conscious it would have to project itself into the portion of frame $K_D$ called Mind, unfolding itself into a helix of world lines, and that it cannot do, just a square cannot decide to project itself into a cube. The machine so constructed is the zombie alluded to at the end of **Chapter 18**. It is odd to think that the Libet experiment, in the accepted naturalist interpretation, does not distinguish between a conscious being and a zombie.

*19.4.3. Minimum time between successive wishes to act.* The helix pitch *AL* corresponds to a single willed action, viz., the one launched at Understanding point *A* and whose motor objective appears at *B*. This may be the first in a sequence of willed actions. If so, the next one will be launched at *L*, with a new motor objective appearing at pitch distance *AL* above *B*, and ending at a new Understanding point appearing at distance *AL* above the first understanding point, *L*. In other words, a sequence of willed actions may be periodic—a helix with many turns.

And so we have a third prediction: *there exists a minimum time between successive wishes to act; and clearly this minimum time is just the pitch, AL.* Its value will depend on whether the willed actions are considered to be spontaneous or pre-planned. As the individual wishes to act follow quickly one after the other, it seems unlikely that they could each be pre-planned. Thus from **Fig. 19.10**, and assuming the readiness potential for the single wish begins at –550 ms, we have

$$\begin{aligned}&\text{minimum time}\\ &\qquad\text{between successive wishes to act} = 550\text{ ms} + 150\text{ ms} \\ &\qquad\qquad\qquad\qquad\qquad\qquad\quad = 700\text{ ms}\end{aligned} \tag{19.6}$$

(This interval corresponds to a metronome setting of 86 beats/min.) On personally testing this prediction, the interval seems too large. That is, it seems that one can make independent wishes at higher rate than suggested by (19.6). On the other hand, it may be that what one assumes to be independent wishes are actually unconscious elements of a single wish to perform a rapidly-performed sequence of actions. A sharpened version of the number—as well as a better understanding of will itself—could come from a study of the RP profiles of subjects asked to perform rapid sequences of willed actions.

*19.4.5 Is there a ghost in the machine?* We have defined the self as all that lies within the boundary of her being, the machine *in toto*. So the answer to the question is "no," for the term "ghost" refers to a knowing presence, or homunculus, separate from the machine as such. There is no homunculus, nor is one required. The self—the internal observer who perceives and chooses to act—is present already in the form of a continuous feedback loop, the complicated helical structure depicted in **Fig. 19.10**, spanning both frames $K$ and $K_D$. Note that a ghost *would* be required if it were not for the existence of Understanding world lines *AB* and *KL*. These world lines connect via longitudinal currents to points $E$ and $H$ of the cortex in frame $K$, bringing brain computation into the process of knowing. The presence of world lines *AB* and *KL* enables percepts and willed actions to participate as active elements of the feedback loop, thereby creating an integrated, self-referential structure, obviating the need for a ghost in the machine.

The idea that the knowing self emerges as a consequence of the whole machine interacting with itself is not new. An intriguing version of this idea put forward by Antonio Damasio asserts that core consciousness consists of a continuous flow of stories, and the self arises as a story within the story[25]—something like perceiving oneself in the act of perception.[26] The spacetime helix in our dualist account similarly suggests a continuous flow of stories—a circulation of actions and percepts. However, in that account, the self is not simply one story among others. It is the entire helix.

Of course, the greater difference between these two conceptions of self is that one is formulated in ordinary spacetime, the other in dual spacetime. The upshot is that, in ordinary spacetime, the self can only be an illusion,[27] as there is no room in 4-space for the first-person perspective. In dual spacetime, however, the self is physically real, occupying a first-person realm inaccessible to third-person observers. We know that the self is physically real, because the percepts she observes and the motor objectives she creates are physically real, and these are the essential elements of the helix. Furthermore, the grasping of percepts in past block time, and the shaping of motor objectives in future block time, are both fashioned by the actions of the moving, slit-bearing screens depicted in **Figs. 19.4** and **19.7**. Such actions are hardly illusory, attached as they are to the moving present moment, and can be nothing but physical.

However, it would be wrong to conclude that we now understand fully the physics of self. We do not. We understand schematically how the conscious self is structured, and we have empirical corroboration of the validity that scheme. But that is as far as it goes. For the fundamental grasping and shaping operations are not just hidden from the external viewpoint, but doubly so: first by their occupation of frame $K_D$ and second by their participation in past and future block time. Their hiddenness is not unlike that of the Kantian noumena: one can picture but not observe them. In essence we are stuck with the mysterian view: human consciousness is perhaps not meant to understand itself, or at any rate not entirely. That being the case, it remains, too, a mystery how the natural world could have given birth to something we cannot understand—the conscious self.

## Notes and references


[1] M. A. Goodale and K. J. Murphy, "Space in the Brain: Different Neural Substrates for Allocentric and Egocentric Frames of Reference," in *Neural Correlates of Consciousness*, T. Metzinger, ed. (MIT Press, Cambridge, MA, 2000), pp. 189-202.

[2] S. Blackmore, *Consciousness: An Introduction* (Oxford U, Press, Oxford, UK, 2004), pp. 41-43.

[3] C. Koch, *The Quest for Consciousness* (Roberts & Company Publishers, Englewood, Colorado, 2004), Ch. 7.

[4] S. Zeki, "The Visual Image in Mind and Brain," *Sci. Am.*, Sept. 1992, pp. 68-76.

[5] R. Carter, *Mapping the Mind* (U. California Press, Berkeley, 2010), pp. 112-113.

[6] J. R. Searle, *The Mystery of Consciousness* (A New York Review Book, New York, 1997), pp. 33-34, 40.

[7] K. R. Popper and J. C. Eccles, *The Self and Its Brain* (Routledge & Kegan Paul, London, 1977), p. 534.

[8] P. S. Churchland, *Neurophilosophy: Toward a Unified Science of the Mind/Brain* (MIT Press, Cambridge, MA, 1886), pp. 120-125.

[9] A. R. Damasio, "How the Brain Creates the Mind," *Sci. Am.*, December, 1999, pp. 112-117.

[10] P. S. Churchland and T. J. Sejnowski, *The Computational Brain* (MIT Press, Cambridge, MA, 1992), p. 155.

[11] R. Scruton, *Modern Philosophy* (Penguin, New York, 1994), pp. 234-236, 248-250.

[12] H. H. Kornhuber and L. Deeke, "Hirnpotential ändrungen bei Willkürbewegungen und passive Bevegungen des Menschen: Bereitschaftpotential und reafferente potentiale," *Pflügers Archiv* **284**, 1-17 (1965).

[13] B. Libet, C. A. Gleason, E. W. Wright and D. K. Pearl, "Time of conscious intention to act in relation to onset of cerebral activities (readiness-potential): the unconscious initiation of a freely voluntary act," *Brain* **106**, 623-642 (1983).

[14] C. S. Soon, M. Brass, H-J Heinz and J-D Haynes, "Unconscious determinants of free decisions in the human brain," *Nature Neuroscience* **11**, (543-545 (2008).
[15] S. Bode, A. H. He, C. S. Soon, R. Trampel, R. Turner and J-D Haynes, "Tracking the Unconscious Generation of Free Decisions Using Ultra-High Field fMRI, " PLoS ONE 6(6):e21612. doi10.1371/journal.pone.0021612.
[16] I. Fried, R. Mukamel and G. Kreiman, "Internally generated preactivation of single neurons in human medial frontal cortex predicts volition," *Neuron* **69**, 548-562 (2011).
[17] W. R. Klemm, "Free will debates: Simple experiments are not so simple," *Advances in Cognitive Psychology* **6**, 47-65 (2010). doi: 10.2478/v10053-008-0076-2
[18] See **Ch. 18**, Ref. 22.
[19] Sam Harris, *Free Will* (Simon and Schuster, New York, 2012).
[20] For a contrary view, see J. Bagginni, *Freedom Regained* (U. Of Chicago Press, Chicago, 2015).
[21] B. Libet, "Do we have free will?" in *The Volitional Brain*, B. Libet, A. Freeman and K. Sutherland, Eds. (Imprint Academic, Thorverton, UK, 1999).
[22] K. R. Popper, *Realism and the Aim of Science* (Rowman and Littlefield, Totowa, New Jersey, 1956, 1983), p. 145.
[23] S. Thorpe, D. Fize and C. Marlot, "Speed of processing in the human visual system," *Nature* **381**, 520 (1996).
[24] E. Fermi, Z. Phys. **88,** 161 (1934); translated in F. L. Wilson, Am. J. Phys. **36**, 1150 (1960).
[25] A. R. Damasio, *The Feeling of What Happens* (Harcourt Brace & Co., New York, 1999), pp. 189-192; "How the Brain Creates the Mind," *Sci. Am.* **281**, 112-117 (Dec., 1999).
[26] This memorable phrase comes from L. Barnett, *The Universe and Dr. Einstein* (Bantam Books, New York, 1957), p. 117.
[27] See this very interesting and pertinent account of the self from an expert in digital circuit design: Masakazu Shoji, *Self-Consciousness: The Hidden Internal State of Digital Circuits* (iUniverse LLC, Bloomington, Indiana, 2013), pp. 7-9, 216-217.

# PART VIII

# CONCLUSION

# Chapter 20

## From noumena to qualia

In his collection of essays *Conjectures and Refutations*, Karl Popper argues against what he calls *observationalism*—the idea that science advances by accumulating observations. [1] A notebook of observations may form the basis for a library or museum exhibit, but not much else. The way science *really* advances, he says, is by looking for conflict between observations and expectations. These are called *problems*. Advance comes through the attempt to solve the problems, leading, if all goes right, to a heightened understanding of the natural world.

The present volume could be cited an example of how, according to Popper, one tries to make scientific progress. In the first place, there is no shortage of problems. They are dotted across the entire landscape of the world of nature—problems with fundamental particles, gravitation, cosmology and neuroscience. In fact, the weight of them is so great, and so serious, one could be forgiven for thinking that fundamental physical science has come to an end. How is one to tackle all this? One approach would be to deal with the problems one by one, approaching each separately with a physical model deemed appropriate to its solution. That is an understandable if exhausting way to go, but it is not what we have attempted here. Instead, we chose to start with a formal criterion of truth, the Principle of True Representation, which led to a comprehensive Law of Laws and thence systematically to a portfolio of separate laws—new laws true by definition—applicable to a fair selection of the most elusive of open problems. In such an approach, the solutions to the problems emerge spontaneously as we walk through the formalism. One might say they find us, rather than the other way round.

What makes the set of new laws unique (besides being automatically true) is that it reveals the presence of, and is defined with respect to, an expanded spacetime continuum. The basic Minkowski $x^0, \mathbf{x} \in M^4$ remains intact, of course. It's from there that measurements are made. But in addition there is found to exist a 2-D Minkowskian $x^5, x^6 \in \bar{M}^2$, where $x^5$ and $x^6$ are flat and infinite, $x^5$ playing the role of an extra spatial coordinate with time-like metrical signature, and $x^6$ an extra temporal coordinate with space-like signature. The two Minkowskian continua combine to form the 6-D product manifold $M^{4+2} = M^4 \times \bar{M}^2$. And that is not all. A study of the symmetry properties of the Dirac equation in six dimensions reveals the existence of a second 6-D product manifold $M_D{}^{4+2} = M_D{}^4 \times \bar{M}_D{}^2$ related to $M^{4+2}$ by a $-90°$ rotation in the $x^0$-$x^5$ plane: $x^5, \mathbf{x} \in M_D{}^4$ and $-x^0, x^6 \in \bar{M}_D{}^2$. The physical world is thus described twice, once in $M^{4+2}$ and once in $M_D{}^{4+2}$, the two manifolds together forming dual spacetime $M^{4+2} + M_D{}^{4+2}$. And there is still more. The negative probabilities in the *Zitterbewegung* imply the existence of a 2-D inner spacetime $u^0, u \in N^2$, where $u^0$ is the (infinite) inner temporal dimension and $u$ a bounded spatial dimension. Total spacetime thus has the form $\left(M^{4+2} + M_D{}^{4+2}\right) \times N^2$.

The $2+2$ dimensions forming $\bar{M}^2$ and $N^2$ are invisible. As a result, although the majority of particular laws proceeding from the Law of Laws are straightforward generalizations of known laws (such as the Dirac and Maxwell equations), two have not been seen before (the field equations of Riemann-Cristoffel $F_{\alpha\beta}{}^{\mu\nu}$ and Casimir $F_{\alpha}{}^{\mu\nu}$). With these latter two are associated fifth and sixth forces of nature, with serious implications for geometrodynamics. Also invisible to observers in $M^{4+2}$ are phenomena that take place in $M_D{}^{4+2}$, this manifold dual being separated from $M^{4+2}$ by a finite rotation. If Nature is correctly described in terms of these hidden dimensions and laws, as the writer thinks it is, small wonder the open problems have avoided solution. **Table 20.1** summarizes the main problems addressed in this book by the Principle of True Representation. For each is stated where in the book it is discussed, the pre-solution nature of the puzzle, the means of solution (theory) and final result.

The highlighted problems in this table are ones found in everyone's list of unsolved problems in physics. If Kant were alive today, his list would certainly include problem (1); and Eccles' and Popper's list would include problems (30) and (31). Fittingly, these problems begin and end the set. Fitting, because (1) entails the invisible, external world of noumena, whose unification with phenomena initiates access to the source of physical law; and (30) and (31) entail the invisible, internal world of conscious experience, by any measure the crowning, end-point achievement of creation.

| Table 20.1 | Problems addressed by the Principle of True Representation | | | |
|---|---|---|---|---|
| Problem | Where dis-cussed | Conven-tional under-standing | Theory/means of solution | Result P=prediction, E= explanation |
| | | General | | |
| (1) Is there a real, external world? | **Ch. 2** | Self evi-dent, but not prov-able | Treat external fact as received message ⇒ Law of Laws | (P) External world exists. Corroborated by tests of dualism. See accompanying text. |
| (2) Are there extra dimen-sions? | **Secs. 3.5-9** | Unknown No evidence for them. | Analogy with optical self-imaging. | (P) There are four extra dimensions. See sum-mary in text above. |
| (3) Origin of wave-particle dualism | *Sec. 3.9.3* | Origin unknown | Self-imaging in $M^{4+2} = M^4 \times \bar{M}^2$ | (E) Descriptor $\chi$ is wave; propagator $\hat{R}$ is particle |
| (4) Origin of charge $Q$ quanti-zation | *Sec. 3.9.9* **Sec. 4.9** | Origin unknown | Self-imaging in $M^4 \times \bar{M}^2$. Modal expan-sion index $n$. | (E) $Q = \begin{cases} -n, & \text{leptons, bosons} \\ -n - B, & \text{quarks} \end{cases}$ $B$ baryon number |
| | | Fermions | | |
| (5) Origin of flavor | **Secs. 4.4, 4.11** | Origin unknown | Self-imaging in $M^4 \times \bar{M}^2$ | (E) Defined jointly by 5-momentum shell radius $G$ and charge $Q$ |
| (6) Proton decay | **Secs. 4.13-14** | Allowed in SU(5) GUT | Baryon and lepton #s separately conserved in all processes | (P) Proton decay forbidden |
| (7) Neutrino-less double $\beta$ - decay | **Secs. 4.13-14** | Allowed if neutrinos are Majorana | Baryon and lepton numbers separately conserved in all processes | (P) Neutrinoless double $\beta$ - decay forbidden |
| | | Bosons | | |
| (8) What is dark matter? | *Secs. 5.4.3, 5.5.3* | WIMPs | Maxwell-Proca in $M^4 \times \bar{M}^2$ | (P) Two dark particles: Pseudoscalars $Z^5(m = m_Z)$ $W^5(m = m_W)$ |

| | | | | |
|---|---|---|---|---|
| (9) Dark radiation | **Sec. 6.4** | Unanticipated | Maxwell in $M^4 \times \bar{M}^2$ | (P) Pseudoscalar potential $A_5$; Pseudovector field $\left(\frac{E_5}{c}, \mathbf{B}_5\right)$ |
| Geometrodynamics | | | | |
| (10) Black hole singularity | **Sec. 8.9** | GR breakdown: 0 diameter, $\infty$ density | Rank 4 field equations of Riemann-Cristoffel (R-C) | (P) Repulsive R-C force drives the 'singularity' to finite diameter ~ Planck length |
| (11) Force of Big Bang explosion | **Sec. 8.10** | GR breakdown: force unknown | Rank 4 field equations of Riemann-Cristoffel (R-C) | (P) Repulsive R-C force drives explosion: $\frac{F_{\text{R-C}}}{F_{\text{GRAV}}} \sim 10^{40}$ |
| (12) The graviton. Does it exist? | **Sec. 8.12** | Generally assumed to exist. | Rank 4 field equations of Riemann-Cristoffel (R-C) | (P) Graviton does not exist. |
| (13) Quantum gravity: unify gravity and QM | **Sec. 8.12** | Leading theories: String theory, Loop quantum gravity. | Rank 4 field equations of Riemann-Cristoffel (R-C) | (P) String theory predicts the graviton, which according to the above does not exist. String theory contradicted. |
| (14) Shape of the Universe | **Sec. 9.1** | Flat by WMAP | Structure of $M^4 \times \bar{M}^2$: $x^5 \perp M^4$ | (P) Universe $M^4$ is flat. |
| (15) Origin of cosmic energy density ratio $\frac{\rho_{\text{QFT}}}{\rho_{\text{CRIT}}} \sim 10^{124}$ | *Sec. 9.6.5* | Cosmological constant problem: unresolved | Rank 3 field equations of Casimir | (P) Repulsive Casimir force drives cosmic expansion. $\frac{\rho_{\text{QFT}}}{\rho_{\text{CRIT}}} \sim \left(\frac{D_{\text{U0}}}{\ell_{\text{P}}}\right)^2$ $D_{\text{U0}}$ = current radius of physical universe $\ell_{\text{P}}$ = Planck length |

| | | | | |
|---|---|---|---|---|
| (16) Radius of Universe ratio $\frac{D_{U0}}{D_{OBS}}$ | **Sec. 9.7** | Ratio $\sim 10^{23}$ (Guth inflation) | Rank 3 field equations of Casimir | (P) From above equation, if $D_{OBS} \sim 4.35\times 10^{28}$ cm, then $\frac{D_{U0}}{D_{OBS}} \sim 3.5$! |
| Internal Structure of Leptons | | | | |
| (17) Baryon asym-metry: Where is the missing antimatter? | *Sec. 10.7.2* | No promising expla-nation | Existence of inner space-time $N^2$. Negative probabilities in Zitter-bewegung. | (P) Parallel universes: matter on one side of $N^2$, antimatter on other side, separation distance $W = h/2m_e c$ . |
| Neutrino Mixing and Oscillation | | | | |
| (18) Neutrino: Dirac or Majorana? | **Sec. 12.3** | Unknown | Self-imaging in $M^4 \times \bar{M}^2$ | (P) Neutrino is Dirac particle. |
| (19) Neutrino mass | **Secs. 4.7, 12.4** | Because they oscillate, neutrinos must be massive | Flavor neutrino $m_\nu^2$ propor-tional to elec-tric charge $q_\nu$. | (P) Since $q_\nu = 0$, all flavor neutrinos are massless. Refutes standard oscillation theory. |
| (20) Why do neutrinos oscillate? | **Sec. 12.5** | Analog of quark mix-ing, if $\nu$ s massive. | Structure of mixing matrix in $M^4 \times \bar{M}^2$ | (E) Neutrinos must oscillate to exist. Nevertheless, lepton number $L_\ell$ is conserved. |
| (21) Why is mixing between $\nu_\mu$ and $\nu_\tau$ nearly maximal? | *Sec. 12.13. 6* | Unex-plained | Structure of mixing matrix in $M^4 \times \bar{M}^2$ | (E) Consequence of the masslessness of the two flavor neutrinos. |
| The Family Problem | | | | |
| (22) Mass spectrum of charged leptons (relative to $m_e$) | **Ch. 13** | Genera-ted by coupling to vac-uum Higgs, but not predicted. | Energy spectrum of electron trapped in $N^2$. | (P) Masses predicted within 0.5% of experimental values. Hidden lepton $\sigma^\mp$ predicted, $\frac{m_\sigma}{m_e} = 421.071$ MeV . |

| | | | | |
|---|---|---|---|---|
| (23) Origin of the electron mass $m_e$ | *Sec. 4.8.4* **Secs. 10.7, 13.7** | SM coupling to vacuum Higgs field. | Wave function reëntrant in $N^2$. | (E) $m_e$ determined by spacetime geometry, not coupling to Higgs field |
| (24) Mass spectrum of bare quarks | **Ch. 14** | Generated in SM, same way as leptons | Quark-lepton universality | (P) Reasonable agreement with experimental values. c and t coupled to $\sigma^{\mp}$. |
| Structure of spacetime | | | | |
| (25) Why are there three spatial dimensions? | **Ch. 15** | E.g., required for orbital stability of planets | Calculate *Zitter.* of electron in *N* spatial dimensions | (E) Only in a world with $N = 3$ can the electron exist. |
| (26) What is the origin of electric charge? | **Ch. 16** | Unknown | Self-imaging in $\bar{M}^2 \times N^2$ of lattice structure of internal time $u^0$ | (E) All integer and fractional charges resident in $\bar{M}^2 \times N^2$. Fractional charges grouped as found in $p/\bar{p}$ and $n/\bar{n}$. |
| (27) How does prime number 137 arise physically? | **Ch. 16** | Unknown | Self-imaging in $\bar{M}^2 \times N^2$ of lattice structure of internal time $u^0$ | (E) $\frac{D}{\lambda} = 137$, where $D = \hbar / m_e c$ = charge pattern repeat interval, $\lambda$ = propagation w. l. |
| (28) Non-locality and reduction of the state vector | **Ch. 17** | Un-explained | Dual spacetime $M^{4+2} + M_D^{4+2}$ and moving present moment | (E) Owing to moving present moment, space-like intervals in frame *K* of $M^4$ are in causal contact in frame $K_D$ of $M_D^4$. |
| The hard problem of consciousness | | | | |
| (29) Can a man-made machine be conscious? | **Ch. 18** *Sec. 19.4.2* | Monistic material-ism (natur-alism) as-sumes "yes" | Dual spacetime $M^{4+2} + M_D^{4+2}$ and moving present moment | (E) Conscious experience occurs in frame $K_D$ of $M_D^4$, inaccessible to 3rd-person builders of machines in frame *K* of $M^4$. |

| | | | | |
|---|---|---|---|---|
| (30) Free will vs. determinism | **Sec. 19.2** | Naturalistic neuroscience claims free will to be an illusion. | Dual spacetime $M^{4+2} + M_D^{\;4+2}$ and moving present moment | (P) Free will is real, no illusion. Determinist view of will refuted. |
| (31) Mind-brain dualism vs. monistic naturalism | *Secs. 19.3-4* | Naturalistic neuroscience assumes dualism refuted by non-conservation of energy. | Dual spacetime $M^{4+2} + M_D^{\;4+2}$ and moving present moment | (P) Dualism prevails over monistic naturalism. Dualism corroborated empirically. Naturalism falsified. |

There is another, more significant connection between problems (1) and (30)/(31). We recall from *Sec. 19.4.1* that two tests were performed on our theory of mind-brain dualism. The tests were severe, exposing the theory to falsification should certain predicted effects be absent. The effects appeared, however, exactly as predicted, corroborating our version of dualism and the reality of free will. But our theory of dualism depends on the existence of dual spacetime; and *that* structure is a direct outcome of the unification of appearance and reality. It follows that our two tests, designed to test a theory of dualism, end up corroborating as well the existence of the external world. We may say, with some irony, that the existence of the privileged, *internal* world of conscious experience corroborates the existence of an unobservable, *external* world of noumena.

The tabulated results divide into two types: explanatory (E) and predictive (P). The explanatory ones are obviously of interest. Everyone, for example, seeks an explanation for the phenomenon of non-locality in quantum mechanics [entry (28)]; and welcomes an existential reason for the existence of three spatial dimensions [entry (25). Yet the explanatory results are of no value unless the theory as a whole holds good. And whether or not it does depends on the outcomes of tests of its predictions. Some of these are untestable. For example, in (10), how is one to measure the diameter of what resides at the center of a black hole? And in (17), how is one to detect the presence of a universe parallel to the one we live in? However, some of the predictions are perfectly testable. For example: whether or not the proton is stable [entry (6)]; whether or not double $\beta$-decay can occur unaccompanied by two neutrinos [entry (7)]; and whether or not flavor neutrinos are truly massless [entry (19)]. Such tests pose a severe threat to the theory as a whole. If the results of any of these tests should go against prediction, the whole—including its explanations and remaining predictions—would probably have to be discarded. For in the event it is falsified, there is almost no way to revise the theory, given its very specific structure, to keep it viable.

However, suppose that the Principle of True Representation survives the tests unmolested. What might it be applied to next? A first suggestion might be to ask it to give a better account than is available presently of the creation and annihilation of particles; specifically, with a view to doing away with the unwanted infinities in quantum field theory. Another would be to have another look at the problem of quantum gravity, as string theory appears not to be viable. As mentioned at the end of **Chapter 8**, it should not be hard to unify the Riemann-Cristoffel force with the electromagnetic and weak forces. Since the Einstein field equations are already embedded in the equation, of Riemann-Cristoffel, it might then be possible to see directly the connection between gravitation and the other two forces and the common origin of all of them.

## Reference

[1] K. R. Popper, *Conjectures and Refutations: Growth of Scientific Knowledge* (Harper Torchbooks, New York, 1963, 1965), pp. 127 ff.

## A



## B



## C



## D



## E

## F



## G



## H



## I



## K



## L

## W



## Z

Printed in the United States
By Bookmasters